Lecture Notes in Networks and Systems   778

The series "Lecture Notes in Networks and Systems" publishes the latest developments in Networks and Systems—quickly, informally and with high quality. Original research reported in proceedings and post-proceedings represents the core of LNNS.

Volumes published in LNNS embrace all aspects and subfields of, as well as new challenges in, Networks and Systems.

The series contains proceedings and edited volumes in systems and networks, spanning the areas of Cyber-Physical Systems, Autonomous Systems, Sensor Networks, Control Systems, Energy Systems, Automotive Systems, Biological Systems, Vehicular Networking and Connected Vehicles, Aerospace Systems, Automation, Manufacturing, Smart Grids, Nonlinear Systems, Power Systems, Robotics, Social Systems, Economic Systems and other. Of particular value to both the contributors and the readership are the short publication timeframe and the worldwide distribution and exposure which enable both a wide and rapid dissemination of research output.

The series covers the theory, applications, and perspectives on the state of the art and future developments relevant to systems and networks, decision making, control, complex processes and related areas, as embedded in the fields of interdisciplinary and applied sciences, engineering, computer science, physics, economics, social, and life sciences, as well as the paradigms and methodologies behind them.

Indexed by SCOPUS, INSPEC, WTI Frankfurt eG, zbMATH, SCImago.

All books published in the series are submitted for consideration in Web of Science.

For proposals from Asia please contact Aninda Bose (aninda.bose@springer.com).

José Manuel Machado · Javier Prieto ·
Paulo Vieira · Hugo Peixoto · António Abelha ·
David Arroyo · Luigi Vigneri
Editors

Blockchain and Applications, 5th International Congress

 Springer

Editors
José Manuel Machado
Universidade do Minho
Braga, Portugal

Paulo Vieira
Guarda Polytechnic Institute
Guarda, Portugal

António Abelha
Universidade do Minho
Braga, Portugal

Luigi Vigneri
Pappelallee
IOTA Foundation
Berlin, Germany

Javier Prieto
BISITE Research Group
University of Salamanca
Salamanca, Spain

Hugo Peixoto
Universidade do Minho
Braga, Portugal

David Arroyo
Científicas, ITEFI
Instituto de Física Aplicada, Consejo
Superior de Investigaciones
Madrid, Spain

ISSN 2367-3370 ISSN 2367-3389 (electronic)
Lecture Notes in Networks and Systems
ISBN 978-3-031-45154-6 ISBN 978-3-031-45155-3 (eBook)
https://doi.org/10.1007/978-3-031-45155-3

Preface

The 5th International Congress on Blockchain and Applications 2023 was held in Guimarães (Portugal) from 12 to 14 of July. This annual congress reunites blockchain and artificial intelligence (AI) researchers who share ideas, projects, lectures, and advances associated with those technologies and their applications.

Among the scientific community, the advance in next-generation blockchain and distributed ledger technologies is seen as a promising combination that will transform the production and manufacturing industry, media, finance, insurance, e-government, agriculture, energy, or even medicine. Besides, the combination of DLT with other cutting-edge disruptive technologies (AI, metaverse, and quantum computing) will open a multitude of amazing opportunities that entail an interesting challenge of research at the same time.

The BLOCKCHAIN'23 congress is devoted to: promoting the investigation of next-generation blockchain and DLT technologies; exploring the latest ideas, innovations, guidelines, theories, models, technologies, applications, and tools of blockchain/DLT combined with other cutting-edge technologies; and identifying critical issues and challenges that researchers and practitioners must deal with in future research. We want to offer researchers and practitioners the opportunity to work on promising lines of research and to publish their developments in this area.

In this 5th edition, the technical program has been diverse and of high quality, focused on contributions to both well-established and evolving areas of research. The congress had an acceptance rate close to 50%, with more than 110 full papers submitted and 54 of them accepted. The volume includes three papers from the WEB3-TRUST workshop, seven papers from the DAG-DLT workshop, two papers from the FRAMS4CLIMATE workshop, and seven papers from the Doctoral Consortium.

Submissions come from more than 30 different countries (Spain, USA, Sweden, UK, Saudi Arabia, Ireland, Morocco, France, Italy, India, Australia, Portugal, Qatar, Greece, Vietnam, Malta, China, Luxembourg, Lithuania, Switzerland, Canada, Germany, Turkey, Japan, Tanzania, Croatia, Brazil, Israel, Romania, Indonesia, Bangladesh, Singapore, Slovenia, Russia, and Austria).

This conference is organized by the LASI and Centro ALGORITMI of the University of Minho (Portugal). We would like to thank all the contributing authors, the members of the program committee, the reviewers, the sponsors and the organizing committee for their hard and highly valuable work. Their work contributed to the success of the BLOCKCHAIN'23 event. Finally, special thanks to the local organization members and the program committee members for their hard work, which was essential for the success of BLOCKCHAIN'23.

José Manuel Machado
Javier Prieto
Paulo Vieira
Hugo Peixoto
António Abelha
David Arroyo
Luigi Vigneri

Organization

General Chairs

Javier Prieto Tejedor   University of Salamanca, Spain, and AIR
Institute, Spain

José Manuel Machado   University of Minho, Portugal

Advisory Board

Abdelhakim Hafid   Université de Montréal, Canada
Ashok Kumar Das   IIIT Hyderabad, India
António Pinto   Instituto Politécnico do Porto, Portugal
Paulo Leitao   Technical Institute of Bragança, Portugal
Francisco Luis Benitez Martínez   Fidesol, Spain
Pedro Nevado   University of Salamanca, Spain

Technical Program Committee Chair

Paulo Vieira   Guarda Polytechnic Institute, Portugal

Local Chairs

Hugo Peixoto   University of Minho, Portugal
António Abelha   University of Minho, Portugal

Local Organizing Committee

Paulo Novais (Chair)   University of Minho, Portugal
José Manuel Machado (Co-chair)   University of Minho, Portugal
Hugo Peixoto   University of Minho, Portugal
Regina Sousa   University of Minho, Portugal
Pedro José Oliveira   University of Minho, Portugal
Francisco Marcondes   University of Minho, Portugal
Manuel Rodrigues   University of Minho, Portugal

Filipe Gonçalves   University of Minho, Portugal
Dalila Durães   University of Minho, Portugal
Sérgio Gonçalves   University of Minho, Portugal

Organizing Committee

Juan M. Corchado Rodríguez   University of Salamanca and AIR Institute, Spain
Fernando De la Prieta   University of Salamanca, Spain
Sara Rodríguez González   University of Salamanca, Spain
Javier Prieto Tejedor   University of Salamanca and AIR Institute, Spain
Ricardo S. Alonso Rincón   AIR Institute, Spain
Alfonso González Briones   University of Salamanca, Spain
Pablo Chamoso Santos   University of Salamanca, Spain
Javier Parra   University of Salamanca, Spain
Liliana Durón   University of Salamanca, Spain
Marta Plaza Hernández   University of Salamanca, Spain
Belén Pérez Lancho   University of Salamanca, Spain
Ana Belén Gil González   University of Salamanca, Spain
Ana De Luis Reboredo   University of Salamanca, Spain
Angélica González Arrieta   University of Salamanca, Spain
Angel Luis Sánchez Lázaro   University of Salamanca, Spain
Emilio S. Corchado Rodríguez   University of Salamanca, Spain
Raúl López   University of Salamanca, Spain
Beatriz Bellido   University of Salamanca, Spain
María Alonso   University of Salamanca, Spain
Yeray Mezquita Martín   AIR Institute, Spain
Sergio Márquez   AIR Institute, Spain
Andrea Gil   University of Salamanca, Spain
Albano Carrera González   AIR Institute, Spain

Technical Program Committee

Mahmoud Abbasi   University of Salamanca, Spain
Rim Abdallah   University of Le Havre, France
Mansoor Abdulhak   Consultant & Researcher, Malaysia
Ermyas Abebe   USA
Imtiaz Ahmad Akhtar   XLENT Link, Sweden
Morteza Alaeddini   Department of Computer Engineering and
                        Information Technology, Amirkabir University
                        of Technology, Iran

Francisco Javier Estrella Liébana	University of Jaén, Spain
Sante Dino Facchini	DISIM - Università degli Studi dell'Aquila, Italy
Xinxin Fan	IoTeX, USA
Wenjun Fan	Xi'an Jiaotong-Liverpool University, China
José Álvaro Fernández Carrasco	Vicomtech, Spain
Christof Ferreira Torres	Interdisciplinary Centre for Security, Reliability and Trust (SnT), University of Luxembourg, Luxembourg
Ernestas Filatovas	Vilnius University, Lithuania
Marco Fiore	Politecnico di Bari, Italy
Nikos Fotiou	AUEB, Greece
Muriel Franco	University of Zurich, Switzerland
Jean-Philippe Georges	University of Lorraine, France
Shahin Gheitanchi	Senior consultant, UK
Raffaele Giaffreda	CREATE-NET, USA
Seep Goel	IBM Indian Research Labs, India
Hélder Gomes	Escola Superior de Tecnologia e Gestão de Águeda, Universidade de Aveiro, Portugal
Dhrubajyoti Goswami	Concordia University, Canada
Volker Gruhn	Universität Duisburg-Essen, Germany
Ivan Gutierrez	TECNALIA, Spain
Abdelatif Hafid	University of Montreal, Canada
Christopher G. Harris	University of Northern Colorado, USA
Yahya Hassanzadeh-Nazarabadi	Ferdowsi University of Mashhad, Turkey
Pham Hoai Luan	NAIST, Japan
Hsiang-Jen Hong	University of Colorado Colorado Springs, USA
Yining Hu	University of Technology Sydney, Australia
Maria Visitación Hurtado	University of Granada, Spain
Shahid Hussain	National University of Ireland, Ireland
Fredrick Ishengoma	The University of Dodoma, Tanzania
Hans-Arno Jacobsen	University of Toronto, Canada
Syed Muslim Jameel	Sir Syed University of Engineering & Technology, Pakistan/National University of Ireland Galway, Ireland
Marc Jansen	University of Applied Sciences Ruhr West, Germany
Zakwan Jaroucheh	Edinburgh Napier University, UK
Eder John Scheid	University of Zurich, Switzerland
Christina Joseph	National Institute of Technology Karnataka, India
Raja Jurdak	QUT, Australia
Chandrasekaran K.	NITK, India
Christos Karapapas	Athens University of Economics and Business, Greece

Francisco Moya	University of Jaen, Spain
Daniel-Jesus Munoz	ITIS Software, Universidad de Malaga, Spain
Pavas Navaney	Oracle America Inc; University of Southern California, USA
Mikhail Nesterenko	Kent State University, USA
Binh Minh Nguyen	Hanoi University of Science and Technology, Vietnam
Luca Nizzardo	Protocol Labs, India
Francesco Nocera	Polytechnic University of Bari, Italy
Suhail Odeh	Bethlehem University, Israel
Joseph Oglio	Kent State University, USA
Andrea Omicini	Alma Mater Studiorum—Università di Bologna, Italy
Kazumasa Omote	University of Tsukuba, Japan
Arindam Pal	Data61, CSIRO, Australia
Andreea-Elena Panait	University of Bucharest, Romania
Gaurav Panwar	New Mexico State University, USA
Nohpill Park	Oklahoma State Univ., USA
Alberto Partida	URJC, Spain
Remigijus Paulavičius	Institute of Data Science and Digital Technologies, Lithuania
Jan Pennekamp	RWTH Aachen University, Germany
Teresa Pereira	Universidade do Minho, Portugal
Cristina Pérez-Solà	Universitat Rovira i Virgili, Spain
Miguel Pincheira	Fondazione Bruno Kessler, Italy
Karl Pinter	TU Vienna, INSO, Austria
Pedro Pinto	Instituto Politécnico de Viana do Castelo, Portugal
António Pinto	ESTG, P.Porto, Portugal
Iakovos Pittaras	Athens University of Economics and Business, Greece
Steven Platt	Universitat Pompeu Fabra, Spain
Pedro Plaza	UNED, Spain
Matthias Pohl	Otto-von-Guericke-Universität Magdeburg, Germany
George Polyzos	Athens University of Economics and Business, Greece
Hauke Precht	University Oldenburg, Germany
Wolfgang Prinz	Fraunhofer, Germany
Yuansong Qiao	Athlone Institute of Technology, Ireland
Venkatraman Ramakrishna	IBM Research—India, India
Paul Reaidy	CERAG, France
Cristina Regueiro	Tecnalia Research & Innovation, Spain
Richard Richard	Bina Nusantara University, Indonesia

Antonio Robles-Gómez	UNED, Spain
Bruno Rodrigues	University of Zurich, Switzerland
Ivan Rodriguez-Conde	University of Arkansas at Little Rock, USA
Thomas Rose	Fraunhofer, Germany
Abiola Salau	University of North Texas, USA
Gernot Salzer	Vienna University of Technology, Austria
Georgios Samakovitis	University of Greenwich, UK
Altino Sampaio	Instituto Politécnico do Porto, Escola Superior de Tecnologia e Gestão de Felgueiras, Portugal
Elio San Cristóbal Ruiz	UNED, Spain
Ricardo Santos	ESTG/IPP, Portugal
Henrique Santos	University of Minho, Portugal
Helena Saraiva	Instituto Politecnico da Guarda, Portugal
Vishal Saraswat	Robert Bosch Engineering and Business Solutions Pvt. Ltd., (RBEI/ESY)
Dominik Schmelz	Research Group for Industrial Software (INSO), Austria
Jongho Seol	Middle Georgia State University, USA
Gokarna Sharma	Kent State University, USA
Pramod Shashidhara	Chainlink Labs, USA
Chien-Chung Shen	University of Delaware, USA
Ajay Shrestha	University of Saskatchewan, Canada
Mirza Kamrul Bashar Shuhan	bKash Limited, Bangladesh
Manuel Sivianes	Universidad de Sevilla, Spain
Pamella Soares	Universidade Estadual do Ceará, Brazil
Benfano Soewito	Bina Nusantara University, Indonesia
Denis Stefanescu	Ikerlan, Spain
Burkhard Stiller	University of Zurich, Switzerland
Marko Suvajdzic	University of Florida, USA
Chamseddine Talhi	École de Technologie Supérieure, Canada
Teik Guan Tan	pQCee, Singapore
Keisuke Tanaka	Tokyo Institute of Technology, Japan
Subhasis Thakur	National University of Ireland, Galway, UK
Tuan Tran	University of California, Santa Cruz, USA
Sergio Trilles	UJI, Spain
Muhamed Turkanović	University of Maribor, Faculty of Electrical Engineering and Computer Science, Slovenia
Aitor Urbieta	IK4-Ikerlan Technology Research Centre, Spain
Julita Vassileva	University of Saskatchewan, Canada
Massimo Vecchio	Fondazione Bruno Kessler (FBK), Italy
Andreas Veneris	University of Toronto, Canada
Luigi Vigneri	IOTA Foundation, Germany

Chenggang Wang	University of Cincinnati, USA
Lei Xu	Kent State University, USA
Yury Yanovich	Skoltech, Russia
Amr Youssef	Concordia University, Canada
Kaiwen Zhang	École de technologie supérieure de Montréal, Canada
Haofan Zheng	UC Santa Cruz, USA
Mirko Zichichi	Universidad Politécnica de Madrid, Spain
Francesco Zola	Vicomtech, Spain
Avelino F. Zorzo	PUCRS, PUCRS
André Zúquete	University of Aveiro, Portugal

BLOCKCHAIN'23 Sponsors

Contents

**Workshop on Beyond the Promises of Web3.0: Foundations and
Challenges of Trust Decentralization (WEB3-TRUST)**

**Beyond the Chain: Workshop on DAG-Based Distributed Ledger
Technologies (DAG-DLT)**

**Workshop on Building the Potential of Blockchain in Farming
(FARMS4CLIMATE)**

Doctoral Consortium

Main Track

Implementation and Evaluation of Blockchain-Based Applications in SMEs' Supply Chains: Proof-of-Origin and Process Automation

Lambert Schmidt[1]([✉]), Marc Hübschke[2], Vanessa Carls[1], Eugen Buss[2], Stefan Lier[2], Elmar Holschbach[2], and Marc Jansen[1]

[1] Institute of Computer Science, University of Applied Sciences Ruhr West, 46236 Bottrop, Germany
Lambert.Schmidt@hs-ruhrwest.de

[2] Faculty of Engineering and Economic Sciences, South Westphalia University of Applied Sciences, 59872 Meschede, Germany

Abstract. Blockchains are utilized in supply chains of small and medium enterprises in addition to conventional technologies. As part of a research project, two use cases have been implemented, tested and evaluated in pilot projects using two demonstrators. The first use case tested is a transparent proof of origin for wood products in order to reduce illegal logging and to be able to prove certifications. The second use case focuses on the automation and validation of an order process to eliminate errors due to manual activities and standardize interfaces. The paper provides a detailed overview of the use cases, as well as the technical implementation and economical evaluation of the solutions. Both qualitative and quantitative aspects have been analyzed.

Keywords: Blockchain · SMEs · Supply Chain · Process Automation · Proof of Origin

1 Introduction

The use of blockchain technology in supply chain management has gained significant attention in recent years [12]. While much of the focus has been on the application of this technology in large corporations, there is a growing interest in exploring its potential for small and medium-sized enterprises (SMEs).

The research project 'Blockchain for Supply Chain' has the goal, to integrate the blockchain technology into the supply chains of SMEs. A technical wholesaler and a manufacturer of wooden-materials are participants of the research project and have two different use cases for a blockchain-based solution: First the transparent proof of origin for wood products and second the automation and validation of an order process. Transparency and traceability of wood products

© The Author(s), under exclusive license to Springer Nature Switzerland AG 2023
J. M. Machado et al. (Eds.): BLOCKCHAIN 2023, LNNS 778, pp. 3–12, 2023.
https://doi.org/10.1007/978-3-031-45155-3_1

are crucial to prevent illegal logging and ensure sustainable sourcing. Automation and validation of order processes can streamline the process, reduce errors, and improve efficiency. In this paper, we provide a detailed overview of these use cases, the technical implementation of the solutions, and the evaluation of their effectiveness.

2   State of the Art

Blockchain technology has gained significant attention in recent years due to its potential to transform various industries, including supply chain management, finance, healthcare, and other industries [7]. While many of the applications of blockchain technology have been focused on larger corporations, there is growing interest in its potential for SMEs [15]. The benefits of blockchain technology for SMEs are indeed numerous and extensively examined. It can enhance transparency and traceability in supply chains, reduce costs related to manual record-keeping, and improve efficiency by streamlining processes [13]. Furthermore, blockchain can significantly enhance security by providing an immutable ledger of transactions, thus reducing the risk of fraud and cyber-attacks [2].

While there are many potential benefits of blockchain technology for SMEs, such as smart contracts, traceability and immutability, there is often confusion about whether to use a public or private blockchain [4]. Public blockchains are decentralized networks that allow anyone to participate and contribute to the code, offering transparency and security. Examples include Bitcoin and Ethereum, which can be used for various applications like supply chain management. On the other hand, private blockchains are permissioned networks operated by a central authority or consortium, providing greater control and customization. They offer flexibility and efficiency but may be less transparent. The choice between public and private blockchains depends on the specific needs of the SME, considering factors like transparency, security, scalability, and customization [16].

The process of deciding whether to choose a private or a public blockchain solution is described in a previous paper [5] and has been carried out with a QOC analysis, which resulted in choosing a public blockchain. The specific public blockchain that has been chosen for these use cases is the WAVES blockchain, that is designed primarily for building scalable, user-friendly apps. In case of the two pilot projects these are two relevant aspects. The scalability is important to connect new customers/suppliers easily and the user friendliness is inevitable for users with a low level of IT know how.

3   Use Cases

The following chapter presents the two use cases that are implemented in the research project and compares the current situation with the future situation at the project partners.

3.1   Proof of Origin

The first project partner is a manufacturer of wood-based materials, who wants to provide the proof of origin of the wood used in a transparent and secure way, over the whole supply chain with selected suppliers and logistics providers. Since it is important to work sustainably it should be clarified that the processed wood originates only from legal logging. This use case describes how blockchain technology can be utilized to enable traceability in the wood supply chain, from the initial harvesting of the wood to its processing. The process involves the following steps: The supplier of the wood logs into a website provided by the manufacturer and uploads an eldatSmart file [1] containing information about the timber and the stacked-up wood. This file contains important details such as the origin of the wood, the type of wood, and other relevant information. The supplier saves the information in the blockchain, a decentralized platform that enables the creation and management of custom digital assets. The information is saved without any additional cost or effort for the supplier, making the process seamless and efficient. The manufacturer then arranges for a logistics provider to go out and gather the stacked-up wood. The order for the logistics provider is sent via the blockchain, ensuring that the transaction is secure and tamper-proof. The logistics carrier then picks up the wood and can mark the wood-stacks on the blockchain as collected. This step enables the logistics carrier to provide proof of delivery, which is recorded in the blockchain, and allows the manufacturer to track the location of the wood as it moves through the supply chain. The wood is tracked again when it arrives at the manufactures site, where it undergoes further processing. The blockchain entries enable the manufacturer to trace the wood back to its origin, providing complete transparency and traceability in the supply chain. This use case exemplifies how blockchain technology can establish a transparent and secure supply chain for wood products. By leveraging blockchain, the manufacturer can verify that the wood used in their products meets high-quality and environmental standards, ensuring sustainability and responsible sourcing. Additionally, blockchain technology simplifies and streamlines the supply chain process, reducing costs and enhancing efficiency for suppliers and carriers. The adoption of blockchain technology brings significant benefits to the wood industry, ensuring sustainable practices and providing a trustworthy system for all stakeholders involved.

3.2   Process Automation

The second project partner is a technical wholesaler that wants to automate the process of order transmission with its customers to optimize costs and save time. Since the interfaces between the project partner's ERP system and its customers are diverse, a blockchain-based solution is proposed to simplify and expedite the connection of new customers, replacing the current clearing house method. Previously, orders were sent via email as PDF files, manually processed, and entered into the target ERP system. The implementation of blockchain technology offers two distinct approaches for customers to interact with the wholesaler's system. Firstly, customers can utilize a user-friendly website interface to enter their orders, ensuring a seamless order placement experience. This enables customers to conveniently submit their orders, which can then be recorded and processed on the blockchain. Secondly, the wholesaler can provide a backend system that converts incoming orders into the format used by their ERP system. This backend system acts as an intermediary, bridging the gap between customers and the wholesaler's system. By converting orders into a compatible format, it facilitates the order process and ensures smooth integration with the ERP system. The integration of blockchain in the customer order process offers several benefits. By connecting all customers to the wholesaler's ERP system through a unified solution, it enhances efficiency and transparency in the supply chain. This centralized approach simplifies the order placement process for customers and streamlines order management for the wholesaler. Moreover, the blockchain functionality includes a smart contract that checks the validity of orders, such as verifying them against a price sheet uploaded by the wholesaler. This automatic validation process ensures that customers are using the correct pricing, minimizing the risk of errors. Consequently, the order process becomes more accurate and trustworthy, guaranteeing that customers are charged correctly based on the prices defined by the wholesaler. It's worth noting that the order process described here serves as an example to demonstrate the plausibility check functionality, which can be customized for other use cases as well. In conclusion, the utilization of blockchain technology in the customer order process offers two approaches, namely website-based order placement and backend mapping, to streamline the order process and connect customers to the wholesaler's ERP system. The additional functionality of checking order validity against a price sheet enhances accuracy and trust in the order process, making it a promising solution for improving supply chain efficiency and transparency.

4   Technical Implementation

This chapter provides an detailed explanation of the technical implementations of the use cases, as well as the used architecture. Both use cases have been implemented in the Waves Testnet.

4.1   Proof of Origin

The use case of storing certificates of wood origin runs through multiple steps, in which data is accumulated (see Fig. 1). Before exchanging the data, all relevant participants need to create accounts on the blockchain and the wood manufacturer needs to accept these accounts in order for them to upload data. The first step is the wood provisioning notification, that gives information about wood piles that are ready to be processed. For the test use case, this notification has been provided in form of an .eldat file, which contains information about the certification of the wood as well as the geographic location. All relevant information from the .eldat file are stored on the blockchain. In the next step, the manufacturer creates a transport order for logistics providers in form of an excel sheet and also stores its information on the blockchain. The process of transportation is also stored on the blockchain, enabling seamless tracking at any time. Via a web interface, the logistics providers can record the current process step like 'loaded' or 'delivered'.

The following steps will be extended and are not yet implemented in the demonstrator: The project partner generates a production order for the processing of the supplied wood. In the final step, the wood products are delivered to the corresponding purchaser and again

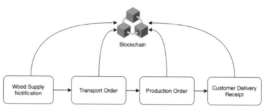

Fig. 1. Certificates of Wood Origin

all relevant information needs to be stored. It is important to mention, that the data that is sensitive is encrypted before storage. The purchaser will be provided with a QR code that forwards to an overview of the wood that has been processed, including its origin and certificates.

4.2   Process Automation

For the use case of process automation the process of an order has been chosen. Therefore a smart contract has been implemented in Ride (Waves' smart contract language) that performs plausibility checks on the orders from the customers. Since every customers could possibly have its own prices - for example depending on the ordered quantity of an item - the technical wholesale first needs to upload a list to the blockchain, that includes the prices for all customers, items and quantities. The upload of this list is enabled via a web interface and all sensitive data are encrypted beforehand [6] (Fig. 2).

The test case has been implemented with one special customer, that uses SAP as an ERP system. Since the project partner uses an ERP system called Gevis [11], the orders and the corresponding fields in it needed to be mapped first. Before sending the

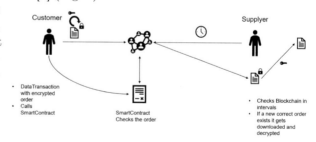

Fig. 2. Automation of order transmission

order, the customer needs to create a Waves account and a request for accepting the customer account is send to the technical wholesaler. Without accepting the customer account, the customer can not interact with the smart contract and the application. Then, the customer is able to send the order via e-mail in form of an IDOC file. The e-mail is then processed automatically and the IDOC is converted to a gevis file. The fields of the resulting gevis file are uploaded to the blockchain and the script of the smart contract gets invoked. Afterwards the price that is deposited in the order is compared to the price, that has been uploaded by the project partner. If these prices match, the check passes and the order is forwarded to a FTP server hosted by the project partner. From there, the order can be imported to the target ERP system.

5   Evaluation

This chapter describes how the evaluation of the blockchain implementation has been conducted using process cost accounting according to Drury [8].

5.1   Evaluation Strategy

The method of process cost accounting begins with the identification of relevant processes [8] that are of importance in the context of the blockchain introduction across companies in the pilot projects. Here, all steps should be mapped at the highest level of detail possible. In the use case "Proof-of-Orign" with the wood manufacturer, the process steps from wood preparation, transport order and production order to manufacturer declaration are considered. In the use case "Process Automation" of the technical wholesaler, it is the customer's order process steps as well as the order processing until payment receipt on the dealer's side. The approach includes a comprehensive analysis of all process steps to gain insights into costs and potentials. The resulting insights should contribute to cost savings and the targeted use of the blockchain to increase transparency and efficiency of the relevant processes of each use case [14]. In the second step of process description and modeling, a detailed recording of the selected processes for both use cases is carried out to gain a comprehensive understanding of the overall process. Subsequently, potentials and risks are identified within the process steps to achieve transparency and efficiency improvements through targeted application of the blockchain at these processes. The introduction of a blockchain in cross-company data exchange processes is a complex undertaking that requires a modeling language that enables clear and intuitive representation. Therefore, in this project, the modeling language OMEGA has been chosen over other modeling languages such as UML, EPK, and BPMN. OMEGA allows for the representation of processes at different levels of abstraction, provides a clear structure, and offers graphical symbols to depict complex relationships. In addition, the language is easy to learn, intuitively understandable, and enables high level of detail and flexibility through the use of free text. OMEGA allows for the modeling of data flows, resources, and events within a process, and is therefore suitable for the modeling of blockchain introductions in cross-company data exchange processes [10]. In the third step of the process analysis, a comprehensive process cost analysis has been conducted to determine the cost structure of the examined business processes. The costs for each process step are identified by calculating cost center costs, dividing them into direct and indirect costs, and allocating them to the process drivers [3]. Direct costs include personnel costs that are directly related to the process, while indirect costs, such as licensing and software costs, mainly target the framework conditions of the process and has been assigned to the process on a proportional basis. The time for each process step has been extracted from the respective ERP systems of the project partners. Subsequently, the costs for each step have been calculated by multiplying the identified process times with the underlying hourly rates of the involved employees and adding indirect costs proportionally. The comprehensive analysis of process costs forms the basis for comparing the costs before and after the introduction of blockchain technology in the project. The fourth step of process cost accounting involves the analysis of process costs to investigate the cost structure of a business process and identify weaknesses and potential savings. This includes examining the costs for each process step as well as the total costs to gain a comprehensive overview of the process costs. Particularly costly steps are examined more closely to determine if they can be optimized or eliminated. As part of process cost accounting, process cost optimization is a crucial component for increasing the efficiency of business processes and reducing costs. After analyzing processes and costs, targeted measures must be developed to address weaknesses and exploit optimization potential. Possible measures include automation, technology use, process revision, employee training, and product and service adaptation [8]. In both use cases the blockchain technology is used to automate manual process steps and reduce costs. This can lead to faster and more efficient process execution, minimize errors, and improve output quality. In the present study, process cost controlling has been considered as a crucial element for monitoring cost development and the success of optimization measures. To identify potential cost savings, an actual-to-target cost comparison has been performed, in which actual and planned costs have been compared. Subsequently, the total savings potential is extrapolated for all relevant customers and suppliers of the project partners, which is part of the evaluation strategy. This approach enables the derivation of conclusions about specific process optimization measures, which allows the estimation of the total cost savings potential through the scaling of the blockchain solutions.

5.2   Results

For privacy reasons, realistic and plausible data provided by the participating companies are used for this paper's evaluation, but exact internal operational information is not disclosed.

Proof of Origin: As already described in Sect. 4.1, the process of the manufacturer of wood-based materials can be subdivided into the four relevant process steps. Here, the blockchain adds value in terms of data exchange, data management and especially on the transparency of the data trough the whole supply chain. In the current project, only the implementation of the process step "Timber provision notification" has been implemented in the WAVES test network, so the evaluation focuses on this process step. According to the evaluation strategy, this project is in the third phase and data collection. The process times have been determined so far, but no associated costs have been assigned, so conclusions regarding the time savings are only possible. A preliminary extrapolation for all suppliers is also not yet possible and will follow in a further publication. Since different suppliers execute this process step, it is heterogeneous and shows the greatest potential for improvement. As sustainability is important in the wood-based materials industry, the admixture of recycled material is around 80% depending on the product (chipboard or medium density fiberboard). The approximately 20% percent fresh wood used for wood products is sourced from over 130 suppliers in 2022. According to a preliminary legislative proposal of the European Parliament, operators in the industry will be obliged to retrace the exact origin of wood [9]. This includes the harvesting and its geolocation data as well as product descriptions and information about the supplier. Consequently, the focus of the blockchain application is the creation of a valid information base that is transmitted in a forgery-proof manner throughout the entire supply chain to the customer. This will enable the origin of the wood product to be proven in the future. The application considers both the notification of the wood in the forestry and by rangers in the forest via mobile devices. In the long term, customers will be able to trace the origin of wood products from this manufacturer directly by using identifiers (e.g. QR code). In this use case, about 10% of the process time can be saved in terms of process optimization at the local forestry enterprise when notifying the provision of the wood because the total process is reduced from 9 process steps to 8. Savings in process time of about 6.5% are possible for the wood manufacturer in the verification and securing of certificate information because a check is no longer necessary and the process steps are reduced from 8 to 7. This is also in line with the main objective of this project, a transparent data-based supply chain from forest to wood product.

Process Automation: Within the scope of the second pilot project, an analysis of process steps with the greatest manual time expenditure in the order and order processing has been conducted. The investigation revealed that manual order data entry, manual order and inventory checks, manual order confirmation creation and transmission, and manual verification of order confirmations, invoices, and payment receipts are particularly time- and cost-intensive. These manual, non-value-added processes have the greatest potential for automation using blockchain technology. While the current process requires 28 process steps on the part of the customer and the dealer from order initiation to invoice verification, the use of a blockchain, including smart contract for verification, order creation, and file exchange, reduces the number of steps to just 15. The 13 automated process steps streamline the previous process by about 46%. Looking at each step individually, the time reduction is 34 min from 58 min previously (exported from the ERP system) for the entire process to 24 min for the optimized process. This corresponds to a time saving of around 59% per order. The time reduction per incoming order using blockchain technology is about 20 min for the technical wholesaler and about 14 min for the customer. The current cost per order is estimated at €53.33, assuming an average hourly rate of €50 for the involved employees and cumulated fixed costs. The price of the cryptocurrency of the used blockchain WAVES averaged €2.01 in March/April 2023. The invocation of the smart contract that performs the plausibility check on the prices costs 0.009 WAVES and the data transactions to upload the order information to the blockchain have an estimated price of 0.027 WAVES. This results in an approximate transaction cost of €0.07 per order. With the use of blockchain technology, the cost per order is reduced to €22.82. This specific customer placed 375 orders with the technical wholesaler last year, resulting in current process costs of approximately €20,000 on both sides. With the implementation of blockchain technology, these costs could be reduced from around €11,500 to approximately €8,500. At the current stage of the project, transferability to other customers could not yet be investigated. In analogy to the use case "Proof of Origin", the blockchain solution is designed in such a generic way that a transferability to other customers is recommendable.

6 Conclusion and Future Work

In summary, the implementation of a blockchain solution can lead to savings in both process costs and process times, and can contribute to greater transparency and trust between collaborating parties. The uniformization of heterogeneous interfaces in the area of ordering processes leads to a more efficient and simplified process. The savings in process costs in the use case of process automation are mainly caused by the replacement of manual activities. General digitization activities would certainly have achieved similar results in time and cost savings and those are therefore not specifically attributable to the Blockchain technology, although it can be said that the Blockchain is also definitely suitable for achieving those results and offers the advantages of a low investment and a high level of scalability. Due to the development stage of the solutions, which are still in the demonstration phase, the analysis in this paper does not yet include all costs of using the blockchain. The potential for scaling through transferability to other customers and suppliers will also be covered by a further publication when the solutions have reached a more advanced stage and gone live. Future work is planned to look at the results in more detail. For this purpose, the investigated processes must be evaluated repeatedly over a longer period of time in order to verify whether the optimization measures have a long-term effect. In addition, the use cases of the pilot projects are to be further developed and expanded in a planned subsequent research.

Acknowledgements. This research was supported by the European Regional Development Fund 2014–2020 within the BC4SC (Blockchain for Supply Chain) project (EFRE 0200617).

References

1. Kuratorium für Waldarbeit und Forsttechnik e.V.: Kwf-ev/eldat-schema:eldat is the data standard for communication between the partners in the wood supply chain. https://github.com/kwf-ev/eldat-schema

2. Abu-elezz, I., Hassan, A., Nazeemudeen, A., Househ, M., Abd-alrazaq, A.: The benefits and threats of blockchain technology in healthcare: a scoping review. Int. J. Med. Inform. **142**, 104246 (2020). https://doi.org/10.1016/j.ijmedinf.2020.104246, https://www.sciencedirect.com/science/article/pii/S1386505620301544
3. Bhimani, A., Horngren, C.T.: Management and Cost Accounting, vol. 1. Pearson Education, London (2008)
4. Bhutta, M.N.M., et al.: A survey on blockchain technology: evolution, architecture and security. IEEE Access **9**, 61048–61073 (2021). https://doi.org/10.1109/ACCESS.2021.3072849
5. Carls, V., Schmidt, L., Jansen, M.: Evaluation and comparison of a private and a public blockchain solution for use in supply chains of SMEs based on a QOC analysis. In: Prieto, J., Benítez Martínez, F.L., Ferretti, S., Arroyo Guardeño, D., Tomás Nevado-Batalla, P. (eds.) BLOCKCHAIN 2022. LNNS, vol. 595, pp. 388–397. Springer, Cham (2023). https://doi.org/10.1007/978-3-031-21229-1_36
6. Carls, V., Schmidt, L., Jansen, M.: Overview of multiple user encryption for exchange of private data via blockchains. In: Prieto, J., Benítez Martínez, F.L., Ferretti, S., Arroyo Guardeño, D., Tomás Nevado-Batalla, P. (eds.) BLOCKCHAIN 2022. LNNS, vol. 595, pp. 447–453. Springer, Cham (2023). https://doi.org/10.1007/978-3-031-21229-1_41
7. Casado-Vara, R., Prieto, J., la Prieta, F.D., Corchado, J.M.: How blockchain improves the supply chain: case study alimentary supply chain. Procedia Comput. Sci. **134**, 393–398 (2018). https://doi.org/10.1016/j.procs.2018.07.193, https://www.sciencedirect.com/science/article/pii/S187705091831158X
8. Drury, C.M.: Management and Cost Accounting. Springer, Cham (2013). https://doi.org/10.1007/978-1-4899-6828-9
9. European Union: Green Deal: EU agrees law to fight global deforestation and forest degradation driven by EU production and consumption (2022). https://ec.europa.eu/commission/presscorner/detail/en/ip_22_7444
10. Gausemeier, J., Dumitrescu, R., Pfänder, T., Steffen, D., Thielemann, F.: Innovationen für die Märkte von morgen: strategische Planung von Produkten, Dienstleistungen und Geschäftsmodellen. Carl Hanser Verlag GmbH Co KG (2018)
11. GWS Gesellschaft für Warenwirtschafts-Systeme mbH: Gevis ERP Business (2023). https://gws.ms/en/gevis-erp-business/
12. Holschbach, E., Buss, E.: Blockchain in Einkauf und Supply Chain. Springer, Wiesbaden (2022). https://doi.org/10.1007/978-3-658-36967-5
13. Keresztes, E., Ildikó, K., Horváth, A., Zimányi, K.: Exploratory analysis of blockchain platforms in supply chain management. Economies **10**, 206 (2022). https://doi.org/10.3390/economies10090206
14. Vanderbeck, E.J.: Principles of Cost Accounting. Cengage Learning, Boston (2012)
15. Wong, L.W., Leong, L.Y., Hew, J.J., Tan, G.W.H., Ooi, K.B.: Time to seize the digital evolution: adoption of blockchain in operations and supply chain management among Malaysian SMEs. Int. J. Inf. Manag. **52**, 101997 (2020). https://doi.org/10.1016/j.ijinfomgt.2019.08.005, https://www.sciencedirect.com/science/article/pii/S0268401219304347
16. Yang, R., et al.: Public and private blockchain in construction business process and information integration. Autom. Constr. **118**, 103276 (2020). https://doi.org/10.1016/j.autcon.2020.103276, https://www.sciencedirect.com/science/article/pii/S0926580520301886

Report on Decentralized Identity Models and Study on Main Privacy Concerns

María Ruiz Molina(✉) ⓘ

Universitat Oberta de Catalunya, Av. del Tibidabo, 39, 08035 Barcelona, Spain
maria-ruiz-molina@proton.me
https://mashenka.substack.com/p/portunus30

Abstract. Decentralized Identity is a expanding topic regarding privacy matters which is strongly developed on blockchain technologies. It allows the users of the model to become the real and only owners of their data, being able of choosing how and with whom to share their information. Due to the growing interest in the topic, several proposals have been created, such as the Verifiable Credentials from the W3C, and recently the SoulBound Tokens from the Ethereum Foundation. The objective of this study is to show the impact on Web3.0 of both schemes, with important contributions to the ever-growing ecosystem. After said analysis, showed the importance of both paradigms due to their consolidated works and strong support by different communities, the necessity of a privacy analysis for their right use, mainly under the GDPR compliance, is highlighted. For the correct use and maintenance of both of them, a definition of their limits and most suitable use cases is made. To this end, the privacy concerns of both main models is studied, and a selection of the most appropriate use cases for each of them is proposed alongside based on their characteristics and necessities.

Keywords: Blockchain · Blockchain applications · Privacy · Web3.0 · Decentralized Identity

1 Introduction

Digital Identity is becoming everyday a more important topic in our society, to the point most of our information is stored virtually. This increasing exposure on the web led to the development of a lot of techniques to obtain data from the users, creating a market focused on their information. Data is used to create profiles of everyone using Internet today and then, sold to different interested parties, making users to lose the control of their identity and more important, of who has access to their information, sometimes involving even personal or sensitive data.

Moreover, the centralization of the identity, makes users to have to remember a different specific set of credentials, usually a username and a password, for accessing every single service on the web. This leads to users repeating and

J. M. Machado et al. (Eds.): BLOCKCHAIN 2023, LNNS 778, pp. 13–22, 2023.
https://doi.org/10.1007/978-3-031-45155-3_2

recycling passwords, or the use of considered weak ones. These main reasons, among others, such as the difficulties to maintain data of users true to themselves or homogeneous among different databases, led to the creation of new identity models [1], such as Identity Management Architectures (where a user can identify themself with the same pair of name-password in a set of trusted domains) or Federated Identities (where the previous concept is expanded and now the user after being identified within the federation, it is given a token that they can use to identify themselves in the parties conforming the group). As it can be appreciated, the main focus was to make the process more friendly to the users, so they do not have to remember a lot of passwords. Now the focus is in the decentralization of this information [2], giving back to the users their own identity.

This Decentralized Identity is one of the main topics where Blockchain Technology has a promising future, reason why several proposals have been suggested in order to establish a standard for Self-Sovereign Identity that allows developers to build on a uniformed ecosystem, such as the Verifiable Credentials (VC) model from the W3C, which has a strong impact in the European Blockchain Service Infrastructure (EBSI) [3].

Now with the introduction of the SoulBound Tokens, a new paradigm must be taken into account. The practical implementation of said tokens has already started under the EIP-5484: Consensual Soulbound Tokens, an interface that allows the creation of tokens under a similar structure than to the referred in the ERC-721 standard. Given the starting high level of acceptance around it in Web3.0 communities, this new paradigm makes the whole ecosystem to be reconsidered. Due to the differences regarding privacy concerns, actors and interactions within the ecosystem, it is necessary to define their limits and find the most suitable use cases for each of them, highlighting too, those where both can coexist and where an interoperability would be needed to communicate both models.

2 Privacy Matters Regarding Blockchain

The European Union regulates individual's data through the General Data Protection Regulation (GDPR), being it applicated since 2018 [4]. Nonetheless, this regulation was structured implying that both data controllers and data processors can be identified. This idea collides sometimes with the intrinsic implementation of blockchain technologies, where transactions can be written in the ledger by the data subject, or clear definitions over what participants should be considered as such roles have not been officially established by the European Data Protection Board (EDPB) neither in court [5].

Another main issue that blockchain technologies bring regarding the GDPR compliance is the immutability of the information written on it. This collides with the right to erasure (articles 17 and 19), and the right to rectification (articles 16 and 19). Also, personal data should be stored anonymized and never in clear or plain text, neither using techniques that are not considered by the

GDPR as correct for said purpose. This means that the anonymized data should not be obtainable only with the original one, and of course the original data cannot be deduced from the anonymized one [6].

For this, information that may be inscribed on it has to be done in a way that the deletion of a component used for said anonymization can be technically equal to the erasure of the original information, since the original data cannot be retrieved anymore without the existence of said element. A way of achieving this condition, proposed by several reports before, as in *Blockchain and GDPR, a thematic report prepared by The European Union Blockchain Observatory and Forum* [7], imply the use of a hashing function alongside a cryptographic key. This way we make sure that anonymized data is not obtainable only with the original one, given the necessity of using the algorithm along said key. Also, we can achieve a state similar to the erasure of the information the moment the key is completely deleted, because if so, there would not be a possible way of retrieving the information from the anonymized data written on-chain.

3 Verifiable Credentials

A Verifiable Credential refers to information and data related to a specific entity, that can be proved to be original and issued by a specific trustworthy organization or unit. This purpose is achieved through digital signatures, used to give validation to their content. Moreover, the Verifiable Credentials can be presented through Verifiable Presentations to different Service Providers for them to validate before accepting them. These Verifiable Presentations allow to submit several Verifiable Credentials in one unique bundle, signing the total of the data to certify the validity of the credentials as a group, but also keeping intact the single signatures of every single Verifiable Credential. Several specifications have been created around the Verifiable Credentials idea [8], one of them being from the W3C [9], a big referent due to have been accepted by the European Union as the model to use within the reference framework for Digital Identity wallet architecture [10].

As it can be seen in this model, we have three different entities [11] interacting on it.

- **Issuer.** Entities that are allowed to emit credentials and sign them. These signatures are demonstrable through the blockchain, where these Issuers are registered along with their public keys, so anyone with access to that ledger can check if they are valid. Usually, this registration is made by a main authority or authorities that select and allow the Issuers to emit Verifiable Credentials in the model to certain degree of proof and sensitivity.
- **Service Provider or Verifier.** Entities that can request or receive data in the form of Verifiable Credentials from different users in the model and verify through the blockchain the veracity of the signatures, and so, of the data, included in them.
- **User or Holder.** The owners of the data represented in the Verifiable Credentials. They are the main and only owners of it, having full control on to whom they want to present their full or partial information.

These Verifiable Credentials are technically represented on JSON Web Tokens, so they are not a data introduced necessarily on-chain. This is why they have intrinsically specified a timestamp scope, both for the time of the issue and where the token does not have validity anymore and expires. Of course, if minted on a blockchain, they would have the first timestamp written permanently on-chain the moment the Verifiable Credencials are emitted.

3.1   Privacy Regarding Verifiable Credentials

Verifiable Credentials do not have to be fully written on-chain to work and operate as valid tokens, due to the signature attached to them being the blockchain evidence of their integrity and legitimacy.

This way, when data is presented to a Service Provider, the users can select what attributes they want to disclose and which ones to keep to themselves. On the flip side, this means that the information will be stored off-chain. For it, some possibilities for this are peer-to-peer systems [12] like IPFS (InterPlanetary File System [13], a decentralized protocol that enables the storage and sharing of files on a P2P network), a centralized server or locally by the user. While this may not be the best approach for a fully decentralized model, it is more user friendly since data can be backed-up.

Verifiable Credentials should be accessible only by the owner and the Issuer during the initial generation. Once generated and signed, the users should have exclusive control and be able to revoke access to other actors by any time. To achieve this, the Verifiable Credential should be stored securely, granting only the holder access to prevent unauthorized use or access by others.

After it, a user should be able of choosing what data within the credentials wants to disclose to the different Service Providers. For this, they can use different Zero Knowledge Proofs [14], by means of which they can show to the Verifiers that they own certain information without fully disclosing it, or generate micro-credentials, prompts out of the content of a Verifiable Credential where only part of the data is disclosed and the remaining is kept private.

Removing Access to Verifiers. Nonetheless, once a Verifiable Credential is presented to an entity for its corroboration, the user should be allowed to remove said access at any point. For this, it is needed to keep an authorization control where the user can easily remove this access. This control can be done forcing the Verifier to access first an environment to request or check an information needed to validate the data in the Verifiable Credential. Given the previous necessity of keeping a key used for hashing the data, this could be the necessary asset. This way, the Verifiers could hash the received data with the accessed key and compare it with the information on-chain. This way we would keep track of the times a Verifier checks the validation of a piece of data.

To make the user in control of the revocation of said access, it could be requested to the Verifier to have a token with the following characteristics to access the key used for hashing.

– Created by the user the moment it is sent to the Verifier along with the data for first time. It must be signed by the user to avoid the creation of undesired copies.
– The system where the Verifier access to get the key requires this token and checks the signature of the user.
– The user can revoke the validity of the token at any time, and when that happens the Verifier will not be able of accessing the system where the keys are hosted. This event will be registered on-chain too with its respective timestamp.

While the system does not prevent Verifiers from copying data or keys upon initial access, in the case of them checking the on-chain data after a revocation, an event could be immediately triggered. This event would detect that the access to the data happened after the timestamp of the revocation, and so, it could alarm to an authority about the misuse. This way, the user can keep their full ownership on their identity even after disclosing some information to different parties.

Different Levels of Issuance. Given the wide variety of data, issuers should be allowed to emit Verifiable Credentials according to a level of capacity and privacy. An entity such as an online unofficial academy should be allowed of issuing data about those certificates, but never personal data related to other fields that are not the unofficial academic ones, for example a University Degree or a Passport.

For this to be achieved, when a Issuer is registered along their signature on-chain to allow Verifiers the validation of the credentials, they should get first the permission to do so and there should be kept a clear difference among what they can emit and what they cannot and should not be accepted by Service Providers even if the signature is valid. To this end, there should be an entity or group of entities that verify and do this registration, not only storing on-chain the signatures but also a code for what kind of data the Issuers should be allowed to create. Additionally, the registered signatures should be revocable by the overseeing entities in case of irresponsible behavior, with on-chain attributes indicating the state of the signature for the Service Providers to check the moment of a verification.

4   SoulBound Tokens

SoulBound Tokens (SBTs) are a ERC standard (Ethereum Request for Comments [15], a standardization proposal used to propose and discuss improvements, protocols, and guidelines for the Ethereum blockchain [16]) specifically the ERC-5484 [17]. This means that SBTs are not exclusive of Ethereum main net, but also feasible of developed in other chains or layers that allow the execution of Smart Contract in the same way that the main net does.

This standard inherits from the ERC-721 [18], the Non Fungible Tokens (NFTs) one. This means that they include all the characteristics of an NFT and will add others. The main functionalities of NFTs are the following, making them a good representation for collectibles or tickets and passes:

- **Unique and fungible.** This means that they are differentiable of each other, having their own code on-chain, set of characteristics, address and identifier if part of a collection (this means a set of NFTs minted under the same Smart Contract). Each of them can have a different value, independent of other tokens created using the same interface.
- **Representation of attributes.** These tokens allow the specification of arbitrary attributes, allowing customization depending on the NFT needs, that will be written on-chain the moment they are minted. Among these attributes, media can be attached. It is true that embedding big size files inside of a token on-chain has high costs, both computational and on gas fees (a payment that must be done for every written action on-chain). This is why most NFTs attach a URL instead, pointing to a storage system, preferable a P2P one to avoid centralization.
- **Transferable.** These tokens are not attached to the address that receives them the first time, allowing users to exchange, sell and buy, create auctions and bid them. Everytime they change of owner, a transaction is written on-chain specifying costs, sender and receiver, among other specifications.
- **Ownership.** They are linked to an address, the owner of the token. They are the only ones with the rights to interact with them via their wallets, such as Metamask [19] or others.

Nonetheless, SBTs modify one of these characteristics and add another one, making them suitable for the concept of identity instead of digital collectibles or tickets, items that can be exchanged and are not intrinsic to oneself.

- **Nontransferable.** The main overwritten characteristic is that SBTs are not transferable once they are minted and sent to the holder. This way there is a correlation one-on-one between the holder and the token itself. Only burning is allowed and under certain circumstances, sending the token to the Null Address and making it unusable and unrelated to the owner [20] (but if tracked to the previous transaction).
- **Setting of Burning Authorization.** The moment the SBT is issued and so, created, there is a establishment of who will be allowed of burning the token if necessary. This entity can be either only the issuer (*IssuerOnly*), only the holder of it (*OwnerOnly*), both of them through a consensus (*Both*) or none of them (*Neither*).

These tokens are Smart Contracts, and so, intrinsic to the blockchain. This means that as everything that is created on-chain, they will have a timestamp embedded the moment a transaction happens. This includes, the moment of the issue, but also if burning happens.

4.1   Privacy Regarding SoulBound Tokens

SoulBound Tokens' information is on-chain, but linked media to reduce gas fees. This means that anyone with reading access to the ledger can see the information regarding any SBT, due to their Smart Contract nature. This property, if it makes the token more immutable thanks to the characteristics of the ledger, it also makes the information completely accessible to everyone on a public ledger, and to every kind of participant on a private one.

If SBTs were used to create tokens regarding personal data, this would not only be against the GDPR compliance, but can surely bring serious issues to the owners, who can find out that their information is being showed publicly. With this approach using *vanilla* SBTs for storing personal information, now the data is accessible to everyone making it much less private than before. This is why this implementation has to be done along with the use of other techniques to keep privacy.

A Solution for Proper Data Anonymization. It is proposed on *Decentralized Society: Finding Web3's Soul*, where the concept of SBTs is formally presented for the first time, to use hashing algorithms to anonymize the data before uploading it on-chain [20]. This could be a good approach because this technique allows to get the cyphered data from the original one, but not the other way, keeping the hashed information private by itself. Nonetheless, the GDPR does not accept this technique as a proper anonymization technique. The reason behind this is that for it to accomplish this definition, the anonymized data cannot be obtained using only the original data. This is the case for a hashing algorithm, which will always generate the same output for a specific input.

Because of this, to achieve said compliance with the regulation an external element has to be included in the algorithm so for getting the anonymized data is needed more than just the original file. The use of a cryptographic key alongside the algorithm alters the result from the use of it without any kind of key. This way, for getting the cyphered result it would be needed not only the original data, but also the key. This way, the user could also request the erasure of their data, referring to the Article 17 in the GDPR, by the elimination of the key; and in a similar way also modify the information by generating after it the corrected token.

This key should be secured and kept in a private place where only allowed participants in a Decentralized Identity model can access to it. This may include the issuer for generating the information before sending it to the holder, and the users if it is required the recovery of the information, including the on-chain one, by themselves. It must be made clear that this technique will only be mandatory when personal data is involved, this being physical entities data, not legal ones, since the GDPR only applies to individuals and their data that could reveal their identity or part of it either by itself or in combination with others piece of information.

Use Case: Airdrops to SBT Wallets. Application of the GDPR when the data can identify an individual in combination with other traits can enter in dispute with some use cases where the information being public can be beneficial, such as the membership to communities or attendance to certain congresses or conferences. The information by itself could not be enough for finding out who is the person owning those SBTs, but if an individual collects several tokens from different events, gathering them publicly in their wallet, the anonymization group will become smaller with each addition, and it may be possible to retrieve from lists of attendance or similar datasets who is the common person among all of them. While keeping this information public to allow entities to airdrop tokens to possibly interested parties based on what places they attend as a way of promotion, it is important to keep privacy in mind and the issues that could arise from models like this, where spamming techniques could arise too and make the users victims of it. A solution for this kind of situations may be the creation of a model where SBTs are hashed with a proper key and mechanism, and stakeholders that would like to airdrop tokens to certain individuals should be first verified, and after that users should accept within the ecosystem what information they are happy to publicly share. A Trusted Execution Environment (TEE) could be used where the execution of the airdrop happens in the following way, where a TEE is an area in the main processor separated from the Operative System that can receive commands and execute them isolated from every user, even the ones with root permissions.

1. Users select what kind of airdrops they would like to receive in their wallets and in what kind of tokens (fungible and/or non-fungible, transferable and/or nontransferable, and related to Layer 2 Protocols/NFTs/DeFi... being some of the possible categories). They send the data related to the information of their hashed SBTs.
2. The keys used for hashing the data on-chain are sent to the TEE.
3. The original data is hashed with the received key and compared to the information on-chain, the SBT. This way the data is authenticated and verified.
4. The interested parties, after verification within the model, register what kind of tokens they would like to airdrop and to what kind of users.
5. The execution takes place inside of the TEE. Data stored in the SBTs and interests of both holders and parties are contrasted and matched.
6. The tokens of the airdrop are sent to only those holders that meet the specified requirements for the airdrop.

This way, the entity making the airdrop would not have access to the keys neither to the private information of the users, but the execution could happen within the trusted environment ensuring that the promoted tokens will be airdropped only to interested users that fulfill the requirements for being picked.

5   Conclusions

It is important to find proper technical solutions to make sure that privacy concerns regarding the General Data Protection Regulation are achieved on blockchain technologies and decentralized identity models based on these new technologies. In the following steps for evolving towards a more private world, the use of distributed ledgers alongside other solutions, such as hashing or peer-to-peer networks for storage, can serve as ways of addressing the issues regarding proper anonymization and granting full control to the users of the model over their own data, even after disclosing information to different entities.

For it, a private and user-centric model can be achieved using the Verifiable Credentials system or a SoulBound Tokens one. The first has been already accepted within the European Union for issuing identity related data. Some of the use cases that are already being proposed are ID documents, such as passports, health or medical related data, education and academic related data. As it can be seen, currently they are more oriented to services provided or possibly required by public administrations. On the other hand, SoulBound Tokens count with a bigger exposure of the data due to their on-chain nature. However, this approach can be more suitable for community membership proof, or events attendance. Some of the stored information, however could need from other technologies to keep it private.

This need of treating data properly and GDPR-compliant makes this scheme, whatever model is chosen, to be used in combination with proper private systems. This includes proper anonymization using hashing algorithms with keys and secure storage of said keys used for it; a system, such as the one mentioned using tokens issued by the holder, to prevent Verifiers from keeping copies of the data; and using Trusted Execution Environments where interactions with data are completely isolated from unwanted actors or too-invasive actions. This way it is possible to keep the nature of SoulBound Tokens as readable for everyone on-chain and the privacy and anonymization of the data stored in them.

While these implementations are not fully accepted by the official organisms, the development and establishment of blockchain technologies in some territories will be slowed down as proper standards and regulation around them could not be developed. These reasons are why these steps and further studies must be taken in order to implement a more private and user-focused data scheme. It is important to note the necessity of making this approach enough user-friendly too, finding a balance between decentralization and complete reliability of data by the users.

Acknowledgements. I would like to thank my University teacher Mercedes for showing me the importance of privacy in a connected world, to my mother Fany for being always supportive with my choices in life whether they are related to studies or not, to my partner Nat for believing in my capabilities and her encouragement throughout this project, and to my friend Juan for enduring all my technical explanations.

References

1. Soltani, R., Nguyen, U.T., An, A.: A survey of self-sovereign identity ecosystem. Secur. Commun. Netw. **2021**, 1–26 (2021)
2. Chang, S.S., Park, J.Y.: Understanding of blockchain-based identity management system adoption in the public sector. J. Enterp. Inf. Manage. **34**(5), 1481–1505 (2021)

3. EBSI W3C Verifiable Credentials (VCs) and W3C Verifiable Presentations (VPs). https://ec.europa.eu/digital-building-blocks/wikis/pages/viewpage.action?pageId=555222155. Accessed 11 Apr 2023
4. Council of European Union: Regulation (EU) 2016/679 - General Data Protection Regulation (2016)
5. Commission Nationale de l'Informatique et des Libertés: Solutions for a responsible use of the blockchain in the context of personal data, Official Guidelines (2018)
6. European Parliament, Directorate-General for Parliamentary Research Services, Finck, M.: Blockchain and the general data protection regulation: can distributed ledgers be squared with European data protection law? (2019). https://doi.org/10.2861/535
7. Lyons, T., Courcelas, L., Timsit, K.: Blockchain and the GDPR, a Thematic Report Prepared by The European Union Blockchain Observatory and Forum, 1st edn. (2018)
8. Kudra, A., Lodderstedt, T., Bastian, P., Mollik, M., van Leuken, M., Roelofs, C.: Credential comparison matrix. https://docs.google.com/spreadsheets/d/1Z4cYfjbbE-rABcfC-xab8miocKLomivYMUFibOh9BVo/edit#gid=422610777. Accessed 19 Apr 2023
9. Stone, M., et al.: Verifiable credentials data model v1.1. W3C Recommendation (2022). https://www.w3.org/TR/vc-data-model/. Accessed 14 Apr 2023
10. The European Digital Identity Wallet Architecture and Reference Framework: The Common Union Toolbox for a Coordinated Approach Towards a European Digital Identity Framework (2023). https://www.intesigroup.com/en/wp-content/uploads/sites/4/2023/02/ARF-v1.0.0-final.pdf
11. Liu, Y., He, D., Obaidat, M.S., Kumar, N., Khan, M.K., Choo, K.K.R.: Blockchain-based identity management systems: a review. J. Netw. Comput. Appl. **166**, 102731 (2020)
12. Daniel E., Tschorsch F.: IPFS and friends: a qualitative comparison of next generation peer-to-peer data networks (2022). https://arxiv.org/abs/2102.12737
13. Protocol Labs: InterPlanetary File System. https://docs.ipfs.tech/. Accessed 19 Apr 2023
14. Dieye, M., et al.: A self-sovereign identity based on zero-knowledge proof and blockchain. IEEE Access (2023). https://doi.org/10.1109/ACCESS.2023.3268768
15. Ethereum: Ethereum Improvement Proposals: ERC. https://eips.ethereum.org/erc. Accessed 19 Apr 2023
16. Wood, G.: Ethereum: a secure decentralised generalised transaction ledger. Ethereum Proj. Yellow Pap. **151**, 1–32 (2014)
17. Cai, B.: ERC-5484: consensual soulbound tokens. Ethereum Improvement Proposals, no. 5484 (2022). https://eips.ethereum.org/EIPS/eip-5484
18. Entriken, W., Shirley, D., Evans, J., Sachs, N.: ERC-721: non-fungible token standard. Ethereum Improvement Proposals, no. 721 (2018). https://eips.ethereum.org/EIPS/eip-721
19. ConsenSys Formation: Metamask. https://metamask.io/. Accessed 19 Apr 2023
20. Weyl, E.G., Ohlhaver, P., Buterin, V.: Decentralized society: finding web3's soul. SSRN (2022)

Blockchain-Based Decentralized Autonomous Organizations (DAOs) for Democratization in the Environmental, Social, and Governance (ESG) Rating Process

Richard[1]($\boxtimes$) ⓘ, Alexander Agung Santoso Gunawan[2] ⓘ, Felix Irwanto[1],
Gabrielle Peko[3] ⓘ, and David Sundaram[3] ⓘ

[1] School of Information Systems, Bina Nusantara University, West Jakarta, Indonesia
`richard-slc@binus.edu`
[2] School of Computer Science, Bina Nusantara University, West Jakarta, Indonesia
[3] Business School, The University of Auckland, Auckland, New Zealand

Abstract. The Environmental, Social, and Governance (ESG) initiative has recently become an emerging research area. ESG rating is a critical activity in ESG initiatives that still holds some issues. The ESG rating process needs significant transparency, democratization, and standardization improvement. This research aims to propose the technical architecture of a blockchain-based ESG rating system that applies the concept of DAO as the democratization enabler. The Design Science Research (DSR) method is used for observation and theory building of the technical architecture. This research's outcome illustrates how the proposed technical architecture can fit into the typical ESG rating process. The technical architecture can also become the guideline for future experiments and implementing blockchain-based ESG rating systems.

Keywords: Environmental, Social and Governance (ESG) Reporting · Blockchain · Decentralized Autonomous Organization · Design Science Research

1 Introduction

Environmental, Social, and Governance (ESG) rating has become a crucial consideration for investors as they seek to invest in companies that promote sustainability values on top of the company's financial performance. The ESG rating is often acclaimed as the mirror of a company's sustainability and ethical actions, albeit the process of generating the rating is siloed. There are various problems in the ESG rating system, such as inconsistent rating techniques, a lack of transparency, and limited access to information [1]. These problems might result in flawed assessments, which makes it challenging for investors to make wise selections [2].

Since transparency is one of the main problems in the ESG rating process, blockchain technology is one potential solution that can be applied to solve the problem. By enabling

J. M. Machado et al. (Eds.): BLOCKCHAIN 2023, LNNS 778, pp. 23–31, 2023.
https://doi.org/10.1007/978-3-031-45155-3_3

more precise reporting and verification of sustainability activities, blockchain can give ESG programs transparency and immutability [3]. Smart contracts and Decentralized Autonomous Organizations (DAOs) are two examples of blockchain components that can work with the ESG rating system [4]. DAOs are gaining popularity as a potential answer to the transparency problem in ESG ratings as it runs on a public blockchain network, allowing for total transparency and decision-tracking of all organization activities. With a community-driven decision-making process, smart contract-based governance, and token-based incentives, DAO participants can ensure that the organization supports sustainability and adheres to the values of its constituents.

This study aims to provide a model to implement a blockchain-based DAO that can address the ESG rating by promoting transparency, democratic decision-making, smart contract-based governance, and token-based incentives. The first section will define the problems in ESG Rating and how blockchain can become the enabler. The second section will define the general ESG rating process, DAO, and the related works in this area. The third section will explain the methodology used in this research. Finally, the fourth and fifth sections will discuss the proposed model and conclude this study. The results of this research will help to fill in any gaps in the scope of existing knowledge and offer insights into the potential limitations of blockchain technology in the ESG rating. Additionally, this study will serve as a guide for creating blockchain-based ESG rating systems that apply the concept of DAO.

2 Literature Review

2.1 ESG Ratings

ESG indicator looks at a company's performance in three key areas: corporate leadership, social responsibility, and environmental responsibility. The techniques used to create ESG ratings may vary significantly amongst service providers [2]. However, they almost all include a grading system that awards points to companies based on how they behave regarding various ESG indicators.

In increasing the integrity of the ESG ratings, third-party agencies were assigned to assess the ESG metric and data provided by the company. Third-party agencies increasingly provide ESG ratings as a service to investors [5]. ESG rating agencies will match the assessment metric and the proposed ESG Data. These agencies use various methodologies to assess the ESG performance of companies, including proprietary scoring systems that consider a range of environmental, social, and governance factors. Some of the most well-known third-party ESG rating agencies include MSCI [6], Sustainalytics [7], and ISS ESG [8].

ESG ratings have come with criticism because the criteria tended to be arbitrary, challenging to measure consistently, impacted by a company's disclosure procedures, and excessively centered on short-term financial performance [9]. Despite the contras, ESG ratings grow more significantly over time for investors who want to align their assets with their principles. Therefore, these people must consider the benefits and drawbacks of such evaluations by including them in their decision-making process and other analytical techniques.

2.2 Decentralized Autonomous Organization (DAO)

A decentralized autonomous organization (DAO) is a virtual organization governed by code and runs on a decentralized network [4]. Technically, DAO is often linked with the blockchain as their place of existence. With smart contracts, the rule and agreements between DAO working groups can be accessed and examined transparently.

DAOs aim to be autonomous, transparent, and democratic, with a division of the working group that makes the decisions. Researchers have investigated the potential of DAOs in various fields, including governance, investment, and crowdfunding [10]. DAOs permit direct participation by stakeholders, lowering the possibility that businesses will manipulate their ESG ratings and evaluation. Using a secure blockchain ledger, DAOs can improve accountability and transparency by storing all decisions and acts.

Despite the benefits, there are worries regarding the possibility that DAOs could be manipulated by a small number of stakeholders or that they are not accountable to the larger society [11]. The necessity for a consistent framework for ESG criteria and possible manipulation or bias in the decision-making process are two obstacles to implementing DAOs for ESG rating.

2.3 Related Works

Since the issue of ESG rating is growing, there is not much similar research in this area—one of the research related to this article is a paper from Liu et al.. The article introduces a data-driven ESG assessment approach using blockchain technology and stochastic multicriteria acceptability analysis (SMAA-2) to address the data opaqueness and assessment subjectivity problems [12]. Another article related to this area is written by Jiang et al., which proposes a blockchain-based life cycle assessment to enhance the ESG reporting and assessment process [13].

3 Research Method

The Design Science Research (DSR) method is used for creating and analyzing artifacts to address challenging real-world issues [14]. This study involves making observations and developing theories to produce a novel output that solves the identified problem. During the observation phase, a problem is found, existing remedies are examined, and the problem's underlying causes are examined. A theoretical framework is developed in the theory-building stage to design an artifact to address the issue. This study's observation object is the applicable ESG rating process in several countries. Furthermore, the model of blockchain-based DAO for ESG rating is developed and proposed to address the problem identified.

3.1 Observation Stage

In the DSR method, the observation stage is critical. The stage entails problem identification and information collection to fully comprehend the issue and its underlying causes and pinpoint current solutions. It lays the groundwork for original solutions that deal with the underlying causes of the issue. In understanding the problem, the typical ESG rating assessment process is extracted from several rating agencies. Figure 1 shows the typical ESG rating process used as the baseline of the problem defined in this research.

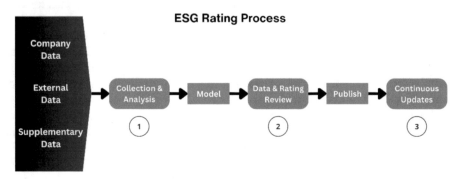

Fig. 1. Typical ESG Rating Process

There are several steps in the ESG rating process mentioned in Fig. 1. The ESG rating process starts with multiple data feeds, such as company, external, and supplementary data. The agency acquires these data to be processed and standardized in maintaining consistency across many businesses and industries. The first step in Fig. 1 shows that data is collected and analyzed before being converted into the assessment model. A model is the baseline of the rating and review process in ESG rating. The ESG ratings are then calculated using weightings and criteria unique to the industry. The final overall ESG rating, which combines the individual ESG scores to get an overall rating for each company, is then published. The last phase is ongoing monitoring and updating of the ESG ratings as new data becomes available. The agencies seek to give investors an unbiased and consistent evaluation of companies' ESG performance through this method.

There are difficulties in the ESG rating procedure. The lack of standardization and consistency among various ESG rating firms is one of the main problems. Investors may become confused, and comparing ESG scores between various companies may be challenging. Additionally, the ESG ratings may be inaccurate or inconsistent because of the dependence on self-reported data. Investors may find it challenging to comprehend how ESG ratings are determined and what variables are considered due to the lack of openness in the ESG rating process, which is another difficulty. Finally, keeping up with shifting ESG trends and best practices can be challenging due to the quickly growing nature of ESG concerns and the absence of industry standardization.

3.2   Theory Building Stage

The theory-building stage is a critical component of the DSR methodology. In this stage, the researcher uses the insights gained from the observation stage to develop a theoretical framework for designing an artifact to solve the problem. The theoretical framework provides a basis for developing a prototype or solution, which can be tested and refined in subsequent iterations. The theory-building stage involves synthesizing existing knowledge from multiple disciplines and creating new knowledge by developing the theoretical framework. By building a solid theoretical foundation, DSR researchers can create innovative solutions grounded in a deep understanding of the problem and its underlying causes.

4   Discussion

4.1   Blockchain Technology in ESG Rating Process

Based on the observed ESG rating process in the previous section, the following Fig. 2 will illustrate how blockchain can involve in each sub-processes of ESG rating.

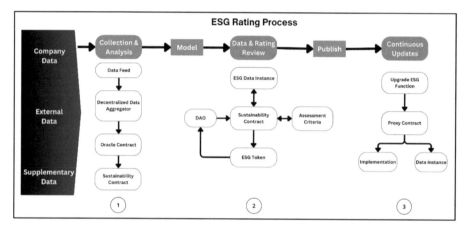

Fig. 2. Blockchain in ESG Rating Process

The blockchain role in ESG rating starts from the data collection and analysis. In the data collection and analysis process, blockchain uses a decentralized data aggregator to verify and aggregate the data feed from multiple sources. Data is then forwarded and stored in an oracle contract while being used by the sustainability contract in the future. A sustainability contract is a smart contract that holds the ESG reporting and rating rule. The contract is publicly accessible, so the ESG data and rating is transparent and traceable.

In the data rating and review process, the sustainability contract is called and executed by the DAO working groups that act as the ESG rating assessors and verifiers. The data rating and review process interact with ESG data instance and assessment criteria for the match and assessment process. After the ESG rating is assessed and published, the DAO working groups will be incentivized for their contributions to the assessment process.

The continuous updates stage exists to maintain the quality of the ESG rating process. A smart contract is an immutable object that cannot be changed once stored on the blockchain. The immutability is a counter fact of the need for ESG rule adjustment in the future. In addressing the issue, the upgradeable pattern is used to maintain and update the ESG rating rules in the sustainability contract. The proxy contract is the umbrella and pointer to implementing the sustainability contract and its data instance. Finally, the ESG rating systems can potentially increase data security, reliability, and transparency with blockchain-based DAO and smart contracts.

4.2   The Proposed Architecture of Blockchain-Based ESG Rating System

The technical architecture is conceptualized and created based on the literature review on the typical implementation of blockchain, smart contracts, and DAO in various fields. The common practice was then applied to the ESG rating process with the help of the author's experience in this research. Figure 3 shows the proposed technical architecture of Blockchain-based ESG rating systems. The proposed architecture consists of incentive, identity, privacy, interface, and blockchain layers used by the ESG mechanism.

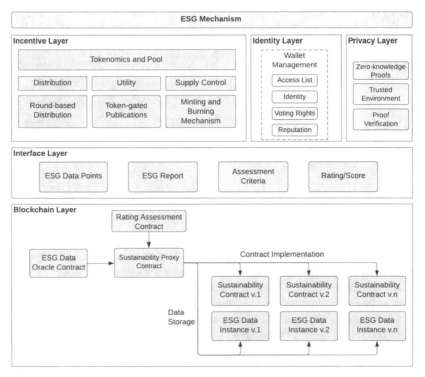

Fig. 3. Technical Architecture of Blockchain-based ESG Rating System

ESG Mechanism. This section represents the interaction between parties in ESG rating mechanisms. The involved parties include ESG-responsible organizations, investment organizations, decentralized autonomous organizations (DAO), and regulators. The DAO has a role as the rating agency that works decentralized by democratizing the assessment process through multiple working groups. The ESG-responsible organizations are companies responsible for preparing and publishing their sustainability and rated by the rating agencies. Investment organizations manage investments and portfolios based on the ESG score. The ESG rating is essential for the investment organization as they typically maintain the sustainability of the managed portfolio. Last, the regulators oversee the respective countries' entire ESG rating and investment process. Figure 4 shows the involved parties in the ESG mechanism.

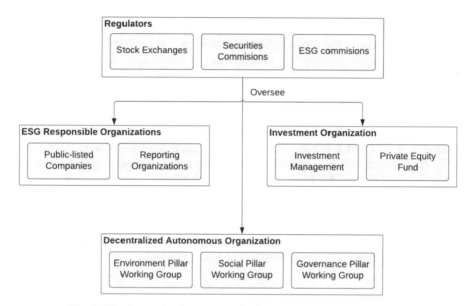

Fig. 4. The interaction between parties involved in the ESG mechanism

Incentive Layer. This layer is the main engine driving the economics of ESG rating systems. Tokenomics is the core mechanism that orchestrates the circulation of ESG tokens. The ESG token is initially stored in the token pool, while the rest is distributed to the investment organization. The ratio between pool and distribution is ruled and defined in tokenomics. Mechanisms such as token-gated publication [15] can be applied to create the utility of ESG tokens. The investment organization must hold and spend ESG tokens before accessing the ESG score. The token spent in accessing the ESG score will be used as an incentive for the DAO working group.

Identity Layer. The identity layer rules the integrity of involving parties in the ESG mechanism. The Blockchain-based identity (or self-sovereign identity) refers to a set of attributes stored in the blockchain owned and controlled by a single entity [16]. The common attributes in the identity layer are access list, identity, voting rights, and reputation. These attributes are wrapped and accessed through wallet management before being used by the involved parties in the ESG mechanism.

Privacy Layer. Privacy is a vital aspect of the ESG rating mechanism. One of the most crucial problems in ESG rating is transparency. However, transparency can enormously decrease the privacy of ESG-responsible organizations. Without non-disclosure agreements, the company will never agree to share its internal data with public and decentralized rating agencies (DAO). The privacy-preserving mechanism rescues this problem with cutting-edge technology such as Zero-knowledge proofs, trusted environments, and proof verification. Companies can state that their ESG data is valid without revealing the original value. The DAO can use verifiable proof as the ESG assessment and rating baseline.

Interface Layer. The interface layer is a front-end or back-end application involving parties interacting with the smart contract stored in the blockchain. The interface layer holds several business and programmatic functions in ESG rating systems. In Fig. 4, the business functions include ESG data points, ESG reports, assessment criteria, and rating/score. The typical connections and interactions to the blockchain use JSON-RPC protocol.

Blockchain Layer. The blockchain layer represents the smart contracts stored in the blockchain network to host and execute the ESG rating rules and data (Sustainability Contract and ESG Data Instance). In addressing the issue of ESG continuity, an upgradeable pattern is used in this design—the proxy contract act as the gateway that points any programmatic request to the active contract implementation. The rating process will use a rating assessment contract while interacting with other data sources through the ESG data oracle contract.

5   Conclusion

The involvement of technology as the enabler for better ESG rating systems has become a hot issue in recent years. This research explains blockchain technology's potential implementation in the ESG rating process. Some blockchain components highlighted in this research are smart contracts and DAO. This research's first output shows how blockchain can fit into the ESG rating process (Fig. 2). The second output is the proposed technical architecture of blockchain implementation in the ESG rating process that contains several layers, such as incentive, identity, privacy, interface, and blockchain layers. The proposed architecture is developed based on the literature review and study of the typical blockchain implementation in various fields.

The future direction of this research is to adjust the architecture to different ESG processes in other countries. Even though the architecture is designed by looking into the typical ESG rating process by MSCI, there is a possibility that the rating process is different in other cases. Moreover, the technical architecture proposed in this research can be used as the reference for the proof-of-concept of blockchain implementation in the ESG rating process. The possible experiment that can be done to extend this research is an experiment related to data consistency and performance in ESG metrics. The other experiment is related to implementing zero-knowledge-proof technology to preserve the privacy of the company's ESG data.

References

1. Tayan, B., Larcker, D., Watts, E., Pomorski, L., ESG Ratings: A Compass without Direction. Rock Cent. Corp. Gov. Stanford Univ. Work. Pap., pp. 1–16, Aug. 2022, Accessed: 28 Apr 2023. [Online]. Available: https://corpgov.law.harvard.edu/2022/08/24/esg-ratings-a-compass-without-direction/
2. Escrig-Olmedo, E., et al.: Rating the raters: evaluating how ESG rating agencies integrate sustainability principles. Sustain 11(3), 915 (2019). https://doi.org/10.3390/su11030915
3. Liu, X., Haoye, W., Wei, W., Ye, F., Huang, G.Q.: Blockchain-enabled ESG reporting framework for sustainable supply chain. In: Scholz, S.G., Howlett, R.J., Setchi, R. (eds.) Sustainable Design and Manufacturing 2020: Proceedings of the 7th International Conference on Sustainable Design and Manufacturing (KES-SDM 2020), pp. 403–413. Springer Singapore, Singapore (2021). https://doi.org/10.1007/978-981-15-8131-1_36
4. Wang, S., Ding, W., Li, J., Yuan, Y., Ouyang, L., Wang, F.-Y.: Decentralized autonomous organizations: concept, model, and applications. IEEE Trans. Comput. Soc. Syst. 6(5), 870–878 (2019). https://doi.org/10.1109/TCSS.2019.2938190
5. Del Giudice, A., Rigamonti, S.: Does audit improve the quality of ESG scores? Evidence from corporate misconduct. Sustainability 12(14), 5670 (2020). https://doi.org/10.3390/su12145670
6. MSCI. MSCI – Powering better investment decisions. 2021. https://www.msci.com/. Accessed 28 Apr 2023
7. Sustainalytics. Home – Sustainalytics. https://www.sustainalytics.com/. Accessed 28 Apr 2023
8. ISS. Home | ISS. https://www.issgovernance.com/. Accessed 28 Apr 2023
9. Macmahon, S.: The challenge of rating esg performance. Harvard Business Review, vol. 2020, no. September-October. pp. 17–18, 2020. Accessed 28 Apr 2023. [Online]. Available: https://hbr.org/2020/09/the-challenge-of-rating-esg-performance
10. Kypriotaki, K.N., Zamani, E.D., Giaglis, G.M.: From bitcoin to decentralized autonomous corporations: extending the application scope of decentralized peer-to-peer networks and blockchains. In: ICEIS 2015 – 1 International Conference on Enterprise Information Systems, Proceedings, vol. 3, pp. 284–290 (2015). https://doi.org/10.5220/0005378402840290
11. Morrison, R., Mazey, N.C.H.L., Wingreen, S.C.: The DAO controversy: the case for a new species of corporate governance? Front. Blockchain 3, 25 (2020). https://doi.org/10.3389/fbloc.2020.00025
12. Liu, X., et al.: Data-driven ESG assessment for blockchain services: a comparative study in textiles and apparel industry. Resour. Conserv. Recycl. 190, 106837 (2023). https://doi.org/10.1016/j.resconrec.2022.106837
13. Jiang, L., Gu, Y., Yu, W., Dai, J.: Blockchain-based life cycle assessment system for ESG reporting. SSRN Electron. J. (2022) https://doi.org/10.2139/ssrn.4121907
14. Peffers, K., Tuunanen, T., Rothenberger, M.A., Chatterjee, S.: A design science research methodology for information systems research. J. Manag. Inf. Syst. 24(3), 45–77 (2007). https://doi.org/10.2753/MIS0742-1222240302
15. Lens Protocol. Gated publications. https://docs.lens.xyz/docs/gated. Accessed 28 Apr 2023
16. Liu, Y., He, D., Obaidat, M.S., Kumar, N., Khan, M.K., Choo, K.-K.R.: Blockchain-based identity management systems: a review. J. Netw. Comput. Appl. 166, 102731 (2020). https://doi.org/10.1016/j.jnca.2020.102731

FeDis: Federated Learning Framework Supported by Distributed Ledger

Rafael Barbarroxa⬥, João Silva⬥, Luis Gomes⬥, Fernando Lezama⬥,
Bruno Ribeiro⬥, and Zita Vale(✉)⬥

GECAD – Research Group on Intelligent Engineering and Computing for Advanced Innovation
and Development, LASI – Intelligent Systems Associate Laboratory, Polytechnic of Porto, R.
Dr. António Bernardino de Almeida, 431, 4249-015 Porto, Portugal
{rroxa,dasil,lfg,flz,brgri,zav}@isep.ipp.pt

Abstract. Despite the several advantages that distributed ledgers provide to end-users and the system, it also gives access to all data in the chain to every user with a participating ledger node, even if encrypted. While modifying data in the blockchain is a difficult and complex process, the sole existence of data in this way raises some security concerns. To address this issue, this article proposes FeDis, a framework that combines federated learning with a distributed ledger. In the proposed FeDis framework, federated learning is used to train a global model by aggregating local gradient models trained by the participants in the learning process. One of the key features of the framework is the possibility of handling heterogeneous data and devices from different participants. By combining federated learning with a distributed ledger, local model weights are saved in the ledger after encryption and then fetched by the server to be aggregated and to create a global model. The use of a distributed ledger in FeDis increases user trust by providing an additional layer of security and traceability. In addition, the framework is tested using the NASA Bearing dataset showing that the extra layer of security does not compromise the accuracy of the local models.

Keywords: Blockchain · Decentralized Machine Learning · Distributed Ledger · Federated Learning · Privacy-Preserving Machine Learning

1 Introduction

The decentralized nature of distributed ledgers (DL) means no central point of control manages its users. Instead, every participant has access to the whole distributed database via their node. Once data is added to a block in the ledger, it cannot be changed or deleted, guaranteeing the ledger has immutability. Additionally, transparency is guaranteed in the ledger because tampering with the blocks that make up the chain is a process that takes a lot of time and computing power [1]. Because the ledger operates in a distributed way, the data in it might be exposed to malicious actors compromising the security of the users. Current machine learning (ML) models require accurate data, usually in the form of large datasets for training [2]. In certain areas, such as finance, healthcare, and energy

J. M. Machado et al. (Eds.): BLOCKCHAIN 2023, LNNS 778, pp. 32–41, 2023.
https://doi.org/10.1007/978-3-031-45155-3_4

[3], these datasets might contain user information that needs to be kept private for safety reasons [4]. If malicious actors acquired the users' information, they could figure out investment habits, learn someone's disease history [5], or know when a person is out of their home [6].

To address the above-mentioned issues, federated learning (FL) can be used with traditional machine learning models [7]. FL is a decentralized machine learning method where client devices train their model with their private local data and afterward, the trained model gradients are sent to a server that has the function of aggregating the weights into a global model which is distributed to all the clients after.

In that line of research, this article proposes FeDis, a framework that combines distributed ledgers with federated learning to save intelligent model weights instead of users' data in the DL, diminishing the privacy issues raised by centralized solutions, and decentralized solutions with data sharing mechanisms. By replacing the Remote Procedure Call (RPC) between clients and the server with a distributed ledger used for saving and retrieving the model gradients in FL, there is an increase in the traceability and accountability of the platform. In addition, FL allows the collective use of DL to promote an intelligent distributed solution.

In the literature, there exist some solutions that utilize DL and FL in conjunction. In the Internet of Things (IoT) area, [8] utilized both technologies to provide privacy in the data-sharing process made by IoT devices. In healthcare, [9] has combined FL and DL to create a method of delivering authentic information on COVID-19.

In the proposed solution, the federated component was developed using the Flower framework [10]. The DL component was developed utilizing the Substrate blockchain framework [11]. Finally, the Message Queuing Telemetry Transport (MQTT) protocol was used to enable notifications between FL clients and the server.

The rest of the article is organized as follows. Section 2 presents work related to the areas of FL and DL. Section 3 describes the proposed FeDis framework combining FL and DL. Section 4 presents the case study and the results achieved after the implementation of FeDis. Finally, Sect. 5 provides the main conclusions obtained from this study.

2 Related Works

This section describes related works concerning federated learning, and distributed ledgers. After addressing these two concepts, related works combining both concepts are also explored.

In [12], the Flower Federated Learning Framework is presented, and its capability for executing large-scale FL experiments as well as being able to work on highly heterogenous FL device scenarios is highlighted. In [13], it is presented an exploration of the Flower Framework, concluding that despite algorithmic advancements in FL, on-device training of FL algorithms on edge devices remains poor. The authors propose a new federated optimization algorithm, a modified version of Federated Averaging (FedAvg), in which each client is assigned a processor-specific cutoff time to send its model parameters regardless of having finished the local epochs.

On a different line of research, [14] discusses the concept of blockchain-based DL technology (DLT) in transactive energy models, proposing a novel transactive management infrastructure for peer-to-peer energy transactions based on Smart Contracts, JavaScript Object Notation (JSON), and the MQTT protocol. A Proof of Energy function for energy exchanges via a DLT was also proposed and demonstrated as a way to reach consensus among the nodes of the distributed ledger. In [15], the use of blockchain technology to enable secure and trustful transactions in local electricity markets (LEM) is proposed. LEMs operate on a smart grid architecture, and so require a distributed and decentralized way for end users to be able to negotiate and transact electricity (different from the current and typical centralized approach in energy systems).

In [16], issues related to the misallocation of energy storage demands and resources were exposed and the authors proposed an energy-sharing mechanism that uses a Substrate private blockchain to enable trust and transparent trading. On the topic of smart home systems, [17] proposes a system using FL to create models able to predict customers' requirements and consumption from the data collected by their home appliances.

The combination between FL and DLT is also a possibility that combines decentralization, traceability, and privacy. In [8], it is proposed a system that combines blockchain with FL for data sharing from Internet of Things (IoT) devices while maintaining privacy. While data is still stored locally by the owners, the authors affirm that the data retrieval process is not a privacy threat. Additionally, a new type of consensus protocol, a Proof of Training Quality (PoQ), was enabled by the use of FL. Considering applications in health care, [9] proposes the FedMedChain, a system that is a combination of FL and blockchain technology applied to the Internet of Medical Things (IoMT) in a COVID-19 context. The solution proposed by the paper attempts to resolve, through the usage of FL, the issues of data privacy and security that appear with the usage of the data collected by the IoMT devices on machine learning and deep learning models. The paper proposes using blockchain technology as a means to increase trust in public media in COVID-19 information dissemination by taking advantage of a distributed ledger's consensus protocols.

The solutions available in the state of the art demonstrate the potential of combining FL and DLT, creating complex solutions enabling data sharing while compliant with data privacy requirements. The solutions mentioned have applications based on IoT, one for industry and one for health, however, they do not focus on a community-based learning approach, where multiple machines share knowledge, using deep learning models, to collaboratively create a combined and unified learning model. The proposed solution, of this paper, addresses the combined knowledge among machines using a decentralized approach for deep learning, supported by FL, and data storage, using DLT.

3   Proposed FeDis Framework

The proposed solution aims to solve the privacy issue of data sharing in the DL by utilizing FL's feature of sharing model weights instead of raw data. This way, the data that is shared among all the participants in the DL still contains all the knowledge but with fewer data privacy risks. Likewise, by utilizing the DL as the mechanism for model weight sharing in the FL training process, there's an increase in the transparency of the platform and accountability of each user. This is because the model weight uploads to the ledger are immutable, so it would be easier to track down and identify malicious actors attempting to sabotage the FL training process via their model weight uploads.

Regarding the architecture, the FeDis framework is composed of 4 different components: the FL component which can be a server or a client (the former is responsible for aggregating model weights and the latter is responsible for local model training), the DL node that connects to other nodes and forms the blockchain where the model weights are stored, and a real-time data stream component that handles the communication between various FeDis nodes. Finally, each node has the FeDis client API, responsible for managing the interactions between the components described above.

4   Implementation of FeDis Framework

The DL of the FeDis framework was implemented in the Substrate framework from Polkadot. Substrate was chosen due to its open-source nature, extensive documentation, active community, and ease of development. The framework simplifies the setting up process of a distributed ledger by providing a solid foundation on which developers can expand and build on top of it. A distributed ledger developed in Substrate can be defined by one or more pallets. Pallets consist of code that defines the structure of the data to be saved on the ledger, as well as establish the functions that operate on said data that can be called by users.

The Application Programming Interface (API) used to interact with the DL was Polkadot JS, which already provides the methods to connect to the ledger (via WebSocket (WS)), execute functions on the ledger, and listen for events, among other functionalities. The RESTFul API for end-users to interact with the ledger was developed in NodeJS, it provides HTTP/S endpoints that pertain to different functionalities users can utilize, such as saving data on the ledger or reading information from the blockchain that has a certain address or belongs to a certain user.

The federated component was developed using Flower FL, an open-source user-friendly Python framework that simplifies the process of setting up FL servers and clients and the communication between them.

Finally, to ensure that each node/client is notified when data was uploaded to the ledger, and to avoid too many requests to the blockchain, FedDis integrates the MQTT protocol to publish messages to the broker.

Overall, the frameworks and tools used were chosen due to the explained characteristics and the fact that they allowed integration with each other. It is worth noting that the described implementation of FeDis can be later modified or expanded to allow other blocks with other functionalities (or improvement over the already established ones). Figure 1 displays the architecture of a single node/client in the FeDis framework, and how the components interact.

A FeDis client cannot operate in a standalone mode. Thus, it must be connected to the other FeDis clients, as represented in Fig. 2. The MQTT broker is the server responsible for transmitting messages from/to clients in the FL Flower framework.

The general functionality of FeDis can be described as: FL clients use a FeDis endpoint to save the trained model weights on the DL. At the same time, the FeDis

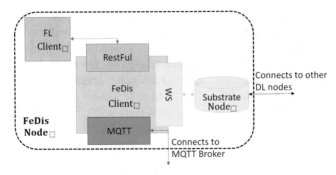

Fig. 1. FeDis implemented client architecture.

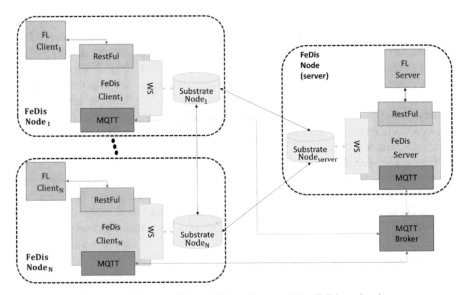

Fig. 2. Interconnection of different FeDis clients and the FeDis end-point server.

endpoint publishes an MQTT message on the topic provided by the user, informing the topic subscribers that the model weights were saved to the blockchain. On the other hand, the FeDis endpoint with the FL server receives the information that the model weights were uploaded, enabling the retrieval of local models to perform their aggregation to create the updated global model. Afterward, the updated global model is published to the DL, and an MQTT message to the FeDis endpoint with FL clients will be sent. Regarding the federated learning process (the blue blocks in Fig. 2) in more detail, each of the devices has an incorporated model as well as all necessary functions to register and save data and train the model with the newly acquired data.

Using the after mentioned solution, the various devices can upload the resulting weights of the training to the blockchain from which the server will extract these weights by applying the FedAvg [18] strategy. After that, the aggregated weights are published in the blockchain, and the clients are notified through the MQTT broker allowing the clients to update their models with the aggregated weights.

In terms of the ledger, it is on the DL where data, such as model weights generated by the FL training, are saved, and retrieved by other nodes (including the server endpoint that aggregates all the model weights from the clients). Likewise, data on the distributed ledger can be retrieved not only by the address attributed when the data was added to the chain but also can be retrieved via a user address (which in fact can return all data added to the ledger by that user). End users are not supposed to interact with the ledger directly. Instead, the FedDis framework provides HTTP/S endpoints to simplify the interactions and utilization of different applications.

5   Case Study and Results

To test and validate the FeDis framework, the NASA Bearing dataset was used along with a Jupyter Notebook describing the implementation of a supervised learning system [19]. The goal of the case study is to apply the approach proposed in [19] using three clients with the FeDis framework. The selected FL classification model is the XGBoost, having clients training different portions of the dataset.

Thus, in the first step, the dataset is divided into three folders, each representing a time frame of data extraction of vibration signals recorded at 20 kHz. The samples of the first folder contain eight input channels, two from each of the four bearings. The recording interval of vibrations was 10 min (except for the first 43 entries which were 5 min apart). The other two folders have only one input channel per bearing resulting in four inputs [20]. Table 1 summarizes the characteristics of the dataset.

Table 1. Dataset description.

Folder	Data Size	Input Channels	Samples	Data Interval	Data Format
Set1_timefeature	2.30 GB	8	2156	22/10/2003 to 25/11/2003	ASCII
Set2_timefeature	519 MB	4	984	12/2/2004 to 19/2/2004	ASCII
Set3_timefeature	3.26 GB	4	6324	4/3/2004 to 4/4/2004	ASCII

To test the FeDis framework, each of these folders was assigned to a different client. With that, each client can apply the FL process locally. The training and evaluation percentages of the Jupyer notebook were utilized, 70% of the dataset was used for training while the remaining 30% was for testing and validation.

From these folders, the raw inputs, stored in the ASCII files, are extracted and converted into a list of relevant signal-based features to be fed to the classification model. Thus, from the vibration signals, a list was obtained, containing the mean (absolute mean), std (standard deviation), skew (Skewness), kurtosis, entropy, rms, max, p2p (peak to peak), crest (crest factor), clearance factor, shape factor and impulse of each file in the folder to feed our model. These features are then used as input to an artificial neural network with a support vector machine to extract the conditions of each bearing, as depicted in Fig. 3.

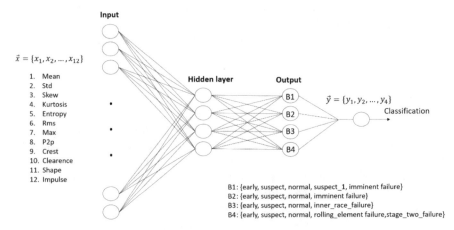

Fig. 3. Input features and Outputs of the classification process in FeDis.

To assess FeDis's performance, two experiments were performed. In the first experiment (base case), the learning process was implemented locally, without implementing interactions with the server and the DL, and without the use of FL. In the second experiment, the clients are allowed to communicate with the global server and use FL with FeDis in the classification process. The tests were performed on a single laptop with separate executions, one for the FL server and one per each client execution. The machines were run on a laptop with Windows 10 with an AMD Ryzen 7 5800H with integrated graphics, 32 GB RAM, and an NVIDIA GeForce RTX 3070.

Table 2 presents the results of the two experiments performed to assess the effectiveness of the FeDis framework. It can be observed in experiment 1 that the accuracy of the XGBoost classifier is around 70%–80% for the three clients without the use of FL. In experiment 2, the accuracy of the global model is only slightly different (with a gap <3% w.r.t. the experiment 1), demonstrating the effectiveness in the accuracy of FeDis. The same conclusion can be reached with the loss, with an even lower difference between the model without FL and FeDis. The table also shows the training time by the client. It can be noticed that the training time is related to the dimension of the data handled by each user. In FeDis, this could represent a bottleneck in the execution since clients that finish first their training process have to wait for others, having their resources locked during that time. This is an issue that should be studied and addressed in future works (for instance, by finding different approaches for the communication and implementation of the framework). The observed drop in accuracy of some of the clients is derived from the FedAvg strategy that aims to have all the clients tend to median value as an attempt to increase the overall performance of the system.

Table 2. Results of the two experiments performed.

Client	Experiment 1 (base case without FL)		Experiment 2 (using FeDis with DL and FL)		Training Time (minutes)
	Accuracy	Loss	Accuracy	Loss	
Client 1 (Set1_timefeature)	84.2%	5.6%	81.7%	5.5%	45
Client 2 (Set2_timefeature)	76.8%	5.8%	77.4%	5.6%	25
Client 3 (Set3_timefeature)	73.3%	5.4%	75.9%	5.5%	55
Average value	78.1%	5.6%	81.7%	5.5%	41

6 Conclusion

After developing and implementing the FeDis framework, it can be concluded that the integration of federated learning with distributed ledgers can benefit both technologies. On one hand, federated learning increases the privacy of the distributed ledger data by replacing it with model gradients that still allow for the knowledge to be shared without impacting data privacy. On the other hand, utilizing distributed ledgers on the federated learning process allows for an additional layer of security and traceability due to the decentralized nature of the ledger and the possibility of permanently tracking all written data (even malicious ones) on the ledger. Furthermore, the framework simplifies user interaction with the distributed ledger by providing an API interface to communicate with the ledger easily. In addition, as was shown by the results, the global models that are a result of the aggregation of the various local models during the federated learning process keep the accuracy compared with traditional classification models without FL, an increase of 3,6% of the average accuracy and drop 0.1% drop of loss. In future work, different and more accurate models can be implemented in FeDis to test the flexibility of the architecture, as well as the usage of more clients with more diversified datasets in terms of sizes. Also, the security and traceability characteristics of the framework should be further explored and validated in different scenarios and with clients of different characteristics.

Acknowledgments. The present work has received funding from European Regional Development Fund through COMPETE 2020 – Operational Programme for Competitiveness and Internationalisation through the P2020 Project F4iTECH (ANI|P2020 POCI-01-0247-FEDER-181419) and has been developed under the EUREKA – CELTIC-NEXT Project F4iTECH (C2021/1-10), we also acknowledge the work facilities and equipment provided by GECAD research center (UIDB/00760/2020) to the project team.

References

1. IBM Blog. https://www.ibm.com/blog/how-transparency-through-blockchain-helps-the-cyb ersecurity-community/. Last accessed 17 Apr 2023
2. Goncalves, C., Barreto, R., Faria, P., Gomes, L., Vale, Z.: Energy community consumption and generation dataset with appliance allocation. IFAC-PapersOnLine **55**(9), 285–290 (2022). https://doi.org/10.1016/J.IFACOL.2022.07.050
3. Macieira, P., Gomes, L., Vale, Z.: Energy management model for HVAC control supported by reinforcement learning. Energies **14**(24), 8210 (2021). https://doi.org/10.3390/en14248210
4. Rieke, Nicola, et al.: The future of digital health with federated learning. npj Digit. Med. **3**(1), 119 (2020). https://doi.org/10.1038/s41746-020-00323-1
5. Kanwal, T., et al.: A robust privacy preserving approach for electronic health records using multiple dataset with multiple sensitive attributes. Comput. Secur. **105**, 102224 (2021). https://doi.org/10.1016/J.COSE.2021.102224
6. Abdalzaher, M., Fouda, M., Ibrahem, M.: Data privacy preservation and security in smart metering systems. Energies **15**(19), 7419 (2022). https://doi.org/10.3390/en15197419
7. Brendan McMahan, H., Moore, E., Ramage, D., Hampson, S., Agüera y Arcas, B.: Communication-efficient learning of deep networks from decentralized data. In: Proceedings of the 20th International Conference on Artificial Intelligence and Statistics, AISTATS 2017 (2016). https://doi.org/10.48550/arxiv.1602.05629
8. Lu, Y., Huang, X., Dai, Y., Maharjan, S., Zhang, Y.: Blockchain and federated learning for privacy-preserved data sharing in industrial IoT. IEEE Trans. Industr. Inform. **16**(6), 4177–4186 (2020). https://doi.org/10.1109/TII.2019.2942190
9. Samuel, O., et al.: IoMT: a covid-19 healthcare system driven by federated learning and blockchain. IEEE J. Biomed. Health Inform. **27**, 823–834 (2022). https://doi.org/10.1109/JBHI.2022.3143576
10. Flower Homepage. https://flower.dev/. Last accessed 21 Apr 2023
11. Substrate Homepage. https://substrate.io/. Last accessed 21 Apr 2023
12. Beutel, D.J., et al.: Flower: a friendly federated learning research framework (2020) [Online]. Available: http://arxiv.org/abs/2007.14390
13. Mathur, A., et al.: On-device Federated Learning with Flower. (2021), [Online]. Available: http://arxiv.org/abs/2104.03042
14. Siano, P., De Marco, G., Rolan, A., Loia, V.: A survey and evaluation of the potentials of distributed ledger technology for peer-to-peer transactive energy exchanges in local energy markets. IEEE Syst. J. **13**(3), 3454–3466 (2019). https://doi.org/10.1109/JSYST.2019.290 3172
15. Santos, G., Faia, R., Pereira, H., Pinto, T., Vale, Z.: Blockchain-based local electricity market solution. In: International Conference on the European Energy Market, EEM, vol. 2022, September (2022). https://doi.org/10.1109/EEM54602.2022.9921035
16. Luo, B., Shen, X., Ping, J.: Energy storage sharing mechanism based on blockchain. In: 2020 IEEE Student Conference on Electric Machines and Systems, SCEMS 2020, pp. 913–917 (2020). https://doi.org/10.1109/SCEMS48876.2020.9352347
17. Zhao, Y., et al.: Privacy-preserving blockchain-based federated learning for IoT devices. IEEE Internet Things J. **8**(3), 1817–1829 (2021). https://doi.org/10.1109/JIOT.2020.3017377
18. Sun, T., Li, D., Wang, B.: Decentralized federated averaging. IEEE Trans. Pattern Anal. Mach. Intell. **45**(4), 4289–4301 (2023). https://doi.org/10.1109/TPAMI.2022.3196503

19. Cavalaglio Camargo Molano, J., Strozzi, M., Rubini, R., Cocconcelli, M.: Analysis of NASA Bearing Dataset of the University of Cincinnati by Means of Hjorth's Parameters. Last accessed: 11 Apr 2023. (2020). Available: https://iris.unimore.it/handle/11380/1203704
20. Qiu, H., Lee, J., Lin, J., Yu, G.: Wavelet filter-based weak signature detection method and its application on rolling element bearing prognostics. J. Sound Vib. **289**(4–5), 1066–1090 (2006). https://doi.org/10.1016/J.JSV.2005.03.007

Enhancing Privacy Protection in Intelligent Surveillance: Video Blockchain Solutions

Kasun Moolika Gedara$^{(\boxtimes)}$, Minh Nguyen, and Wei Qi Yan

Auckland University of Technology, Auckland, New Zealand
Kasun.moolikagedara@autuni.ac.nz

Abstract. Blockchain has emerged as a contemporary innovation that ensures secure operations in distributed networks, including decentralized applications, finance, logistics, and cross-border organizational control. In this paper, we introduce "Video Blockchain" as a novel method to store and manage visual data in smart cities, due to the lack of tamper resistance in existing systems. A relationship is established between video frames from surveillance videos and blockchain technology, integrating the visual data into a decentralized storage platform. A unique approach is leveraged to extract hash values and signatures from video blockchains using cryptographic functions, thereby enhancing surveillance data security. A decentralized blockchain prototype was developed, and appropriate cryptographic algorithms were selected to create a sustainable video blockchain. The contributions of this research project are to enhance blockchain security and minimize privacy-preserving gaps in intelligent surveillance, which lead to more secure, robust and reliable surveillance systems for smart cities.

Keyword: Video Blockchain · Cryptography · Intelligent Surveillance · Smart Cities

1 Introduction

Blockchain was introduced in 2009 and has been employed in large-scale businesses across industries such as global trade, insurance, finance, distributed energy, and healthcare. It allows for transactions and processes to occur without the involvement of third parties, effectively solving complications related to data integrity and verification in the areas such as medical records, electricity, and gas systems in smart cities [4, 6, 7].

As smart cities become more complex and the technologies are more integrated, the idea "smart cities" has gained traction worldwide, with a focus on improving safety and security for citizens while reducing crimes and accidents. Surveillance systems are an integral part of this need, but traditional mechanisms such as centralized client-server methods and network security measures do not always satisfy the necessary requirements [18].

One critical aspect of data storage in smart cities is privacy. With sensitive data being transmitted and stored on blockchains, there is a need to protect the privacy of citizens. Therefore, privacy-preserving techniques need to be employed so as to ensure that personal data remains confidential and secure[20].

J. M. Machado et al. (Eds.): BLOCKCHAIN 2023, LNNS 778, pp. 42–51, 2023.
https://doi.org/10.1007/978-3-031-45155-3_5

Fortunately, blockchain technology provides an ideal platform for privacy-preserving data storage. For example, zero-knowledge proofs can be harnessed to enable secure transactions without revealing the actual data being transferred. Additionally, homomorphic encryption can be accommodated to encrypt data while still allowing computations to be performed on it without the need for decryption by incorporating privacy-preserving methods into the blockchain-based surveillance system, the smart city can ensure that the privacy and security of its citizens are adequately protected. This, in turn, will foster greater trust and confidence in the system, leading to its wider adoption and success[9].

In this paper, our aim is to find a solution for selecting the best cryptographic functionalities, combining privacy-preserving data storage – one of the main limitations in blockchain implementations – and enhancing the immutability of the blockchain to provide a secure mechanism for video surveillance in smart cities. Overall, our research aim is to contribute to the development of a new method of video blockchain for securing surveillance data in a smart city, which can improve the efficiency, effectiveness, and security of surveillance systems.

2 Background and Related Work

Securing surveillance [5] is a crucial way to face detection, human behaviour analysis and traffic rule violation detection. These tools have shown plenty of contributions in notably preventing crimes, anomalous incidents, and privacy policy violations. Also our previous works [10, 16, 17, 26–28] related to blockchain and computer vision lead to enhance the more robust method to address the malicious attackers and hackers can illegally manipulate video repositories and surveillance cameras, thereby rendering the recorded footage unusable in criminal cases. To prevent tampering and attacks [12], Blockchain is employed as a solution to handle a diversity of attacks that occur within surveillance systems. For instance, attackers may manipulate or tamper with video footage, which leads to compromised integrity.

Tamper-resistant and immutability of blockchain were employed to protect stored data and ensure data integrity. Hashing is a reliable method for creating confidentiality between two blocks in the chain [8]. Cryptographic hash functions convert confidential data into a random string having a size, security requirements of one-witness and collision-resistance are necessary. Various blockchain-based systems have been proposed, such as the BlockSee method [9], which provides validation and immutability to surveillance videos.

Moreover, research trends in intelligent surveillance and blockchain include the use of dashboard cameras mounted on vehicles to capture vehicle accidents in smart cities, connected IoT devices and monitor air and water quality[13, 14], as well as food delivery tracking [19]. Decentralization, filtering, and privacy features of blockchain are applied to ensure the authenticity, time-lapse features are employed to transfer unmodified data to a shared repository.

The uniqueness of blockchain methods makes it suitable for large-scale industries such as global trade, insurance, banking, distributed energy, and healthcare. Blockchains have been adapted to smart transportation systems, food supplier management, government, identity verification, and smart cities. The blockchain's ability to solve problems related to data integrity has been offered in medical record verification systems and intelligent gas monitoring systems [2].

To develop a computational method of video blockchain for intelligent surveillance in smart cities, [7] selecting cryptographic functions is an effective method for connecting surveillance videos and blockchains to enhance the resistant against different kinds of attacks that can improve digital surveillance systems, the design of combining different algorithms together and implementing new methods is effective, as a slew of attacks have been identified based on the existing blockchain platform [1, 22, 23].

Blockchain proves to be an effective tool for securing surveillance and ensuring data integrity in intelligent surveillance. Its tamper-resistant and immutable nature protects stored data from malicious hackers. However, the development of new methods and the integration of different algorithms to enhance resistance against attacks on existing blockchain platforms must continue. Emphasizing the importance of selecting appropriate cryptographic functions and implementing new methods is crucial to improve the privacy of blockchain-based systems in the realm of intelligent surveillance.

3   Our Proposed Solutions

An effective method for connecting blockchain function can be identified by analysing recent findings of blockchain. According to [25] myriads attacks on the blockchain platform have been reported, in order to avoid these identified and unidentified attacks, the design of combining different algorithms together and implementing a new method can enhance resistance against multiple kinds of attacks that can happen to surveillance systems recorded data repositories.

3.1   The Best Cryptographic Functions to Implement a Video Blockchain

Choosing the right cryptographic function is crucial for a secure and efficient video blockchain system. It must provide strong security guarantees against attacks, with options like SHA-256, SHA-3, and BLAKE2 [8] being commonly used for blockchain applications. Efficiency is also key, optimizing the function's performance on the specific hardware and software architecture of the blockchain network. Compatibility is another consideration, ensuring seamless integration and interoperability with the existing blockchain infrastructure. For instance, using Ethereum-compatible cryptographic functions like Keccak-256 or SHA-3 [14] is recommended for a video blockchain built on Ethereum.

In this paper, we choose cryptographic features and take advantage of the proposed combination to create a robust mechanism for a blockchain-based computing solution. One of the most important requirements for establishing a blockchain application solution is to ensure data integrity and confidentiality. We have also explored the methods such as Merkel tree [16], hash list [24], H-tree [20] and SM-Tree [3] methods. After comparing different technologies, we will determine which one is most suitable for the required level of security.

The method [2] of creating a blockchain solution for the Dubai government will be employed for comparative analysis of blockchain industrial solutions. According to previous research work [11] the requirements for choosing cryptographic functions and algorithms for scalability. Moreover, energy consumption is a significant consideration for blockchain-based implementations, it is crucial to find the best algorithm that meets these requirements. Overall, the integration of selected solutions requires an in-depth investigation to build a strong and reliable computer approach. This will assist us to deliver a secure solution for intelligent surveillance in smart cities. Table 1 shows the comparisons of our selected algorithms for blockchain.

We selected our proposed method. First, identify the most suitable algorithm from Table 1. Assess new trends and gaps in the selected methods to construct an effective and efficient solution. Additionally, connect the algorithms to design computational methods for video blockchain.

Table 1. The comparisons between conventional databases.

Characteristics	Algorithms/Methosd	Conventional Databases
Authority	Decentralized	Centralized
Architecture	Peer-to-peer model	Client-server model
Performance	Relatively slow	Fast
Cost	Costly	Cheap
Data Handling	Only read and write	Create, Read, Update, Delete
Data Integrity	Has data integrity	Doesn't have data integrity
Transparency	Transparent	Non-transparent
Cryptography	Yes	No

3.2 Solution for Privacy Preserving Over the Blockchain

The proposed method for enhancing surveillance data integrity in smart cities involves using a Merkle tree, hashing function, and peer-to-peer data storage. The verification process detects changes in image frame order and identifies the specific image modifications. A Merkle tree is generated for each block, and its root is stored in the blockchain to ensure integrity. The experimental design utilizes selected cryptographic algorithms, generating output from video frames to validate the blockchain implementation. An interface is designed to test functionality and address video frame-related issues.

The solution emphasizes lightweight functions for inter-block communication to enhance security and prevent attacks like man-in-the-middle interception. Selected cryptographic features create a robust mechanism for blockchain-based computing. The initial computational method incorporates connection hashing and block matrix functions for video blockchain, bridging gaps between frames and improving security. However, adversaries may still estimate the amount of legal data, even with blocked public information equity for verification rate, hash value, and sibling path size.

For experimental analysis, multiple datasets with sample videos were created to be loaded into the system. The surveillance videos, usually recorded at 25 fps, were increased to 30 fps in this project to include more content in the experiments. Using a Samsung S7 (G930F) smartphone, our own dataset of 7,000 video frames was created, focusing on the Auckland city. The objective of this research is to generate hash values for video frames, enhancing resistance against potential attacks.

This paper establishes a connection between surveillance video footage and blockchains, storing the data in a decentralized repository. The main contribution is enhancing the security of observational data through the use of cryptographic algorithms for hashing and signature, distinguishing it from other works. These algorithms ensure accurate connection of video frames and enable detection and location of any frame changes. The verification procedure, using Merkle trees and hashing functions, further strengthens the security measures.

Privacy-preserving problem exists in blockchain implementation [15]. We propose a blockchain-based solution for ensuring and improving the integrity of surveillance data in smart cities, aiming for increased loyalty, reliable results, and controlled disclosures. Combining computational approaches and video blockchain regulates data security, reducing unauthorized access. This enables close monitoring of law enforcement, insurance firms, and traffic management systems, facilitating necessary modifications for improved security and compliance in smart city video surveillance.

To verify frame integrity, a Merkle tree is constructed from the block matrix hash values. The Merkle tree's root hash is stored in the blockchain, allowing detection of any modifications by comparing block and Merkle tree hashes. This tamper-resistant approach enhances system security.

In addition, block matrix operations [21], such as matrix multiplication and matrix inverse, can be employed for video processing tasks, such as compression, filtering, and restoration. These operations can be performed on the block matrices stored in the blockchain, which allows for highly efficient and secure video processing.

Overall, storing video frames in a blockchain-based system using block matrices provides a secure and efficient method for video storage. The use of hash values, Merkle trees, and block matrix operations enhances the tamper-resistance, integrity, and reliability of the system.

In this implementation, the Merkle Tree function takes in an array of data and recursively constructs a Merkle tree. In the base case (when there is only one data item left), the function returns the data item itself. Otherwise, it recursively constructs the left and right subtrees, hashes them together using SHA-256 algorithm, and returns the resulting hash. The resultant hash is the root of Merkle tree.

Algorithm 1: Merkle Tree

Input: A list of data blocks.
1)   Divide the data blocks into fixed-size chunks (usually 1-2KB).
2)   Compute the hash of each data chunk using a cryptographic hash function.
3)   Pair up adjacent data chunk hashes and compute the hash of each pair.
4)   Repeat step 4 until there is only one hash left, which is the Merkle root hash.
5)   Store the Merkle root hash as the identifier of the data blocks.

Algorithm 2: Block Matrix

Input: A video file consisting of frames.
1)   Divide each frame into fixed-size blocks (16x16 pixels).
2)   Store the blocks of each frame in a matrix, where each row represents a block, and each column represents a frame.
3)   Apply compression algorithms (JPEG) to each block to reduce the amount of data.
4)   Store the compressed block matrix as a binary file.
5)   To access a specific frame, load the compressed block matrix and retrieve the corresponding column of blocks.
6)   To access a specific block within a frame, retrieve the corresponding row of the block matrix and decompress the block.

The block matrix function takes in an array of data and a block size, and constructs a matrix where each row represents a block of data. The matrix is filled in by iterating over the data array, slicing it into blocks of the given size, and placing each block in the appropriate row of the matrix. If the length of the data array is not a multiple of the block size, the last row of the matrix will contain padding to fill out the remaining space.

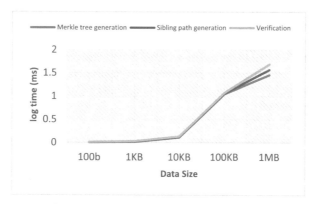

Fig. 1. Average computational time (millisecond) for authentication based on Merkle tree by data size.

Together, these algorithms can be employed to store video frame data in a secure and efficient manner. The video frames can be split into blocks, and a Merkle tree can be constructed over the blocks to provide integrity and authentication for the data. This method supports the distributed storage facility to be store data transferring from the surveillance systems.

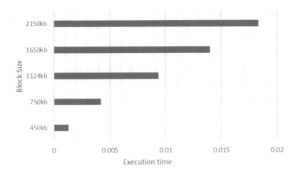

Fig. 2. Computational time for different input sizes and block sizes.

4   Result Analysis and Discussion

This project aims to explore the use of video blockchain for surveillance systems. The method includes converting recorded videos into video frames, each frame has 50 KB–1024 KB, which was employed to implement a private blockchain system based on a Windows 11 64-bit operating system; an experimental setup was conducted to test the effectiveness of the newly proposed method against various attacks. The experimental results in the creation of new computational methods of video blockchains, integrating selected cryptographic algorithms and video blockchains together. Overall, this research provides a technological contribution to the field of video blockchain, as it presents a new method of securing video data for surveillance systems. The findings of this research project can be offered as a foundation for future work in video blockchain and cryptographic algorithms.

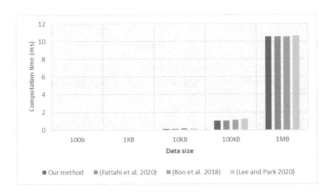

Fig. 3. Comparisons of computational time between ours and other similar projects

Each root structure in the Merkle tree guarantees the connection between video frames and hashing order, preventing changes to the order of images without changing the entire root structure of the tree. Our future work aims to add a real-time change detection feature to the implementation, enhance its reliability and resistance against privacy-pervasive and quantum computer attacks. Overall, the result provides insights into the development of web interfaces for video blockchain systems, and its findings can be utilized to enhance the reliability and security of such systems in the future.

In this paper, we focus on Merkle tree-based approaches for data verification. The evaluation was conducted by measuring the computational time and data size of each experiment, with each experiment being repeated 100 times to minimize errors caused by outliers.

Figure 1 shows an upward trend in computational time related to the increase in data size for the three Merkle tree-based approaches. However, after 100 KB of data, these approaches exhibit only a slight difference, as generating a Merkle tree for a 1 MB data file accounts for 99.9% of the computational time required by the prover. We plot the results to determine the time complexity. The time is subject to different input sizes and block sizes, this results in determining the function's time complexity.

In Fig. 2, the computational time on y-axis and the input size or block size on x-axis for different input sizes or block sizes to compare the performance of the function. In Fig. 3, we compare the computation time and data size metrics of the study with those of other similar works. The results show a significant change with the data size 100 KB, which indicates that the study's outcomes are comparatively reliable. These findings can be applied to improve the efficiency and accuracy of Merkle tree-based approaches for data verification.

The use of blockchain technology in intelligent surveillance faces a many of challenges such as scalability, interoperability, and regulatory compliance. Scaling blockchain to handle large volumes of data and transactions, integrating it with the existing systems, and navigating regulatory frameworks are crucial issues. In our future work, we will plan to solve these problems.

5   Conclusion and Future Work

In this project, our primary objective is to establish the relationship between video frames captured by intelligent surveillance systems and blockchain, integrating the data into a decentralized storage platform for video surveillance. Our approach is distinct from the existing studies as it leverages cryptographic functions to extract hash values and signatures from video blockchains, thereby augmenting the security of surveillance data. Furthermore, this research work primarily targets the enhancement of tamper-resistant data storage within surveillance systems rather than focusing on mitigating the risks posed by considering quantum computer attacks on blockchains. Nevertheless, in future, we will explore the solutions outlined in Sect. 3.2 to bolster the resilience of blockchains against quantum threats. Privacy concern is one of the main problems in blockchain implementation. However, in the future, there is a need to further address limitations of scalability, interoperability, and regulatory issues.

This research aims to propose a blockchain-based approach that enhances the security and integrity of surveillance data while fostering significant levels of trust, reliability, and controlled disclosure in smart cities. By integrating computer vision with video blockchain, we concentrate on fortifying the security of surveillance data, offering a solution that deters tampering and unauthorized access by external parties. The contributions of this project pave the way for necessary advancements to achieve heightened security and adaptability for video surveillance in smart urban environments.

References

1. Aldairi, A., Tawalbeh, L.: Cyber security attacks on smart cities and associated mobile technologies. Procedia Comput. Sci. **109**, 1086–1091 (2017)
2. Alketbi, A., Nasir, Q., Abu Talib, M.: Novel blockchain reference model for government services: Dubai government case study. Int. J. Syst. Assur. Eng. Manag. **11**, 1170–1191 (2020)
3. Becker, G.: Merkle Signature Schemes, Merkle Trees and Their Cryptanalysis. Master's Thesis. Seminararbeit Ruhr-Universit¨at Bochum (2008)
4. Chen, X., Xing, Z., Karki, B., Li, Y., Chen, Z.: Blockchain simulation: a web application for IT education. In: Annual Computing and Communication Workshop and Conference (CCWC), pp. 486–491 (2021)
5. Deepak, K., Badiger, A.N., Akshay, J., Awomi, K.A., Deepak, G., Harish Kumar, N.: Blockchain-based management of video surveillance systems: a survey. In: International Conference on Advanced Computing and Communication Systems (ICACCS), pp. 1256–1258 (2020)
6. Engelhardt, M.A.: Hitching healthcare to the chain: an introduction to blockchain technology in the healthcare sector. Technol. Innov. Manag. Rev. **7**, 22–34 (2017)
7. Fill, H., Haerer, F.: Knowledge blockchains: applying blockchain technologies to enterprise modelling. In: Hawaii International Conference on System Sciences, pp. 4045–4054 (2018)
8. Fu, J., Qiao, S., Huang, Y., Si, X., Li, B., Yuan, C.: A study on the optimization of blockchain hashing algorithm based on PRCA. Secur. Commun. Netw. **2020**, 1–12 (2020). https://doi.org/10.1155/2020/8876317
9. Gallo, P., Pongnumkul, S., Nguyen, U.Q.: BlockSee: Blockchain for IoT video surveillance in smart cities. In: IEEE International Conference on Environment and Electrical Engineering and IEEE Industrial and Commercial Power Systems Europe (EEEIC/I and CPS) (2018)
10. Gedara, K., Nguyen, M., Yan, W.: Visual blockchain for intelligent surveillance in a smart city. Blockchain Technologies for Sustainable Development in Smart Cities, IGI Global (2021)
11. George, R.V., Harsh, H.O., Ray, P., Babu, A.K.: Food quality traceability prototype for restaurants using blockchain and food quality data index. J. Cleaner Prod. **240**, 118021 (2019). https://doi.org/10.1016/j.jclepro.2019.118021
12. Gergely, A.M., Crainicu, B.: Randadminsuite: a new privacy-enhancing solution for private blockchains. Procedia Manuf. **46**, 562–569 (2020)
13. Gřivna, T., Drápal, J.: Attacks on the confidentiality, integrity and availability of data and computer systems in the criminal case law of the Czech Republic. Digit. Investig. **28**, 1–13 (2019)
14. Hartwig, M.: ECDSA security in bitcoin and Ethereum: A research survey. Blog. Coinfabrik, pp. 1–10 (2016)
15. Hasan, O., Brunie, L., Bertino, E.: Privacy-preserving reputation systems based on blockchain and other cryptographic building blocks: a survey. ACM Comput. Surv. **55**(2), 1–37 (2023)
16. Hu, R.: Visual Blockchain Using Merkle Tree. Master's Thesis, Auckland University of Technology, New Zealand (2019)
17. Hu, R., Yan, W.: Design and implementation of visual blockchain with Merkle tree. In: Handbook of Research on Multimedia Cyber Security, pp. 282–295. IGI Global (2020)

18. Khan, P., Byun, Y.-C., Park, N.: A data verification system for CCTV surveillance cameras using blockchain technology in smart cities. Electronics **9**(3), 484 (2020)
19. Khan, M.A., Salah, K.: IoT security: review, blockchain solutions, and open challenges. Futur. Gener. Comput. Syst. **82**, 395–411 (2018)
20. Koo, D., Shin, Y., Yun, J., Hur, J.: Improving security and reliability in Merkle tree-based online data authentication with leakage resilience. Appl. Sci. (Switzerland) **8**, 2532 (2018)
21. Kumar, M., Kaur, G.: High performance scalable recursive block matrix inverse for multi-core architectures. In: International Conference on Parallel, Distributed and Grid Computing, pp. 45–49 (2022)
22. Li, T., et al.: Rational protocols and attacks in blockchain system. Secur. Commun. Netw. **2020**, 1–11 (2020). https://doi.org/10.1155/2020/8839047
23. Majdoubi, D.E.L., El Bakkali, H., Sadki, S.: Towards smart blockchain-based system for privacy and security in a smart city environment. In: International Conference on Cloud Computing and Artificial Intelligence: Technologies and Applications (CloudTech) (2020)
24. Michael, M.M.: High performance dynamic lock-free hash tables and list-based sets. In: Annual ACM Symposium on Parallel Algorithms and Architectures, pp. 73–82 (2002)
25. Mosakheil, J.H.: Security threats classification in blockchains. Culminating Projects in Information Assurance 141 (2018)
26. Shu, Y.: Blockchain for Security of a Cloud-based Online Auction System. Master's Thesis, Auckland University of Technology, New Zealand (2018)
27. Shu, Y., Yu, J., Yan, W.: Blockchain for security of a cloud-based online auction system. Exploring Security in Software Architecture and Design, pp. 189–210. IGI Global (2019)
28. Shu, Y., Yu, J., Yan, W.: Blockchain for security of cloud-based online auction. Research Anthology on Blockchain Technology in Business, Healthcare, Education, and Government. IGI Global (2021)

Energizing Blockchain: Legal Gaps and Power Plays in the Energy Sector's Digital Transformation

Pardis M Tehrani$^{(\boxtimes)}$ [ID]

Associate Head of Law School, University of Sunderland, Sunderland SR6 0DD, UK
pardis.tehrani@sunderland.ac.uk

Abstract. For more than a century, power systems across the globe have relied on centralized electric generation and distribution. Centralized generation denotes intensive and large-scale energy generation at centralized facilities. These centralized power plants operate on fossil fuels such as natural gas and coal. The reliance and dependency on centralized fossil fuel plants have resulted in problems of scarcity and environmental issues. The high reliance on centralized energy servers is likely to result in a single point of failure where a single flaw in the energy market design and configuration can potentially affect the functioning of the entire energy system. The conventional energy system is highly susceptible to power failures and outages. Centralized energy systems can result in greater risk of data breaches, lower transparency and traceability, trust gap and higher operating costs. These deficits can be easily removed with a decentralized and distributed energy system to verify transactions and ensure validity and reliability of the data entered onto the blockchain ledger. This article explores the challenges of the application of blockchain technology in the energy sector with a focus of peer-to-peer energy trading. It investigates the lack of proper legal and regulatory framework to facilitate the adoption of blockchain technology in the energy sector.

Keywords: Blockchain · Energy Sector · Regulatory Framework

1 Blockchain Used Cases in the Energy Sector

Blockchain technology has gained significant traction in the energy industry as it is clearly an important innovation to be leveraged upon, more so with the usage of renewable energy sources. Blockchain technology has been perceived by scholars to be the "driver of energy revolution," attracting new market players to produce electricity, consume it and trade the excess electricity with others. Blockchain technology with the features of decentralization, digitalization, automation, security, and transparency can swiftly facilitate the energy revolution [1]. While the application of blockchain technology offers promising solutions, there are major issues faced by blockchain in the energy sector such as the consumption of exorbitant amount of energy, privacy concerns and security concerns. In what follows, the researchers then explore the lack of proper legal and regulatory framework to facilitate the adoption of blockchain technology in the

© The Author(s), under exclusive license to Springer Nature Switzerland AG 2023
J. M. Machado et al. (Eds.): BLOCKCHAIN 2023, LNNS 778, pp. 52–60, 2023.
https://doi.org/10.1007/978-3-031-45155-3_6

energy sector. Therefore, the researchers define and analyze the legal challenges that might hinder the adoption and application of blockchain-based peer to peer energy trading in the domain of energy law, tort law, property law, contract law, competition law, and data protection law, by looking at these areas in general terms, and specifically to the legal and regulatory.

1.1 Key Issues of Adopting Blockchains Technology in the Energy Sector

Consumption of Exorbitant Amount of Energy

There are numerous challenges with the Power of Work (PoW) consensus mechanism. This consensus mechanism is applied in various peer-to-peer energy trading schemes and projects as portrayed by the tabulation above. In order to generate a new block, the "consensus nodes" are required to solve a "computationally expensive puzzle" [2]. This puzzle is challenging to solve, and once solved, the solution is added to the new block. PoW requires a great deal of energy to mine a block due to the rigorous calculation costs. Therefore, due to the great calculation load, it makes the PoW an inappropriate consensus mechanism to be applied in many blockchain applications, more so for the purposes of peer-to-peer energy trading. Scholars have mentioned that replacing the PoW consensus mechanism with Proof of Stake (PoS) might not be a better solution because it would be difficult to reach a "sustainable carbon footprint" due to the decentralized nature which requires a great deal of communication between peers [3]. Based on the PoS consensus mechanism, the possibility of validating the next new block is proportional to the shares it owns. This is indeed unfair as only the wealthiest individuals can validate new blocks and administer the blockchain.

Privacy Concern

Transactions on a blockchain are "cryptographically sealed" and the Public Key (PK) linked to each transaction amounts to the identity of the individual generating that transaction [4]. Besides that, the individual can vary the PK for every transaction. Nevertheless, scholars have mentioned that transactions via different PKs can be linked with each other, resulting in the deanonymization of the individual. Transactions of individuals can be observed, including their energy preferences and consumption patterns [4].

Energy trading platforms such as Slock.it and SolarCoin have not addressed privacy concerns. Nevertheless, scholars have recommended solutions to attend to this issue. One scholar suggested that energy trading information and documents regarding digital assets are put into a "consortium blockchain" [5]. The trading information is pseudonymized to protect data privacy. Besides that, transaction records are encrypted, and time stamped. To ensure accuracy, trading information is signed with digital signatures [5]. Other scholars have proposed a "noise-based privacy-preserving approach" to conceal the trading distribution patterns. This is vital to prevent data mining attacks and "linking attacks" due to the transparent and traceable trading environment present in the blockchain system. The mechanism is designed to achieve different privacy and for that purpose, "dummy accounts" and "dividing accounts" are introduced [6]. Nevertheless, as propounded by Ye-Byoul Son et al., this design merely protects statistical information and not transaction records [7]. The critical component is to decipher and propose methods to protect transaction records during the pairing of prosumers and consumers on a blockchain [7]. It was proposed by the scholar that the peer-to-peer trading system can be conducted on a private blockchain. Prosumer and consumer matching are performed via "encrypted bids" and by utilizing a "functional encryption (FE)-based smart contract". Without encryption of bids, blockchain nodes can obtain the details of the transaction and related data. The bids are encrypted using function-hiding inner product encryption with a secret key (SK). To ensure that this design is workable, the scholar proposed numerous goals that have to be achieved. Firstly, the identities of users participating in peer-to-peer energy transactions remain anonymous to each other and are also kept confidential from blockchain nodes. Secondly, the contents of each bid by the participating users must be kept confidential from blockchain nodes. Seller and buyers who have been paired for the energy transaction should not be allowed to retract their bid. The transactions should be verifiable, and the "integrity of bids" should be maintained by disallowing any modifications towards the bids. As for "functional encryption (FE)-based smart contract", the smart contract for "peer matching" is generated on a blockchain with "open parameters" (op) and it is publicly verifiable. Therefore, the data is encoded in the "encrypted bids" and not opened to the nodes, but the "peer matching" transactions are carried out by blockchain nodes that can be verified publicly via smart contracts. Nevertheless, it has been elucidated by other authors that utilizing a private blockchain and auction or bidding methods of trading may be unfeasible for numerous users [7]. Many prosumers and consumers cannot partake in auctions and bidding due to time limitation and lack of expertise in technology [7].

The application of blockchain technology in the energy sector demands the application of multiple privacy safeguards. The following section introduces various primary privacy protection approaches in blockchain. These cryptographic solutions and technologies can be adapted in the blockchain network as a foundation towards facilitating privacy preservation, namely Secure Multiparty Computation, Zero Knowledge Proof, Commitment schemes, zkSNARK (zero-knowledge Succinct Non-Interactive ARgument of Knowledge), Homomorphic encryption, Ring signatures, Group signatures. In the energy sector, certain startups/companies, such as Greeneum and Solara have integrated privacy-preserving features when utilizing blockchain.

Security Concerns

There are numerous emerging security issues from the application of blockchain technology. These security issues are present in blockchains' application layer, smart contract layer, incentive layer, consensus layer, network layer and data layer.

Blockchain *Application Layer* includes various attacks such as DDoS attack, Initial Coin Offering Tampering and Phishing attack and 51% attack.

The Smart Contract Layer dealt with Reentrancy Attack, Unauthorized Access Attack and Solidity Development Security Issues.

Consensus Layer covers Long-Range Attack and Sybil Attack. Network Layer also dealt withEclipse attack and BGP Hijacking. Finally, Data Layer which deals with Malicious information attack.

1.2   Lack of Proper Regulatory Framework – Peer to Peer Energy Trading

Energy Law

Peer to peer energy trading may well be technically and economically possible but may prove to be an impossibility if a proper regulatory framework is not devised to address a potential norm in the decentralized world. Regulatory barriers dissuade and deter potential participants from participating in peer-to-peer energy trading networks. It is pertinent to understand the primary role of prosumers in a peer-to-peer energy trading market. Energy prosumers are expected to increase the generation capacity, which will complement the current energy systems, more so with the increase in the demand of energy [8]. Besides that, by utilizing various smart technologies, such as blockchain technology and smart meter, the energy prosumer can actively partake in demand management. With the heightened number of energy suppliers, there will be greater competition amongst the participants in the energy markets [8]. The European Union Directive 2018/2001 postulates on other benefits of the growth of decentralized renewable energy production, namely decrease energy transmission losses and enhance the "local security of energy supply", amongst others. The European Commission boldly expressed that "passive consumers" are becoming "active prosumers". The conventional understanding of the role of the consumer is shifting in that consumers not only consume energy but are involved in the production, storage and sale of energy [9].

There are numerous barriers faced by energy prosumers on peer-to-peer electricity trading platform in relation to their integration within the energy landscape. The primary challenge is that peer to peer energy trading is only possible when parties to the transaction are allowed access to the electricity grid. In the past, large energy companies were reluctant to allow prosumers to have access to the grid but have circumspectly become more accepting towards allowing such access. Regulatory instruments can facilitate prosumers' access to the electricity grid, thus liberalizing the entire energy system and preventing large companies from restricting such access [10].

The second challenge is to enable access to information and communication platforms in order to participate in peer-to-peer energy trading networks which utilize smart technologies to operate. In the communication domain, various communication devices and applications are used [11], and for a prosumer to effectively participate in the peer-to-peer energy network, they should have access to communication domains and services.

The third challenge by Tushar et.al elucidates that regulations are necessary to determine the market design that should be permitted, either fully decentralized market, community-based market or composite/hybrid market. Besides that, a suitable regulatory framework must be devised for a suitable taxation and fee/charge structure. It is also vital to elucidate the manner in which peer to peer trading systems can be integrated with the existing energy supply system. There is a dearth of literature on these areas, and it still remains ambiguous.

The fourth challenge is that the requirement of retail supply licenses to engage in the supply of electricity is shrouded with complexities and constrictions especially to the peer-to-peer energy trading market [12]. As discussed earlier with reference to German Energy Law, the current legislation stipulates requirements on the form of the contract, balancing responsibility, notification of energy delivery, data storage obligations. Currently, energy is sourced from a single supplier which presents hurdles to peer-to-peer trading. The existing retail contracts require these single suppliers to pay system costs on behalf of consumers. Peer to peer energy trading on the other hand is principally "swapping their energy supplier on a minute-by-minute basis" and in accordance with the preference of consumers [12].

The fifth challenge is that the role of Distribution Service Operator has not been clearly delineated. In a peer-to-peer energy trading landscape, the role and responsibilities of other market actors such as DSO's arise. Scholars stipulate that DSO's can take on a system operator or coordinator's role. With the development of peer-to-peer energy trading, it is perceived that the role of DSOs is likely to be augmented further.

The imposition of network charges in peer-to-peer energy trading has yet to be addressed. While the recast to the Renewable Energy Directive and the Electricity Directive stipulates that prosumer are not to be subject to cost-reflective, transparent, and non-discriminatory network charges that account separately for the electricity fed into the grid, there is no clarification on whether the standard network charges are applicable to peer-to-peer energy trading. If network charges are to be imposed, it should be a non-discriminatory and proportionate charge to market peers to ensure the viability of peer-to-peer energy trading.

A potential issue that is not widely discussed is the balancing obligations of prosumers with the imbalances that surface in the electricity grid. In the event that prosumers engaging in peer-to-peer trading are considered energy suppliers, they will be subject to the same obligations as those imposed on large scale energy suppliers. Article 15(2)(f) of the Internal Market Electricity Directive states that active customers are "financially responsible for the imbalances they cause in the electricity system; to that extent they shall be balance responsible parties or shall delegate their balancing obligations" [13]. Besides direct blockchain-enabled peer to peer energy trading between different households, a conventional energy supply company can provide additional services to its customers with the deployment of peer-to-peer energy trading platforms. Prosumers who have energy generation facilities such as solar panels may utilize the platform deployed by the conventional energy suppliers to trade surplus energy with other consumers. In the event of insufficient energy supply by prosumers considering the volatility of renewable energy, the energy suppliers can always guarantee energy supply to consumers [13]. As these conventional energy supply companies act as intermediaries in providing prosumers with platforms to trade energy, the technical and bureaucratic requirements are also complied by these companies, particularly that associated with licensing and registration requirements, taxation obligations and balancing obligations ordinarily applicable to an energy supply company.

Data Protection

Data protection could be undermined because of the lack of legal clarity on whether and how data protection laws are applicable to blockchain technology, especially the role and obligations of data users, and the rights of data subjects participating in peer-to-peer trading. In peer-to-peer energy trading, a smart metering solution is required to facilitate users to actively participate in the smart grid infrastructure necessary to enable peer to peer trading. The information stored on smart meters include (a) demand of power consumption of the user; (b) available power budgets for trading in the grid; (c) length of contract; (d) tariff data; (e) the cost of utilization/rent of the infrastructure; and (f) the amount of contracted power [14]. From the above, the data collected and stored in a smart meter relates generally to the consumption and generation behaviour of the participants in the peer to peer energy network. Another important element in the context of peer-to-peer energy trading is the data quality, and the smart meter infrastructure which is able to provide reliable and "higher temporal-resolution" data [15]. As such, the smart meter infrastructure can be regarded as an essential prerequisite towards the workings of peer-to-peer energy trading. As a significant amount of data becomes available, complemented by other forms of data such as network data, the risks of data protection breaches arise. Besides that, there are numerous other privacy issues that arise from the application of blockchain technology. Firstly, with the advent of blockchain technology, there might be privacy threats through "transaction linkability". The participants' identity can be deciphered by way of the transactions entered by them. An intruder or hacker may be able to ascertain the identity of the participant by way of a "blockchain-transaction-graph analysis". Privacy breaches can also occur via network analysis through "IP address network analysis" and by the utilization of "statistical analysis" from transaction data [16].

As mentioned, the GDPR seems to pose challenges to blockchain technology, and the incompatibilities of blockchain might hinder widespread blockchain-based peer to peer energy trading adoption. The prominent issues are immutability of the blockchain ledger leading towards a potential breach of the right to be erasure and right to rectify data subject's personal data, under Article 16 and Article 17 of GDPR respectively. While these rights form a foundational bearing of data protection, in terms of blockchain technology, the application of these rights have not been succinctly clarified. Therefore, scholars have suggested the need for further clarification of the applicability of these rights in a decentralized setting.

Contract Law

The conventional electricity market is grounded on a bilateral agreement between parties, the buyer and seller. Nevertheless, in a peer-to-peer energy trading market, it is based on "multi-bilateral transaction exchanges of heterogeneous energy commodities or assets among participants" [17]. The legal question of the contractual relationship of prosumers and consumers in relation to multi-bilateral trading arises, and the issue of the legal design of peer-to-peer trading platforms has to be addressed via a regulatory framework. The application of blockchain technology and smart contracts, which automate and digitalize energy trading, pose various contract law issues. Technological innovation makes it challenging to reconcile traditional contractual law principles between buyers and seller, to the multi-bilateral trading arrangements amongst peers. The first issue is contractual liability, in that who should be liable for the breach of contract. The question that arises is whether it is possible to identify the defaulting party in the peer-to-peer energy transaction. Besides that, it is also important to evaluate whether the energy trading platforms are contractually liable in a decentralized energy trading network. To this end, it is pertinent to devise a suitable regulatory framework to ensure legal certainty in a decentralized peer to peer energy network, to compensate the affected network participants.

From academic literature, there are two ways that smart contracts can be leveraged in decentralized energy trading:

The contractual terms are confirmed via traditional contract negoatiation between the parties before the trading of energy and subsequently outlined and executed in a coded programming language [18].

The smart contract deployed on a blockchain network contains business logic for the transactions on the blockhain, in the absence of a traditional contract [18].

In relation to (a), a smart contract for peer-to-peer energy trading involves a three-step process. Firstly, the participants must reach an agreement on the transaction of energy. This involves the negotiation of terms until there is the "meeting of the minds" and a legally binding contract is entered into by the parties. The smart contract is "encoded" to comprise of requirements and instructions as agreed upon. Secondly, upon satisfying the requirements and instructions, the "first 'block' unlocks" and triggers the process of distribution of energy by way of the encoded instructions. Nevertheless, if the requirements are not complied with, the "block will remain locked" and energy would not be distributed, such as non-payment which is a contractual obligation of the participant. There are instances where, depending on the model of the smart contract, the units of energy that is traded is represented by tokens. This is known as tokenization, symbolizing the energy commodity to be traded on the trading platform which is attached to the smart contract [18]. The instructions that specifies the performance of the smart contract are not executed in a "coded programming language" which is stored on the blockchain, instead of the standard legal language contained in traditional contracts. The coded language functions as executable clauses, and the pre-defined requirements and preconditions must be satisfied before the energy tokens and units of energy are transferred from the prosumer to the consumer. The coding and reading of the language coded onto a smart contract must be carried out by a person who is able to comprehend the usage of smart contract. According to scholars, a smart contract executes a legal agreement, but it is not in and of itself legally enforceable due to the decentralized feature of blockchain. Therefore, it is more practicable if a hybrid system is put in place to address this concern. For a more "time-dependent actions", a smart contract can be utilized, but "context-sensitive provisions" should be encompassed within a traditional contract [389].

As for (b), the trading procedure is implemented as a smart contract. Prosumers and consumers put forth their request to sell or purchase energy to the smart contract, and all the request is stored on the blockchain ledger which can be easily assessed by the parties. Subsequently, matching and clearing of transactions is automatically executed by the smart contract [19]. An offer can be made towards one specific individual or more than one individual (public). The initial stages of a smart contractual agreement and traditional contractual agreement is the similar where the parties to the transaction must be agreeable to the same contractual terms. The offer and acceptance of the parties are viewed objectively. If a participant of a transaction post the request to buy or request to sell to the smart contract on a blockchain platform, and the other party accepts such communication using his cryptographic key, the posting of the request to the "on-chain smart contract" on a blockchain platform constitutes an offer. The "blockchain in a binary computer code which specifies precisely the terms of the transaction", will generally be held as an offer instead of a mere invitation to treat [20].

The UK has a legislative footing based on the common law system. Smart contracts are not explicitly recognized in the UK nor EU law. Nevertheless, the Renewable Energy Directive issued by the EU in 2018 defined peer to peer energy trading as taking place "by means of a contract with pre-determined conditions governing the automated execution and settlement of the transaction, either directly between market participants or indirectly through a certified third-party market participant". Scholars have interpreted it to encompass smart contracts and have elucidated that national contract law is the "most reliable source" to access whether smart contracts are valid [21]. While smart contracts may pose several contract law issues, the difficulties in the application of smart contracts are exacerbated as smart contracts are usually in a form of a computer code which needs to be translated into human language. Smart contracts are relatively inflexible unlike contractual law provisions which are relatively flexible in light of the circumstances. Therefore, the functioning and legal recognition of smart contracts may present a legal barrier for smart contracts.

In a peer-to-peer energy trading setting, it has to be determined if electricity platforms should be held liable for the non-performance in the energy supply of individual prosumers. While scholars have raised joint liability, it is crucial to determine if direct or indirect control needs to be exerted in the performance of energy supply service by a prosumer before prosumer can be subjected to such liability. In the event the entire liability is allocated to the prosumer for the prosumers' non-performance, it might deter consumers from participating actively in blockchain-based energy sector, more so since the technical and financial ecosystem is provided by platform operators.

2 Conclusion

It is envisaged that the findings will contribute towards creating a liberalized market structure and facilitate blockchain-enabled peer to peer energy trading. In determining the commercial viability of the deployment of blockchain in the realm of the energy landscape, various technical and regulatory aspects are explored.

There are major technical challenges faced by the adoption of blockchain in the energy sector. On the consumption of exorbitant amounts of energy, currently the consensus mechanisms applied require a great amount of energy to mine a block. Scholars have discussed various technical solutions that can be adopted to provide a more sustainable avenue for the application of blockchain technology. Besides that, as further mentioned, there are numerous privacy and data protection concerns that can arise from the blockchain-enabled applications specifically the disclosure of energy production and consumption data, and energy preferences. While most start-ups and pilot projects have not thoroughly addressed privacy concerns in their White Paper, scholars have suggested various solutions that can be utilized by blockchain engineers in creating blockchain-enabled energy applications and solutions.

On the regulatory aspect, there is a consensus among scholars on the presence of a regulatory layer which is essential to facilitate the adoption of blockchain technology. As our focus is on blockchain-based peer to peer energy trading, and the associated legal and regulatory aspects, the researchers explore and analyze the domain of energy law, data protection and contract law. In the energy domain, regulations should adopt provisions for both, monetary aspects such as the payment of network charges, and other technical aspects such as privacy concerns, licensing requirements, contractual and tortious liability, amongst others.

References

1. Mika, B., Goudz, A.: Blockchain-technology in the energy industry: Blockchain as a driver of the energy revolution? With focus on the situation in Germany. Energy Syst. **12**, 285–355 (2021)
2. Mollah, M.B., et al.: Blockchain for future smart grid: a comprehensive survey. IEEE Internet Things J. **8**(1), 18–43 (2020)
3. Besanger, Y., Tran, Q.T., Le, M.T.: On the applicability of distributed ledger architectures to peer-to-peer energy trading framework. In: 2018 IEEE International Conference on Environment and Electrical Engineering and 2018 IEEE Industrial and Commercial Power Systems Europe (EEEIC/I&CPS Europe). IEEE (2018)
4. Dorri, A., et al.: Peer-to-peer energytrade: a distributed private energy trading platform. In: 2019 IEEE International Conference on Blockchain and Cryptocurrency (ICBC). IEEE (2019)
5. Kang, J., et al.: Enabling localized peer-to-peer electricity trading among plug-in hybrid electric vehicles using consortium blockchains. IEEE Trans. Ind. Inf. **13**(6), 3154–3164 (2017)
6. Gai, K., et al.: Privacy-preserving energy trading using consortium blockchain in smart grid. IEEE Trans. Ind. Inf. **15**(6), 3548–3558 (2019)
7. Son, Y.-B., et al.: Privacy-preserving peer-to-peer energy trading in blockchain-enabled smart grids using functional encryption. Energies **13**(6), 1321 (2020)
8. Milčiuvienė, S., et al.: The role of renewable energy prosumers in implementing energy justice theory. Sustainability **11**(19), 5286 (2019)
9. Council, T.E.P.A.O.T., Directive (EU) 2018/2001, The European Parliament
10. Butenko, A.: User-Centered innovation and regulatory framework: energy prosumers' market access in EU regulation (2016)
11. Zhang, C., et al.: Peer-to-Peer energy trading in a Microgrid. Appl. Energy **220**, 1–12 (2018)
12. Poudineh, R.: Liberalized retail electricity markets: What we have learned after two decades of experience? (2019)
13. Klein, M.: Die Blockchain-Technologie: Potentziale und Herausforderungen in den Netzsektoren Energie und Telekommunikation. Bundesnetzagentur (2019)
14. Sigl, C., et al.: The role of smart meters in P2P energy trading in the low voltage grid (2018)
15. Shishido, J., E.U. Solutions: Smart meter data quality insights. In: ACEEE Summer Study on Energy Efficiency in Buildings (2012)
16. Thukral, M.K.: Emergence of blockchain-technology application in peer-to-peer electrical-energy trading: a review. Clean Energy **5**(1), 104–123 (2021)
17. Science, F.o., Publications–Trusted Networks Lab. corporate (2020)
18. Lee, J., Khan, V.M.: Blockchain and smart contract for peer-to-peer energy trading platform: legal obstacles and regulatory solutions. UIC Rev. Intell. Prop. L. **19**, 285 (2019)
19. Song, J.G., et al.: A smart contract-based p2p energy trading system with dynamic pricing on ethereum blockchain. Sensors **21**(6), 1985 (2021)
20. Finocchiaro, G., Bomprezzi, C.: A legal analysis of the use of blockchain technology for the formation of smart legal contracts. MediaLaws (2020)
21. Schneiders, A., Shipworth, D.: Community energy groups: can they shield consumers from the risks of using blockchain for peer-to-peer energy trading? Energies **14**(12), 3569 (2021)

Student Certificate Sharing System Using Blockchain and NFTs

Prakhyat Khati[1]([✉]) [iD], Ajay Kumar Shrestha[2] [iD], and Julita Vassileva[1] [iD]

[1] University of Saskatchewan, Saskatoon, SK S7N 5C9, Canada
prakhyat.khati@usask.ca, jiv@cs.usask.ca
[2] Vancouver Island University, Nanaimo, BC V9R 5S5, Canada
ajay.shrestha@viu.ca

Abstract. In this paper, we propose a certificate sharing system based on blockchain that gives students authority and control over their academic certificates. Our strategy involves developing blockchain-based NFT certifications that can be shared with institutions or employers using blockchain addresses. Students may access the data created by each individual institute in a single platform, filter the view of the relevant courses according to their requirements, and mint their certificate metadata as NFTs. This method provides accountability of access, comprehensive records that are permanently maintained in IPFS, and verifiable provenance for creating, distributing, and accessing certificates. It also makes it possible to share certificates more safely and efficiently. By incorporating trust factors through data provenance, our system provides a countermeasure against issues such as fake and duplicate certificates. It addresses the challenge of the traditional certificate verification processes, which are lengthy manual process. With this system, students can manage and validate their academic credentials from multiple institutions in one location while ensuring authenticity and confidentiality using digital signatures and hashing for data protection against unauthorized access. Overall, our suggested system ensures data safety, accountability, and confidentiality while offering a novel approach to certificate distribution.

Keywords: Blockchain · Smart contract · Non-fungible token · NFT · viewNFT · Certificate sharing · Data Trust

1 Introduction

Over the past years, the use of blockchain technology has gained immense popularity across various domains. One of the areas where the application of blockchain technology has the potential to revolutionize is educational data sharing. Educational data such as academic certificates, and student transcript data are all important personal data which are unique as it identifies each student. The uniqueness of the data makes it fit to be represented as Non-fungible token (NFT). NFT is considered a piece of data stored on a blockchain. It certifies the uniqueness of an asset. NFTs are seen to be used to prove the authenticity, legitimacy of digital assets and real assets that have a digital footprint [1]. Numerous studies have explored the use of blockchain-based technologies for managing

© The Author(s), under exclusive license to Springer Nature Switzerland AG 2023
J. M. Machado et al. (Eds.): BLOCKCHAIN 2023, LNNS 778, pp. 61–70, 2023.
https://doi.org/10.1007/978-3-031-45155-3_7

certificates. However, the utilization of Non-fungible Tokens (NFTs) in the domain of educational certificate management is still at an incipient stage.

In recent years, there has been a significant increase in the number of students studying abroad or applying for jobs in foreign countries. These students often need to provide proof of their academic credentials through certificates. However, the traditional paper-based certificate system requires multiple steps for certification, translation, and authentication, which can be time-consuming and costly. Additionally, students with multiple certificates must repeat the process for each certificate they hold, causing further delays and expenses. Also, the certificate once shared cannot be retrieved back, this could lead to misuse and data trust issues of such important credentials. To address certificate sharing, some institutions have implemented web2-based centralized systems for sharing certificates. However, they still face challenges such as a lack of global standardization and are often limited to a group of universities or institutions. Another major drawback of these approaches is that they lack data provenance, as the history and origin of the certificates cannot be traced in such implementations. This can lead to issues with trust and legitimacy when it comes to verifying the authenticity of certificates and the institute that issued the certificate. As a result, there is a growing need for a more secure and reliable global system for managing ownership and sharing academic certificates which has led to the exploration of blockchain-based solutions, such as the use of NFTs.

This paper proposes an NFT-based certificate sharing framework that provides an immutable and secure platform for academic institutions and students to manage ownership and share certificates. The framework leverages blockchain technology and smart contracts to ensure transparency, traceability, and authenticity of certificates. On top of that NFTs are used for assigning ownership, which makes it easier to track and verify the authenticity. By having metadata linked to them, NFTs can offer additional information about the certificate, including details such as who issued it, when it was issued, and a description of the achievement. This valuable information can be utilized to provide more context around the certificate, and to establish varying levels of access rights to the credentials, enabling individuals to grant or revoke access as necessary.

The rest of the paper is organized as follows. Section 2 describes the overview of blockchain, smart contracts and NFTs. A brief analysis of the existing architecture with their limitations is given in Sect. 3. After that Sect. 4 presents the solution architecture, our implemented model and describes our implementation details with the proposed approach of evaluation of the implemented system. Finally, the Sect. 5 concludes and summarizes the paper with future directions.

2 Background

The section provides background information about blockchain, smart contracts, non-fungible tokens (NFTs) and some of the terminology that is going to be used in the proposed framework.

Blockchain, also known as distributed ledger technology (DLT), is a data structure that allows for the creation of both private and public digital transactions by maintaining a shared ledger of transactions among network nodes [2]. The Ethereum community built an automation layer on top of a public permissionless blockchain using smart contracts managed by the decentralized network.

Smart contracts can be thought of as "if/then" conditional statements kept on the distributed ledger; they are pieces of executable code that are executed when triggered by an authorized or agreed-upon occurrence. The applications created on top of smart contracts are supported by the state-transition mechanism [3]. All participants share the states that contain the instructions and parameters, ensuring the accuracy of the directives.

Non-fungible tokens are distinct and non-interchangeable tokens stored on a blockchain. The standard for creating and maintaining non-fungible tokens using Ethereum smart contacts is governed by the ERC-721 standard. The standard defines how each NFT can represent someone's ownership of a specific digital and physical asset cryptographically. NFTs applications rely on blockchain technologies with smart contracts to ensure ownership, provenance, and exclusivity of the asset.

ERC-4361(SignIn with Ethereum) [4] is a standard proposed by the Ethereum community for enabling a decentralized authentication process with blockchain based wallets such as MetaMask. With this standard, users can sign in to decentralized applications (dApps) using their Ethereum wallets instead of creating separate login credentials for each dApp. The standard is based on the OAuth 2.0 framework and aims to provide a simple and secure way for dApps. Here we will be using MoralisSDK [5] to integrate with firebase, the backend, and the database. The centralized portion of our framework is handled using firebase and fire store.

IPFS (InterPlanetary File System) enables decentralized file sharing and storage [6]. It uses a peer-to-peer network of nodes to store and retrieve content, in contrast to conventional web protocols that depend on centralized servers. The unique content-addressing method used by IPFS allows files to be identified by their hash. This makes it possible to retrieve files even when the original source is unavailable, and it also makes file verification and versioning simple. It is possible to use hybrid encryption to prevent unauthorized access to stored content on IPFS. Here we have used pinata as a platform to store our NFTs metadata off-chain. These data are addressable using the hash, also known as Content-Identifiers (CIDs).

3   Related Works

Blockchain technology has the potential to revolutionize the way we issue and verify certificates and credentials. In the early stages of blockchain, various institutes presented several platforms and research solutions for generating certificates and badges. These platforms allowed universities to issue and backup certificates for their students or supported self-generated certificates and badges within their own private blockchain ecosystem. NFTs were not utilized by any of these applications. The author in [7] presented a detailed comparison of what are the benefits and challenges faced by the existing blockchain based education certificate sharing systems and proposed a new framework called NFTCert [7].

In late 2021 and early 2022, we saw the development of using NFTs for managing and defining ownership of digital arts. This spiked the exploration of the potential benefits of using NFTs for adding uniqueness to certificates, and badges to define ownership. Some proposed certificate systems, such as NFTCert [7] and "Ethernal Digital Certificates"[8] either incorporated NFTs to their existing blockchain frameworks or proposed new ones to solve problems in the existing ones.

NFTCert is a platform that allows the institute to create their student NFT-based certificates and transfer the ownership to the student. It also provides hash value for verifying the authenticity of NFT-based certificates. The control over data is only accessible to the owner who mints the NFT, here in this case it's the University. The platform also incorporated blockchain Oracle to add an online payment gateway for the necessary payment to retrieve the certificate.

While these NFT-based systems have played a role in reducing fraudulent activities by taking advantage of the offered technologies, they still have several limitations. For example, in NFTCert, the institute only has the right to mint the NFT. From the student point of view, this approach is time consuming as it involves more interaction with the institute, and the paper does not clearly state whether the ownership or the right to access the NFT is transferred to the student address; this make the student completely dependent upon the institute Also, here student cannot create purpose centric view of certificates, for example, if a student is applying for a software developer job, not all subjects or credits that he studied might be relevant to that particular job, in such as case, it would have been convenient if the student could be able to give access only to certain subjects that are relevant. However, this should not compromise the authenticity of the credential issued by the university. Furthermore, this platform focuses on NFT certificate management rather than sharing the NFT certificate and is centered towards private network implementation. This may limit the scope of the system and may not meet the needs of all stakeholders.

In contrast to these approaches, in this paper, we propose a novel NFT-based certificate-sharing system which not only overcomes the weaknesses of these systems but also introduces new features for managing and sharing NFTs.

4   Solution Framework and Discussions

The proposed solution framework for certificate issuance and ownership offers a decentralized approach that grants students maximum ownership of their certificates. Under this framework, university administrators can register their institutions and create university and student profiles containing metadata pertaining to the issuer address and signature for verification purposes. This metadata is then shown to the student's dashboard, which they can access by logging into the system using decentralized authentication via a MetaMask wallet.

Decentralized authentication involves using the Ethereum account to authenticate with an off-chain service by signing a standardized message format parameterized by scope, session detail, and security mechanisms. This can play a crucial role in the context of NFT certificate and credit sharing systems that provide ownership to students and implement access control. By leveraging the Ethereum blockchain, decentralized authentication can enable students to create self-sovereign identities using decentralized identity (DID) protocols like Decentralized Identifiers (DIDs). This can help to ensure that students have full control over their personal information and sole access to their certificates and credit information which is only granted to authorized parties.

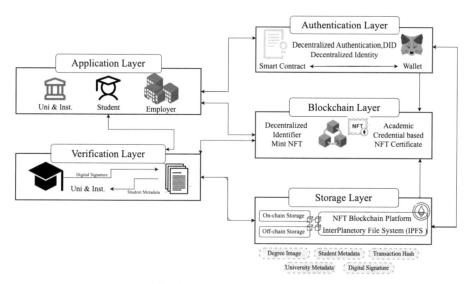

Fig. 1. A general system overview

With reference to this proposed general system overview in Fig. 1, the actual implementation is portrayed with the solution framework in Fig. 2. Here, the University generates a hash of the metadata and signs the hash using its private key to create a digital signature. The digital signature is then added to the metadata along with the original metadata. The university creates the NFT metadata, mints and then transfers the NFT to the student wallet giving the student full ownership of the NFT.

Students have the flexibility to decide how many viewNFTs they want to mint from their original certificate. ViewNFT is a novel term introduced in this framework to reference the NFT that are created by students. As per the smart contract rules, these viewNFTs will have flag that enables student to filter only courses and does not allow to alter any other information of the credentials. So now a student can just put relevant courses as a view transcript with degree completion certificate and mint these data as viewNFT certificate. The minted viewNFT will have student as the owner and the student can then share and allows access to these certificates and credential wherever necessary. Most importantly, even after sharing the certificate and transcript, the ownership remains with the student due to the rules set in the smart contract. This allows the student to take back ownership of their certificate even after sharing it with others. The NFT can be shared using the receiver's wallet address, and the student can fetch the NFT back to their own dashboard by changing the access rights.

Upon receiving the viewNFT generated by the student, the receiver will receive a notice mentioning that it is a viewNFT, and the student chose to hide certain subjects in this viewNFT. The receiver can use the university public key to decrypt the digital signature attached to the viewNFT, the hash obtained afterwards represents the original NFT from where the viewNFT is made. The function verifies the ownership and authenticity by comparing the hashes of the original data attached to the viewNFT and other supporting elements in the metadata.

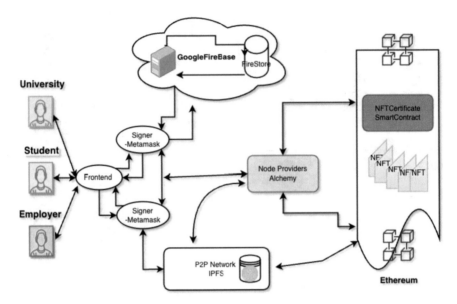

Fig. 2. System architecture

In summary, the solution framework grants students full ownership of their certificates, simplifies the transaction process, provides greater control and customization options, and offers an added layer of security to ensure that the student's data always remains under their control. The framework utilizes decentralized authentication, allowing students to access their profile information without the need for a centralized authority. University administrators play a crucial role in registering their institutions and creating student profiles, while employers can view and verify the certificates issued to potential employees.

4.1 Schema

According to the ERC721 standard, we define the metadata JSON schema to represent the certificate. It includes institute details, degree details, student details, course issued to the students and the respective wallet address details as shown in Fig. 3.

Here, the course in the schema is dynamically set by the university/institute that is creating the profile for the student. The viewSubjects are the subjects that students filtered from the original subject to create a viewNFT. The smart contract is deployed just once for each node on the Ethereum blockchain which stores the access control and tokenID to identify the user who minted their certificate issued to them by their University/Institute. Here the receivers can view the course subject that the student has set, and the smart contract keeps track of the changes made by the student. If the receiver wants to verify the authenticity of the academic credential, the receiver can view the original metadata validated by the university and decrypt the signature using the university public key to get access to the hash. After receiving the hash, the receiver can verify the authenticity

```
{
  "Title": "Certificate Metadata",
"Description": {                      "OriginalNFT": {
  "Creator": String,                    "TokenId": uint256,
  "Owner": String,                      "TokenURI": String,
  "StudentID": String,                  "Subjects": [{"mark": String,
  "StudentWalletAddr": String,          "subject": String},
  "StudentName": String,                {"mark": String, "subject":
"StudentDescription": String,         String}]},

                                        "UniversityName": String,
"ViewNFT": {                            "UniversityID": String,
"TokenId": uint256,                     "UniversityAddress":
"TokenURI": String,                   String,
"ViewSubjects": [                       "UniversityMetadata":
          {"mark": String,            String,
"subject": String},                     "Degree": String,
  {"mark": String, "subject":           "CertificateHash": String
String}
      ]}}
```

Fig. 3. ViewNFT certificate metadata

of the data by comparing the hash value. Since the NFT allows verification of its history regarding ownership, it is easier to verify the owner's address.

We have used OpenZeppline smart contracts to create our NFTCertificate sharing smart contact. The smart contract developed with Solidity contains the following functions (Fig. 4).

Here, the createToken () method takes the tokenURI as parameter, the tokenURI is hash value generated after uploading the metadata to IPFS. The data from the Firestore is given in JSON format with all the necessary metadata. The metadata is first uploaded to IPFS using Pinata API. This function mints an NFT and assigns ownership to the student wallet address. The CreateListedToken () function is used solely to keep track of the number of NFTs issued via the smart contract. To display data on the frontend of the application, we need to create additional methods, such as getAllNFTs (), which lists all NFTs issued using the smart contract, and viewNFT (), which displays a particular NFT. Similarly, the getMyNFTs () method returns a list of all the viewNFTs owned by the current user, while the getTransferredNFTs () method retrieves a list of viewNFTs that have been transferred to universities and employers. The executeTransfer () method in the smart contract is used to transfer view rights to an Ethereum address. The owner can set a default time period after which the transferOwnership function is executable to reclaim the viewing rights of the NFTs. The owner can revoke view rights by calling the transferOwnership () method from the application frontend.

```
Contract NFTCertificate is ERC721URIStorage {
                                    Struct TokenTransferScheduler
Struct ListedToken                  {
{                                   uint256 transferBackTime
uint256 tokenId,                    address transferBackTo
address payable owner,              }

  address payable viewer,             Event TokenListSuccess
  uint256 transferBackTime,           (
  bool currentlyListed,               uint256 indexed tokenId,
  bool currentViewer                  address owner,
  }                                   address viewer,
                                      uint256 transferBackTime,
                                      bool currentlyListed,
                                      bool currentViewer )


function createToken (string memory tokenURI, address Studen-
tAddress) public payable returns (uint)
function createListedToken (uint256 tokenId) private
function getAllNFTs () public view returns (ListedToken []
memory)
function getMyNFTs () public view returns (ListedToken []
memory)
function getTrasnferedNFTs () public view returns (ListedToken
[] memory)
function executeTransfer (uint256 tokenId, address receiver,
uint256 transferBackTime) onlyOwner
function transferOwnership (uint256 tokenId) onlyOwner
```

Fig. 4. Smart contract of NFT based certificate sharing system.

4.2   Performance Evaluation

This section covers the evaluation of the performance of the smart contract. The evaluation focuses on two aspects: transaction latency and resource consumption. The network setup used by default on Kaleido consists of two nodes, each running on 0.5vCPU with 1024 MB of memory. Figure 5. Shows the CPU resource consumption of both nodes and the system. The transaction latency recorded was 4.16 s. Here, y-axis of the graph represents the percentage of CPU usage, while the x-axis represents the duration.

Fig. 5. CPU utilization of nodes

During the experiment, the nodes consumed only 2% of the CPU to maintain operations. However, there was a temporary increase in CPU usage for a few minutes, up to 14%, when we triggered the minting of an NFTCertificate using the deployed smart contract.

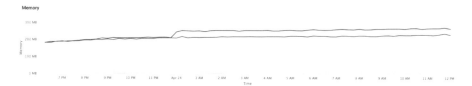

Fig. 6. Memory used of nodes.

In Fig. 6, the value on the y-axis denotes memory usage in MB and the x-axis denote the time in minutes within the day. The data indicates that each node utilized less than 253 MB of memory. At 12.15 AM on the test day, Node 2 on the graph used the smallest amount of memory, which was recorded as 227 MB. On the other hand, Node 1 used the largest amount of memory, approximately 252 MB, at around the same time. The memory consumption pattern remained steady throughout the day with no significant spikes observed. Finally, Fig. 7. Illustrates the number of transactions and blocks per hour. On average, there were 0.02 transactions (Tx) and 11 blocks per hour. It took an average of 5 s for a new block to be added to the blockchain.

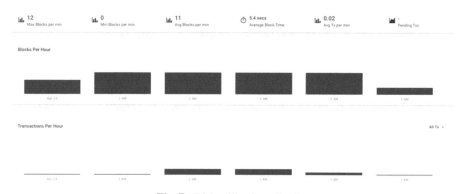

Fig. 7. Disk utilization of nodes

5 Conclusion

In summary, our proposed solution framework provides a decentralized approach to certificate issuance and ownership, giving students full control over their credentials. The use of Ethereum currency for transactions simplifies the process, while the ability to mint viewNFT from certificate credentials allows for greater customization and control over their academic certificate. Our future work involves improving the current model by studying users' attitudes to academic certificate sharing with NFT and blockchain where students have ownership and access control even after sharing their degree. We are also planning to evaluate the usability, trust perceived security, usefulness of the proposed framework using the validated constructs of the Technology Acceptance Model (TAM). We will also be investigating the possibility of introducing incentive for universities through this framework.

References

1. Khati, P., Shrestha, A.K., Vassileva, J.: Non-fungible tokens applications: a systematic mapping review of academic research. In: 2022 IEEE 13th Annual Information Technology, Electronics and Mobile Communication Conference, IEMCON 2022, pp. 323–330 (2022). https://doi.org/10.1109/IEMCON56893.2022.9946500
2. Shrestha, A.K., Vassileva, J.: User data sharing frameworks: a blockchain-based incentive solution. In: 2019 IEEE 10th Annual Information Technology, Electronics and Mobile Communication Conference, IEMCON 2019, pp. 360–366 (2019). https://doi.org/10.1109/IEMCON.2019.8936137
3. Wang, Q., Li, R., Wang, Q., Chen, S.: Non-Fungible Token (NFT): Overview, evaluation, opportunities and challenges. ArXiv (2021)
4. ERC-4361: Sign-In with Ethereum. https://eips.ethereum.org/EIPS/eip-4361. Accessed 8 Apr 2023
5. Web3 Firebase Authentication – Create a Web3 Sign-In with Moralis – Moralis Web3 | Enterprise-Grade Web3 APIs. https://moralis.io/web3-firebase-authentication-create-a-web3-sign-in-with-moralis/. Accessed 9 Apr 2023
6. What is IPFS? | IPFS Docs. https://docs.ipfs.tech/concepts/what-is-ipfs/. Accessed 9 Apr 2023
7. Zhao, X., Si, Y.-W.: NFTCert: NFT-based certificates with online payment gateway; NFTCert: NFT-based certificates with online payment gateway. In: 2021 IEEE International Conference on Blockchain (Blockchain) (2021). https://doi.org/10.1109/Blockchain53845.2021.00081
8. Nikolic, S., Matic, S., Capko, D., Vukmirovic, S., Nedic, N.: Development of a blockchain-based application for digital certificates in education. In: 2022 30th Telecommunications Forum, TELFOR 2022 – Proceedings (2022). https://doi.org/10.1109/TELFOR56187.2022.9983672

Transparency Disclosure for End Consumers in Private Food Supply Chains - A Systematic Literature Review

Hauke Precht[(✉)] , Ella Mahnken, and Jorge Marx Gómez

Carl von Ossietzky University of Oldenburg, Ammerländer Heerstraße 114-118, 26129 Oldenburg, Germany
{hauke.precht,ella.charlott.mahnken,jorge.marx.gomez}@uol.de

Abstract. End consumers state a rising interest in reliable information about food products. Blockchain technology is considered a beneficial technology to ensure data authenticity and provide traceability and transparency in the food supply chain. In this paper, we, therefore, analyze the integration of end consumers in the blockchain-based system within the food supply chain. As often private permissioned systems are used, we especially focus on this area of technology. To identify the current state of the art and approaches for transparency disclosure for end consumers, we conduct a systematic literature review, following Webster and Watson. We queried well-known databases ACM Digital Library, IEEE Xplore, Web of Science and The DBLP Computer Science Bibliography to obtain our initial set of literature. Based on the set of 111 papers, we performed a two-step process to narrow the set down to 19 relevant papers, from which only seven focused on end consumers. Our findings, based on the analysis of these seven papers, show that the aspect of transparency disclosure for the end consumer in the food supply chain does not go beyond the idea of QR code or tag scanning. Thus, the integration of end consumers into blockchain systems in the food supply chain is largely ignored, and their integration into the system is underrepresented. This raises the question of why end consumers are not considered relevant stakeholders and if private blockchains are suitable when end consumers should access information, opening new discussions around the integration of end consumers.

Keywords: End Consumer · Private Blockchain · Transparency Disclosure

1 Introduction

In the view of climate change, the need for action regarding sustainability is becoming increasingly urgent. End consumers are becoming more aware that their consumption behavior has an impact in ecological and social terms. Accordingly, it becomes increasingly important for them to be able to buy sustainably

J. M. Machado et al. (Eds.): BLOCKCHAIN 2023, LNNS 778, pp. 71–81, 2023.
https://doi.org/10.1007/978-3-031-45155-3_8

and fairly produced food [6]. In this respect, the origin of the food, on the one hand, and the cultivation and production, on the other, play a decisive role. Information about where food comes from and how it was grown and processed therefore becomes increasingly important for the end consumer. Further, the interest in fair trade and organic production methods is increasing as well [17]. Overall, consumers have a growing interest in the food they consume while demanding truthful information about the quality of products [10]. But at the same time, in Germany, for example, around one-third of the population do not trust that the food they consume is checked in detail [11]. This implies distrust of existing food inspectors and producers. However, trust in the information on quality and sustainability provided by food companies is essential for consumers to be willing to spend more money on sustainably and fairly produced food. Therefore, It is also in the interest of food producers to ensure that consumers trust the information on their products.

The usage of novel technologies in the area of Food Supply Chain (FSC) comes a long way and is subject to an active field of research, see, for example, [2,7,15,18,22]. Considering the enhancement of transparency, similar to general supply chain management, the usage of blockchain technology is researched as a promising addition.

2   Related Work

As mentioned, blockchain technology in the context of FSC is already subject to several scientific studies where the end consumer integration is, at least roughly, discussed as well. In [1], for example, the authors describe the potential of using a public permissioned blockchain to provide the end consumer with detailed information and the possibility to trace the food. They leveraged the Hyperledger Sawtooth blockchain along with a REST API to enable relevant parties to access information about the FSC, also providing a QR code for end consumers that can be scanned [1]. A similar approach can be found in the works by [3,14] that uses a REST API provided by Hyperledger themselves to connect end consumers to the blockchain layer. But also, the web3.js library, which is commonly used to communicate with Smart Contracts running on the Ethereum blockchain, is used on approaches within the FSC, as shown, for example, in [12].

However, these works do not describe in detail how the end user is connected to the REST API and how this, in turn, is integrated into the blockchain layer. Often, the usage of QR codes to be scanned by the end consumer is described. But while [12] and [16] describe the scanning to be done via a smartphone app and [3] and [14] propose the usage of a web application, works like [1,5,8] do not provide any detail at all. Further, the transparency disclosure in a private permissioned blockchain is, due to its restricted nature, more complicated than on public blockchain systems and would require further investigation. In public blockchains, there are search tools that allow anyone, not just members, as in a private blockchain, to have insight into all transactions made so far [21]. The limited access to a private blockchain means that non-authorized stakeholders of a FSC, such as the end consumer, do not have direct access to blockchain information, posing challenges for transparency disclosure.

3 Research Question and Methodology

As shown, end consumer integration is, at best, only casually discussed, and there is not yet a comprehensive study of approaches for transparency disclosure for end consumers in FSC. With this paper, we aim to close this gap and analyze the current approaches of transparency disclosure for end consumers in the context of private blockchain supported FSC. As shown, the current scientific literature discusses the usage of blockchain in general but lacks a detailed analysis/concept for transparency disclosure for end consumers. We focus on private systems especially, as they are often used within the context of FSC [19]. We, therefore, pose the following research questions:

What approaches are currently used to achieve transparency disclosure for end consumers in the food supply chain that leverages private blockchain systems?

By answering these questions, we contribute to the existing scientific literature by providing an in-depth analysis of potential transparency disclosure for end consumers in the FSC. Further, we highlight current shortcomings and challenges that require further attention in future research.

In order to obtain the current state of the art concerning the integration and transparency disclosure for end consumers in the FSC, a literature review is conducted, following [20]. Before starting the actual search process, the scientific databases to be used must be determined. For our literature, we rely on the commonly used and well-known databases ACM Digital Library, IEEE Xplore, Web of Science and The DBLP Computer Science Bibliography. In order to obtain the most comprehensive state of the art possible, as recommended by ([20]), we do not limit ourselves to one type of literature but consider all papers, regardless of their journal type, research method, or geographic region. The search terms to be used are derived from the defined research questions while also including potential synonyms for specific search terms. Thus we derived the search terms by extracting the four central nouns of the before-stated research questions: "private blockchain", "food supply chain", "consumer", and "transparency". Further, we integrated the synonyms for consumers, namely "endconsumer" and "customer", into the search string via a linkage with the OR operator. In addition, the term "end-to-end" is included in the search string, as this is also increasingly used to describe the integration of all stakeholders up to the end consumer of a FSC within a system. Thus the final constructed search string is: "private blockchain" AND "food supply chain" AND "transparency" AND ("consumer" OR "endconsumer" OR "customer" OR "end to end"). By performing a backwards search, further papers were identified that might not focus on the FSC itself but on an adjacent area and thus were also taken into consideration for further analysis. This is justified by the fact that end-user integration in another supply chain sector is likely to translate to FSC accordingly. The following table

1 shows for each database the initial number of found papers, papers found by backward search, duplicates and the number of papers identified as relevant.

Table 1. Overview of used databases and number of identified papers

Database	ACM	IEEE	WoS	Total
Number of Papers	27	64	20	111
Papers identified via backwards search	4	53	5	62
Duplicates	-	10	3	13
Number of papers after abstract screening	7	15	2	24
Number of relevant papers	**3**	**14**	**2**	**19**

After gathering the initial papers based on our keyword search, a two step-process is conducted to identify the relevant paper. Note that only papers written in English were considered. First, an abstract screening for each paper was carried out. During the abstract screening, the general relevance of the paper in the context of our stated research question is checked. If the abstracts show that the paper deals, at least loosely, with transparency for end consumers in FSC leveraging blockchain technology, it is subject to the next step of identifying relevant papers. Otherwise, the paper is discarded at this step of analysis. In the second step, the selected papers, after the abstract screening, are subject to an in-depth analysis. The papers are now analyzed in greater detail, focusing on end consumer integration in the respective systems. By reading the conclusion and screening the paper under review, the final set of 19 papers was identified. From this set, only seven papers highlight the end consumer integration within their research. The other 12 papers only casually mention the potential integration or propose loose ideas in this area.

4   Discussion of Identified Paper

In the following, the seven identified papers are presented and described in greater detail, focusing on the used technologies and patterns with respect to end consumer integration.

[8] describe the Food Safety Quick Response Block (FoodSQRBlock) framework that uses blockchain and QR codes to enhance the traceability of food/products in the FSC. Their framework is deployed to the Google Cloud Platform to facilitate its integration and scalability [8] Further, the authors identified two main challenges when introducing blockchain technology to the FSC: the limited storage capacity of data and the integration of end consumers into the blockchain system to improve the accessibility of the track and trace system [8]. The latter, they aim to tackle by leveraging QR Codes. Their described system architecture consists of three layers: (1)The physical layer that contains the different food products of the different manufacturers, (2)the digital layer where products from

the physical layer are linked with digital data that can be used for traceability and other purposes and (3) the cloud layer which is used to process the digital data on the blockchain to enable traceability of products and accessibility to food information [8] When a manufacturer produces a new product, the relevant information generated is digitized and stored on the blockchain [8]. Within the FoodSQRBlock framework, the authors describe two modules that are used, namely, an encoding module and a decoding module. The encoding module digitizes the product information and creates new blocks as well as QR codes (that can be generated at any point in time of the FSC)that contain product information (e.g. Product name, manufacturer ID, product size). The decoding module is a open-source software that enables the end consumer to retrieve information about products and verify them [8] End consumers can access information from the previous block while the decoding module performs hash computations on the retrieved information from the blockchain and compares these hashes with the hash values that are provided via scanning the QR codes [8] If these values match, the information is considered authentic. However, the paper does not describe how the information is made available online to the end user, nor does it describe how the end user can access the necessary decoding module or blockchain. Further, the authors do not use an established blockchain system but implemented their own version in Python [8]. Thus, common tools from existing ecosystems can not be used, making the integration even more difficult and not suitable at a large scale. Furthermore, the blockchain is solely hosted on the Google Cloud Platform, counteracting the idea of decentralization.

In [3], the pilot project from Bytable Inc is described that leverages Hyperledger Sawtooth and IoT technology within the context of FSC. By using IoT sensors and therefore leveraging automated data collection, the authors aim to further ensure data integrity [3]. The authors chose Hyperledger Sawtooth as underlying blockchain technology as specific REST APIs can be leveraged [3]. To ensure that not everyone has access authorization to sensitive corporate information, different stakeholders have different read access to blockchain data [3]. This is ensured by requiring users to have a valid private key to interact directly with the blockchain via a proxy microservice [3]. Since individual, private end users are not considered stakeholders in the FSC by [3], they do not have a valid key to access the authentication proxy server [3]. Therefore, they need to access information from the blockchain via a public REST API and a public proxy client service [3]. It is described that the end consumer scans a QR code on the egg package and thereby gains access to a web application, which can then be used to obtain information about the product [3]. Next to the scanning of the QR code that leads the end consumer to a web application, the end consumer must enter additional information that is printed on the package to obtain product-specific information. The paper does not explain how the web application through which the end consumer receives information is connected to the blockchain.

[4] present AgriBlockIoT, a traceability system leveraging a combination of Hyperledger Sawtooth and IoT sensors. Similar to [3], the IoT sensors are leveraged for automated data collection [4]. Within their research, they compared the

Hyperledger Sawtooth blockchain with the Ethereum blockchain and found that
Hyperledger Sawtooth is more suitable for a traceability System than Ethereum
as it provides better performance in terms of, for example, latency [4]. Overall, they define three main modules for their software: (1) a REST API with
a high level of abstraction enabling other applications to communicate easily
with the system, (2) a Controller component that sits between the API and
the blockchain module, translating the high-level requests from the API to low-
level request towards the blockchain and vice versa and (3) the blockchain as
the main component, handling business logic in Smart Contract (SC)s [4]. It
is notable that the authors see the blockchain module as independent of the
actual chosen blockchain implementation. However, as mentioned, they chose
Hyperledger Sawtooth over Ethereum. It is also described that all users must
be registered on the blockchain and have an asymmetric key pair in order to
sign digital operations on the ledger [4]. End users can verify information about
products by scanning a tag previously attached to the product [4]. However,
it is not described how exactly they then get the blockchain information, for
example, via a web application or mobile application.

[16] presents a whitepaper discussing the usage of blockchain within the FSC
by attaching a digital identity to a physical product. They describe their systems as being constructed based on a set of independent yet coupled programs
that take over specific responsibilities. Overall, they distinguish four programs,
namely the registration program, the standards program, the production program, as well as manufacturing program [16]. The registration program is used
to identify real-world actors such as certifiers, auditors or producers and create
a dedicated digital identity using a Public Key Infrastructure. The standards
program ensures that seals and certificates on food packaging are linked to the
information required for their requirements via the blockchain. New manufacturers and food information are added to the blockchain via the production
program. In the manufacturing program, the transformation of input goods into
end products is implemented. Tagging establishes a secure link between physical
products and their digital representation. In addition, products are given tags
(e.g., RFID tags) so that they can be scanned. The whitepaper mentions that a
DApp is presented. However, the DApp is not discussed in greater detail. It is,
therefore, possible that the smartphone application mentioned is a DApp that
has direct access to the blockchain. In general, [16] only provides an abstract
description of their system, and they do not elaborate much on the specific technical implementation. Therefore, the exact end-user integration in the system is
not clear.

[13] describe the framework ProductChain governed by officials to enable
information availability for end consumers. Instead of a single blockchain managing all occurring transactions, several parallel blockchains (shards) are used
in the system as side-chains [13]. Sharding is based on the concept of parallel processing of transactions, so transaction verification can be shared between
multiple nodes in parallel [13]. Overall, they propose a three-tier network architecture, where tier once consists of participants that are divided into FSC par-

ticipants and non-FSC participants. Tier 2 consists of local validators running side chains, and tier three includes the blockchain query manager along with a product ledger [13]. Note that in tier one, the end consumer is not considered to be a FSC participant [13] even though the main focus of the proposed system serves the goal of food transparency. End users can nevertheless query the product ledger, e.g. to obtain the origin information about a product [13] but do not have access to specific side chains. The global validator provides a query interface that returns complete or incomplete information depending on the read permission [13]. However, the global validator in tier three does not show the end consumer the entire transaction flow of a product, but only the information he or she has requested (e.g., on the origin of a specific product) by scanning a QR code [13]. Further, it is not explained exactly how the connection from the end consumer to the global validator is established.

[9] present a blockchain-based framework for product traceability using Hyperledger Fabric. The authors present their framework as a double-layer framework, with the goal of addressing existing disadvantages in other product traceability systems, such as a lack of government participation, performance bottlenecks or insufficient protection of private enterprise data [9]. The main layer consists of a consortium blockchain where also government regulatory agencies participate, while the sub-layer consists of multiple private blockchains that are deployed in and by the respective companies [9]. However, note that the main and sub-layer are not independent, but rather each company defines a key node that is part of the consortium blockchain as well as part of the private blockchain [9]. Thus, the consortium chain is used to share data between the respective companies and also functions as the interface towards the end consumer [9]. The different stakeholders of a supply chain are located in a different layers depending on their role [9]. Authorities are responsible for controlling the supply chain, and end users are in the main layer, while companies operate mainly in the sub-layer but partake in the main layer via their key nodes, as already mentioned [9]. The main layer is responsible for queries from end users, as it can provide information about products, while the sub-layer is responsible for storing product information [9]. However, even though the authors acknowledge the importance of end consumer integration and fast processing of their product information queries, they do not explain in detail how they interact with the main layer.

In [1], the authors describe a traceability system based on Hyperledger Sawtooth with the goal of enabling end consumers to trace the complete food history. They propose a multi-layer architecture, currently consisting of three layers. The first layer, the physical flow, contains various products throughout different organizations [1]. Digital data represents the second layer, where data is linked to a physical asset from within the first layer [1]. As the third layer, the authors define the blockchain system where the digital data from layer two is stored for traceability enabling consumers to track and trace specific products [1]. A QR code is attached to each product that contains information about the product, such as the manufacturing company and the temperatures and locations along the FSC [1]. The end user can access this information by scanning with the provided mobile application. Calls are made on the state of the Sawtooth REST API after scanning to get to the addresses with the requested information[1]. It is not made clear whether the mobile application might be a DApp that has direct access to the REST API of the Hyperledger Sawtooth blockchain. But [1] mentioned in a sentence that a DApp was developed, but this was not further elaborated.

In order to present and highlight similarities and differences in the described paper, Table 2 presents the specific approaches, used technologies and patters as well as the general view point of end consumer integration.

Table 2. Overview of the end consumer consideration and used architecture approaches

Authors	ID	Information access for end consumer	Architecture consideration	Used Blockchain
Dey et al.	[8]	QR code scanning	A layered architecture (3 layers) along with encoding and decoding modules	Self-developed Blockchain Solution
Bumblauskas et al.	[3]	QR code scanning plus web application where additional information must be entered	Usage of (Proxy) Microservices and a web application for end consumer	Hyperledger Sawtooth
Caro et al.	[4]	Tag scanning	Three main modules: REST API, Controller, Blockchain	Hyperledger Sawtooth
Provenance .org	[16]	QR code / Tags, dApp (smartphone app)	4 interconnected programs, namely registration, standards, production, manufacturing	-
Malik et al.	[13]	QR code scanning	Three tier with tier 1 being participants, tier 2 local validators and side chains (sharding)and tier 3 blockchain query manager and product ledger	-
Ding et al.	[9]	Querying via Main Layer	Double-layer framework with a main layer (consortium blockchain) and sub-layer (private blockchain)	Hyperledger Fabric
Baralla et al.	[1]	QR code scanning	Multi-layer; Layer one physical flow, layer two digital data linked to layer one, layer three blockchain system	Hyperledger Sawtooth

As shown in table 2, the overall dominant approach to enable end consumers to access information within the FSC is the scanning of either QR codes or tags. Further, most proposed architectures are designed as multi-layer or multi-tier architecture. Within the usage of private permissioned/consortium blockchain, solutions provided by the Hyperledger Foundation are commonly used, highlighting the potential for scalability and transaction speed. Therefore, it is no surprise that blockchain systems provided by Hyperledger are the most used. Note, however, that [16] and [13] only describe a high-level system concept without any implementation; thus, they do not name any specific blockchain. But [13] state that they aim to move beyond the sole concept and implement their design by using a Hyperledger blockchain.

Furthermore, only [3] and [4] consider the potential of IoT sensors for automated data collection as a way to increase data integrity by mitigating the need for manual data collection that poses potential threats to data integrity.

5 Conclusion

In this paper, we were able to show that the benefits of blockchain technology are widely recognized within the context of FSC. We further highlighted the fact that even though the traceability and transparency feature of blockchain technology is of interest to the end consumer that state requirements for reliable food data and information, they are sometimes not even considered stakeholder of the FSC [13]. Considering this perspective, it is no surprise that from the initial 19 relevant identified papers, only seven have a somewhat focus on the end consumer. As shown in the discussion in Sect. 4, even these seven papers do not explain the (technical) integration of end consumers in greater detail. This leaves room for speculations if, for example, the integration is done via a central server, which could lead to a single point of failure (SPoF), further contradicting the initial idea of transparency and traceability of blockchain.

The lack of consideration of end consumer integration clearly indicates the need for further research. This is most likely not limited to the FSC but rather affects other areas as well where the usage of private blockchain systems is preferred from an enterprise point of view but making it, therefore, hard for proper end consumer integration. This can raise the question of the actual feasibility of private blockchain systems in areas where end consumers have a valid interest in the managed data. For future work, we plan to further analyze end consumer integration in blockchain-based systems in other areas to check if similar challenges can be identified. Further, the connection to Web3 usability will need to be made as well.

References

1. Baralla, G., Pinna, A., Corrias, G.: Ensure traceability in European food supply chain by using a blockchain system. In: 2019 IEEE/ACM 2nd International Workshop on Emerging Trends in Software Engineering for Blockchain (WETSEB), pp. 40–47. IEEE (2019). https://doi.org/10.1109/WETSEB.2019.00012

2. Ben-Daya, M., Hassini, E., Bahroun, Z.: Internet of things and supply chain management: a literature review. Int. J. Prod. Res. **57**(15–16), 4719–4742 (2019). https://doi.org/10.1080/00207543.2017.1402140

3. Bumblauskas, D., Mann, A., Dugan, B., Rittmer, J.: A blockchain use case in food distribution: do you know where your food has been? Int. J. Inf. Manage. **52**, 102008 (2020). https://doi.org/10.1016/j.ijinfomgt.2019.09.004

4. Caro, M.P., Ali, M.S., Vecchio, M., Giaffreda, R.: Blockchain-based traceability in Agri-Food supply chain management: a practical implementation. In: 2018 IoT Vertical and Topical Summit on Agriculture - Tuscany (IOT Tuscany), pp. 1–4. IEEE, Piscataway, NJ (2018). https://doi.org/10.1109/IOT-TUSCANY.2018.8373021

5. Chan, K.Y., Abdullah, J., Shahid, A.: A framework for traceable and transparent supply chain management for Agri-Food sector in Malaysia using blockchain technology. Int. J. Adv. Comput. Sci. Appl. **10**(11), 1–8 (2019). https://doi.org/10.14569/IJACSA.2019.0101120

6. Connolly, J., Shaw, D.: Identifying fair trade in consumption choice. J. Strateg. Mark. **14**(4), 353–368 (2006). https://doi.org/10.1080/09652540600960675

7. Dabbene, F., Gay, P., Tortia, C.: Traceability issues in food supply chain management: a review. Biosys. Eng. **120**, 65–80 (2014). https://doi.org/10.1016/j.biosystemseng.2013.09.006

8. Dey, S., Saha, S., Singh, A.K., McDonald-Maier, K.: FoodSQRBlock: digitizing food production and the supply chain with blockchain and QR code in the cloud. Sustainability **13**(6), 3486 (2021). https://doi.org/10.3390/su13063486

9. Ding, Q., Gao, S., Zhu, J., Yuan, C.: Permissioned blockchain-based double-layer framework for product traceability system. IEEE Access **8**, 6209–6225 (2020). https://doi.org/10.1109/ACCESS.2019.2962274

10. European Comission: (2020). https://de.statista.com/statistik/daten/studie/1196049/umfrage/hauptsorgen-bei-lebensmittelbetrug-in-eu-und-deutschland/

11. GfK: Ich vertraue darauf, dass die verfügbaren lebensmittel ausführlich kontrolliert worden sind und gesundheitlich unbedenklich sind. (2013). https://de.statista.com/statistik/daten/studie/253989/umfrage/verbrauchervertrauen-hinsichtlich-lebensmittelkontrollen-nach-altersklassen/

12. Kim, M., Hilton, B., Burks, Z., Reyes, J.: Integrating blockchain, smart contract-tokens, and IoT to design a food traceability solution. In: 2018 IEEE 9th Annual Information Technology, Electronics and Mobile Communication Conference (IEMCON), pp. 335–340. IEEE (2018). https://doi.org/10.1109/IEMCON.2018.8615007

13. Malik, S., Kanhere, S.S., Jurdak, R.: Productchain: scalable blockchain framework to support provenance in supply chains. In: 2018 IEEE 17th International Symposium on Network Computing and Applications (NCA), pp. 1–10. IEEE (2018). https://doi.org/10.1109/NCA.2018.8548322

14. Miatton, F., Amado, L.: Fairness, transparency and traceability in the coffee value chain through blockchain innovation. In: 2020 International Conference on Technology and Entrepreneurship - Virtual (ICTE-V), pp. 1–6. IEEE (2020). https://doi.org/10.1109/ICTE-V50708.2020.9113785

15. Pal, A., Kant, K.: Smart sensing, communication, and control in perishable food supply chain. ACM Trans. Sens. Netw. **16**(1), 1–41 (2020). https://doi.org/10.1145/3360726

16. Provenance.org: Blockchain: the solution for supply chain transparency (2015). https://www.provenance.org/whitepaper

17. Squires, L., Juric, B., Bettina Cornwell, T.: Level of market development and intensity of organic food consumption: cross-cultural study of Danish and New Zealand consumers. J. Consum. Mark. **18**(5), 392–409 (2001). https://doi.org/10.1108/07363760110398754

18. Verdouw, C.N., Wolfert, J., Beulens, A., Rialland, A.: Virtualization of food supply chains with the internet of things. J. Food Eng. **176**, 128–136 (2016). https://doi.org/10.1016/j.jfoodeng.2015.11.009

19. Vo, K.T., Nguyen-Thi, A.T., Nguyen-Hoang, T.A.: Building sustainable food supply chain management system based on hyperledger fabric blockchain. In: 2021 15th International Conference on Advanced Computing and Applications (ACOMP), pp. 9–16. IEEE (2021). https://doi.org/10.1109/ACOMP53746.2021.00008

20. Webster, J., Watson, R.T.: Analyzing the past to prepare for the future: writing a literature review. MIS Q **26**(2), xiii-xxiii (2002). http://dl.acm.org/citation.cfm?id=2017160.2017162

21. Xu, X., Weber, I., Staples, M.: Architecture for Blockchain Applications. Springer, Cham (2019)

22. Zhu, Z., Chu, F., Dolgui, A., Chu, C., Zhou, W., Piramuthu, S.: Recent advances and opportunities in sustainable food supply chain: a model-oriented review. Int. J. Prod. Res. **56**(17), 5700–5722 (2018). https://doi.org/10.1080/00207543.2018.1425014

Repositioning Blockchain Technology in the Maritime Industry for Increased Accountability

Rim Abdallah[1,2(✉)], Cyrille Bertelle[1], Jérôme Besancenot[2], Claude Duvallet[1], and Frédéric Gilletta[2]

[1] Université Le Havre Normandie, LITIS EA 4108, 76600 Le Havre, France
{cyrille.bertelle,claude.duvallet}@univ-lehavre.fr
[2] HAROPA PORT, 71 Quai Colbert, 76600 Le Havre, France
rim.abdallah@etu.univ-lehavre.fr,
{rim.abdallah,jerome.besancenot,frederic.gilletta}@haropaport.com
https://www.haropaport.com/fr

Abstract. Digitalization has transformed the maritime industry, but accountability and transparency remain challenging due to the centralized nature of electronic data exchange platforms. To address this issue, this article proposes a blockchain-based solution that leverages legal identifiers and smart contracts to enhance accountability and transparency in portuary information systems. Extensive research indicates that a permissioned blockchain infrastructure is optimal for establishing a secure and decentralized environment among authorized parties. The proposed solution utilizes a modular and distributed blockchain infrastructure to notarize actions within the existing system, thereby increasing transparency and accountability. The article presents a proof of concept as the initial step in showcasing the capabilities and significance of this novel architecture. By tackling current challenges in the maritime industry, this proposed architecture offers a secure, efficient, and effective means of integrating portuary information systems.

Keywords: Smart port · Accountability · Transparency · Blockchain

1 Introduction

The digitalization of the maritime industry has led to the adoption of centralized platforms like Port Community Systems (PCS) and Cargo Community Systems (CCS) for efficient data exchange[7]. However, managing user accounts for the diverse range of actors within the network becomes challenging in a centralized system that covers the entire network.The International Maritime Organization (IMO) advocates for standardized Legal Identifiers (LIs) to authenticate organizations in these platforms, promoting security and trust. LIs are generally considered unique and secure as they are issued and approved by government

J. M. Machado et al. (Eds.): BLOCKCHAIN 2023, LNNS 778, pp. 82–92, 2023.
https://doi.org/10.1007/978-3-031-45155-3_9

authorities, adhering to various regulations and standards to guarantee accuracy and authenticity. A widespread approach in the maritime sector is the utilization of LIs to establish a brokered authoritative transitional (BAT) access. Each organization uses its legal identifiers to create an administrator account and manages its own hierarchy in the centralized system using different roles, permissions and access rights for its different users while adhering to system oversight. Through LIs, organizations represent their users on the network and satisfy legal obligations to treat other parties honestly. This approach strikes a balance between centralized control and local autonomy, enhancing efficiency and regulatory compliance. By utilizing LIs and administrative accounts, the central system is relieved of excessive administrative burdens, while LIs enhance security and trust in the network. However, limitations linked to accountability and transparency have sparked a need for decentralization in maritime digital networks. Blockchain technology offers a solution by providing decentralization, immutability, and transparency. Leveraging blockchain can overcome centralization challenges and establish a distributed network where multiple stakeholders participate in transaction validation and verification.

2   Research Rationale

The BAT approach is commonly used in information systems and particularly in maritime electronic platforms and PCS [9,11,17]. It provides numerous benefits such as improved scalability, enhanced security and compliance.

Using legal identifiers and BAT access in information systems can bring numerous benefits, such as improved scalability, enhanced security, and compliance [12]. However, this approach can also compromise accountability and transparency.

Accountability refers to the ability to track and trace individual and organizational interactions within the system. When organizations manage their users and create administrative accounts on the central system, it becomes challenging to maintain a complete and accurate record of all user activities. This can make it difficult to hold individual users and organizations accountable for their actions within the system.

Transparency refers to the visibility throughout the system and the recording of decision-making processes. If entities manage their hierarchy and user permissions within the PCS, it complicates monitoring and governance for the central network resulting in a lack of transparency that is crucial for effective governance and trust. This could potentially lead to miscommunications or intentional misconduct, making it more difficult to detect and hold entities accountable.

To address security compromises, PCS should implement measures that promote transparency, along with robust auditing and reporting mechanisms to ensure accountability. However, this can be challenging in the complex ecosystem of the maritime sector. With a large number of parties, diverse interests, and different legal frameworks, it is difficult for PCS to maintain accountability and openness across the entire marine ecosystem. Furthermore, the variety in

PCS infrastructure and superstructure further complicates the development of a versatile solution that can comply with all requirements across the global supply chain. Nevertheless, promoting accountability and transparency remains vital to the efficient operation of PCS networks and may promote the expansion and development of the marine industry.

Based on a thorough review of relevant literature and applications recommended strategic measures to achieve optimum performance and maintain accountability and transparency can be grouped into the following pillars:

- Prioritize focus areas where transparency and accountability are most significant [5]
- Promote collaboration and transparency amongst network entities [16]
- Acknowledge the importance of transparency and accountability[6]
- Leverage technological advancements [13]

Anchoring the last suggestion, we ascertain the need for alternative solutions that can enhance fluidity and security and propose an innovative blockchain-based architecture as a well-suited proposition to improve transparency and accountability that aligns with the remainder of the suggestions while maintaining traditionally used technologies such as LIs.

3   An Introduction to Blockchain

Blockchain technology is characterized by being secure, transparent and tamper-proof. It consists of two security layers [4,8] to enact autonomous trust within a network. The first is distribution. A blockchain ledger is distributively replicated across the network's user pool. The users' pool, access rights and control within the network determine the blockchain type: public, permissioned or private. The second layer of security is the use of cryptographic algorithms to chain blocks of data. Each transaction has a fixed-length digital signature on the ledger associated with its owner's public key and is incorporated in its containing block's hash. And each block contains the hash of its previous forming a hard-to-reverse-engineer and alter chain as detailed in 1. Together, these measures ensure data integrity and immutability. Our hypothesis envisioned blockchain as a viable solution to accommodate enhanced security and transparency for the sector while preserving current business manners. The technology's decentralised architecture provides access to a shared collectively maintained verifiable tamper-proof single source of truth. The integration of this technology into the sector enables additional security and helps achieve a transparent system that promotes accountability.

4   Background Work

On a blockchain network, identity representation can be achieved through the use of Decentralised identifiers (DID).

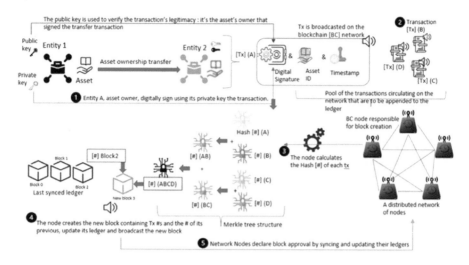

Fig. 1. Double the Security: Exploring the Anatomy of Blockchain

4.1 What Are DID?

DIDs are a relatively new form of digital identification frequently linked with blockchain because they allow for autonomous digital identity.

Our research was not primarily focused on the technical aspects of DIDs. Instead, what piqued our interest was their characteristics. DIDs, based on blockchain and key pair technologies, offer a decentralized and tamper-proof system for recording and verifying identity information, while their main benefit lies in enhancing transparency and accountability in online interactions [14]. They promote self-sovereign identity (SSI) by giving users control over their own identities and addressing trust issues in centralized systems. SSI can be effectively used in single-sign-on (SSO) scenarios, requiring only identity proof without sharing private data, and in authentication contexts where portability of secure verifiable credentials is advantageous.

These characteristics can further elevate technological development, particularly in a global industry such as the maritime as they can enhance operational efficiency and convenience. They converge towards IMO recommendations such as the 2017 STCW amendment that necessitated identification documents' standardization and encouraged secure and verifiable electronic certificate use for enhanced security and efficiency.

4.2 A Challenge Statement

The implementation of DIDs in the maritime sector faces significant challenges due to the lack of standardized governance mechanisms and a globally spread infrastructure. The decentralized and distributed trust models of DIDs conflict with existing hierarchical governance structures in the sector. Despite ongoing efforts in standardization, regulations, and digital identity projects, the adoption of DIDs in the maritime sector is still a distant prospect. The sector must first address underlying infrastructure and superstructure challenges before embracing DIDs [2]. Additionally, the maritime sector is known for its technological conservatism, and resistance to DID implementation may arise due to the presence of established Legal Identifiers (LIs) that serve a similar purpose in maritime information systems.

4.3   A Solution Strategy

The exploration of the benefits and challenges of adopting Decentralized Identifiers (DIDs) in the maritime industry has prompted a strategic shift in our approach. Implementing a full blockchain solution in the maritime sector may be complex and face various obstacles, making it challenging to achieve widespread adoption. Therefore, we have focused on a specific use case that is more manageable and feasible, allowing us to showcase the added advantages of the technology. This specific use case involves addressing accountability and transparency concerns through the concept of brokered authoritative transitional (BAT) access. Our shift in strategy also considers the functional similarities between DIDs and Legal Identifiers (LIs). Both serve as unique digital identifiers within information systems, promoting trust and authenticating the identity of the entities involved. However, it is important to note that we do not use these identifiers interchangeably.Additionally, we have drawn inspiration from the existing balance between centralization and autonomy, aiming to innovate and address the industry's needs and challenges. We aim to build upon pre-existing technologies and criteria while introducing new and beneficial advancements in the maritime sector.

4.4   Conceptualization

In BAT information systems, balancing operational efficiency with centralized oversight and control poses challenges for accountability. Legal Identifiers (LIs) provide administrative access, granting organizations autonomy over their users and access permissions, but this can result in a lack of transparency and accountability. Blockchain-based DIDs are promoted as a promising solution for accountability by simplifying authentication and identification processes[10]. In the maritime industry, we propose the novel combination of LIs and smart contracts as a revolutionary approach. This approach maintains current operational practices while introducing blockchain technology to enhance transparency and accountability. By leveraging smart contracts, administrative access to the centralized system can be notarized in a decentralized environment, leading to increased transparency and effectiveness while improving security. The combination of LIs and smart contracts has the potential to significantly transform transparency, accountability, and overall system performance.

5   Proof of Concept: An Architectural Proposition

5.1   An Entry

To demonstrate and further elaborate on our proposed approach, we suggest the following architectural proposition that seamlessly integrates portuary information systems and promotes collaboration. Based on the extensive technological research of blockchain in the maritime sector [1–3], we identified permissioned blockchain as the most adept infrastructure for our secure decentralized environment for the following reasons:

- Compliance: restricting access to the network to only authorized parties' relevant rules and regulations can be enforced in the environment
- Security: prevent unauthorized access to sensitive maritime data
- Privacy: network operators can implement privacy-preserving measures and access controls
- Scalability: network can be designed to meet the specific needs of the sector such as the number of participants and the data volume to be stored and processed.

The proposed architecture based on permissioned blockchain specifically Hyperledger Fabric (HF) does not take away from the governance within the centralized system but adds to it. Through Membership service, access control and certificates the environment remains secure. The information system previously governing LI creation and access to the system will also govern our blockchain environment where LI will also be used to regulate the access and permissions within the system as prior. This can be easily achieved via Fabric CA, a fabric component, used to easily manage (issue, revoke and manage) certificates for environment participants such as nodes and users within the environment. Additionally, organizations may use their own certificate authorities (such as Microsoft active directory certificate services or open SSL) for increased scalability and more seamless integration.

Thus we create a blockchain environment that revolves around the currently used information systems and seamlessly integrate them.

5.2   Workflow

Once the participant is identified and authorized within the information system. The PCS itself will communicate with the blockchain infrastructure and correspondingly tie a certificate to the organization's LI. Thus we place the PCS within a blockchain environment, making organisations' integration seamless. We change nothing to the current business process thus overcoming the obstacle of varying technical and technological capabilities. Stakeholders will participate within the secure digital blockchain-based environment without having to dive into blockchain complexities.

The IS will communicate with the blockchain infrastructure via API calls. We chose API calls because they are already widely common and familiar within the industry and easy to integrate. The blockchain infrastructure can be shared and maintained amongst IS governing, IS participating stakeholders and other trusted intermediaries within the maritime environment (Fig. 2).

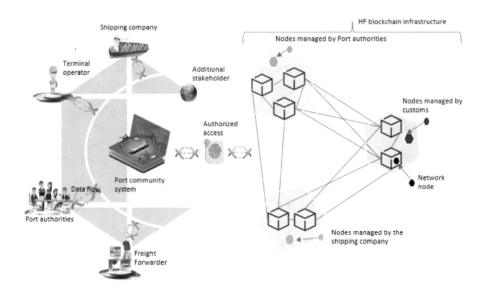

Fig. 2. Conceptual Blockchain based architecture for maritime IS

The maritime environment is reigned by competitiveness factors from different organizations present (for example several shipping companies). This makes the concept of a shared distributed ledger more challenging to implement. Moreover, IS superstructure in the maritime sector varies widely as it can be governed by one or more entities from the public sector, and/or one or more entities from the private sector. Hence, developing a blockchain-based solution that complies and adapt to distinct IS is complex. However, through HF's modularity and different node types, this can be achieved. We will demonstrate our architecture reasoning through a hybrid IS example governed by two different entities:

- Entity A is a public sector governmental entity actively involved in decision-making and possessing endorsing nodes on the network.
- Entity B, a private sector entity, oversees and ensures the competitiveness and maintenance of the information system through validator nodes, preserving the integrity of the ledger on the network.

We will further elaborate our developed proof of concept using a typical scenario with the maritime industry that first, elaborates on the use of BAT access and second, demonstrates part of our system's capabilities and importance Fig. 3.

Our Proof of Concept (POC) focuses on demonstrating how a shipping company, XYZ, can utilize a Port Community System (PCS) to manage operations

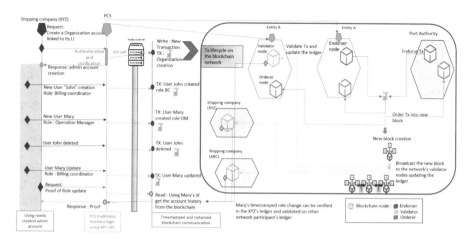

Fig. 3. PoC Worflow illustration

at Port AB. The PCS serves as a crucial platform for shipping companies to interact with various actors in the maritime industry, streamlining operations, reducing costs, and enhancing competitiveness.

The POC utilizes a modular and distributed blockchain infrastructure to add transparency and accountability to the traditionally used PCS. By combining Legal Identifiers (LIs) for authentication and identification with smart contracts, relevant actions and events within the system are notarized on the blockchain, creating a secure and transparent record.

The POC introduces shipping companies XYZ and ABC to the network, where XYZ gains access through LI authentication and the PCS acts as a liaison for authentication and verification. By recording all account management activities and updates on a shared distributed ledger, the POC addresses the need for transparency and accountability in the maritime sector. It tackles challenges such as tracking access privileges, preventing unauthorized changes, and establishing clear timelines for actions. Valid transactions created by XYZ are endorsed by endorser nodes, ordered in blocks, and validated by validator nodes, while a gossip protocol efficiently shares updates across the network, enhancing scalability and fault tolerance. Timestamped records on the blockchain enable a secure and transparent audit trail, reducing errors and instilling trust among stakeholders.

In the PCS system, John, an XYZ Billing Coordinator (BC), had his account deleted when he moved to another branch, and Mary took over the BC role. A dispute arises later over unpaid invoices, leading to ambiguity and compromised accountability as it is unclear who handled the document in question. The lack of a proper record of access privilege changes within the PCS makes it challenging to establish a clear timeline and identify unauthorized changes, hindering transparency and accountability. We propose a solution that preserves conventional practices while addressing these vulnerabilities, emphasizing the need to revoke John's privilege and delete his account to prevent unauthorized access and ensuring proper record-keeping of role changes. This improved solution enhances transparency, accountability, and effective operations management under the oversight of XYZ's nodes and the portuary governing entities A and B. Since XYZ and ABC are competitors the communication concerning John's actions remain hidden from ABC through the concept of channels [15]. The POC preserves conventional business practices while leveraging blockchain technology to enhance data protection, validation, and certification. It aims to reinforce existing trust within the maritime sector and facilitate the introduction of decentralized approaches for increased transparency, accountability, and competitiveness.

6   Discussion and Future Work

The innovative aspect of combining legal identifiers with smart contracts in the maritime industry lies in its ability to address the industry's technological conservatism and facilitate the adoption of decentralized identifiers. By implementing a permissioned blockchain and a node server, decentralized identifiers can be seamlessly integrated into existing systems, reducing implementation complexity. The use of smart contracts ensures secure and tamper-proof transaction execution, providing a reliable solution for the industry's valuable goods transportation sector. Furthermore, this approach enables the industry to leverage decentralized identifiers without requiring a significant technological overhaul, accommodating its slow adoption of new technologies.

The solution offers numerous benefits, including secure transaction execution, seamless integration, and the potential for wider use cases beyond accountability and transparency. However, further development is required, particularly in the area of sophisticated smart contracts that can handle complex transactions. Performance testing is also crucial to optimize system efficiency for high transaction volumes.

In summary, the innovative aspect lies in how legal identifiers and smart contracts are combined to overcome the maritime industry's technological conservatism, promoting the adoption of decentralized identifiers. The solution provides a secure and familiar approach, with potential for expanded use cases. Continued research and development are necessary to fully unlock the benefits of this solution.

References

1. Abdallah, R., Bertelle, C., Duvallet, C., Besancenot, J., Gilletta, F.: Blockchain potentials in the maritime sector: a survey. In: Daimi, K., Al Sadoon, A. (eds.) Proceedings of the ICR 2022 International Conference on Innovations in Computing

Research. ICR 2022. Advances in Intelligent Systems and Computing, vol. 1431, pp. 293–309. Springer, Cham (2022). https://doi.org/10.1007/978-3-031-14054-9_28

2. Abdallah, R., Besancenot, J., Bertelle, C., Duvallet, C., Gilletta, F.: Assessing blockchain challenges in the maritime sector. In: Prieto, J., Martínez, F.L.B., Ferretti, S., Guardeño, D.A., Nevado-Batalla, P.T. (eds.) Blockchain and Applications, 4th International Congress. BLOCKCHAIN 2022. Lecture Notes in Networks and Systems, vol. 595, pp. 13–22. Springer, Cham (2023). https://doi.org/10.1007/978-3-031-21229-1_2

3. Abdallah, R., Besancenot, J., Bertelle, C., Duvallet, C., Gilletta, F.: An extensive preliminary blockchain survey from a maritime perspective. Smart Cities **6**(2), 846–877 (2023)

4. Abdallah, R., Abdallah, R.: A blockchain based methodology for power grid control systems. In: Daimi, K., Al Sadoon, A. (eds.) Proceedings of the ICR 2022 International Conference on Innovations in Computing Research, ICR 2022. Advances in Intelligent Systems and Computing, vol. 1431, pp. 431–443. Springer, Cham (2022). https://doi.org/10.1007/978-3-031-14054-9_40

5. Christie, P., et al.: Why people matter in ocean governance: incorporating human dimensions into large-scale marine protected areas. Marine Policy **84**, 273–284 (2017). https://doi.org/10.1016/j.marpol.2017.08.002, https://www.sciencedirect.com/science/article/pii/S0308597X17300532

6. de la Peña Zarzuelo, I.: Cybersecurity in ports and maritime industry: reasons for raising awareness on this issue. Transp. Policy **100**, 1–4 (2021). https://doi.org/10.1016/j.tranpol.2020.10.001

7. Dewulf, W., Dekker, R., Notteboom, T., Van Hooydonk, E.: Digitalization of port logistics: a review of the port of Rotterdam. Marit. Econ. Logistics **17**(1), 1–18 (2015)

8. Dib, O., Brousmiche, K.L., Durand, A., Thea, E., Hamida, E.B.: Consortium blockchains: overview, applications and challenges. Int. J. Adv. Telecommun. **11**(1), 51–64 (2018)

9. Fransoo, J.C., de Koster, R.: PortBase: enabling logistics the success of the Dutch port community system. Transp. Res. Part E: Logistics Transp. Rev. **77**, 1–14 (2015). https://doi.org/10.1016/j.tre.2015.01.007

10. Maldonado-Ruiz, D., Torres, J., El Madhoun, N., Badra, M.: An innovative and decentralized identity framework based on blockchain technology. In: 2021 11th IFIP International Conference on New Technologies, Mobility and Security (NTMS), pp. 1–8. IEEE (2021)

11. of Rotterdam, P., of Amsterdam, P.: Portbase. https://www.portbase.com/. Accessed 18 Apr 2023

12. Saeed, A.: Legal identifiers in the maritime industry: case study on the Iranian port system. J. Marit. Law Commerce **48**(2), 225–238 (2017)

13. Simmons, E., McLean, G.: Understanding the paradigm shift in maritime education: the role of 4th industrial revolution technologies: an industry perspective. Worldwide Hospitality Tourism Themes **12**(1), 90–97 (2020). https://doi.org/10.1108/WHATT-10-2019-0062

14. Stockburger, L., Kokosioulis, G., Mukkamala, A., Mukkamala, R.R., Avital, M.: Blockchain-enabled decentralized identity management: the case of self-sovereign identity in public transportation. Blockchain Res. Appl. **2**(2), 100014 (2021)

15. Surjandari, I., Yusuf, H., Laoh, E., Maulida, R.: Designing a permissioned blockchain network for the halal industry using hyperledger fabric with multiple channels and the raft consensus mechanism. J. Big Data **8**(1), 1–16 (2021)

16. Tran, T.M.T., Woo, S.H., Yuen, K.F.: The impacts of sustainable inter-firm collaboration on business performance of shipping companies. Int. J. Logistics Manage. **32**(3), 766–789 (2021). https://doi.org/10.1108/IJLM-11-2020-0453
17. Valenciaport: Valenciaportpcs. https://www.valenciaportpcs.com/. Accessed 18 Apr 2023

IPFS and Hyperledger Fabric: Integrity of Data in Healthcare

Pedro Silva⬤, Tiago Guimarães$^{(\boxtimes)}$⬤, Ricardo Duarte⬤, and Manuel Filipe Santos⬤

Algoritmi Research Center, School of Engineering, University of Minho, Azurém Campus, 4800-05 Guimarães, Portugal
tsg@dsi.uminho.pt

Abstract. Since the Information of Things (IoT) arrival, one of the main problems we encounter daily is data breaches and data integrity. Now more than ever, the expertise needed to develop an attack is decreasing. Developers are creating software that does the same thing as an expert, requiring less knowledge from the attacker. Also, ill-intended professionals within the institutions can compromise and access information without authorization.

Healthcare Information Technologies must be aware of and have a proactive approach to this problem within each sector. With this in mind, researchers and developers must propose and study solutions and architectures to store and query sensitive files and information more securely.

When we think about security, there are more dimensions to consider other than the technology itself, but it does remove some constraints. This article presents an architecture that relies on Blockchain through HyperLedger Fabric and Interplanetary File System (IPFS) to securely host sensitive documents such as contracts within Healthcare.

Keywords: Blockchain · Healthcare Industry · Data Integrity · Ipfs · Hyper Ledger Fabric · Kubernetes · Security · Permissioned Blockchain · Web3 · Linux Foundation

1 Introduction

The healthcare industry handles vast amounts of sensitive information, from patient records and contracts to clinical trial data. As such, ensuring the security of this data is crucial. In this paper, we aim to propose a solution for securely storing sensitive information in the healthcare industry [12], based on permissions.

Previous research has emphasized the importance of secure data storage in healthcare, but there are gaps in the literature regarding the use of combined centralized and decentralized mechanisms for this purpose. Our proposed solution combines a master node and slave nodes for managing communications between containers, an ordering node for message production in the hyper ledger channel, and a local network of IPFS peers for file storage and hash generation.

© The Author(s), under exclusive license to Springer Nature Switzerland AG 2023
J. M. Machado et al. (Eds.): BLOCKCHAIN 2023, LNNS 778, pp. 93–102, 2023.
https://doi.org/10.1007/978-3-031-45155-3_10

To assess the feasibility and potential benefits of our proposed solution, we conducted a SWOT analysis. Strengths include the use of both centralized and decentralized mechanisms, as well as permissions-based access control. Weaknesses include the need for a secure authentication system to manage permissions, as well as potential compatibility issues with existing healthcare infrastructure. Opportunities include the potential to enhance data security and streamline data management processes. Threats include potential resistance to change and the need for additional resources to implement the proposed solution. In conclusion, our proposed solution offers a potential solution for securely storing sensitive information in the healthcare industry, based on permissions. By utilizing a combination of centralized and decentralized mechanisms, as well as permissions-based access control, we are able to ensure data integrity and increase security. Future works could include testing the proposed solution in a healthcare setting, refining the architecture to address any compatibility issues, and exploring additional mechanisms for enhancing data security.

2   Materials and Methods

2.1   Decentralized Distributed File System

IPFS(Interplanetary File System) [6,17,18] is a peer-to-peer solution to store files in a decentralized manner and it was built to offer high availability and immutability of files across the world. In a nutshell, is a combination of bittorrent, git and also kademlia [16] that offers a interplanetary file system solution peer-to-peer. Examples of nowadays usages for this tech are NFTs(Non Fungible Tokens) [20] and private immutable files. Every non ordering node will have one as you will see further.

2.2   Analysis of a Permissioned Blockchain Framework for Enterprise Applications

Hyper Ledger Fabric [8,10,11], on the other hand, is a technology that builds trust between organizations. It is a peer-to-peer role based tech, that nowadays is used to connect organizations such as banks, created by Linux foundation it has gain massive adoption because it mixes centralized technologies with decentralized, providing immutable and yet relevant information between parties that need information and because all organizations agree upon the same policies on that system and the data is immutable, this system becomes very reliable to use. You will see hyper ledger with this roles containerized. You should note that, for example Accenture, one of the biggest companies on the market is betting in this technology as prove of value.

2.3   Related Works

In order to make clear to the reader which cases we can apply this kind of tech, we grouped some interesting projects where some of them are runned by matured organizations. We even tought ahead and divided it, to achieve comprehension about single mentioned techs, but also how they are integrated together.

2.3.1 IPFS

In what concerns IPFS, the projects that we think are valuable to speak of are:

- Arbore [13] - Ipfs application for storing files in a decentralized way.
- FileNation [5] - Filenation lets you send files as a link to an email. The files cannot be censored as they are sent stored and transmitted decentralised because it uses the InterPlanetary File System (IPFS).
- Snapshot [3] - Web3 project that enables the user to vote in proposals. Normally this votes are saved in a public blockchain [7] like Ethereum [19], but this project is more economic, because instead of saving the data into a public chain it saves on the hyper ledger fabric that does not require gas fees.
- Berty.tech [1] - Non required mobile internet messaging system, this project has the objective to exchange messages between mobile phones that will be able to run their IPFS nodes locally without internet, using any proximity protocol like Bluetooth.

2.3.2 Hyper Ledger Fabric

In the other hand for hyper Ledger Fabric we have projects such as:

- IBM Food Trust [15] - In this project, IBM ensures the quality of the food and the chain it self by creating a network of connecting growers, processors, distributors, and retailers through a permissioned, permanent and shared record of food system data.
- NIIT Technologies: Chain-m Blockchain [2] - Project to airlines, to keep track of: number of tickets sold to fare amounts, commissions, taxes collected and more.
- Hitachi Streamlines and Secures Procurement [9] - Project that uses Hyper Ledger Fabric to process contracts.

2.3.3 Both

- Starling Project [4] - This is a social media project, that uses multiple decentralized technologies including IPFS, GUN, Hyperledger Fabric, and Hedera Hashgraph throughout its process. It uses IPFS for storing images and files and uses HyperLedger Fabric to store those hash's, that can also be stored in Hedera Hashgraph(Public blockchain).
- Decentralized Data Access with IPFS and Smart Contract Permission Management for Electronic Health Records [14] - This article examines the use of smart contracts in blockchain for electronic health record (EHR) permission management. Benefits include traceability, transparency, and improved security. However, EHRs are still stored locally despite decentralized permission management, allowing undetected queries within hospital infrastructure.

Smart contracts pose challenges for verifying actor identity and handling lost or stolen private keys. Our proposal integrates blockchain technology with the InterPlanetary File System (IPFS) for decentralized EHR storage. IPFS ensures peer-to-peer distributed network storage, enhancing security and availability. A governing body is introduced to handle public key governance and identity verification. Our solution combines blockchain, decentralized storage, and robust identity verification to enhance EHR permission management.

3   Architecture

3.1   Basic Explanation

In order to explain our idea, we tough in showing a basic layer representation of it, which contains the tech, from the lowest level until the highest. According to this ideology, what we will have is the operative system, multiple containers, CA Authority for authenticate who's communicating with the components, the hyperledger, Ipfs, Kubernetes, the API and finally the front office in React.

3.2   Node Composition

Since we will have multiple machines, this is the composition of a single node. This one is a Master node that will be the one responsible for orchestrating all the nodes in the network, but the slave nodes will be the same exactly thing, the only difference is that they will be commanded by this one. IPFS will not be in a container in every node, in order to use the much resources it can get and also because it does not make sense to put it inside a container.

3.3   Global Representation

For example purposes this is our global conception of the system, which will consist in multiples nodes, which can be represented in 3 different purposes. First one will be responsible for ordering the information shared between all the nodes. Second will be responsible for orchestrating the slaves. The third will be a slave. Note that the administrator of the network will be the one which have access to the Master node, and will be responsible for maintaining both Hyper Ledger Fabric, IPFS and also the Kubernetes Network. Hyper Ledger is a peer-to-peer and role-based system that is typically implemented for collaboration between multiple organizations. However, in this particular case, our focus is on a single hospital environment. Instead of multiple organizations, we envision representing different departments within the hospital as separate entities within the network (Fig. 1).

Despite the current single hospital setup, the network structure remains adaptable and can be linked to other organizations if the need arises. Currently,

Fig. 1. Layer Architecture

our approach is centered around a partnership-type organization within the hospital. However, in the future, if necessary, we can transition to a more pervasive type of organization, enabling seamless integration with external entities.

Similar to the standard implementation, each peer in the network is still represented as a container. However, in this context, each container represents a specific department within the hospital, rather than a separate organization.

Just like in the usual setup, different departments within the hospital can have their own dedicated channels within the network. This allows for secure and isolated communication between departments while maintaining the overall integrity of the system.

Every machine or peer within the network can be assigned different roles based on the certificate authentication of individuals using them. This role-based approach ensures that authorized personnel have appropriate permissions and access levels according to their assigned roles.

The flexible network design also enables connections to different nodes within the network. This facilitates seamless communication and collaboration between peers, enhancing the overall efficiency of data sharing and interaction within the hospital environment (Figs. 2, 3 and 4).

3.3.1 Security Measures The security of storing files in IPFS (InterPlanetary File System) can be strengthened through various measures. Storing the hash of the file and its versions in a Hyperledger Fabric network provides additional security and transparency. Managing interactions using certificates, main-

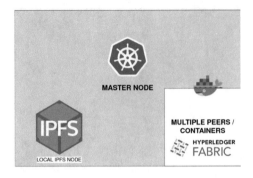

Fig. 2. Master Node

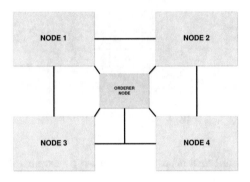

Fig. 3. Master Node

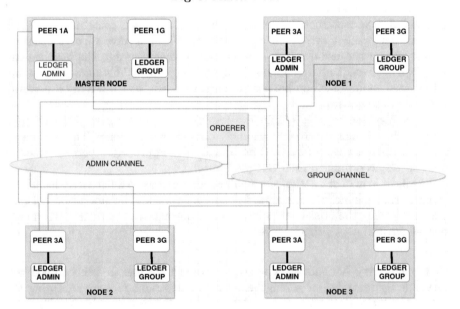

Fig. 4. Master Node

taining logs of every operation performed by certificates, conducting regular analysis of sensitive files and corresponding logs, and leveraging Kubernetes for network management and high availability further enhance the overall security of the system.

By storing the hash of a file and its versions in a Hyperledger Fabric network, you establish a tamper-evident record of the file's integrity and version history. This ensures that any modifications or updates to the file can be traced and audited. The Hyperledger Fabric network provides a trusted and immutable ledger that securely stores the file hashes, adding an extra layer of security and facilitating regulatory compliance and dispute resolution.

Managing interactions using certificates is crucial for securing the system. Certificates, issued by a trusted Certificate Authority (CA), validate the identity of network participants and ensure secure communication. By utilizing certificates, you can authenticate and authorize users, preventing unauthorized access and ensuring that only trusted entities can interact with the system. Furthermore, maintaining logs of every operation performed by certificates provides an audit trail for accountability and forensic analysis, enabling you to track and investigate any suspicious activities.

Regular analysis of sensitive files and their corresponding logs is essential to identify unexpected behavior and potential security threats. By conducting periodic reviews of the logs and analyzing the activities related to sensitive files, you can detect any unusual patterns, unauthorized access attempts, or data breaches. This proactive approach allows you to promptly respond to security incidents and mitigate potential risks before they escalate.

Managing the network using Kubernetes brings additional security benefits. Kubernetes is a container orchestration platform that provides robust management capabilities, including workload isolation, automatic scaling, and service discovery. By leveraging Kubernetes, you can ensure that the network components running IPFS, Hyperledger Fabric, and other related services are deployed in a controlled and scalable manner. Additionally, Kubernetes offers high availability features, such as automatic failover and load balancing, which help mitigate the impact of potential failures or disruptions, ensuring the continued availability and reliability of the file storage system.

By combining these security measures, including storing file hashes in Hyperledger Fabric, managing interactions with certificates, maintaining comprehensive logs, analyzing sensitive files and logs regularly, and leveraging Kubernetes for network management and high availability, you establish a robust and secure environment for storing files in IPFS. These measures enhance the integrity, confidentiality, and availability of the stored files, protecting them from unauthorized access, tampering, and potential system failures.

3.3.2 Workflow The process begins when a doctor logs into the system using their unique credentials and digital certificate. This initial authentication ensures the identity and authorization of the doctor. Once logged in, the doctor proceeds to upload, for example, a file related to a contract he wishes to store.

At this point, a Rest Service comes into play. It validates the provided credentials and certificate, ensuring the doctor's identity and authenticity. With successful validation, the Rest Service takes the uploaded file and saves it into the InterPlanetary File System (IPFS). IPFS provides a decentralized and secure storage solution for files.

After saving the file, IPFS generates a unique hash that acts as a digital fingerprint representing the content of the file. This hash ensures the integrity and tamper-proof nature of the stored data. The Rest Service then retrieves this hash and saves it into the Hyper Ledger Fabric using a Smart Contract.

Depending on the type of operation, the workflow branches into two paths: file update or simple save. In the case of a file update, the system creates a file record and a version record with a specific sequence. However, for a simple save operation, the system creates a file record and a version record with an initial sequence set to 1. These records maintain a chronological history of file versions.

To maintain an audit trail, every transaction within the system is recorded in a dedicated logs data structure. This structure captures essential information, including the file ID, the version ID, and the data that was inserted during the transaction. These logs serve as a comprehensive record, allowing for traceability and accountability.

Once the Rest Service has completed the necessary steps, it receives confirmation that the file has been successfully stored in the system. This confirmation ensures that the storage process has been completed without any issues. Simultaneously, the doctor who initiated the file storage receives confirmation as well, ensuring they are aware of the successful operation.

This workflow applies not only to file storage but also to other types of operations, following the same principles. Whether it's creating, reading, updating, or deleting files (CRUD), the workflow remains consistent, with the variation lying in the nature of the operations performed.

4   Discussion

As discussion, we will reflect about the bad sides and positives sides of our project. In that behalf, we conceded a SWOT(Strengths,Weaknesses, Opportunities, Threats) diagram:

5   Conclusion and Future Works

In conclusion, the use of InterPlanetary File System (IPFS) and Hyperledger Fabric in the hospital environment is a promising solution to address data security, interoperability, and transparency challenges. By leveraging the distributed nature of IPFS and the blockchain technology of Hyperledger Fabric, hospitals can create a secure, tamper-proof, and decentralized system for storing, sharing, and accessing sensitive patient data (Fig. 5).

However, to assess the effectiveness of this system, it is necessary to measure its performance and compare it with centralized solutions. This will require

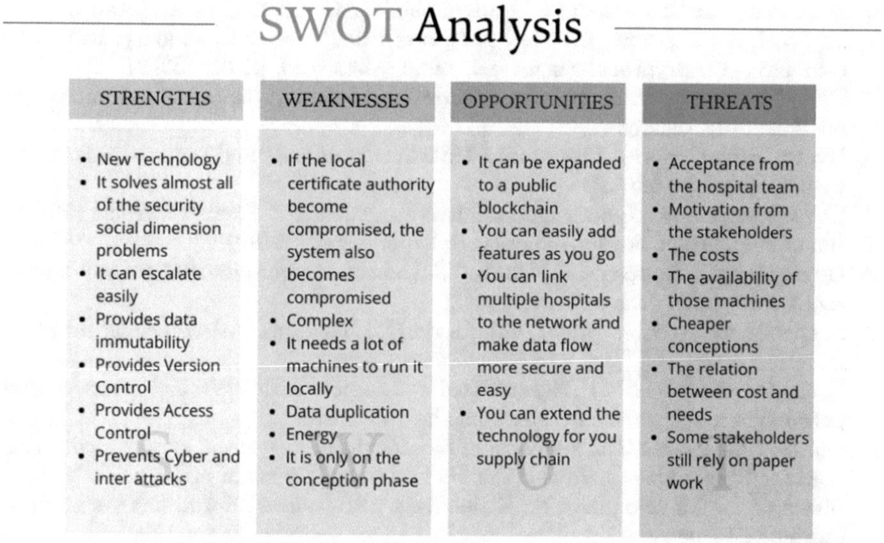

SWOT Analysis

STRENGTHS	WEAKNESSES	OPPORTUNITIES	THREATS
• New Technology • It solves almost all of the security social dimension problems • It can escalate easily • Provides data immutability • Provides Version Control • Provides Access Control • Prevents Cyber and inter attacks	• If the local certificate authority become compromised, the system also becomes compromised • Complex • It needs a lot of machines to run it locally • Data duplication • Energy • It is only on the conception phase	• It can be expanded to a public blockchain • You can easily add features as you go • You can link multiple hospitals to the network and make data flow more secure and easy • You can extend the technology for you supply chain	• Acceptance from the hospital team • Motivation from the stakeholders • The costs • The availability of those machines • Cheaper conceptions • The relation between cost and needs • Some stakeholders still rely on paper work

Fig. 5. Swot analysis

further research and testing to determine factors such as data throughput, scalability, and latency.

Overall, the combination of IPFS and Hyperledger Fabric has the potential to revolutionize the way hospitals handle patient data, staff data and improve the quality of care for patients. Further exploration and experimentation will be essential to validate these claims and ensure that the system meets the unique needs of the healthcare industry.

Acknowledgements. This work has been supported by FCT - Fundação para a Ciência e Tecnologia within the R&D Units Project Scope: UIDB/00319/2020

References

1. bertytech. https://berty.tech/. Accessed 21 Apr 2023
2. Niit technologies introduces chain-m, a blockchain powered solution for airlines and its partners. https://www.prnewswire.com/in/news-releases/niit-technologies-introduces-chain-m-a-blockchain-powered-solution-for-airlines-and-its-partners-693610061.html. Accessed 21 Apr 2023
3. Snapshot. https://snapshot.org/#/. Accessed 21 Apr 2023
4. Starling. https://hedera.com/users/starling. Accessed 21 Apr 2023

5. Alexsicart. Filenation. https://github.com/FileNation/FileNation. Accessed 21 Apr 2023
6. Juan Benet. IPFS - content addressed, versioned, P2P file system (draft 3)
7. Tapscott, A., Tapscott, D.: Blockchain revolution: how the technology behind bitcoin and other cryptocurrencies is changing the world, p. 432 (2018)
8. Bortnikov, V., et al. Hyperledger fabric: a distributed operating system for permissioned blockchains
9. Hyper Ledger Fabric. Case study Hitachi streamlines and secures procurement with hyperledger fabric
10. Linux Foundation. Hyperledger architecture, volume 2: Smart contracts (2018)
11. Linux Foundation. An introduction to hyperledger (2018)
12. Guimarães, T., Santos, M.F., Cunha, J., Duarte, R.: Permissioned blockchain approach using open data in healthcare
13. MichaelMure. Arbore. https://github.com/MichaelMure/Arbore. Accessed 21 Apr 2023
14. Poels, G., Verdonck, M.: Decentralized data access with IPFS and smart contract permission management for electronic health records
15. Do, L., Nguyen, H.: The adoption of blockchain in food retail supply chain: Case: IBM food trust blockchain and the food retail supply chain in Malta
16. Maziéres, D., Maymounkov, P.: Kademlia: a peer-to-peer information system based on the XOR metric
17. Chen, P., Dong, X., Zheng, Q., Li, Y.: An innovative IPFS-based storage model for blockchain
18. Amin, R., Kumar, S., Bharti, A.K.: Decentralized secure storage of medical records using blockchain and IPFS: a comparative analysis with future directions
19. Tikhomirov, S.: Ethereum: state of knowledge and research perspectives
20. Imran, J., Bawany, N.Z., Rehman, W., e Zainab, H.: NFTs: applications and challenges

Advancing the Environmental, Social, and Governance (ESG) with Blockchain: A PRISMA Review

Richard[1]([✉]) [ID], Erwin Halim[1] [ID], Felix Irwanto[1], Gabrielle Peko[2] [ID], and David Sundaram[2] [ID]

[1] School of Information Systems, Bina Nusantara University, Jakarta Barat, Indonesia
richard-slc@binus.edu
[2] Business School, University of Auckland, Auckland, New Zealand

Abstract. The Environmental, Social, and Governance (ESG) initiative has recently become an emerging research focus. ESG mainly focuses on reporting and assessment activities. The reporting and assessment area in ESG requires high visibility and transparency. Blockchain is a technology that can potentially address those issues. This research aims to comprehend the possibilities, opportunities, and challenges of Blockchain implementation in ESG. A systematic literature review was conducted and guided by the Preferred Reporting Items for Systematic Review and Meta-Analysis (PRISMA) Statement standards to achieve the research objective. The result shows that smart contracts in Blockchain can advance ESG data collection, management, and validation. This research also mentions several challenges and future research directions in implementing Blockchain in ESG initiatives.

Keywords: Environmental · Social · And Governance (ESG) · Blockchain · PRISMA

1 Introduction

ESG (Environmental, Social, and Governance) initiatives have become increasingly adopted in various countries' stock markets. Companies listed in the stock market must prepare and publish a Sustainability Report (SR). A third-party auditor will rate the SR, and the ESG rating is published to mirror the company's ESG rating. The ESG rating demonstrates the company's commitment to sustainability and social responsibility. However, there are growing concerns about the effectiveness of these initiatives and whether they are achieving their intended goals.

The urgent issue is the need for more standardization and transparency in ESG reporting, which makes it difficult for investors and other stakeholders to assess the impact of these initiatives [1]. Another area for improvement is the potential of greenwashing, where companies overstate their ESG efforts without significantly changing their practices [2]. Additionally, ESG initiatives prioritize shareholders' concerns over other stakeholders, such as employees and local communities. These issues highlight the need

J. M. Machado et al. (Eds.): BLOCKCHAIN 2023, LNNS 778, pp. 103–112, 2023.
https://doi.org/10.1007/978-3-031-45155-3_11

for greater accountability and oversight in ESG initiatives to ensure they serve the initial purpose.

One potential solution to address the issue in ESG is using blockchain technology. Blockchain can provide transparency and immutability to ESG initiatives by allowing for more accurate reporting and verification of sustainability efforts [3]. Blockchain has several components that can fit with the ESG initiatives, such as smart contracts and Decentralized Autonomous Organizations (DAO). The company can set specific goals and metrics for its ESG initiatives with smart contracts. Smart contracts can provide greater accountability and transparency for investors and stakeholders to prevent greenwashing.

A literature review will identify relevant studies, reports, and publications on blockchain applications in the ESG field. The literature review will be conducted with the following objective: (a) comprehends Blockchain technologies and how they are applied throughout ESG Reporting and Assessment; (b) determine Blockchain opportunities and challenges in ESG initiatives, and (c) acknowledge new concepts, perceptions, and approaches that can lead to a research agenda for prospective studies.

The findings of this research will provide insights into the opportunities and limitations of Blockchain in the ESG domain and help to identify gaps in the existing literature. Furthermore, this research will provide a reference for designing a blockchain-based ESG initiative that regulators and companies can use.

2 Literature Review

2.1 Environmental, Social, and Governance (ESG)

ESG (Environmental, Social, and Governance) factors have become increasingly important in business as investors and stakeholders demand more accountability and transparency from companies. Environmental factors include a company's impact on the natural environment, such as carbon emissions, waste management, and resource consumption. Meanwhile, social factors refer to a company's impact on society, such as employee relations, human rights, and community involvement. Last, governance factors include a company's internal practices and structures, such as executive compensation, board diversity, and risk management.

ESG's effects on several business-related factors, including stakeholder engagement, risk management, and financial performance, have been conducted by researchers. According to previous research, businesses with strong ESG performance typically have better financial performance [4]. ESG has also been shown to impact brand reputation and customer loyalty positively [5]. ESG reporting can increase accountability and transparency [6] while assisting businesses in identifying and addressing risks and opportunities [7]. The lack of consistency and standardization in reporting and the difficulty in assessing and quantifying ESG performance are some challenges of ESG reporting. There are also concerns about "greenwashing," where companies may exaggerate their ESG performance to improve their reputation [2]. Further study is required to efficiently address the issues with ESG reporting and comprehend how it impacts the company.

2.2 Blockchain

Blockchain is a decentralized, distributed ledger technology that enables secure, transparent, and tamper-proof transactions. With cryptography, Blockchain provides unaltered data storage that can be tracked. Several Blockchain platforms, such as Ethereum, have an additional execution layer called Virtual Machine. The virtual machine can interpret and execute code stored in the Blockchain [8]. The program runtime and state are maintained through the blockchain network while forming a decentralized computer network.

Blockchain has been widely explored in various fields, including finance and government. In the finance field, Blockchain has been studied extensively for its potential to transform the industry through increased efficiency, transparency, and security. Researchers have explored the use of Blockchain in various financial applications, such as payments, remittances, and asset management [9]. In government, Blockchain has been studied as a solution to increase transparency, accountability, and efficiency in public services. Researchers have explored the use of Blockchain in various areas, such as voting systems, land registries, and identity management [10]. Blockchain can transform various industries by increasing efficiency, transparency, and security. However, challenges are also associated with adopting Blockchain, such as scalability, interoperability, and regulatory issues [11].

2.3 Blockchain in the ESG Mechanism

The exploration of blockchain technology in ESG mechanisms is considered a new area in blockchain research. As of April 2023, only a few articles were exploring this area. An article from Liu et al. introduces a data-driven ESG assessment approach using blockchain technology and stochastic multicriteria acceptability analysis (SMAA-2) to address the data opaqueness and assessment subjectivity problems [12]. Another article is written by Jiang et al., which proposes a blockchain-based life cycle assessment to enhance the ESG reporting and assessment process [13]. Both articles highlight the capabilities of smart contracts as the baseline for ESG reporting and assessment processes.

Smart contracts are self-executing contracts/agreements that are automated, enforceable, and transparent, with the contents of the agreement written straight into code [8]. It can be used to carry out a variety of transactions without the requirement of an intermediary. Smart contracts can improve efficiency, reduce transaction costs, and increase privacy and security. With smart contracts, an organization can establish a Decentralized Autonomous Organization (DAO) to govern several organization functions. DAO is the concept of an organization that is governed by code and operates on a decentralized network. With DAO, the ESG process.

3   Research Method

This study aims to present a comprehensive overview of blockchain technologies and identify the challenges and opportunities associated with their application in ESG. A systematic literature review was conducted and guided by the standards of the Preferred

Reporting Items for Systematic Review and Meta-Analysis (PRISMA) Statement [14]. Academic peer-reviewed journals were searched to gather literature on both current and emerging technologies. A review of current technologies is crucial for understanding the current state of the blockchain landscape and analyzing the development, adoption, and application of Blockchain in ESG. The study has implemented a three-stage methodological approach by following the PRISMA protocol [15].

3.1 Stage 1 (Planning Stage)

This research aims to identify how Blockchain technology can address the issue in ESG initiatives. The following keywords were used to develop the initial search criterion: "ESG Blockchain", "Blockchain in Environmental, Social, and Governance", "ESG Reporting Blockchain", and "ESG Assessment Blockchain". These keywords defined the boundaries of the research areas and gave an overview of Blockchain technologies currently used in ESG.

The initial search covered multiple bibliographic repositories, including Scopus, Semantic Scholar, and Google Scholar. The ESG was first coined in 2003, and Blockchain was popularized by Satoshi Nakamoto with its Bitcoin in 2008. Based on this, the initial search filtered into January 2008 to March 2023. The keyword search was conducted in March 2023 and obtained 130 results that satisfied the search criteria. After removing the duplicated articles, 124 articles were retained. The article list will be filtered through the primary inclusion/exclusion criteria. The number of articles filtered with the primary inclusion/exclusion criteria is 99. Table 1 shows the inclusion and exclusion criteria developed to effectively decrease the number of screened articles.

Table 1. Primary and secondary inclusion/exclusion criteria

Primary Criteria		Secondary Criteria	
Inclusion	Exclusion	Inclusion	Exclusion
Journal articles Conference articles Peer-reviewed Accessible online Written in English	Duplicate records Textbooks Patent Conference/Journal Abstract	Blockchain in ESG Blockchain in ESG Reporting Blockchain in ESG assessment	Not in ESG-related issue Not in Blockchain technology scope (e.g., cryptocurrency trading)

3.2 Stage 2 (Conducting the review Stage)

The remaining 99 articles are screened and filtered in this stage using the secondary inclusion/exclusion criteria. The screening process involves a two-step filtering process. First, we read the title, abstract, and keyword section and removed the articles that fit into the exclusion criteria. The second step is the reading process. We read the content thoroughly until we find the papers that will be reviewed. The final number of articles in the review process is 14.

The review will be based on the initial research objectives, which include (a) comprehending Blockchain technologies and how they are applied throughout ESG Reporting and Assessment; (b) determining Blockchain opportunities and challenges in ESG initiatives; and (c) acknowledging new concepts, perceptions, and approaches that can lead to a research agenda for prospective studies. The final article list will be reviewed and mapped into those objective sections.

3.3 Stage 3 (Reporting Stage)

In Stage 3 (reporting), 14 articles were analyzed using the descriptive techniques of explanation building. The report will be separated into four sections, including a general description and research objectives. The information should generate a powerful insight for the reader, as the articles chosen in this research have passed the filtering processes (Fig. 1).

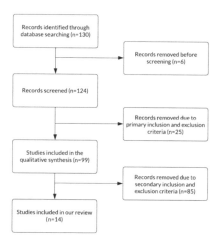

Fig. 1. Screening and filtering process.

4 Discussion

4.1 General Description

The research trend of Blockchain use in ESG initiatives is still growing. Since ESG is a relatively unfamiliar topic in the academic research area, the number of academic research papers found in this research is low. Our search result shows that the study increased from 4 relevant articles in 2021 to 9 in 2022. The article type of our search is relatively even, with 9 for a journal article and 5 for conference articles. Figure 2 shows the publication years and article type data. Since the number of studied papers is low, the summary provided in this paper will be preliminary. However, the output can still be used as a reference for this area (Blockchain for ESG).

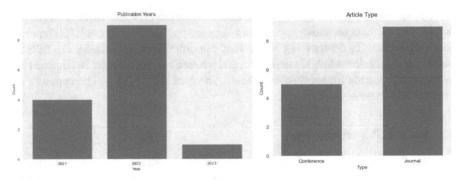

Fig. 2. Publication years and article types.

4.2   How can Blockchain be Applied Throughout ESG Reporting and Assessment?

There are various use cases of blockchain technology implementation in Environmental, Social, and Governance (ESG) assessment and reporting. The primary benefit of using blockchain technology in the ESG area is to ensure a transparent and secure ESG reporting and assessment environment [16]. With the help of a Blockchain-based Life-Cycle Assessment (LCA) system, ESG reporting can be automatically cross-validated [17]. Another benefit is streamlining ESG data/score flow among stakeholders [12] and providing a platform for equal access and sharing of information in organizations [18]. Blockchain can also provide the ability to enable data sharing, tracking, tracing, and verification across multi-party systems [19] (Fig. 3).

Fig. 3. The Overview of Blockchain Application in ESG.

Smart contracts can automate the implementation of agreements and assess the continuity between raw data and final ESG reports [20]. The smart contract can also permanently store and track every version of the ESG procedure [21]. With the right design, a smart contract can control the ESG procedure within applicable content and format [17].

The smart contract can also create a digital asset with a value designed/programmed into it (Tokenization). By using the concept of tokenization, a company can issue a "Carbon Token" that represents the carbon credit [22, 23]. Another case study for blockchain applications in ESG-related initiatives is cobalt sourcing practices. The company can responsibly source, monitor, measure, and report their environmental and social performance in their supply chains to their stakeholders and shareholders [24].

4.3 What are the Opportunities and Challenges in Implementing Blockchain in ESG Initiatives?

The challenges in implementing Blockchain into any industry are very high, and the implementation in ESG initiatives is no different. The stakeholders must deal with many fundamental issues in Blockchain, such as the user experience, scalability, security, performance, and cost efficiency issues [20]. While the technology is growing, implementing Blockchain in ESG still has many opportunities. The advancement of ESG initiatives by using Blockchain still has potential in the future.

Firmly, blockchain technology can ensure the transparency, traceability, and security of ESG data [12, 25]. Blockchain implementation definitely can lead to a positive effect on business ethics, corporate governance, and social sustainability [18]. Blockchain can solve the data reliability and distribution problem by reliably storing and distributing data from various sources without tampering via its decentralized and secure nature [17]. Adopting blockchain technology will result in cost-saving, efficient, and trustworthy ESG reporting [21, 24].

Smart contracts can automate the reporting process, ensuring complete, accurate, and timely reporting [26]. Tokenization through smart contracts [16] also opens an opportunity to create a fair carbon market and more comprehensive carbon asset management [23]. A smart contract's transparency can help combat greenwashing practices [17, 20]. Furthermore, integrating Blockchain with other cutting-edge technology, such as the Internet of Things (IoT), can enhance the efficiency and reliability of environmental data collection and reporting [25].

4.4 What are the New Concepts, Perceptions, and Approaches that Can Lead to a Research Agenda for Blockchain Implementation in ESG?

As mentioned in the previous section, Blockchain can advance ESG initiatives in several ways. However, technology has a broader potential to address other sustainability issues. It is suggested that future research should consider the effect of Blockchain on different dimensions of sustainability [18]. The blockchain-powered carbon market can increase the traceability and integrity of the carbon credit itself [23]. The integrity of the carbon market can be even better by integrating intelligent sensor devices that validate the data collection process in carbon credit issuance [21].

Blockchain is a decentralized and immutable computer that can store a program and state. Any sustainability practices can utilize the technology to increase transparency by democratizing the reporting and controlling process. Beyond the benefit mentioned, Blockchain is still a growing technology that the community is improving. Blockchain was born with several limitations, such as performance, scalability, and interoperability. However, the current blockchain platform development, such as Layer 2, gradually solves the issue. Layer 2 blockchain is an additional layer synchronizing with the existing blockchain network. The synchronization is enabled by applying Zero-Knowledge technology.

Even after the improvement, the overall experience of using a blockchain-based system is still unfavorable for non-technical people. Technical issues such as private key management, user experience, asset bridge, and transaction cost still become a pain for everyday users [22]. The representation of ESG rules also becomes an issue from the development side. The ESG rules can change in the future, while the smart contract program cannot easily be changed. It is recommended to create a modular smart contract code to adjust ESG rule changes in the future [17].

5   Conclusion

There are many possibilities for implementing Blockchain technology in ESG. In the logical domain, the immutability and transparency of the technology can help the ongoing issue in ESG. The typical process that requires visibility and transparency in ESG is ESG reporting and assessment. With the help of smart contracts stored on the Blockchain, ESG rules, and procedures can be directly applied to code and visible to all parties. The convergence of Blockchain and other technology like IoT devices also can benefit ESG reporting and assessment. IoT and Blockchain potentially generate autonomous data collection, governance, and management.

Beyond all the benefits, Blockchain's technical limitation also becomes an issue in Blockchain-ESG implementation. Smart contract modularity, Layer 2 chain, and the privacy-preserving feature can help address those technical problems. For the future research agenda, it is suggested that the Blockchain-ESG implementation should match the current applicable regulations. A comprehensive framework design is also needed as the baseline for Blockchain implementation in ESG.

References

1. World Economic Forum: Business Leaders Call for a Common Standard on 'Onerous ESG Frameworks (2022). https://www.weforum.org/press/2022/05/business-leaders-call-for-a-common-standard-on-onerous-esg-frameworks/
2. Szabo, S., Webster, J.: Perceived greenwashing: the effects of green marketing on environmental and product perceptions. J. Bus. Ethics 171(4), 719–739 (2021). https://doi.org/10.1007/s10551-020-04461-0
3. Dario, C., Sabrina, L., Landriault, E., De Vega, P.: DLT to boost efficiency for financial intermediaries. An application in ESG reporting activities. Technol. Anal. Strateg. Manag. (2021). https://doi.org/10.1080/09537325.2021.1999921
4. Zhao, C., et al.: ESG and corporate financial performance: empirical evidence from China's listed power generation companies. Sustainability 10(8), 2607 (2018). https://doi.org/10.3390/su10082607
5. Reber, B., Gold, A., Gold, S.: ESG disclosure and idiosyncratic risk in initial public offerings. J. Bus. Ethics 179(3), 867–886 (2022). https://doi.org/10.1007/s10551-021-04847-8

6. Ellili, N.O.D.: Impact of ESG disclosure and financial reporting quality on investment efficiency. Corp. Gov. **22**(5), 1094–1111 (2022). https://doi.org/10.1108/CG-06-2021-0209

7. Serafeim, G., Eccles, R.: The performance frontier innovating for a sustainable strategy. https://hbr.org/2013/05/the-performance-frontier-innovating-for-a-sustainable-strategy (2013). Accessed: 11 Apr 2023

8. Richard, R., Prabowo, H., Trisetyarso, A., Soewito, B.: Smart contract development model and the future of blockchain technology. In: ACM International Conference Proceeding Series, pp. 34–39 (2020). https://doi.org/10.1145/3446983.3446994

9. Norta, A., Leiding, B., Lane, A.: Lowering financial inclusion barriers with a blockchain-based capital transfer system. In: INFOCOM 2019 – IEEE Conference on Computer Communications Workshops, INFOCOM WKSHPS 2019, pp. 319–324 (2019). https://doi.org/10.1109/INFOCOMW.2019.8845177

10. Jafar, U., Juzaiddin, M., Aziz, A., Shukur, Z., Wu, J., Wang, H.: Blockchain for electronic voting system—review and open research challenges. Sensors **21**, 5874 (2021). https://doi.org/10.3390/S21175874

11. Benahmed, S., et al.: A comparative analysis of distributed ledger technologies for smart contract development. In: IEEE Intrnational Symposium Personal Indoor Mobile Radio Communications PIMRC, vol. 2019-Sept, pp. 1–6 (2019). https://doi.org/10.1109/PIMRC.2019.8904256

12. Liu, X., et al.: Data-driven ESG assessment for blockchain services: a comparative study in textiles and apparel industry. Resour. Conserv. Recycl. **190**, 106837 (2023). https://doi.org/10.1016/j.resconrec.2022.106837

13. Jiang, L., Gu, Y., Yu, W., Dai, J.: Blockchain-based life cycle assessment system for ESG reporting. SSRN Electron. J. (2022). https://doi.org/10.2139/ssrn.4121907

14. Page, M.J., et al.: The PRISMA 2020 statement: an updated guideline for reporting systematic reviews. Int. J. Surg. **88**, 105906 (2021). https://doi.org/10.1016/j.ijsu.2021.105906

15. Sarkis-Onofre, R., Catalá-López, F., Aromataris, E., Lockwood, C.: How to properly use the PRISMA statement. Syst. Rev. **10**, 117 (2021). https://doi.org/10.1186/s13643-021-01671-z

16. Wu, W., et al.: Consortium blockchain-enabled smart ESG reporting platform with token-based incentives for corporate crowdsensing. Comput. Ind. Eng. **172**, 108456 (2022). https://doi.org/10.1016/j.cie.2022.108456

17. Gu, Y., Jiang, L., Yu, W., Dai, J.: Towards blockchain-enabled ESG reporting and assurance: from the perspective of P2P energy trading. SSRN Electron. J. (2022). https://doi.org/10.2139/ssrn.4121798

18. Ronaghi, M.H., Mosakhani, M.: The effects of blockchain technology adoption on business ethics and social sustainability: evidence from the Middle East. Env. Dev. Sustain. **24**(5), 6834–6859 (2021). https://doi.org/10.1007/s10668-021-01729-x

19. Subramoniam, R., Parameswaran, A., Ramanan, R., Sreekumar, R., Cherian, S.: Generating trust using product genome mapping: a cure for ESG communication. In: 2022 IEEE 1st Global Emerging Technology Blockchain Forum Blockchain Beyond, pp. 1–5 (2022). https://doi.org/10.1109/IGETBLOCKCHAIN56591.2022.10087063

20. Saxena, A., et al.: Technologies empowered environmental, social, and governance (ESG): an industry 4.0 landscape. Sustainability **15**(1), 309 (2022). https://doi.org/10.3390/su15010309

21. Liu, X., Wu, H., Wu, W., Fu, Y., Huang, G.Q.: Blockchain-enabled esg reporting framework for sustainable supply chain. Smart Innov. Syst. Technol. **200**, 403–413 (2021). https://doi.org/10.1007/978-981-15-8131-1_36/COVER

22. Khan, N., Ahmad, T.: DCarbonX Decentralised Application: Carbon Market Case Study. arXiv Prepr. arXiv2203.09508. https://arxiv.org/abs/2203.09508v1 (2022). Accessed 11 Apr 2023

23. Golding, O., Yu, G., Lu, Q., Xu, X.: Carboncoin: blockchain tokenization of carbon emissions with ESG-based reputation. In: 2022 IEEE International Conference on Blockchain and Cryptocurrency (ICBC), pp. 1–5 (2022). https://doi.org/10.1109/ICBC54727.2022.9805516
24. Mugurusi, G., Ahishakiye, E.: Blockchain technology needs for sustainable mineral supply chains: a framework for responsible sourcing of Cobalt. Procedia Comput. Sci. **200**, 638–647 (2022). https://doi.org/10.1016/j.procs.2022.01.262
25. Wu, W., Chen, W., Fu, Y., Jiang, Y., Huang, G.Q.: Unsupervised neural network-enabled spatial-temporal analytics for data authenticity under environmental smart reporting system. Comput. Ind. **141**, 103700 (2022). https://doi.org/10.1016/j.compind.2022.103700
26. Bora, I., Duan, H.K., Vasarhelyi, M.A., Zhang, C., Dai, J.: The transformation of government accountability and reporting. J. Emerg. Technol. Account. **18**(2), 1–21 (2021). https://doi.org/10.2308/JETA-10780

Industry 4.0 Business-Oriented Blockchain Design Decision Tree

Denis Stefanescu[1,2(✉)] , Leticia Montalvillo[1] , Patxi Galán-García[3] ,
Juanjo Unzilla[2] , and Aitor Urbieta[1]

[1] Ikerlan Technology Research Centre, 20500 Arrasate-Mondragon, Spain
{distefanescu,lmontalvillo,aurbieta}@ikerlan.es
[2] University of the Basque Country (UPV/EHU), 48013 Bilbao, Spain
juanjo.unzilla@ehu.eus
[3] Entrii, 46024 Valencia, Spain
patxigg@deusto.es

Abstract. Industry 4.0 integrates IoT and robotics to create smart factories that work together seamlessly. However, this integration requires secure and efficient data management systems. Blockchain is a distributed ledger technology that stores data in a secure and transparent manner via cryptography mechanisms. Since the release of Bitcoin, it has gained popularity in finance, but its potential applications in other sectors such as Industry 4.0 is vast. In Industry 4.0, blockchain can revolutionize businesses by improving transparency, efficiency, and security. However, developing an adequate blockchain for Industry 4.0 business mechanisms requires a specific approach. In this paper, we propose a decision tree for designing a well-suited blockchain solution for business-oriented Industry 4.0 that takes into account the requirements of the aforementioned field.

Keywords: Blockchain · DLT · Industry 4.0 · Decision Tree

1 Introduction

The Fourth Industrial Revolution, known as Industry 4.0, is changing the way companies operate and interact with customers. Industry 4.0 combines various disruptive technologies such as Artificial Intelligence (AI), data analytics or robotics to enable smart digital factories that can communicate and collaborate automatically [1].

However, these innovative technologies also require efficient data management systems [2]. This is where the blockchain technology comes in. Blockchain is a type of DLT that stores and shares data securely and transparently. Its decentralized nature lacks of any central point of failure, making it useful in a variety of different fields.

In the field of Industry 4.0, blockchain has the potential to completely transform business operations by boosting transparency, efficiency, and security.

J. M. Machado et al. (Eds.): BLOCKCHAIN 2023, LNNS 778, pp. 113–123, 2023.
https://doi.org/10.1007/978-3-031-45155-3_12

Nonetheless, designing a suitable blockchain that caters to the specific demands of the industry calls for a more specific approach. Business-focused blockchains are subject to the needs of particular industries, and developing them necessitates a tailored methodology [3].

Therefore, the need for a decision tree to design a suited blockchain for the industry 4.0 business field arises from the fact that different industries have varying requirements for their data management systems. For example, supply chain management may require a blockchain that is capable of tracking goods from production to delivery, while manufacturing may require a blockchain that can integrate with the existing Enterprise Resource Planning (ERP) systems. Furthermore, many use cases might need to include the capacity of executing automated agreements on top of a blockchain (i.e., smart contracts), and in many cases, these automated agreements might need to be provided with external data, which is not a straightforward process [4]. Therefore, a generic blockchain solution is not be suitable for all industries.

Thus, in this paper we explore the use of the blockchain technology in the business-oriented part of Industry 4.0 and discuss the requirements of the aforementioned field. We also highlight some of the existing blockchain solutions for industry 4.0 and their limitations. Finally, we propose a decision tree for designing a well-suited blockchain solution for the Industry 4.0 business field that takes into account its unique requirements and needs. Specifically, we perform a comprehensive exploration of the characteristics that a business oriented blockchain should have according to several input parameters regarding the given needs.

The remainder of the paper is organized as follows. In Sect. 2 we analyze the existing related work in this field and outline our contributions. In Sect. 3 we analyze the requirements that a business oriented blockchain for Industry 4.0 should have. In Sect. 4 we describe the proposed decision tree for Industry 4.0 blockchain design. In Sect. 5 we present an example use case for the proposed decision tree. Finally, Sect. 6 includes the conclusion of the paper and future work insights.

2   Related Work

In this section we analyze the most relevant works regarding blockchain design methodologies and decision frameworks within the business related Industry 4.0 field of applications [5]: supply chains, Enterprise Resource Planning (ERP) and logistics. Finally, we outline our contributions regarding the existing works.

A framework for evaluating blockchain platforms within supply chain networks is proposed in [6], which highlights 45 essential criteria across 10 dimensions. This framework offers valuable guidance for the development of blockchain-based platforms in other organizational systems, despite being initially developed for a specific domain. Therefore, the requirements gathered by the authors can be utilized as important considerations when selecting a blockchain platform.

In [7], an automated decision-making framework is introduced, which aids in determining the most suitable alternative platforms based on the given requirements and preferences. The framework matches the necessary requirements with quality features, such as ISO 25010, and then compares various platforms against the selected features to benchmark their performance.

In [8], a comprehensive framework for selecting blockchain software is introduced. It is based on the Fuzzy Analytical Hierarchy Process (FAHP) to create a hierarchy of selected criteria and related attributes. Weighting criteria is then employed to offer detailed instructions for evaluating a prospective blockchain platform. The primary identified criteria in the framework are cost, speed, privacy, functionality, and developer availability.

The work presented in [9] provides a comprehensive overview of the various applications of blockchain technology in Industry 4.0. It offers insights into how blockchain can be used to enhance supply chain management, improve data security and management, enable secure machine-to-machine communication, and enhance cybersecurity. This work intends to help researchers who are interested in designing a blockchain frameworks for Industry 4.0.

The works that have been analyzed offer an overview of the current state-of-the-art in decision-making processes for designing and selecting a blockchain platform. However, the coverage of this topic is superficial, with only a brief examination of essential criteria such as "security," "performance," "interoperability," and "functionality". Therefore, additional research is needed to explore these factors more profoundly and analyze other relevant considerations that affect the selection of a blockchain platform.

Our main contribution is a deeper exploration of specific blockchain characteristics, such as the implementation of smart contracts, oracles, private channels, etc., to ensure desired criteria like "security" and "interoperability". We also define a decision tree that is intended to be applied in a more focused field, such as the Industry 4.0 business domain, making our decision tree more comprehensive for this particular field.

3   Industry 4.0 Business Requirements

In this section, we enumerate the main requirements of a business-oriented Industry 4.0 environment and the characteristics that it should possess to meet the needs of this field. Thus, we define 11 fundamental requirements based on the current knowledge on the field from Sect. 2 and additional industry focused literature [10,11]:

1. **Security.** One of the most critical requirements for a business Industry 4.0 environment is security. Due to the integration of various technologies and the management of sensitive information, there is a significant risk of cyber-attacks and data breaches. Therefore, businesses must ensure that their systems and networks are secure from external threats. This can be achieved by implementing strong access controls, data encryption, network monitoring for suspicious behavior and trustworthy traceability.

2. **Transparency.** Another key requirement for a business Industry 4.0 environment is transparency. Typically, it is challenging to keep track of data

and processes due the integration of various systems and devices. Therefore, businesses must ensure that their systems provide traceability to enable better decision-making and transparency.

3. **Data Privacy.** Data privacy is a crucial requirement in the Industry 4.0 and most digital fields. Nowadays, businesses collect a great amount of data from various sources. Therefore, there must be an assurance that data is collected, processed, and stored in compliance with data privacy regulations and standards in order to provide trust.

4. **Interoperability.** The Industry 4.0 environment is highly complex, with multiple systems and devices communicating with each other. To ensure seamless communication, businesses need to ensure interoperability between different systems. This requires the adoption of standard protocols and interfaces that enable communication between different systems and devices.

5. **Performance and scalability.** A successful Industry 4.0 environment requires both scalability and performance. As businesses grow and the amount of data and traffic increases, they need systems and networks that can handle the load and perform tasks quickly and accurately.

6. **Reliability.** In Industry 4.0, businesses rely on technology to run their operations. Thus, it is necessary to ensure that the systems and devices used are reliable and available when needed. This requires implementing redundant systems and devices and conducting maintenance to prevent downtime.

7. **Governance.** Governance is a critical aspect of any digital environment. The implemented systems must have a well-defined governance model that ensures the integrity, accountability, and transparency.

8. **Compliance.** Industry 4.0 companies are subject to many regulations and standards, and their systems must be compliant with these regulations.

9. **Automation.** Business process automation involves automating repetitive tasks and workflows, allowing employees to focus on higher-level tasks that require human decision-making and creativity. This approach can be applied to various business processes, including manufacturing planning, supply chain management, customer service, and financial management.

10. **Flexibility.** Finally, businesses in the Industry 4.0 environment need to be flexible so they can adapt to changing market conditions and technologies. This requires the adoption of methods that enable rapid modification, prototyping, implementation and testing of current or new technologies.

11. **Costs.** Implementing Industry 4.0 requirements requires significant investment in technology infrastructure, talent acquisition, data analytics, cybersecurity, and equipment and machinery. Companies need to carefully evaluate the costs and benefits of Industry 4.0 before implementing it.

4   Proposed Decision Tree

In this section, we present a decision tree for designing a blockchain that effectively addresses all the business-related Industry 4.0 requirements identified through an extensive review of existing literature. The proposed decision tree comprises a series of strategically ordered questions aimed at guiding the design process towards implementing specific technical components and mechanisms that meet the blockchain requirements.

The election of the requirements and their order within the decision tree is meticulously curated based on several criteria, including their relative importance, potential dependencies, and most significantly, the logical flow of the decision tree, as shown in [12]. By organizing the aspects in a manner that adheres to established blockchain designing patterns, we ensure a more coherent and structured approach to the design process. This logical progression not only allows for seamless navigation through the decision tree but also facilitates a comprehensive understanding of the relationships between various design elements and their impact on the overall blockchain architecture.

Furthermore, the proposed decision tree ensures that the blockchain system is functional, efficient, compatible with other systems, reliable, secure and maintainable, as specified in the ISO 25010 standard.

Due to space limitations, the visual diagram of the decision tree can be found externally[1].

1. **Security:**
   - Q1: Does the blockchain require secure access mechanisms and trustworthy traceability via identity check?
     - Yes: Consider implementing a permissioned blockchain with identity management and encryption mechanisms to ensure secure transactions and prevent unauthorized access.
     - No: Skip to Q2.
   - Q2: Does the blockchain require traceability for long periods of time?
     - Yes: Consider implementing a blockchain that does not remove old transactions.
     - No: Skip to Q3.
   - Q3: Does the blockchain require regular security audits and updates to ensure the ongoing protection of data against emerging threats and vulnerabilities?
     - Yes: Consider implementing a blockchain that is under continuous development and based on a widely used programming language that includes the possibility of performing vulnerability scanning.
     - No: Skip to requirement 2 - Transparency.
2. **Transparency:**
   - Q1: Is full transparency of all transactions required on the blockchain?
     - Yes: Consider implementing a public blockchain that allows any user to view the ledger and its transactions.
     - No: Consider implementing a private-permissioned blockchain that provides selective access to the ledger and transactions.
3. **Data privacy:**
   - *Answer to this question only if full transparency (requirement 2) is **not** required.*

---

[1] https://tinyurl.com/blockchaindecisiontree.

   – Q1: Is it necessary for the blockchain to provide confidentiality for some transactions?
- Yes: Consider implementing a private-permissioned blockchain with encryption mechanisms or private channels to ensure confidentiality and selective access to transactions.
- No: Skip to requirement 4 - Interoperability.

4. **Interoperability:**
   – Q1: Is it essential for the blockchain to interact with other blockchains or other distributed systems?
   - Yes: If the number of distinct blockchains to interoperate is significant (i.e., more than 3), consider implementing an interoperable blockchain with interoperability protocols like cross-chain atomic swaps or sidechains. If the number of blockchains to interoperate is 3 or less, consider implementing specific blockchain interoperability connectors (gateways, APIs...).
   - No: Skip to requirement 5 - Performance and scalability.

5. **Performance and scalability:**
   – Q1: Is it expected that the number of transactions and participants on the blockchain to grow significantly over time?
   - Yes: Consider implementing a sharded or layered blockchain architecture, off-chain processing mechanisms like state channels or IPFS storage, or use a consensus mechanism with high throughput and scalability capacities.
   - No: Skip to requirement 6 - Reliability.

6. **Reliability:**
   – Q1: Is it essential for the blockchain nodes to be always online and have minimal downtime?
   - Yes: If the chosen blockchain is private-permissioned, consider implementing a Byzantine fault-tolerant (BFT) consensus mechanism, and redundant nodes to ensure high availability and reliability. If it is a public blockchain, no further action should be needed.
   - No: Skip to requirement 7 - Governance.

7. **Governance:**
   – Q1: Is a governance model to manage the evolution of the blockchain needed? Specifically, is the blockchain expected to change its characteristics at some point in time?
   - Yes: Consider implementing a blockchain with a formal governance structure such as a Decentralized Autonomous Organization (DAO) or a voting-based decision-making process.
   - No: Skip to requirement 8 - Compliance.

8. **Compliance:**
   – Q1: Does the blockchain need to comply with regulatory requirements?
   - Yes: Consider implementing compliance mechanisms like regulatory reporting, identity verification, or anti-money laundering (AML) and know-your-customer (KYC) procedures.
   - No: Skip to requirement 9 - Automation.

9. **Automation:**
    - Q1: Is there a need for automatic enforcement or execution of business logic based on predetermined conditions? OR Is there a need for automation of tasks, such as triggering events or notifications based on specific conditions?
        - Yes: Consider implementing a blockchain with smart contracts.
        - No: Skip to requirement 11 - Costs.
10. **Flexibility:**
    - *Answer to these questions only if the elected blockchain includes smart contracts (requirement 9).*
    - Q1: Does the smart contract require access to real-world data or events that are not natively available on the blockchain? OR Does the smart contract need to be able to communicate with off-chain APIs or other external systems?
        - Yes: Consider implementing a blockchain with a flexible smart contract platform that has the capability to access external data via oracle mechanisms.
        - No: Skip to requirement 11 - Costs.
11. **Costs:**
    - Q1: What specific additional costs could potentially arise from using the blockchain network for transactions? Could these include elements like gas fees or transaction fees?
        - Yes: Skip to Q2.
        - No: END - No further action is needed.
    - Q2: Have you analyzed the potential cost savings associated with using blockchain technology for supply chain management, such as reducing the need for intermediaries and improving traceability and transparency?
        - Yes: If the costs associated with blockchain are acceptable compared to the expected savings, no further actions need to be taken. If the costs are not acceptable, consider using a fee-less blockchain.
        - No: Skip to Q3.
    - Q3: Does your business require a high volume of transactions, and if so, will the cost of using the blockchain network for these transactions be feasible?
        - Yes: Consider implementing a blockchain with zero or near zero associated costs.
        - No: END - No further action is needed.

Blockchain Platform Election. In the context of our proposed decision tree for blockchain design, we identify various blockchain platforms that align with the defined choices. These platforms are classified based on the type of blockchain as per Almeshal et al.'s categorization [5]: private-permissioned, public-permissioned, and public-permissionless. This classification system serves as a guiding tool for selecting the appropriate real-world platform.

1. **Private-Permissioned:**
   - Hyperledger Fabric or Sawtooth with private channels, BFT consensus, regulatory mechanisms, smart contracts, and blockchain oracles.
   - R3 Corda with private channels, interoperability, BFT-based consensus, regulatory mechanisms, and smart contracts.
   - IOTA with fee-less transactions, scalable and lightweight architecture, and support for IoT use cases.
   - Private Ethereum with private channels, interoperability, BFT consensus, DAO/voting system, AML/KYC, smart contracts and oracles.
   - Quorum with private channels, interoperability, BFT-based consensus, DAO/voting system, and smart contracts.
2. **Public-Permissioned:**
   - Permissioned Ethereum with private channels, interoperability, BFT consensus, DAO/voting system, AML/KYC, smart contracts, and oracles.
   - Hyperledger Fabric or Sawtooth with private channels, BFT-based consensus, regulatory mechanisms, smart contracts, and blockchain oracles.
   - Corda Enterprise with private channels, interoperability, BFT-based consensus, regulatory mechanisms, and smart contracts.
   - Hedera Hashgraph with private channels, interoperability, BFT-based consensus, and smart contracts.
   - Ripple with private channels, regulatory systems, and smart contracts.
   - Cosmos with interoperability, fee-less transactions, and smart contracts.
3. **Public-Permissionless:**
   - IOTA with fee-less transactions, scalable and lightweight architecture, and support for IoT use cases.
   - Ethereum with smart contracts and blockchain oracles.
   - Polkadot with interoperability, private channels, and smart contracts.
   - Solana with fee-less transactions, smart contracts, and oracles.

5 Example Use Case: Product Manufacturing Traceability

In this section, we elaborate on an illustrative Industry 4.0 use case, demonstrating the practical application of the proposed decision tree. This use case, though hypothetical, is constructed to mirror real-world scenarios, thereby providing a better understanding of the potential implications and benefits of implementing such a decision-making model.

Example Use Case. A manufacturing company called "X" makes a product that contains parts from multiple suppliers. The company wants to use a blockchain to track the movement of the aforementioned parts from the suppliers to the manufacturing facility. Whenever a part passes from one party to another, a new block is added to the blockchain. Each block includes information about the transaction, such as: time, date, location or the involved parties.

Once the product is manufactured, it is shipped to a logistics company for distribution, allowing the involved parties to track the product in real time. Finally, the retailer receives the product and verifies its authenticity and integrity via the blockchain. Any issues, such as missing or damaged parts, can be traced back to their source, allowing the relevant parties to take appropriate corrective measures.

Decision Tree Application

1. **Security:** Yes (Q1, Q2 and Q3). Data security and long time traceability are very important, and the blockchain participants need to be identified. Thus, a permissioned blockchain must be implemented.
2. **Transparency:** No. In this case full transparency is not required, since the traceability of the parts needs to be tracked only by specific actors. Therefore, a private blockchain is the most optimal choice.
3. **Data privacy:** No. In this case no specific privacy requirement is specified. No encryption or private channels mechanisms are needed.
4. **Interoperability:** No. In this case no relevant blockchain interoperability capacity is required.
5. **Scalability:** Yes. In this case we have a product that has many parts, and these parts belong to several suppliers. While it is not clear whether the number of transactions and participants is significant, this use case requires a margin to be left in case the number of parts and suppliers increases.
6. **Reliability:** Yes. Reliability is highly recommended in order to guarantee correct data registering throughout the whole process and achieve the intended data traceability capacity. Thus, a BFT consensus is recommended.
7. **Governance:** No. In this case there is no evidence that the participants need to posses the capacity of managing the blockchain model.
8. **Compliance:** No. In this case no compliance model is needed, since the stored data belongs to industrial machinery parts.
9. **Automation:** Yes. In this case, smart contracts could be used to automatically update the blockchain with each transaction and verify that the product is complete and in good condition at each stage of the supply chain. It could also trigger alerts and notifications to relevant parties if any issues are detected, such as a missing part or a damaged product.
10. **Flexibility:** No. In this case, if the goal is simply to track the movement of the product and ensure its integrity, a basic blockchain with standard features is sufficient. Therefore, no extra features such as oracles are needed.
11. **Costs:** No (Q1). In this case, no additional costs such as gas fees, network fees, or transaction fees are present.

Therefore, according to the answers that are shown above, we need a blockchain that is: private, permissioned, scalable, reliable and with smart contracts capacity. Given these features, we can choose from several blockchain platforms: Hyperledger Fabric or Sawtooth, R3 Corda or Quorum.

6   Conclusions and Future Work

In this work we presented a decision tree to assist in decision making when designing a DLT platform for business-focused industrial applications. The questions have been defined on the basis of an comprehensive analysis of the requirements present in this area. An attempt has been made to develop a clear and straightforward decision tree with a minimum number of conflicts or dependencies. The dependencies (questions 3 and 10) have been solved by defining additional questions or decision paths so that the intended simplicity of the tree is not altered. Although it is not a trivial task, we also suggest names of existing DLT platforms based on the characteristics that have been previously selected. Finally, we describe a realistic use case to validate our proposal.

However, this work still has plenty of room for extension. Specifically, we plan to delve deeper into the complexity of the decision tree and extend it to more industrial uses beyond the business-related area.

Acknowledgements. This work has been financed by The European commission through the Horizon Europe program under the IDUNN project (grant agreement number 101021911).

References

1. Lasi, H., Fettke, P., Kemper, H.G., Feld, T., Hoffmann, M.: Industry 4.0. Bus. Inf. Syst. Eng. **6**(4), 239–242 (2014). https://doi.org/10.1007/s12599-014-0334-4
2. Singh, M.: Blockchain technology for data management in Industry 4.0. In: Rosa Righi, R., Alberti, A.M., Singh, M. (eds.) Blockchain Technology for Industry 4.0. BT, pp. 59–72. Springer, Singapore (2020). https://doi.org/10.1007/978-981-15-1137-0_3
3. George, R.P., Peterson, B.L., Yaros, O., Beam, D.L., Dibbell, J.M., Moore, R.C.: Blockchain for business. J. Investment Compliance **20**(1), 17–21 (2019)
4. Fernandez-Carames, T.M., Fraga-Lamas, P.: A review on the application of blockchain to the next generation of cybersecure Industry 4.0 smart factories. IEEE Access **7**, 45201–45218 (2019)
5. Almeshal, T.A., Alhogail, A.A.: Blockchain for businesses: a scoping review of suitability evaluations frameworks. IEEE Access **9**, 155425–155442 (2021)
6. Herm, L.V., Janiesch, C.: Towards an implementation of blockchain-based collaboration platforms in supply chain networks: a requirements analysis (2021)
7. Six, N., Herbaut, N., Salinesi, C.: Which blockchain to choose? A decision support tool to guide the choice of a blockchain technology. arXiv preprint (2020)
8. Karayazi, F., Bereketli, I.: Criteria weighting for blockchain software selection using fuzzy AHP. In: Kahraman, C., Cevik Onar, S., Oztaysi, B., Sari, I.U., Cebi, S., Tolga, A.C. (eds.) INFUS 2020. AISC, vol. 1197, pp. 608–615. Springer, Cham (2021). https://doi.org/10.1007/978-3-030-51156-2_70
9. Javaid, M., Haleem, A., Singh, R.P., Khan, S., Suman, R.: Blockchain technology applications for Industry 4.0: a literature-based review. Blockchain Res. Appl. **2**(4), 100027 (2021)
10. Drath, R., Horch, A.: Industrie 4.0: hit or hype? [industry forum]. IEEE Ind. Electron. Mag. **8**(2), 56–58 (2014)

11. Oztemel, E., Gursev, S.: Literature review of Industry 4.0 and related technologies. J. Intell. Manuf. **31**, 127–182 (2020). https://doi.org/10.1007/s10845-018-1433-8
12. Xu, X., et al.: A taxonomy of blockchain-based systems for architecture design. In: 2017 IEEE International Conference on Software Architecture (ICSA), pp. 243–252. IEEE (2017)

Performance Evaluation of Quantum-Resistant Cryptography on a Blockchain

Jonas Deterding[✉], Noah Janzen, David Rohrschneider, Philipp Lösch, and Marc Jansen

Computer Science Institute, Ruhr West University of Applied Sciences, Lützowstraße 5, 46236 Bottrop, Germany
jonas.deterding@stud.hs-ruhrwest.de

Abstract. Blockchain technologies are charting a promising trajectory, gaining increasing acceptance across various economic sectors due to their exceptional trust levels, underpinned by sophisticated encryption techniques. However, despite continuous advancements in the development of nearly indecipherable encryption methods, the rapid evolution of quantum computing capabilities poses a potential threat to these systems in the foreseeable future. In July 2022, the National Institute of Standards and Technology (NIST) introduced four quantum-resistant cryptographic algorithms, marking a significant milestone in this area. Nonetheless, with heightened security comes a potential performance drawback, especially when implemented in blockchain projects such as Ethereum. This study empirically demonstrates that the integration of the quantum-resistant cryptographic algorithm "Dilithium"—one of the "to be standardized candidates" announced by the NIST—significantly decelerates operation speeds within a distributed blockchain architecture when compared to the incumbent algorithms, ECC and RSA.

Keywords: Blockchain · Quantum computers · Dilithium · Performance

1   Introduction

Blockchain technology has captured significant attention and widespread popularity in recent years, owing to its distinct features, namely, decentralization, immutability, and transparency. Serving as a distributed ledger, the blockchain secures data in an immutable and indelible manner, offering an ideal solution for many applications, such as finance, healthcare, supply chain management, and beyond. However, the inherent reliance of blockchains on cryptography to guarantee the security and privacy of data and transactions underscores the essential need for the deployment of robust and unassailable cryptographic methodologies within the blockchain ecosystem.

J. M. Machado et al. (Eds.): BLOCKCHAIN 2023, LNNS 778, pp. 124–133, 2023.
https://doi.org/10.1007/978-3-031-45155-3_13

Blockchains utilize an array of cryptographic methods, prominently including public-key cryptography for transaction signatures via private keys. These signatures can be verified using the corresponding public keys, establishing a signature scheme that authenticates the intended sender while maintaining transaction integrity and ensuring non-repudiation.

Cryptographic hash functions play a pivotal role in a multitude of tasks, such as the resolution of cryptographic puzzles. For instance, the Bitcoin mining process necessitates the solving of a cryptographic puzzle by repeatedly computing hashes and altering a variable known as the nonce to meet a specific condition. This consensus mechanism is recognized as Proof of Work (PoW).

Notably, while hashing functions like SHA-256 are regarded as quantum-resistant, the same cannot be affirmed for public-key cryptography. Blockchains predominantly employ public-key cryptosystems, including RSA and ECC, which have demonstrated security against classical computers. However, the advent of quantum computing presents a formidable threat to these algorithms. Quantum computers bear the potential to compromise the public-key cryptosystems used in blockchains until now, threatening the integrity and confidentiality of data and transactions [8].

Recently, the emerging apprehension within the blockchain community concerning the potential threat presented by quantum computers has instigated a wave of research focused on devising countermeasures. One prominent strategy involves the exploration of quantum-resistant cryptography, specifically engineered to safeguard against quantum assaults. These technologies are based on complex mathematical challenges that are posited to be strenuous for both classical and quantum computing systems to unravel.

This paper aims to conduct a thorough performance comparison of two conventional public-key cryptography methods, ECC and RSA, against a quantum-secure public-key algorithm, Crystals-Dilithium. The evaluation's primary objective is to ascertain the suitability of the Crystals-Dilithium algorithm for deployment within a blockchain context, a milieu that necessitates a balance between robust security and optimized performance.

The organization of this paper is as follows: Sect. 2 offers a comprehensive review of the quantum-resistant public-key algorithm, Crystals-Dilithium, and illuminates its superiority over conventional methodologies. In Sect. 3, the architecture of the used blockchain is described upon integrating the selected algorithms. Section 4 outlines the specificities of the blockchain's implementation, including the selection of programming language and associated libraries. Section 5 assesses the performance of the three algorithms by tracking the time required to confirm varying volumes of transactions within the blockchain. Concluding the paper, Sect. 6 consolidates the findings, underscores the limitations of this research, and proposes potential avenues for future investigation.

2   State of the Art

Numerous projects and researchers have posited that by the year 2035, quantum computers will likely be capable of breaching the contemporary "gold standard" encryption algorithms, such as RSA and ECC, in a markedly reduced time frame [1,5,6].

Currently, public-key and digital signature algorithms, such as RSA and ECC—deemed safe in today's context—are extensively employed within blockchains to authenticate transactions. Nevertheless, these may become susceptible to attacks by quantum computers in the future, thereby introducing a risk to all blockchain technologies. Prior research, as referenced in [9], asserts that at the time of their study, concrete countermeasures were lacking, and quantum-safe blockchains were non-existent. They concluded that both domains—blockchain and quantum computing—are dynamic and rapidly evolving areas of study.

In 2022, the United States National Institute of Standards and Technology (NIST) selected Crystals-Dilithium among others as a candidate for post-quantum cryptography standardization of digital signature algorithms. Dilithium represents a cutting-edge, quantum-resistant digital signature algorithm that offers protection against both quantum and conventional attacks. According to the NIST, various security stages are available, specifically achieving NIST levels II, III, and V. These encompass Dilithium-2 with a public-key length of 1312 bytes, Dilithium-3 with 1952 bytes, and Dilithium-5 with 2592 bytes. Nevertheless, the current recommendation for Dilithium is the level-3 parameter set, as it ensures a security level exceeding 128 bits against all known classical and quantum attacks and also surpasses the recommended public-key length of 1472 bytes, as stated in the referenced Crystals-Dilithium study [4]. In [3] a high-performance implementation of Crystals-Dilithium targeting FPGAs, was compared to other post-quantum digital signature schemes with focus on efficiency and a low signature generation time. The lattice-based Dilithium high-performance implementation stood out as the cryptosystem with the best-known latency at all security levels.

3   Architecture

We built upon an existing Python blockchain implementation as a foundational reference. The primary elements retained from the original architecture were the client implementation, while the node underwent significant reimplementation to facilitate an efficient testing flow. Our resultant code[1], which can serve as a basis for further research, leverages containerization to ensure that multiple instances can operate independently. Within this framework, two principal types of containers are distinguished: the Blockchain Container (BC) and the Client Node (CN).

As depicted in Fig. 1, each Blockchain Container (BC) comprises a Blockchain Node (BN), a Mining Node (MN), and a Relational Database (RDB). The BN encapsulates the blockchain logic and offers an endpoint to facilitate inter-instance communication. All running container endpoints are interconnected,

---

[1] https://github.com/jode-reflact/quantumsafe-blockchain-project.

thereby reflecting a fully-networked architecture. The MN, employing the Proof-of-Work (PoW) mechanism, attempts to solve the PoW problem concurrently and dispatches the discovered nonce to the BN. To maintain a global state of the chain and the pending transactions, both nodes within the BC refer to the local RDB. The Client Node (CN) provides an endpoint for submitting transactions to one of the BNs, which then disseminates the incoming data across the network. Once validated, the transaction is recorded in the RDB's pending transaction table, which the MN subsequently references. Furthermore, two significant processes highlighted in Fig. 1 utilize the cryptographic algorithm, thus directly influencing the runtime. These processes include transaction signing within the CN (1) and transaction validation within the BN (2).

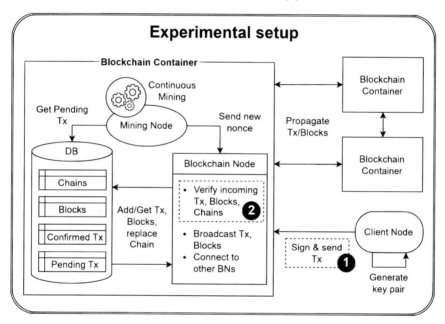

Fig. 1. Illustration of our experimental setup. Point 1 and 2 declare functionality utilizing on the cryptographic algorithm

4   Implementation

The entire implementation is composed in Python. To facilitate communication within and between BCs, we employ the Flask framework to offer API endpoints for transmitting and receiving data. The BCs' local SQLite database interfaces with SQLAlchemy, enabling object-relational mapping for data read-write operations. Consequently, the four classes "chain", "block", "confirmed transaction", and "pending transaction" map to their respective tables in the RDB. To streamline the exchange of cryptographic algorithms, an abstract cipher class is defined. This class comprises three functions: Generating key pairs to simulate a client's wallet, signing transactions, and verifying them based on their signature. For each cryptographic algorithm employed in the ensuing experiment, a distinct class that implements the abstract class is created and utilized in the process delineated in Fig. 1. The Proof-of-Work function employed by the MN strives to discover a new nonce that results in a new block hash possessing n leading zeros, where n denotes the difficulty level. In each iteration, the nonce is assigned a fresh random value between 0 and 100,000,000. If transactions are received during this interval, they are included in the succeeding iteration.

Given that our architecture relies on containers to simulate network participants, we opted for Docker to configure the images and containers, which are subsequently deployed on a Hetzner Cloud server. A discrete Python script is devised to initiate and manage the BCs. This script enables the specification of the number of BCs to start, the port range, the number of initialized transactions, and the selected cryptographic algorithm. Additionally, it connects all active BCs to one another by invoking their API endpoints, and sequentially dispatches the pre-configured quantity of transactions to randomly selected BNs within the network.

5 Evaluation

In Sect. 5.1, we initially delineate the structure of our test iterations. Subsequently, we present the test outcomes and their statistical relevance. Furthermore, we contemplate potential avenues for enhancing performance and undertake an additional test to examine the impact of block size on test duration.

5.1 Evaluation Strategy

To assess the performance of the three algorithms within a blockchain environment, we conducted multiple test cases. In each test, the algorithm employed for signing and verifying transactions was suitably substituted. To maintain comparability, we juxtaposed Dilithium-3 with RSA 3072 and ECC P-256, the latter two of which are deemed equivalent according to the NIST Recommendations for Key Management [7]. Each test was conducted on the same Hetzner Cloud server instance—CPX-51—which boasts 16 VCPUs and 32 GB of RAM. To ensure each test's independent operation and to provide adequate computational power for its two client and ten node containers, a Node.js application, designated as the Test Controller, was established with a test queue.

To evaluate the algorithms under discussion and answer the primary question—whether quantum-safe algorithms like Dilithium can be implemented efficiently within blockchains concerning their performance—we chose a specific time span to assess the test cases. Upon verification, the first node to receive a transaction stamps it with a "received at" timestamp, marking its initial appearance in the network. Consequently, each authenticated transaction within the network carries its unique "received at" timestamp. The earliest of these timestamps, which indicates the first verified transaction in the network, marks the onset of our measurement. Client signing durations are intentionally excluded from the measurement since they occur locally within a real blockchain network, thus not posing a scaling issue. However, our measurement method doesn't account for the duration required for verifying the first transaction, as our implementation only assigns the "received at" timestamp post-verification. While this only affects the first transaction, its impact is negligible. Nonetheless, in future measurements, we plan to set the timestamp prior to verification commencement. The measurement concludes once all test transactions are confirmed within the network. Thus, the end time is the timestamp of the last block encompassing the final transactions of the test. This methodology is advantageous over a simpler "time span between the first and last block" approach, as it also takes the verification time of the transactions (except for the first one) into account, which is dependent on the employed algorithm. To fortify our resistance to outliers, we conducted each test case 10 times, while tests involving 500 and 1000 transactions in the first experiment were conducted 20 times.

5.2   Evaluation Results

In the initial experiment, we compared performances by testing varying transaction counts, maintaining a consistent block size of 10 (comprising 9 client transactions and 1 block reward transaction). During this experiment, an observation was made that the block time increased due to a longer string to be hashed. This extended string resulted from longer transaction signatures (as in Dilithium) as well as from an increased number of transactions, which was then employed to validate the proof-of-work nonce. Consequently, we examined the runtime of this validation function with 10 or 100 transactions in contrast to the runtime without any transactions using a brief test script. The result indicated that even 10 transactions amplified the runtime by nearly 1.5 times (146%) compared to the runtime without transactions, whereas 100 transactions multiplied the duration by more than five times (558%). To counter this, we devised a basic caching system that preemptively holds the transactions as strings and feeds them into the validation function. In our test script, this rudimentary variant reduced the required runtime by 26% for 10 transactions and by 67% for 100 transactions. Consequently, we opted to conduct an additional test run incorporating the caching solution (see 1000_C). In interpreting these results, it's critical to note that the transaction signature is merely a single component of a quantum-safe blockchain solution, as outlined by Allende et al. [2]. For instance, the communication between nodes also presents vulnerability. Our results validate the overhead of quantum-safe algorithms identified in this process, although our overhead is evidenced through extended test duration rather than heightened CPU and RAM usage.

Statistical Significance. The Shapiro-Wilk test, as presented in Table 1, demonstrates that all distributions can be regarded as normally distributed. Consequently, we employed T-tests for the statistical evaluation. Each algorithm

Table 1. Test duration comparison between different public-key algorithms for different numbers of transactions. 1000_C uses an caching implementation. P-values from Shapiro-Wilk-Tests are included as p.

#Txs	RSA 3072			ECC P-256			Dilithium 3		
	mean	std	p	mean	std	p	mean	std	p
100	114 s	33 s	0.20	91 s	23 s	0.23	250 s	68 s	0.62
500	656 s	55 s	0.38	561 s	43 s	0.39	1640 s	171 s	0.89
1000	2032 s	128 s	0.11	1672 s	103 s	0.16	4248 s	451 s	0.25
2000	6564 s	376 s	0.11	6700 s	333 s	0.75	13344 s	871 s	0.28
1000_C	1929 s	83 s	0.55	1644 s	99 s	0.76	4349 s	308 s	0.25

in the test case with 1000 transactions underwent a two-sided test against the other algorithms, culminating in a total of three tests, subsequently adjusted using the Bonferroni correction. The null hypothesis posited that the test times of the algorithms do not vary, signifying that the algorithms demonstrate comparable performance. The tests indicated a significant difference between RSA and ECC ($t(19) = 9.785$, $p = 1.865e{-}11$), between RSA and Dilithium ($t(19) = -21.126$, $p = 4e{-}22$), and between ECC and Dilithium ($t(19) = -24.882$, $p = 1.2e{-}24$). In light of these test results, it can be concluded that the use of Dilithium would entail significant performance losses.

Block Size Experiment. Beyond these findings, we were intrigued by the potential impact of varying block sizes, which we explored in a subsequent experiment. In each test, we employed the straightforward caching solution and authenticated 1000 transactions, while the block size fluctuated between 10 and 100.

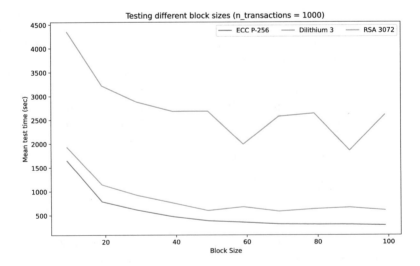

Fig. 2. Mean test time for each algorithm with different block sizes

As shown in Fig. 2, the test time for each algorithm decreases to a lower threshold as the block size increases. Nevertheless, this reduction is lower for Dilithium at 40% than for ECC at 82%. RSA lies between the other algorithms with a reduction of 68%. In order to better compare the algorithms with different block sizes, it is useful to analyze the average block time per test case.

As depicted in Fig. 2, the test duration for each algorithm diminishes to a minimal threshold as the block size expands. However, this reduction is less pronounced for Dilithium, exhibiting a decline of 40%, as compared to ECC which shows a significant reduction of 82%. RSA sits between the two other algorithms with a decrease of 68%. To facilitate a more robust comparison of the algorithms at differing block sizes, it is beneficial to evaluate the average block time per test case.

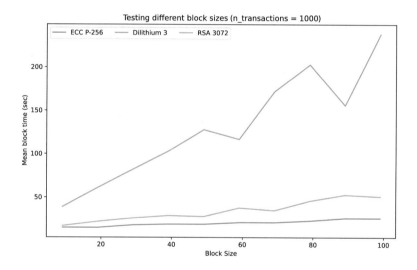

Fig. 3. Mean block time for each algorithm with different block sizes

Figure 3 demonstrates that, as hypothesized, the block time for Dilithium escalates more steeply with increased block size compared to the other two algorithms. This phenomenon might be attributed to the more extensive transaction signatures previously outlined. This trend also elucidates the lesser reduction in test duration as depicted in Fig. 2, indicating that the markedly inferior performance of Dilithium, as seen in Table 1, does not appear to be mitigated by larger block sizes.

6   Conclusion

This study aimed to compare the performance of two conventional (RSA, ECC) and one novel quantum-safe (Dilithium) public-key cryptography algorithm within a blockchain environment. To address the research query, we constructed a blockchain system and executed tests with varying hyperparameters, transaction numbers, and an optional caching setting which cached the string representation of a transaction prior to hashing. The findings from our experiments disclosed significant performance variances across all three public-key algorithms. Our blockchain required more than twice the time to confirm 1000 transactions when employing the Dilithium algorithm in comparison to RSA or ECC. The average block time escalated more dramatically when Dilithium was utilized versus when RSA or ECC algorithms were employed. These outcomes indicate that additional research is needed to ascertain the suitability of Dilithium for blockchain applications, eventually upon specific use cases. The constraints of the study encompass its inability to comment on the performance of the Dilithium algorithm beyond its current implementation or to evaluate all potential quantum-safe algorithms grounded on different mathematical bases. Future research should focus on carrying out experiments on a dedicated GPU or hardware specialized for SHA256 hashing, optimizing the blockchain implementation concerning the caching of the hashing function's input, and administering the experiments with physically distributed mining nodes communicating via the internet. As demonstrated in [3], FPGA-based high-performance hardware can curtail the negative performance impact of larger signature sizes, although the economic and ecological implications require proper assessment. Further examination of how many transactions a blockchain with a more realistic node count can register per second, coupled with a comparison of these results to the existing Bitcoin or Ethereum performance, could provide a clearer understanding of whether Dilithium can meet the expected operational performance. In conclusion, this study offers valuable insights into the performance of quantum-resistant cryptography within a blockchain context, underscoring the necessity for continued exploration of this topic.

References

1. Aggarwal, D., Brennen, G., Lee, T., Santha, M., Tomamichel, M.: Quantum attacks on bitcoin, and how to protect against them. Ledger **3** (2017). https://doi.org/10.5195/LEDGER.2018.127
2. Allende, M., et al.: Quantum-resistance in blockchain networks. Sci. Rep. **13**(1), 5664 (2023)
3. Beckwith, L., Nguyen, D.T., Gaj, K.: High-performance hardware implementation of crystals-dilithium. Cryptology ePrint Archive, Paper 2021/1451 (2021). https://eprint.iacr.org/2021/1451
4. Ducas, L., et al.: Crystals-dilithium - algorithm specifications and supporting documentation (version 3.1). https://pq-crystals.org/dilithium/data/dilithium-specification-round3-20210208.pdf. Accessed 18 Mar 2023
5. Fernández-Caramès, T.M., Fraga-Lamas, P.: Towards post-quantum blockchain: a review on blockchain cryptography resistant to quantum computing attacks. IEEE Access **8**, 21091–21116 (2020). https://doi.org/10.1109/ACCESS.2020.2968985
6. Kearney, J.J., Perez-Delgado, C.A.: Vulnerability of blockchain technologies to quantum attacks. Array **10**, 100065 (2021). https://doi.org/10.1016/j.array.2021.100065. https://www.sciencedirect.com/science/article/pii/S2590005621000138

7. Barker, E.: Recommendation for key management: part 1 - general. https://csrc. nist.gov/publications/detail/sp/800-57-part-1/rev-5/final. Accessed 17 Mar 2023
8. Raikwar, M., Gligoroski, D., Kralevska, K.: SoK of used cryptography in blockchain. IEEE Access **7**, 148550–148575 (2019). https://doi.org/10.1109/ACCESS.2019. 2946983
9. Swathi, P., Dragan, B.: A survey on quantum-safe blockchain system. In: Proceedings of Annual Computer Security Applications Conference (ACSAC) (2022)

An Investigation of Scalability for Blockchain-Based E-Voting Applications

Mohammad Hajian Berenjestanaki[1]([✉]) [ID], Hamid R. Barzegar[1] [ID], Nabil El Ioini[2] [ID], and Claus Pahl[1] [ID]

[1] Faculty of Engineering, Free University of Bozen-Bolzano, Bolzano, Italy
`mhajian@unibz.it`, `hbarzegar@unibz.it`, `cpahl@unibz.it`
[2] School of Computer Science, University of Nottingham Malaysia, Semenyih, Selangor, Malaysia
`elioini.nabil@nottingham.edu.my`

Abstract. In the last years, a number of initiatives have developed and deployed new electronic voting (e-voting) systems, with the goal of improving the paper-based electoral systems. Traditional e-voting systems are vulnerable to manipulation concerns, such as altering election outcomes, due to their centralized nature. Blockchain is a decentralized system technology with a distributed ledger in which every participant in the network accesses the same data source. The immutability of blockchains makes them suited for e-voting systems as no one can alter data. Although blockchain technology has potential for e-voting systems, there are challenges, including the scalability for large-scale elections.
We present a solution to scalability of e-voting systems using the Solana blockchain. Solana provides a fast and scalable blockchain infrastructure utilizing the Proof of History (PoH) consensus algorithm, making it easier for developers to create efficient decentralized applications. The evaluation of factors such as throughput, delay, and cost shows that a Solana implementation can finalize a vote 32 times faster than Ethereum 2.0, with a significantly lower cost per vote. The throughput of e-voting app on Solana indicates that it can deal with a large scale election entirely.

Keywords: e-voting · throughput · DApp · Solana · scalability

1 Introduction

Traditional voting using paper-based methods is the most prevalent type of voting. However, it is known to have various problems such as integrity, including vote tampering, lack of transparency, fairness, or issues with accessibility [1–4].

Blockchain technology has a number of characteristics, including decentralization, immutability, anonymity, and auditability, which make it suitable to address the above challenges. Its decentralized nature eliminates intermediaries,

J. M. Machado et al. (Eds.): BLOCKCHAIN 2023, LNNS 778, pp. 134–143, 2023.
https://doi.org/10.1007/978-3-031-45155-3_14

making transactions secure and transparent. The use of cryptographic algorithms provides privacy and anonymity. Blockchain is being used in financial services, healthcare, supply chain management, and other domains to streamline processes, reduce costs, and enhance security [5].

While blockchain technology offers several advantages in e-voting systems, there are also some challenges that need to be addressed. Scalability is indeed one of the main requirements and challenges of blockchain-based e-voting systems [6–8]. E-voting systems must be able to handle large amounts of data and transactions in order to support large-scale elections and referendums. A definition of scalability in the blockchain domain is as follows:

> **Scalability** is the ability of the blockchain protocol to withstand high transactional throughput and future growth of nodes in the network [9].

However, many e-voting systems are proposed in the domain of blockchain-based voting, but the majority of prior studies merely provided a broad overview of blockchain and did not explore sufficiently the technical aspects of its implementation or analyze its efficiency. Ethereum, Bitcoin, Hyperledger, Quantum, and Zcash are the most popular blockchain platforms currently available [7], but proposed e-voting systems have issues with scalability, and there is an insufficient evaluation of their performance. We contribute to propose an e-voting system on the Solana platform, which is known for its scalability, and address the scalability issue through a comprehensive assessment. In accordance with prior research on scalability, we take into consideration three crucial metrics: cost per transaction, throughput, and finality delay, to assess the scalability of a blockchain-based e-voting system [10,11]. The paper is structured as described below. Section 2 describes the background and concept of the proposed e-voting on Solana platform. Then, Sects. 3 and 4 present the implementation, evaluation procedure and results of the e-voting system for different numbers of voters. In addition, Sect. 5 reviews related work that use various blockchain platforms. In Sect. 6 provides a discussion about the advantages and challenges of using Solana and highlights the need for further research and development in the field. Finally, Sect. 7 presents the conclusions and future work that can be done.

2 Background and Concepts

This section provides an overview of the blockchain concept, introduces the Solana platform, and defines terms and metrics used throughout the paper to evaluate the scalability of e-voting as a distributed application.

2.1 Blockchain and Solana

Blockchain is a digital ledger that records transactions across a decentralized network of computers, using cryptography to secure and verify transactions. Transactions are added to blocks, which are then linked chronologically and protected through consensus mechanisms. The blocks contain information such as the previous block's hash, timestamps, and the Merkle root of transactions, which is the result of hashing all transactions in the block in pairs. This structure allows blockchain data to be protected from tampering and manipulation, as any changes to the data will result in a different hash value in the next block [6].

Regarding the blockchain definition we utilize the Solana platform which is a public-permissionless blockchain and aims to solve the scalability issue in the blockchain based e-voting system. Solana is designed to handle large-scale decentralized applications by providing high transaction processing speed and low latency. Solana's architecture is based on a Proof of Stake (PoS) consensus algorithm and a new strategy called a "Gulf Stream" that enables the process of transaction catching and forwarding even before the next set of blocks for confirmation are finalized. Solana uses the tower Byzantine Fault Tolerance (BFT), which is an algorithm that uses the PoH mechanism. This mechanism leverages a cryptographic technique called Verifiable Delay Functions (VDFs) to create a historical record of timestamps for each transaction and event in the network. This allows Solana nodes to agree on the order and timing of transactions. PoH acts as a global clock that provides a reliable ordering of events and prevents the need for all nodes to agree on the exact order through consensus. This allows the network to process transactions in parallel, significantly increasing throughput. The PoH mechanism works by generating a verifiable proof that demonstrates the passage of a certain amount of time. This proof is created using a computationally intensive process, making it difficult for malicious actors to manipulate the ordering of events. It helps to reach consensus without having to send a flood of communication between the nodes [12,13]. These features allow Solana to process a large number of transactions per second while maintaining a low latency, which is essential for ensuring a smooth user experience in decentralized applications and making it a suitable platform for various use cases, including e-voting systems.

2.2   Technical Concepts

We define the following concepts relevant for the Solana platform:

- SOL: Solana's native token is called a SOL. Nodes in a Solana cluster can accept a SOL in exchange for running an on-chain program or validating its output [13]
- Slot: The duration in which each leader processes transactions and generates a block. A logical clock is made up of slots, and chronologically arranged slots do not overlap with one another [13].
- RPC (Remote Procedure Call) node: A Solana RPC node constructs and administers the infrastructure required for DApps to submit transactions, make requests, and get data from the public Solana blockchain.
- Cost: The SOL based fee that is paid by the network user who performs the transactions to the Solana validators to successfully complete a transaction or smart contract execution.
- Throughput: In general, throughput is the number of valid transactions that can be committed by the blockchain network in a specific period of time. In

this study and e-voting application we defined it as the number of votes that send to the Solana network and their block commitment reach to confirm status per second. Confirmed status means the corresponding block receives votes from 2/3 of the stakes in the cluster.

- Finality Delay: The time duration that a block of a vote transaction is validated by a super majority of the cluster as having reached maximum lockout means that the cluster recognizes the block as finalized.

3 Solana-Based E-Voting Implementation

3.1 E-Voting System Model

This study looks at a scenario in which a group of electors can make a choice among four different options. The proposed scheme includes five different stages. First At the time of registration, the registration authority checks the eligibility of each potential voter. After registering all eligible voters, the registration authority makes a list of voter ID numbers and sends this list to a read-only account on the Solana network. Next, each eligible voter, through authorization, which is made by an ID number, makes a transaction, signs it, and transmits it to the Solana cluster during the voting phase. In detail, the transaction is signed by the voter and sent to the DApp (Decentralized Application), then the e-voting DApp takes the signed transaction and sends it to the RPC servers. After that, the RPC servers send the transaction as a UDP packet to the current and next validator according to the leader schedule in the Solana network. Then the leader validator verifies the signature of the voter and executes it, and afterwards it propagates the transaction to the other Solana validators, known as verifiers for the consensus process [12]. In the next phase, the voters' decisions proceed to the tallying section, and the tallying program in e-voting DApp adds the cast vote to the final result. Verification is the last stage, in which the voters verify their votes and the tally's correctness. Figure 1 illustrates the phases of a voting process.

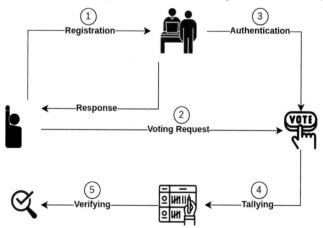

Fig. 1. Voting process

3.2   Architecture

For the Solana platform-based e-voting system, we have two tiers. One is a smart contract and another one is for examination as follows:

1. On-chain program: In this tier, we used the anchor framework, which is a convenient tool for developing smart contracts in Solana using the Rust programming language. In the Solana platform the programs are accounts which are read-only accounts that include their own code and are marked as "executable". There is two type of account in our e-voting DApp. One of them is a construct that aggregates the votes for four different options and another one pertains to individual voters, tracking their eligibility for voting once and voting option in election process.
2. Interact and testing the smart contract: Solana offers a JSON RPC API for this purpose. This API is encapsulated by the @solana/web3.js JavaScript library, which provides a wide variety of useful asynchronous functions.

We implemented the e-voting system based on Solana blockchain through the smart contract. Then we targeted the smart contract for e-voting on the Devnet cluster. Devnet serves as a tested to test drive application development on Solana.

3.3   Evaluation Setup

We examine the e-voting DApp for different numbers of voters (1–1000) and we repeat 10 times each experiment for a specific number of voters. Regarding this repetition we have max, average, and min finality delay. We try to send vote casting transaction through HTTP requests and interact by the Solana RPC using the JSON-RPC 2.0. Solana provides dedicated API nodes for each public cluster to serve JSON-RPC queries. The Devnet public RPC endpoints are available with a per IP request rate limit. Therefore we use proxy pools for multiplexing sent transaction request for evaluating different number of voters through our local machine. Figure 2 indicates the journey of a vote transaction from the local machine until it reaches to the Solana network and starts to be processed [14].

4   Scalability Evaluation

In the following, we present the results of the evaluation for the three scalability metrics cost, throughput, and finality delay that we introduced earlier.

4.1   Cost

To evaluate the cost of the Solana based e-voting system, three aspects should be considered: the cost of deploying the app on the network and initializing the

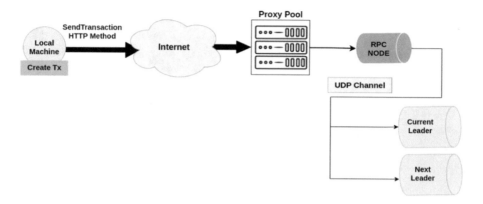

Fig. 2. Transaction of a vote from casting to processing on the Solana network

voting state account, initializing the voters eligibility and choice account, and the cost for each voter to vote.

To store data in Solana, it is necessary to create an account and pay based on the occupied volume. The authority is responsible for covering the deploying and initializing costs of the account used to store voters' decisions, as well as tracking voters' eligibility and choices. There is an option to make the account rent-exempt by depositing a certain amount of SOL. Considering the given rate of 0.000005 SOL per vote transaction sent to the Solana cluster and the approximate price of SOL at 23.95 dollars on February 1st, 2023, the cost for 1000 voters just to cast their votes in total is approximately 0.12 dollars [15]. The total cost of the voting process includes the cumulative costs for the number of voters and the deploying and initializing expenses. The relationship between the cost and the number of voters exhibits a clear linear pattern, as demonstrated in Fig. 3.

4.2   Throughput

For the evaluation of throughput, we increase vote requests rate from 1 voter to 1000 voters per second and then checked the status of the transaction commitment. According to the definition in Sect. 2.2 the number of votes that could entirely reach to confirm status is the throughput of our system.

We found that in all experiments for a different number and rate of vote transaction from 1–1000, the transaction status can be reached to confirm this. It indicates that the e-voting DApp based on Solana can deal with the rate of 1000 casted votes per second.

4.3   Finality Delay

To find the finality delay, we aimed to send vote transaction with the rate 1 to 1000 in a second. First we requested the slot and timestamp that the requests

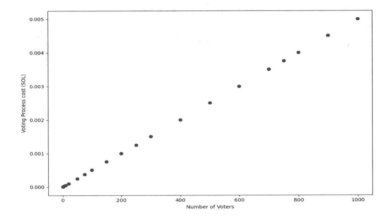

Fig. 3. Cost of voting process

start to be processed in the Solana cluster, then requested using a JavaScript program based on a for loop with a 100 millisecond delay request the vote transaction slot finalizing timestamp. In addition we requested the slot that start is to be processed in the Solana cluster. Therefore, the difference between these two number results in the finality delay.

In Fig. 4 and 5 we present the results of the finality delay experiment. The results indicate the minimum, average, and maximum delay periods according to the number of slots and milliseconds. As the Figs. 4 and 5 show, there is some irregularity for 50 and 400 number of voters. We can attribute this delay to network connection delay as the cause. Overall, however, the general delay average in total is around 12 to 15 s and 30 slots.

5   Related Work

In this section, we briefly review some parts of the recent efforts that provide evaluation of the deployed blockchain based e-voting systems.

Khan et al. [16] provided a detailed analysis of the performance and scalability limitations of an e-voting system. They found a trade-off between different parameters and running several clients simultaneously had a considerable influence on blockchain performance.

Abuidris et al. [17] proposed a secure large-scale e-voting system combining Proof of Credibility and Proof of Stake in a hybrid consensus mechanism with sharding mechanism called PSC-Bchain, which outperforms PoW and PoS in throughput at a network size of 1000 nodes.

Pandey et al. [18] presented VoteChain as a blockchain based voting system. The experiments showed that as the size of the block increases, the amount of time it takes to send blocks between nodes increases. This makes it more challenging to reduce the amount of time it takes to mine each vote. The evaluation showed the time taken for each vote reaches a maximum of 90 s.

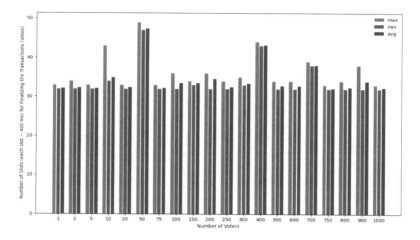

Fig. 4. Finality delay based on slots

In [19] Song et al. proposed a private Ethereum solution was built with Truffle and the evaluation is performed with different numbers of voters and candidates. Furthermore, they indicated the proposed system is cost-efficient, scalable, and reduces the number of computations compared to previous studies.

Sharma et al. [20] proposed a cost-efficient Proof of Stake Voting based blockchain for e-voting systems. The study found that the total cost of the proposed blockchain-based e-voting application with the Proof of Stake Voting protocol was significantly lower than the Proof of Authority and Proof of Work protocols.

6   Discussion

The Solana platform has shown promising results for cost, throughput, and delay metrics, indicating positive outcomes for scalability. While the blockchain trilemma remains a challenge, the use of Solana has demonstrated progress in addressing it. Solana is particularly suitable for decentralized applications that require high transaction throughput and low latency, such as gaming, governance, cryptocurrency exchanges, and marketplaces.

Despite its advantages, Solana's high hardware requirements for validators to handle its high transaction processing speed leads to a lower degree of decentralization compared to other blockchain systems like Ethereum and Bitcoin. When evaluating blockchain-based e-voting systems, it is crucial to consider several factors, including the type of blockchain employed, the number of voters, the complexity of the voting process, and the underlying infrastructure.

In comparison to other platforms e-voting systems, Solana presents a more cost-effective, efficient, and faster election process. While Solana has made progress in terms of scalability, it may come at the cost of some degree of decentralization and security. Thus, it becomes imperative to conduct further research

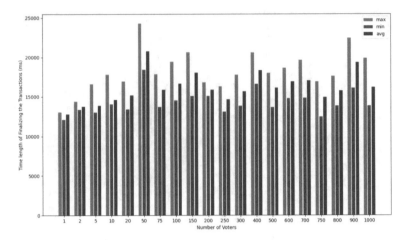

Fig. 5. Finality delay based on milliseconds

to determine whether the balance of scalability, security, and decentralization in Solana makes it a suitable platform for e-voting DApps. This kind of research can involve evaluating the platform's security measures and its ability to maintain decentralization while providing high transaction throughput.

7   Conclusion

The development of new e-voting systems aims to improve traditional paper-based electoral systems and address their limitations. Blockchain technology is seen as a promising solution due to its decentralization and immutable nature, but scalability remains a major challenge.

This study proposes using Solana blockchain to address scalability in e-voting systems, showing that it can handle large-scale elections with high speed and low cost. However, security and privacy must also be considered in the design and implementation of e-voting systems.

Our proposed solution provides a promising approach, but further research and development are needed to ensure security and privacy, and to create fair and transparent elections to safeguard the democratic process. These concerns will be addressed in future work.

Acknowledgements. This work has been supported by the Südtiroler Informatik AG (SIAG) through a PhD Bursary.

References

1. Jafar, U., Aziz, M.J.A., Shukur, Z.: Blockchain for electronic voting system-review and open research challenges. Sensors **21**(17), 5874 (2021)

2. Alvi, S.T., et al.: Classification of blockchain based voting: challenges and solutions. In: 2020 IEEE Asia-Pacific Conference on Computer Science and Data Engineering (CSDE). IEEE (2020)
3. Buyukbaskin, L.A., Sertkaya, I.: Requirement analysis of some blockchain-based e-voting schemes. Int. J. Inf. Secur. Sci. **9**(4), 188–212 (2021)
4. Alvi, S.T., et al.: From conventional voting to blockchain voting: categorization of different voting mechanisms. In: 2020 2nd International Conference on Sustainable Technologies for Industry 4.0 (STI). IEEE (2020)
5. Zīle, K., Strazdiņa, R.: Blockchain use cases and their feasibility. Appl. Comput. Syst. **23**(1), 12–20 (2018)
6. Wang, X., et al.: Survey on blockchain for Internet of Things. Comput. Commun. **136**, 10–29 (2019)
7. Taş, R., Tanrıöver, Ö.Ö.: A systematic review of challenges and opportunities of blockchain for E-voting. Symmetry **12**(8), 1328 (2020)
8. Jafar, U., et al.: A systematic literature review and meta-analysis on scalable blockchain-based electronic voting systems. Sensors **22**(19), 7585 (2022)
9. Khan, D., Jung, L.T., Hashmani, M.A.: Systematic literature review of challenges in blockchain scalability. Appl. Sci. **11**(20), 9372 (2021)
10. Shevkar, R.: Performance-based analysis of blockchain scalability metric. Tehnički glasnik **15**(1), 133–142 (2021)
11. Zhou, Q., et al.: Solutions to scalability of blockchain: a survey. IEEE Access **8**, 16440–16455 (2020)
12. Yakovenko, A.: Solana: a new architecture for a high performance blockchain v0. 8.13. Whitepaper (2018)
13. Solana Docs Blog RSS (2022). https://docs.solana.com/
14. Solana Cookbook: Home to Solana References. Cooking with Solana. Solana Cookbook. https://solanacookbook.com/
15. Solana Price Today, Sol to USD Live, Marketcap and Chart. CoinMarketCap. coinmarketcap.com/currencies/solana/. Accessed 1 Feb 2023
16. Khan, K.M., Arshad, J., Khan, M.M.: Investigating performance constraints for blockchain based secure e-voting system. Future Gener. Comput. Syst. **105**, 13–26 (2020)
17. Abuidris, Y., et al.: Secure large-scale E-voting system based on blockchain contract using a hybrid consensus model combined with sharding. ETRI J. **43**(2), 357–370 (2021)
18. Pandey, A., Bhasi, M., Chandrasekaran, K.: VoteChain: a blockchain based e-voting system. In: 2019 Global Conference for Advancement in Technology (GCAT). IEEE (2019)
19. Song, J.-G., Moon, S.-J., Jang, J.-W.: A scalable implementation of anonymous voting over Ethereum blockchain. Sensors **21**(12), 3958 (2021)
20. Sharma, T., Krishna, C.R., Bahga, A.: A cost-efficient proof-of-stake-voting based auditable blockchain e-voting system. In: IOP Conference Series: Materials Science and Engineering, vol. 1099. no. 1. IOP Publishing (2021)

Data Analysis and Core Findings of Pakistan's National Population Census for a Blockchain-Based Solution

Sana Rasheed[(✉)] and Soulla Louca

Digital Innovation Department, University of Nicosia, Nicosia, Cyprus
rasheed.s@live.unic.ac.cy, louca.s@unic.ac.cy

Abstract. The national population census provides a holistic picture of the country's stability situation in terms of health and economy indexes and sources in policy formation, electoral support and fund allocation. Any flaws in the census process can have serious consequences for national planning and interests. Our previous research revealed six challenges in the national census process, in general, such as population coverage, ethnic and racial discrimination, privacy concerns, etc. Taking into account these challenges, we conducted a case study of Pakistan and interviewed the National Census Bureau to identify their challenges and propose a blockchain-based robust and scalable solution to address them, encapsulating census data sensitivity, privacy and national security concerns. Our research contributes to the field by presenting the challenges and findings of the current system of Pakistan's national population census and outlining the requirements for a new permissioned blockchain system.

Keywords: Pakistan's national population census · structured interviews · data collection & analysis · challenges of nation census system · blockchain-based national census solution · permissioned blockchain

1 Introduction

A national population census collects information on individuals, their demographics and their state of living in their respective jurisdictions. It helps to assess the country's stability and inform policy making, electoral support, fund allocation, etc. Any problems in the census process can compromise the national planning and interests [1]. In our previous research work (discussed in the next section), we explored the issues of the national census process. In this paper, we focus on the case of Pakistan's national census bureau and conduct qualitative research to understand their processes, challenges and perspectives. We contribute to the field by presenting the challenges and findings of the current system of Pakistan's national population census and outlining the requirements for a new system.

J. M. Machado et al. (Eds.): BLOCKCHAIN 2023, LNNS 778, pp. 144–153, 2023.
https://doi.org/10.1007/978-3-031-45155-3_15

2 Previous Work

This paper is a sequel research on the need of a blockchain-based national population census, where we conducted a systematic literature review and have identified six drawbacks of national population census, i.e., population coverage (skip locations due to different identity), ethnic and racial discrimination (amend records to decease headcount to gain minorities benefits or vis-versa), privacy and data distribution concerns, cost inefficiency and cooperation issues. We explored the characteristics of blockchain technology and analyzed potential solutions for these challenges [2] Table 1. Our research focuses on population coverage to provide a solution to the missing person who lives in complex living structures, i.e. under the bridge, on footpaths, in remote areas, etc.

Table 1. Challenges of National Population Census and Potential Solution [2].

Sr	Challenges of Centralized System	Blockchain-based Solution	Potential Solution
1	Population Coverage	improve	Decentralized Non-Government Organizations (NGOs), part of the blockchain, doing the enumeration with smart contracts
2	Ethnic & Racial Discrimination	improve	Immutability of records, timestamped records
3	Privacy Concerns	resolve	Decentralization, cryptographically-secured, immutability
4	Census Data Distribution	resolve	Smart contracts
5	Cost of Census	improve	Decentralization
6	Cooperation & Participation	improve	Transparency and Accountability

3 Related Work

Blockchain technology has features such as immutability and smart contracts that make it suitable for various use cases. Besides supply chain and finance [3], blockchain technology has also been applied to healthcare, energy [4, 5] and academia [6]. Researchers are still exploring possible opportunities in government (e-government services), law and order for improving trust and transparency and promoting accountability of government activities [7–10]. Blockchain technology for the housing and population census enumeration process is yet to be explored.

Recently, the blockchain for population census has been investigated and proposed an infrastructure with the integration of Internet voting applications [11]. Despite the differences between the applications and their requirements, they are discussed together and propose a common infrastructure. The main study focuses on the internet-based voting system and proposes to use it for the census enumeration process. They have considered the census enumeration process only from the cost and security perspective and implied that blockchain provides better security than any existing models, i.e. client–server model, cloud-based technology, or involvement of mobile-based services. They suggest a private blockchain platform over the public blockchain and create a separate peer-to-peer infrastructure. The chain should be managed by the Indian government by making 4,689 constituencies across India is fixed stations which will serve as multi-chain nodes of the blockchain. All these stations must be connected with the nearest constituency station via fiber-optic cable [11].

From a critics' point-of-view, the proposed infrastructure has few drawbacks, i.e.

i. It requires a unique national identity card number to fill in census information. Population with lack a national identity card, have no permanent residence or are illegal immigrants won't be able to login and provide their information. Similar applies to the population living in complex living arrangements or remote areas.
ii. Multiple family members who hold national identity cards can log in to the system and enter census data twice.
iii. Also, voter/voting details are required in the population census survey form, such as voter ID, voter type, voting state/district/subdistrict/town etc.

The domain of national population census has limited research. Considering this and the possible challenges of the existing national population enumeration system, we are exploring the process and the requirements to suggest a suitable solution.

4 Methodology

Besides of the gaps found in the systematic literature review and related work literature, collecting first-hand information on the topic is essential to reveal the current situation and its challenges. For this reason, we have considered structured interviews to collect primary data to perform qualitative research. This method has the advantages of forming relevant questions, ensuring consistency and comparability of interview responses, reducing errors and bias. We have picked a case study of Pakistan and conducted structured interviews at the Pakistan Bureau of Statistics (PBS). The interviews took place face to face during October – December 2022.

The Pakistan Bureau of Statistics is Pakistan's prime official agency responsible for collecting, compiling, and disseminating statistical information on the national population census. We have opted for face-to-face interviews where the researcher and interviewee exchange information and data collected through direct communication. This method helps minimize survey dropout rates and improve the quality of collected data, resulting in tangible outcomes.

5 Structured Interview Questionnaire and Pre-Fieldwork

A list of 34 questions for structured interviews has been prepared and split into five sections (The online version is available here):

   i. Participant demographics – 4 questions
  ii. The existing population census system – 13 questions
 iii. The population missed by the census– 6 questions
 iv. Their Blockchain knowledge – 3 questions
  v. Requirements for a new solution – 8 questions

The logical flow was tested with three participants who had experience with census population data or previously worked with the national census bureau. Later, it was translated into Urdu – Pakistan's national language, so the participants should not feel uncomfortable or disconnected during the session and have a better understanding of what had been asked.

At the pre-fieldwork stage, we identified our target participants from different departments and extracted a list of contact details from the Islamabad head office [12]. The selection criteria were: the participant should have 10+ years of experience serving in the census department and have been involved in planning, data collection, consolidation, and compilation, or should be part of the IT support team. We have set the minimum target bar for 10–15 participants because the national census bureau is a single federal body in the country, and their team will mainly use the proposed system.

6   Data Collection and Data Analysis

We conducted ten interviews with eight officers and two directors. Each interview session took 1–1.25 h approx. To record the responses, *KoboToolbox* has been used [13]. The *Python* programming language with *JupyterNotebook* has been used for data analysis. It is free software, open standards, and web services for interactive computing across all programming languages. Specifically, *Pandas* and *NumPy* libraries are used for data cleaning and transformation, and *Ploty* and *Sweetviz* libraries have been used to plot different data graphs.

We collected 450 data points from 10 participants who responded to 34 questions. As required, we have applied feature engineering techniques on multiple choice questions, transforming them into Yes/No questions and introduced 11 new fields to bring more standpoint to data, which makes it $10 * (34 + 11) = 450$ data points. The analysis shared below is based on five questionnaire sections.

6.1   Participant Demographics

We inquired about their role in PBS, education, experience, involvement in the census data collection and dissemination process and data access.

The participants were three females and seven males, all with a master's degree in Statistics, Economics, Geography, or Information Technology. They were either directors or officers, with two directors and eight officers. They had been working at the organization for 12–28 years. Additionally, 4 out of 10 participants said they had direct data access for purposes such as data correction, verification, consistency, or imputation.

6.2   The Existing Population Census System

This section of questions covered the questions like if the current system is paper-based or paperless, either it is centralized or decentralized; and asks a few direct questions if the existing system is immutable, reliable, ownership, safe and secure; and what protocols are in-placed for the protection of data theft and leakages.

It is confirmed that the existing system is paper-based and centralized in nature. The important aspect is 90% of the respondent said that data is not immutable and can be modified by the supervisors. There is no trace of when and who imputed the data and what has been modified. The data is safe with traditional protocols like antivirus, VPNs, firewall settings, and log maintenance of network security. The system is considered reliable and safe, but data is not readily accessible to the stakeholders. It is available only on demand, and data sharing is only possible via CD, email, printed hardcopy, posted by mail, and USB drives – all are still ancient methods, as revealed in Fig. 1.

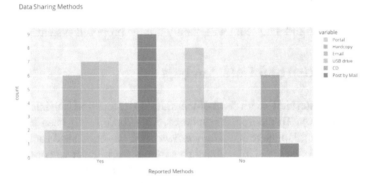

Fig. 1. Data sharing methods

Another key aspect is the ownership of the census results. They only count data they gathered through authorized surveyors and have exclusive rights to the data. They do not use any third-party data or reliable open-source data, even for verification purposes.

Yet, Fig. 2 reveals system issues such as manual effort dependency, data redundancy, and difficulty for a machine to read written text. Other challenges include quality assurance (QA), data provision reluctance in the affluent community, security issues, budget issues, technological issues like network coverage, and software uncertainty.

6.3   The Population Missed During the Enumeration Process

This section focused on their awareness of people who were not counted by the census. All participants either 'strongly disagree' or 'disagree' with this possibility, as the census process has a one-day schedule for the homeless population. A follow-up question on the likelihood of which category may get overlooked during the census conduct. Figure 3 shows that most participants said that no one gets missed, but some mentioned a few groups on a possible basis, such as remote areas of Pakistan, the transgender population, unregistered orphan houses, and congregational settings.

Current System Challenges

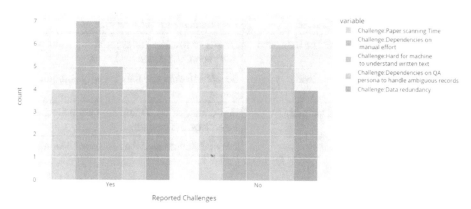

Fig. 2. Challenges in the existing system.

Type of Population get Missed

Fig. 3. Type of missed population

6.4 Blockchain Awareness

Under this section, the collected responses were 'cannot say anything' and 'do not know' about distributed ledger or blockchain technology. We felt a considerable gap between their information technology (IT) knowledge and emerging technology trends. One of the respondents perceived that the Blockchain would be similar to a 'Tablet computer'.

6.5 Requirements of a New Solution

Undoubtedly, all the participants wanted a new digital system that resolves existing challenges. The two new requested features are a collection of Thumbprints and sending confirmation to the interviewee. Open-ended questions were asked to know what challenges they want to resolve and what features they like to have in the new system, shown in Table 2. This information helps us prepare a software requirement specification (SRS) and define a new blockchain-based solution.

Table 2. System issues & requirements.

Issues to address in the new system	Requirements for the new system
Data reliability/accuracy must be improved	De-stigmatization, Rules for accountability must be implemented
Data security and encryption are highly required	Secure software and data security is required
Paper scanning issues & old technology issues	Self-enumeration, Thumbprint collection, Offer incentives and penalties
Data digitization	Provide summarized data access to everyone

7   Core Findings

After consolidating the structured interview data and generating insights in the data analysis section, the following findings have been concluded.

- The participants have direct access to census data for imputation, correction and verification of records, which raises the concern of data not being immutable and not traceable, especially regarding where they were originally captured and modified. Participants confirm that only supervisors can change data because imputing missing values is a manual task requiring an expert (supervisor) review.
- The census bureau only counts data from authorized surveyors, and they do not verify the homeless population, especially those without or with lost national identity cards. This raises another issue that could be solved by a blockchain-based system: self-sovereign identity. This allows the user to control their data and have their own digital identity without relying on a third party [18]. Having a schedule for the homeless population does not guarantee that everyone is counted from homeless areas and communities. Some participants mentioned possible groups that may be missed, such as remote areas, transgender people, unregistered orphan houses, and congregational settings. They have data redundancy issues, but they do not use any reliable data source to verify or update their enumeration for missing populations. Therefore, we need a system that allows authorized entities to register on the census portal and support the census to cover the country. The proposed blockchain system should be a consortium or partially private blockchain.
- Table 2 suggests various requirements, but the main ones are reducing political relevance and increasing public awareness. These suggestions imply missed/undercounts due to racial, ethnic or religious biases that deter these groups from joining census activities. Others are implementing data digitalization, reliability, accuracy, security procedures, and upgrading infrastructure with the latest technology.

8   Blockchain-Based National Census Solution Outlook

We propose a permissioned blockchain solution for Pakistan's national census bureau, which addresses the drawbacks identified in the literature review and considers the data sensitivity, privacy and security issues of the census. Figure 4 shows an abstract level of system design and information flow. We chose the Hyperledger Fabric framework for its performance [14] and proven record in record-keeping applications [15–17]. We will collect thumbprint data from individuals who lack national identity cards, live in remote or complex areas, or face discrimination. To include these people in the census, we will involve Non-Government Organizations (NGOs) in the census process and assign them as enumerators. NGOs can join the census either by self-registration or by invitation from the census bureau. The solution outlook is explained below:

1. When the enumerator completes the census survey form on the client application, the collected data and thumbprint image will be stored on InterPlanetary File System (IPFS) system, with an extra layer of encryption method to encrypt the file content. Storing data as a payload on Hyperledger is not ideal, as it affects the system's performance when processing and querying data.
2. The IPFS system will generate and return the hash code of the content location to the client application.
3. The client application initiates the transaction proposal with its identity to the channel via smart contracts.
4. The peer on the channel checks the identity.
5. Peer generates response events back to the client application.
6. The client application submits the transaction to the orderer peer based on a smart contract.

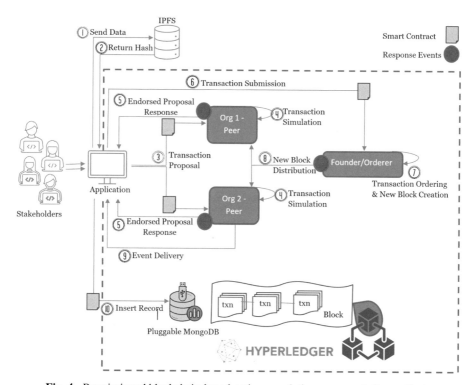

Fig. 4. Permissioned blockchain-based nation population census solution outlook

7. The orderer peer will initiate transaction processing and create a new block.
8. A response event will be generated, and a new block will be distributed to all the peers on the network.
9. Event delivery notification will be further sent to the client application to conclude the transaction.
10. Lastly, the record will also be inserted into the Mongo database for various bureau purposes. This database will be omitted as soon as the Bureau of Census technical team gets first-hand experience with the blockchain client application and peer/query Command Line Interface (CLI).

9 Conclusion

Our research shows that data digitalization, reliability, security, and accessibility stand out with the demand for new features. These challenges arise from the difficulty of reaching some groups of individuals and surveyors who may avoid them due to minority rights or misrepresentation. We propose a permissioned blockchain-based solution that can allocate roles to the humanitarian entities involved and store and maintain data on the individuals who get assistance or aid throughout their lives. Our solution considers national security and data privacy concerns and enhances the accuracy and inclusiveness of the census enumeration by storing the data in a secure and tamper-resistant decentralized storage. Our future work will involve developing a detailed blockchain-based architecture design and covering the technical details of a more robust and scalable solution. We will create and test a pilot solution in some areas of Pakistan. Our objective is to test the system by collecting census data from 250–500 houses and individuals at a block-level of census geolocation and compare the outcomes with the recent population census to assess the quality and coverage of the census data.

References

1. Reynolds, F., John, H. (eds.): The American People: Census 2000, Russell Sage Foundation Census Series (2000)
2. Rasheed, S., Louca, S.: Exploring the need for blockchain-based national population census. In: Themistocleous, M., Papadaki, M. (eds.) Information Systems: 18th European, Mediterranean, and Middle Eastern Conference, EMCIS 2021, Virtual Event, 8–9 Dec 2021, Proceedings, pp. 117–129. Springer International Publishing, Cham (2022). https://doi.org/10.1007/978-3-030-95947-0_9
3. Chen, Y., Bellavitis, C.: Blockchain disruption and decentralized finance: the rise of decentralized business models. J. Bus. Ventur. Insights **13**, e00151 (2020)
4. Scheller, F., Reichelt, D., Dienst, S. et al.: Effects of implementing decentralized business models at a neighborhood energy system level: a model based cross-sectoral analysis. In: 14th International Conference on the European Energy Market (EEM), pp. 1–6. Dresden (2017)
5. Scheller, F., Johanning, S., Seim, S., et al.: Legal framework of decentralized energy business models in germany: challenges and opportunities for municipal utilities. J. Energy Ind. **42**, 207–223 (2018)

6. Themistocleous, M., Christodoulou, K., Iosif, E., Louca, S., Tseas, D.: Blockchain in academia: where do we stand and where do we go? In: The Proceedings of the Fifty-Third Annual Hawaii International Conference on System Sciences, (HICSS 53), 7–10 Jan 2020, Maui, Hawaii, USA. IEEE Computer Society, Los Alamitos (2020)
7. Mattila, J.: The Blockchain Phenomenon – The Disruptive Potential of Distributed Consensus Architectures, ETLA Working Papers 38, The Research Institute of the Finnish Economy (2016)
8. Batubara, F., Ubacht, J., Janssen, M.: Unraveling transparency and accountability in blockchain, In: Proceedings of the 20th Annual International Conference on Digital Government Research, ACM (2019)
9. Alketbi, A., Nasir, Q., Talib, M.: Blockchain for government services – Use cases, security benefits and challenges. In: 15th Learning and Technology Conference (L&T), pp. 112–119. IEEE (2018)
10. BITNATION: Governance 2.0. https://bitnation.co/ (2017). Last accessed 1 July 2021
11. Tirodkar, V., Patil, S.: Proposed infrastructure for census enumeration and internet voting application in digital India with Multichain Blockchain. In: Vasudevan, H., Michalas, A., Shekokar, N., Narvekar, M. (eds.) Advanced Computing Technologies and Applications: Proceedings of 2nd International Conference on Advanced Computing Technologies and Applications—ICACTA 2020, pp. 223–235. Springer Singapore, Singapore (2020). https://doi.org/10.1007/978-981-15-3242-9_22
12. Contacts, Pakistan Bureau of Statistics. https://www.pbs.gov.pk/contacts. Last accessed 7 June 2022
13. Kobo Toolbox. https://www.kobotoolbox.org/. Last accessed 21 June 2022
14. Polge, J., Robert, J., Traon, Y.: Permissioned blockchain frameworks in the industry: a comparison. ICT Express 7(2), 229–233 (2021). https://doi.org/10.1016/j.icte.2020.09.002
15. Walz, A.: Introducing the Alberta Credentials Ecosystem. https://www.evernym.com/blog/alberta-credentials-ecosystem (2019)
16. Ministry of Citizens' Services, BC, British Columbia's Verifiable Organizations. https://orgbook.gov.bc.ca/en/home (2019)
17. Lemieux, V.: Blockchain and public record keeping: of temples, prisons, and the (re)configuration of power. Front. Blockchain 2, 5 (2019)
18. Hendrickson, L.: The Ultimate Guide to SSI: Self-Sovereign Identity Explained, Identity. https://www.identity.com/self-sovereign-identity/ (2023)

Severity Analysis of Web3 Security Vulnerabilities Based on Publicly Bug Reports

Rita Melo[1]($^{(\boxtimes)}$) (iD), Pedro Pinto[1,2] (iD), and António Pinto[2,3] (iD)

[1] Instituto Politécnico de Viana do Castelo, Viana do Castelo, Portugal
`meloana@ipvc.pt`, `pedropinto@estg.ipvc.pt`
[2] INESC TEC, Porto, Portugal
[3] CIICESI, ESTG, Instituto Politécnico do Porto, Porto, Portugal
`apinto@estg.ipp.pt`

Abstract. Web3 has its basis in blockchain and smart contract technologies, supporting secure, distributed, and decentralized applications. Nonetheless, Web3 is still in the process of evolution, and, as with any other software-based product, software bugs, security flaws, and other vulnerabilities are expected to appear. This paper performs a severity analysis of Web3 security vulnerabilities based on publicly available bug reports. Through this analysis, it is feasible to obtain an overview of the evolution and trends regarding the number of reports delivered, growth by platform, severity classification, and the amounts paid for discovering and reporting vulnerabilities.

Keywords: Web3 Security · Survey · Bug bounties

1 Introduction

Web3 is envisioned as a backend revolution that replaces central data storage with a widely distributed environment [9]. As a decentralized web, Web3 can bring advantages such as mitigating problems concerning network vulnerability, misinformation, and information disorder. Web3 allows for creating, identifying, contracting, exchanging, commercializing, and managing public or private content, products, and services [2]. Also, it can provide more security when creating scalable and accessible applications for private information exchange, sharing, and transfer.

Web3 builds on technologies such as Blockchain and smart contracts, among others. A Blockchain is a group of transactions that are part of a shared ledger that has been entered into data storage and has been previously verified by multiple sources. In addition, Blockchain is expected to make data secure and tamper-proof [13]. Ethereum is a decentralized network that can run applications, which can be seen as a set of smart contracts in a distributed environment. The idea is to avoid dependency on a single entity, which must also store and manage users'

© The Author(s), under exclusive license to Springer Nature Switzerland AG 2023
J. M. Machado et al. (Eds.): BLOCKCHAIN 2023, LNNS 778, pp. 154–163, 2023.
https://doi.org/10.1007/978-3-031-45155-3_16

personal and business data [4]. A smart contract is purely electronic, written as code, supported, and enforced by the entire Blockchain system [13].

With the increase of Web3 use, it becomes a more interesting and valuable target for attacks. These attacks are usually targeted against its components, such as the underlying Blockchain or the smart contracts themselves [8]. The problem aggravates if we consider the decentralized finance (DeFI) part of Web3, where financial services and solutions move funds and money. Such funds and money can be stolen by hackers [6]. As with any system and software, users are expected to encounter bugs and flaws that ill-intentioned people might use to exploit it. Examples include the Qubit Finance attack [7] that originated from exploiting a security flaw in a smart contract, which allowed the transfer of a certain amount of money that was never deposited. White hackers upon discovering a software vulnerability or bug, will report them so that the involved entities can correct them. This is widely known as bug-bounties, and some online platforms enable this. The key idea behind bug bounties is to enable better overall security of the online services and platforms that adopt it and for still enable white hackers to be funded by their efforts. These programs offer monetary rewards for discovering and reporting a vulnerability. Reports produced in this context can then be used to improve the security of new or similar existing services.

In this paper, a severity analysis of Web3 vulnerabilities is carried out. This analysis is performed using publicly available bug reports. The contribution of this article involves the collection and classification of the severity of these vulnerabilities, as well as an analysis of their impact over time.

This paper is organized into sections. Section 2 describes related contributions that were made by others. Section 3 describes the adopted methodology, including the platform selection, report collection, analysis and exclusion. Section 4 presents the obtained results and performs its analysis. Section 5 draws the conclusions and identifies future work.

2   Web3 Insecurity

The increased usage of Web3 lead to the growth of related cyber-attacks, with Blockchain and smart contracts being the primary targets. These attacks are primarily based on vulnerability exploitation [8]. Snegireva [12] analysed different vulnerabilities related to Blockchain and smart contracts. A list of recently found vulnerabilities that are publicly disclosed using a Common Vulnerabilities and Exposures (CVE) identifier is also presented. This list includes vulnerabilities in software such as the Bitcoin Core (CVE-2020-14198), the Ethereum virtual machine (CVE-2021-29511), or the Hyperledger product family (CVE-2020-11093), for instance. The analysis concludes that the existing vulnerabilities were either previously in the source codes or have recently emerged due to the development of new systems features, such as the addition of smart contracts.

Sapna et al. [11] examined various types of vulnerabilities and attacks to the Blockchain. They presented real attacks, which occurred between 2016 and 2019, in their explanations of the different vulnerabilities they identified. These real-life attacks had impacts that ranged from tens of thousands of dollars to 60 million dollars. They also compile a list of security solutions and methods that could be used to mitigate security vulnerabilities. They conclude that the analysed attacks demonstrate the insecurity of smart contracts and that it is necessary to continue researching this area to discover new security options and appropriate testing methods. Moreover, they identified a recent increase in attacks in more recent years.

Pise et al. [10] also analysed blockchain-based security vulnerabilities that are specific to smart contracts. They classify the different smart contract vulnerabilities into three main blocks: platform-related, application-related, and code related. The first is related to security flaws that exist in the adopted platform, such as Ethereum or Hyperledger. The second is related to the web applications that usually go along with smart contracts. The third focuses specifically on security flaws in the smart contract code itself. The author specifies each one and presents some safety measures. However, he mentions that with the growth of these systems, it is necessary to develop best practices to combat bugs and future security risks.

Marchesi et al. [5] proposed a security assessment checklist to be used when developing smart contracts. Similarly to Pise, they also classify vulnerabilities into the same three (platform, application, and code). They start their work by identifying and analyzing security patterns for DApps. Next, the authors propose security checklists be applied in different phases of the smart contract development lifecycle. In particular, they propose security assurance checklists for the design, coding, and testing phases. Although this list is being updated over time, this work is based on public sources, focusing on scientific and forum articles. Their work is heavily based on the work of the Open Web Application Security Project (OWASP) and can be summed up to a collection of 48 patterns and best practices for the secure development of smart contracts in Ethereum and Solidity.

Connelly [3] is yet another researcher that addressed smart contract security, in particular, focusing on the Ethereum platform, getting a sense of its current state of security. In his work, Connelly proposes a classification system for smart contracts. Through his analysis, he concluded that smart contracts have immense vulnerabilities, many of which are unknown in their consequences. A key distinction between vulnerabilities and exploits is made, stating that the presence of some sort of insecurity in the code of a smart contract (vulnerability) does not mean that it can be actively exploited, or that its exploitation has a relevant impact. The symbolic execution through Mythril, a code security analysis tool, was adopted in the pursuit of the creation of a digital registry of flawed smart contracts, accompanied by a rating system.

Matulevicius et al. [8] analysed the adequacy of the existing static code analysis tools to Web3 and their smart contracts. To do so, they developed purposely flawed smart contracts and used a set of tools to analyze them. The vulnerabilities in the flawed smart contracts followed a confidentiality, integrity, and availability classification. They concluded that Slither was the best tool in terms of accuracy and one of the best in terms of performance. Nonetheless, the maximum obtained accuracy was 75%, with tools reaching only 25% accuracy. This means that, in the best case, 25% of the vulnerabilities remain undetected by such static analysis tools.

Our analysis differs from the others mainly because of the used sources that consist of publicly available reports of bug-bounties platforms or company audits. This approach made it possible to collect detailed and up-to-date information on security vulnerabilities relating to Web3 that were identified and corrected, as well as to identify the main trends. In addition, data collection and analysis were automated, which enabled the rapid and efficient extraction of accurate data from hundreds of collected reports.

3   Methodology

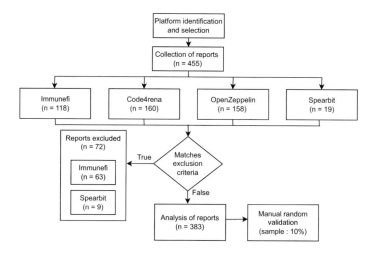

Fig. 1. Adopted methodology

The methodology adopted to perform a systematic collection and analysis of the publicly available reports is shown in Fig. 1. It illustrates all the steps taken, from the platform identification and selection to the final analysis of the reports. First, the identification and selection of relevant platforms were carried out. The selection included bug bounty platforms and platforms that provided reports/audits on Web3 security vulnerabilities. The platform selection was based on the number of available reports and their structure. To be able to perform an automated analysis of such reports, it would be beneficial if such reports had a fixed and predefined structure. This would ease the process of developing code to automate the collection and parsing of data from these reports. The selected platforms were: Immunefi, Code4rena, Spearbit, and OpenZeppelin.

In the next phase, reports were collected from these platforms. This collection was carried out until the end of 2022 since, at the time of collection, there was a low number of results in 2023. There were two data collection approaches for this analysis: manual and automated. The automated approach consists of running a script to analyze and collect the essential data. For the Code4rena and Spearbit platforms, the automated collection was viable, in which a script, adapted per platform, was developed to collect the links of all reports. On the other hand, for the Immunefi and OpenZeppelin platforms, the manual approach was performed due to their non-coherent report structure. After collecting all the links to the reports, the following information was collected: Title, Date, and Amount Paid. Since each platform has a different report format, a script adapted to each source was created to manage the desired data. It should also be noted that the amount paid was only available in the Immunefi platform since it is a bug bounties platform. Their reports state the amount paid for discovering and reporting each vulnerability. In total, 455 reports were collected: 118 from Immunefi, 160 from Code4rena, 158 from OpenZeppelin, and 19 from Spearbit.

Next, all reports were analysed in regards to being within the scope of the analysis, which lead to the exclusion of some reports. Code4rena and Spearbit platforms only report on Web3, so they were all considered within scope. As far as Immunefi and OpenZeppelin are concerned, both platforms include reports that are not relevant, so these were excluded. Data from these sources were collected manually and the report exclusion was also manual. Informative articles, monthly reports, and all types of essays that do not have the objective of explaining/analyzing a vulnerability related to Web3 were excluded. In total, 72 reports were excluded, 63 from Immunefi and 9 from OpenZeppelin.

The remaining 383 reports were then analysed. This step comprises the collection of the assigned severity classification in each report. A script was created to analyze the text and collect the severity classification for each platform. During the development and subsequent execution and test of these scripts, it was feasible to understand that initially, the obtained classification was not accurate due to the presence of different report formats, as well as different word choices for the severity classification, even within the same platform. Such led to the iterative improvement of the scripts. A random manual verification of a sample of 10% of the results was carried out to understand the success rate of the scripts for all platforms and each one individually. In this manual verification, 95.65% of the analysed reports were accurately classified and 4.35% were incorrect. Exploring each source, it was possible to observe that reports from Code4rena, Spearbit, and Immunefi sources had 100%, meaning that all sample reports analysed were correct. As far as the OpenZeppelin platform is concerned, it scored 87%, which means that a small number of reports are incorrect. According to the analysis, this error percentage is likely because this platform presents different report formats and word choices to refer to the severity classification.

Once the data collection phase of the reports was concluded, the severity classification was performed. Each platform uses its scale for severity classification, with most having four levels and Code4rena having only three levels. Table 1 shows the correspondence of the severity classification of each platform to the used severity classification. The severity classifications were normalized, transforming the data to make them comparable.

Table 1. Normalization of severity classifications

Code4rena	Immunefi	OpenZeppelin	Spearbit	Our
Low	Low	Low	Low	Low
Medium	Medium	Medium	Medium	Medium
	High	High	High	High
High	Critical	Critical	Critical	Critical

After the normalization procedure, a script was created and executed that mapped the correspondence according to each platform and each respective classification. Finally, the data was iterated again to leave only the highest severity rating of each report. For instance, if a script initially assigns four levels to one specific report (critical, high, medium, and low), after this final iteration, the only remaining classification level of said report will be critical. This adjustment was made to reduce the number of displayed classifications, highlighting the highest severity level, corresponding to more severe and urgent vulnerabilities.

4 Results and Discussion

After collection and normalization, the data were analysed to perceive the growth, trends, and pattern of the identified vulnerabilities. Despite collecting reports that refer to situations dated until the beginning of 2023, the records from 2023 were not considered in the analysis because these were too few to be statistically significant.

The number of reports per source and severity is shown in Fig. 2a). Considering the Code4rena platform, there are only reports of critical, medium, and low severity, the critical being predominant with 134 reports. Considering the Immunefi platform, there are only reports of high and critical severity, the critical being predominant with 37 reports. Considering the OpenZeppelin platform, there are reports of all levels of severity, the critical being predominant with 35 reports. Finally, considering the Spearbit platform, there are reports of all levels of severity, the critical and high being the predominant ones with 8 reports each. Code4rena has the most significant number of reports, with 158 reports in total, and Spearbit has the smaller number of reports, with 19 reports in total. Immunefi has a total of 39 reports, and OpenZeppelin has 79 reports. It is also noticeable that in all platforms, the predominant severity level is the critical one, so it is possible to conclude that most problems encountered on the analysed platforms have critical levels of severity.

The number of reports by date and severity is shown in Fig. 2b). The dates are organized in semesters and this figure has the purpose of depicting its evolution and trend over time. The period analysed comprises reports dated from the second half of 2016, up to the second half of 2022. Analyzing the low severity level, it has never had a very high number over the years, with its peak being in the first half of 2020, 2021, and 2022 when it reached 8 reports. The reasoning

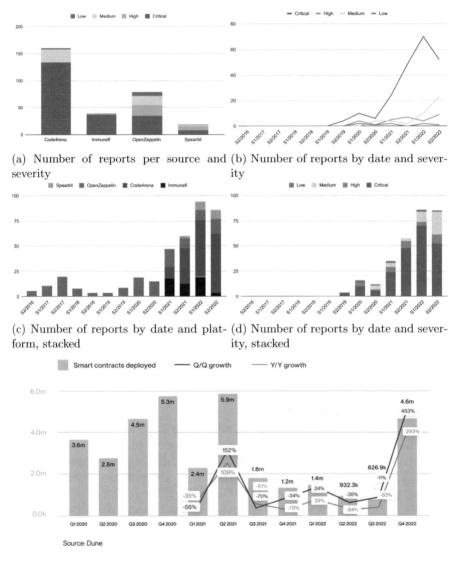

(a) Number of reports per source and severity

(b) Number of reports by date and severity

(c) Number of reports by date and platform, stacked

(d) Number of reports by date and severity, stacked

(e) Smart contracts deployed in the Ethereum [1]

Fig. 2. Trends and statistics regarding Web3 vulnerabilities

is that low-level vulnerabilities might not be appealing to be reported from the security researchers' point of view. Analyzing the medium level, it has been increasing in recent years, and its peak is in the second half of 2022 when it reached 23 reports. Analyzing the high level, it shows an increasing trend over the years. Particularly in 2022, it is possible to observe a steady increase, peaking in the second half of 2022 when it reached 9 reports. Analyzing the critical level, it is noticeable that, over the last few years, there has been significant growth in the number of reports, peaking in the first half of 2022 with 70 reports. Overall, from 2020 onwards, there was a significant increase, which may have been triggered by the forced digitalization brought about by the COVID-19 pandemic. In summary, a significant growth trend in the number of critical reports can be identified.

The stacked number of reports by date and severity is shown in Fig. 2d). It shows the total number of reports per semester over the years. There was a peak in the first semester of 2022, in which there was a more significant number of critical reports, precisely 70. The total number of reports made in the two semesters of 202 is very similar and it can be concluded that there is a steady increase in the total number of reports per year, which is also true if we only consider the critical reports.

It is important to note that OpenZeppelin was founded in 2015, Spearbit in 2021, Code4rena in 2021, and Immunefi in 2020, only the oldest reports refer to OpenZeppelin, given that the other platforms were founded more recently. The total number of reports by date and the platform is shown in Fig. 2c). It shows that the OpenZeppelin platform has the longest database of records, starting in the first semester of 2016 and up to the end of 2022. Code4rena, Spearbit, and Immunefi sources have only recently begun to receive reports. Comparing this Fig. 2c) with Fig. 2d), it is possible to identify a peak in the same period, the first semester of 2022. Worthy of note is the fact that Fig. 2d) does not include reports between 2016 and the first semester of 2019, while Fig. 2c) does. In particular, the reports collected from the OpenZeppelin platform in such a period, do not include a severity classification. The OpenZeppelin platform only more recently started to include the classification in their reports. In summary, there is a clear growing trend over the years and the Code4rena platform has amassed more reports in recent years when compared to the other platforms.

Security researchers receive bounties when they report real security vulnerabilities. The total amount paid for the reports analysed was only available in the Immunefi platform and the values obtained were in dollars, 3,162,347 in S1/2021, 3,608,515 in S2/2021, and 24,215,042 in S1/2022. These values show a steep increase in the revenues paid, more precisely, an increase of more than 20,000,000\$ in the first half of 2022. Concerning the second half of 2022, there is a significant decrease in the total amount paid, contrasting with a similar total number of reports. Nonetheless, the number of reports on the Immunefi platform also decreases significantly, justifying the results.

Considering the results presented, there has been a significant increase in reports over the last few years, with a peak in the first half of 2022, when all platforms received reports. In addition to the high number of reports, the growth of critical reports is also visible, demonstrating that more and more vulnerabilities emerge that pose a risk to the security of smart contracts. Concerning the amounts paid for discovering and reporting vulnerabilities, these peaked in the same period as the remaining amounts, with a notable rise.

The Alchemy's report [1] includes the number of recently deployed smart contracts, per quartile, as shown in Fig. 2e). Considering only the last quartile of 2022, it shows an increase of more than 453%, corresponding to a total of 4.6 million smart contracts deployed. Still, the numbers show a significant increase in smart contract deployment in 2021 and a significant overall decline in 2022, when compared to 2021. On the other hand, there was a substantial increase in the detection of vulnerabilities in 2022. This leads us to conclude that, proportionately, there are more smart contracts with security concerns.

In short, there is a growing trend regarding the number of reports that transpires into reports with critical classification. Consequently, the amounts paid regarding the reporting of such vulnerabilities are also increasing.

5   Conclusions

Web3's purpose is to decentralize the network and allow users to control their online activities. The drawback is the vulnerabilities and their respective consequences, which can seriously affect this entire system and have a strong negative impact. By carrying out this work, it was possible to analyze the number of reported vulnerabilities in smart contracts, presenting total values and trends regarding the number of reports, severity, and financial return. Through the analysis, it was feasible to verify that over the last few years, there has been a significant increase in the total number of reports and critical reports. The growth in amounts paid to security researchers is also visible. It was possible to understand that this increase is also due to the high delivery of smart contracts in the market. Web3 security vulnerabilities are being constantly discovered and reported and this demonstrates the continuous evolution of Web3 and the need to explore and find solutions to combat its vulnerabilities.

In the near future, a better categorization of the collected reports, now considering the used code vulnerability, is expected to be pursued to develop a classification of the main pitfalls, helping Web3 developers regarding security. In addition, it would be interesting to produce an audit checklist for smart contract development based on these reports.

Acknowledgements. This work was partially supported by the Norte Portugal Regional Operational Programme (NORTE 2020), under the PORTUGAL 2020 Partnership Agreement, through the European Regional Development Fund (ERDF), within the project "Cybers SeC IP" (NORTE-01-0145-FEDER-000044).

References

1. Alchemy: Web3 Developer Report Q4 2022 (2022). https://www.alchemy.com/blog/web3-developer-report-q4-2022
2. Cao, L.: Decentralized AI: edge intelligence and smart blockchain, metaverse, Web3, and DeSci. IEEE Intell. Syst. **37**(3), 6–19 (2022). https://doi.org/10.1109/MIS.2022.3181504
3. Connelly, D.S.: Smart contract vulnerabilities on the ethereum blockchain: a current perspective. Ph.D. thesis, Portland State University (2020)

4. Gupta, R.: Hands-On Cybersecurity with Blockchain: Implement DDoS Protection, PKI-Based Identity, 2FA, and DNS Security Using Blockchain. Packt Publishing (2018). https://books.google.pt/books?id=upBiDwAAQBAJ
5. Marchesi, L., Marchesi, M., Pompianu, L., Tonelli, R.: Security checklists for ethereum smart contract development: patterns and best practices (2020). https://arxiv.org/pdf/2008.04761.pdf
6. Lucas, G.: Poly Network attack: here's what happened to biggest DeFi hack in history. CoinGeek (2021). https://coingeek.com/poly-network-attack-heres-what-happened-to-biggest-defi-hack-in-history/
7. Malwa, S.: DeFi protocol qubit finance exploited for $80M. CoinDesk (2022). https://www.coindesk.com/markets/2022/01/28/defi-protocol-qubit-finance-exploited-for-80m/
8. Matulevicius, N., Cordeiro, L.C.: Verifying security vulnerabilities for blockchain-based smart contracts. In: 2021 XI Brazilian Symposium on Computing Systems Engineering (SBESC), pp. 1–8 (2021). https://doi.org/10.1109/SBESC53686.2021.9628229
9. Park, A., Wilson, M., Robson, K., Demetis, D., Kietzmann, J.: Interoperability: our exciting and terrifying Web3 future. Bus. Horiz. **66**(4), 529–541 (2022). https://doi.org/10.1016/j.bushor.2022.10.005. https://www.sciencedirect.com/science/article/pii/S0007681322001318
10. Pise, R., Patil, S.: A deep dive into blockchain-based smart contract-specific security vulnerabilities. In: 2022 IEEE International Conference on Blockchain and Distributed Systems Security (ICBDS), pp. 1–6 (2022). https://doi.org/10.1109/ICBDS53701.2022.9935949
11. Sapna, Prashar, D.: Analysis on blockchain vulnerabilities & attacks on wallet. In: 2021 3rd International Conference on Advances in Computing, Communication Control and Networking (ICAC3N), pp. 1515–1521 (2021). https://doi.org/10.1109/ICAC3N53548.2021.9725403
12. Snegireva, D.A.: Review of modern vulnerabilities in blockchain systems. In: 2021 International Conference on Quality Management, Transport and Information Security, Information Technologies (IT&QM&IS), pp. 117–121 (2021). https://doi.org/10.1109/ITQMIS53292.2021.9642862
13. William, J.: Blockchain: The Simple Guide to Everything You Need to Know. CreateSpace Independent Publishing Platform (2016). https://books.google.pt/books?id=xYauDAEACAAJ

Virtuous Data Monetisation Cycle: A Hybrid Consensus Substrate Automotive Consortium Blockchain Solution

Cyril Naves Samuel[1,2]([✉]), François Verdier[1], Severine Glock[2],
and Patricia Guitton-Ouhamou[2]

[1] Université Côte d'Azur, LEAT/CNRS UMR 7248, Sophia Antipolis, Nice, France
`francois.verdier@univ-cotedazur.fr`,
`cyril-naves.samuel@etu.univ-cotedazur.fr`
[2] Renault Software Factory, Sophia Antipolis, Valbonne, France
{`cyril-naves.samuel,severine.glock,patricia.guitton-ouhamou`}`@renault.com`

Abstract. We propose a data monetization architecture based on the Substrate blockchain framework for leveraging automotive radar data. The architecture is designed to enable a virtuous economic cycle involving the Radar automotive data for the consortium members of Automotive enterprises, Vehicle owners, and Radar Equipment manufacturers. The data represented as Non-Fungible Token (NFT) is sourced from the vehicle owners by the Radar component manufacturers and enhanced upon them. It is offered as a token road signature service (NFT) back to the vehicle owners completing the virtuous economic loop. All the participant interactions are monetized and incentivized by dynamic pricing and commission sharing. Data interactions are certified for integrity, and reviews for each NFT are maintained in the blockchain. Architecture respects the fairness, privacy concerns, as well as fidelity aspects. Architecture is implemented in the Substrate Blockchain framework and tested for hybrid consensus scalability of Proof of Authority algorithms. The algorithms evaluated are Aura and BABE, along with GRANDPA for block authoring and finalization, respectively, in a cloud-based implementation that suits the enterprise consortium networks.

Keywords: Data Monetisation · Non Fungible Token · ERC 721 · Blockchain · Consensus · AuRa · BABE · GRANDPA

1 Introduction

"Data is the new oil" or "Information is the oil of the 21st century, and analytics is the combustion engine" [1] are breakthrough analogical phrase notions adopted by various organizations like Apple, Amazon, Facebook, Alphabet or automotive innovators like Waymo, Tesla, Renault as they embark the digital age of autonomous driving (ADAS). ADAS systems learn and predict using data generated from a fleet of vehicles where each autonomous vehicle can generate upto

© The Author(s), under exclusive license to Springer Nature Switzerland AG 2023
J. M. Machado et al. (Eds.): BLOCKCHAIN 2023, LNNS 778, pp. 164–174, 2023.
https://doi.org/10.1007/978-3-031-45155-3_17

300 Terabytes of data per year comprising of different sensors like Radar, Light Detection and Ranging (LIDAR), Camera, Ultrasonic, Vehicle motion, Global Navigation Satellite System (GNSS) and Inertial Measurement Unit (IMU) [2]. Mckinsey has analyzed incentivization through data monetization, which states it as a differentiator and is still nascent stage [3]. They also explore that with the available data, it is necessary to engage with other partners to create new business ecosystems, and it is essential to dissolve sectoral borders. This is one of the business models we would approach in our architecture to create an ecosystem of interconnected businesses.

1.1 Blockchain Framework by Design: Substrate

We choose Substrate [4], a third-generation blockchain like Cosmos, Polkadot, Cardano, or Avalanche. But to be precise, it is not a systemized blockchain node but a 'modular, extensible framework' which we can utilize to create our custom blockchain node and generate its binary. It can be interoperable and scalable in parachains and building smart contracts directly onto the blockchain node instead of external deployment on EVM. All the signed transactions executed in the smart contract pallets are invariably called Extrinsic in Substrate terminology.

2 State of Art

In this section, we compare and contrast two decentralized Mobility Data solutions already in production with a modest market share thriving in an ecosystem of partners for offering "Return on Data" shared with these networks.

- **Digital Infrastructure for Moving Objects (DIMO)** [13]: It is a data-driven decentralized public IoT platform that requires the vehicle owners to install an AutoPi telematics unit in their vehicle, which then submits the data at frequent intervals to the cloud infrastructure. The compromise we notice with this System is that sharing the data is a benefit. Still, the data lifecycle is a question regarding who procures it, data destruction, data obfuscation by removing sensitive details, and the overhead of installing a device for protocol communication. It was envisaged to solve the problems of centralization and as an alternative to prominent vehicle data aggregators Otonomo and Wejo. It has privacy concerns that need more understanding as the data is ported to centralized actors in the ecosystem.
- **Ocean Protocol**: Ocean Protocol [5] is a Data marketplace built on parity Ethereum Proof of Authority, an EVM-based public Blockchain protocol. It is a decentralized data exchange where individuals or enterprises share the data shared as ERC721 data NFT tokens by the Ocean smart contracts. Then the ERC20 data tokens are generated for data service to access the published data for a dynamic or fixed price. This protocol is quite comprehensive, with no additional hardware requirements, but there is a problem with the data certification. It is an issue as data shared needs to be tracked for its provenance, quality, and genuineness with pre-emptive checks.

A credible data trading system for IoT data minimizing the risk of fraud and transactions is proposed in [14] by twentyfive. The System permits the data producers and consumers to agree on data and settle the payment on the chain. A Credit mechanism is developed to lower the fees incurred during the private Ethereum network implementation participation process. This System's objectives are Consumer Fairness, where customers do not pay for data not received; Producer Fairness ensuring the minimal risk of data loss; and Privacy, limiting data visibility at a minimal operational cost. The System is evaluated regarding gas consumption and incurred blockchain transactions for the trading scheme, as the idea is to minimize the transaction cost related to transaction Gas complexity. They estimate around 35000 units of gas for fund deposit transactions and higher for receipt transactions as it needs data signature checking. Our data-monetization architecture is designed to solve the abovementioned concerns of data provenance, certification streamlined data flow with no additional hardware required. It will also ensure data privacy within a consortium blockchain of agreed and verified partners with mutual benefit for everyone in the network.

3   Data Monetisation Use Case

In this section, we elucidate the use case of creating an automotive radar data value chain by incentivization and fidelity mechanism. Automotive RADAR furnishes the essential range and speed data for driver assistance systems, including Long Range RADAR (LRR) adaptive cruise control, automatic emergency braking, cross traffic alert, lane change assist, and Short Range RADAR (SRR) parking aid, as well as pedestrian detection. Road signature offered as a subscription service back to the vehicle user community [6] is a crowd-sourced localization service for autonomous vehicles to detect the relative position of other vehicles and objects in its environment for accurate and reliable localization. The vehicle generates RADAR and video sensor data while in motion, which will be collated at determined intervals via the Telematics Control Unit of the vehicle to the Cloud of RADAR OEM (Original Equipment Manufacturer). Radar OEM observes all these received data of RADAR and video to extract the object details and regenerate the environment based on the data. Video and RADAR sensor data complement the rendering of a localization environment, and the RADAR OEM, e.g., Continental and Bosch, integrates this data with map data to produce an updated and globally consistent map.

Virtuous Cycle of Road RADAR Signature Data: As represented in Fig. 1 the Road Radar signature workflow is facilitated and augmented with acknowledgment, data certification, transparency, privacy, dynamic pricing mechanism accompanied by fidelity and incentivization mechanism. Our data monetization use case will involve the following components realized through the development of pallet smart contracts on the Substrate blockchain framework.

- **Asset Component:** This component creates an asset vehicle OEM as Non-Fungible Ethereum Request for Comments (ERC) 721 token in the network.

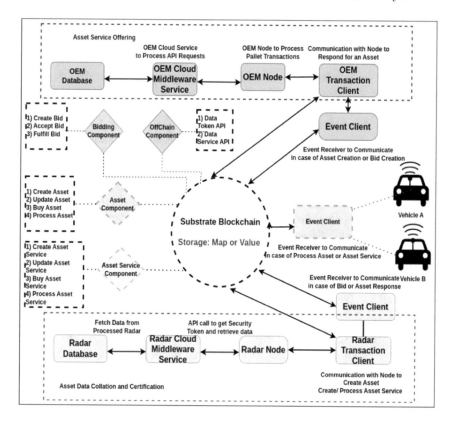

Fig. 1. Automotive Data Monetisation Use Case Flow

RADAR Data of the vehicles for a defined condition of localization, time period, error tolerance, and other miscellaneous meta details is represented as an asset along with its floor and ceiling price.

- **Asset Service Component:** The component is similar to the asset component but consecrated towards the Road Radar Signature Map, which is offered by the RADAR OEM to the vehicle users for enhanced safety, autonomous and optimized localization information which is accompanied by a ceiling and floor price

- **Bidding Component:** Each Asset or Asset Service created as an ERC 721 token has its price determined based on a transparent bidding process between the RADAR OEM, Vehicle OEM, and Vehicle Users consented to the transaction.

- **OffChain Component:** As soon as the Asset or Asset Service is sold, then data has to be offloaded from each vehicle to the RADAR OEM via Vehicle OEM acting as a guarantor or mediator in the blockchain for initial data collation, data cleaning and obfuscating any sensitive or private data which is done off-chain in a cloud middleware but close looped with the blockchain for certification of the process and the data.

- **Asset Data Collation and Certification:** Offloaded data to the RADAR OEM is then processed, analyzed, and interpolated with Map data to be offered as a service later completing the asset finalization along with certification along its entire process.
- **Asset Service Offering:** Road RADAR Map signature is offered to the consenting vehicle back in the blockchain ecosystem with the interest of subscriptions collated in the ledger by the Vehicle OEM and transaction completion post the fund transfer.
- **Incentivisation for Everyone:** For Each transaction of Asset or Asset Service involving RADAR data monetization and Road Radar Signature Service, respectively, the monetary transaction attributed the commission to the Vehicle OEM and the Vehicle Users and the Radar OEM the necessary monetary benefit given that the data certification, anonymity, and quality is maintained. Moreover, the RADAR data or RADAR signature service represented as tokens ascend in higher price appreciation based on its usage review is submitted to the smart contract.

This completes the explanation of the virtuous data cycle since the inception as an asset from the vehicle to its collation in the RADAR cloud and close looping back as an asset service to the vehicle users based on the shared data imbibing the fidelity as well a virtuous mutual sustained benefit for everyone in the monetization ecosystem.

4   Data Monetization Solution Workflow

This section looks at the data monetization solution flow using a smart contract transaction sequence diagram for the Non-Fungible Token-based Asset and Asset Service proposition scenarios. The Blockchain ecosystem network is composed of nodes representing the Vehicle_OEM and Radar_OEM as validators with the possibility of extending the membership to other members interested in data like for example, the map provider or the government service as it deals with traffic and road, toll maintenance. The blockchain network constructed is of a private consortium type where each participant is aware of the public key of the other actors, as each one has a crypto wallet secured by a public cryptography system.

Tokenized Non-Fungible Asset Data-Set Component: The Smart Contract Transaction involving the Offchain Sequence for NFT Asset component involving RADAR data transfer is represented in Fig. 2 for the radar asset and radar asset service process is explained as follows:

1. **Data Demand Phase:** The initial phase when the Radar_OEM needs the Radar data publishes the demand as an event notification transaction onto the network with the location, vehicle speed, and acceptable price limit requirements. The Vehicle_OEM can then respond to the event by broadcasting the demand to its vehicle clients for participation consent.

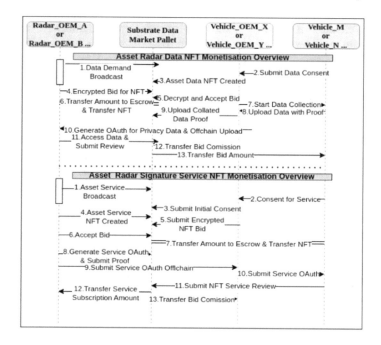

Fig. 2. Data Monetisation Workflow Overview

2. **Data Asset Offering and Bid Phase:** Then in response Vehicle_OEM_X or Y responds individually by creating an ERC 721 token as an Asset with the criteria as well as floor and roof price. Then to maintain the bid privacy, the Radar_OEMs submit an encrypted bid with the public key of Vehicle_OEM_X or Y respectively individually. Then among the bids, the one of RADAR_OEM_A is accepted by the Vehicle_OEM_X as the encrypted bid of RADAR_OEM_B is not an acceptable price. The fund of RADAR_OEM_A for the bid is transferred to the escrow account as an intermediate transfer which will be transferred to the Vehicle_OEM_X when the criteria of the bid and the asset are respected along with necessary certification proof submissions.

3. **Data Aggregation and Certification:** The Vehicle_OEM_X starts the data collection job from its set of vehicle clients with the necessary criteria. Vehicles start recording Radar data and submitting it with its signature hash as proof to the data pool and smart contract. Then Vehicle_OEM_X processes the data removing the user's sensitive identification data, submits the final hash proof to the ledger, and generates a data access token to a middleware service to be retrieved later.

4. **OffChain Data Transfer and OAuth Access:** The blockchain Off-chain worker component then retrieves the OAuth [7] access token from the middleware and submits as an API POST request to the RADAR_OEM_A if the Vehicle_OEM_X submits all the certification proofs to the blockchain.

5. **Fund Transfer including Finalisation of Asset Transfer:** Then the bid amount is transferred by the smart contract from the Escrow to the Vehicle_OEM_X account and Vehicle_Accounts who participated in this monetization process such that the above condition of the access token and proof are respected

6. **Review of transferred Asset:** The RADAR_OEM_A who retrieves the processed data from the data pool utilizing the OAuth access token reviews the data. The result of the review process is submitted to the blockchain against the Asset ERC721 token issued, which is a fidelity action to augment the price of the data issued by the Vehicle_OEM_X in a later transaction or diminish if the review score is bad.

Tokenized Non-Fungible Asset Data-Service Component: The Smart Contract transaction sequence for the Asset Service token is represented in Fig. 2 for the bidding process similar to the asset component and Asset Service finalization for the Radar Signature Map subscription service. This is similar to the earlier Asset Data transaction workflow explained but is in the other sense where the vehicles subscribe to the service offered by the Radar_OEM_A *or* B.

5   Functional Evaluation

Our virtuous cycle of data monetization since the generation from the vehicle traversing Vehicle_OEM to the Radar_OEM and offering a close loop Road Radar Signature service based on the received data improvisation has the functional properties as represented below:

- **Global Fairness:** Fairness in the case of any application is defined by the work of FairSwap [8] "A fair exchange protocol allows a sender S to sell a digital commodity x for a fixed price p to a receiver R. The protocol is secure if R only pays if he receives the correct x." We extend this work in our above design as in [9] for a Global Fairness where every participant in our ecosystem will have the following guarantees: 1) Each Participant is ensured that fund transfer for any $B_{x,y,S}$ happens only when the certification conditions are satisfied either for C_D or C_{Rs}. 2) Fund transfer is not fulfilled immediately between Vo, Ro, or V as it is placed in an escrow account ϵ and then transferred to the destined participant account on verification. In either case of the virtuous cycle from Radar Data conversion to Road Radar Signature, the facilitators who provision the cloud infrastructure or other value adders can benefit from the commission or fidelity rewards.
- **Full Interoperability:** Our solution achieves two levels of interoperability which are intra-chain as well as inter-chain, as follows: 1) **Intra-Chain:** The above monetization solution can integrate with external agnostic (centralized or decentralized) systems through OffChain worker for API requests or responses. Also, as we utilize a balanced mix of events, signed and unsigned transactions (extrinsic), to differentiate the priority of message passing between each actor in the system, it avoids unnecessary overhead in the

distributed system, especially of consensus. 2)**Inter-Chain:** As it is based on the Substrate framework, this implementation can be extended as a para-chain to integrate with other blockchains or para-chains through Polkadot.

– **Chained Data Certification:** For Asset Radar data or Asset Service Road Radar Signature, we ensure the maintenance of the history of the data prove-nance along with its hash signature-based certification we avoid any counter-feit and validate the health of the data as well as the service.

– **Privacy By Design:** The solution has granularised the privacy at each level by the following mechanisms:

  • **Account Pseudonymization:** All the accounts are pseudonymized con-cerning vehicle participants as it is necessary to identify Radar_OEM's and Vehicle_OEM's. A vehicle's original identity is another intermediate identity that can be unrelated in case of necessity. This is to respect the right to forgetting GDPR [10] if the vehicle owner decides to remove his information.

  • **Privacy Data Concealment:** All the bids for either Asset or Asset Ser-vice are sealed using encryption which can be decrypted only by the bid receiver, ensuring competitiveness, transparency, and privacy. Radar Data from the origin in the vehicle until the data reception by the Radar_OEM is privacy treated with the making of sensitive data, and the hash signa-ture generated is added with controlled noise like location, time, etc., to ensure authenticity.

– **Dynamic Pricing:** As discussed earlier, pricing is sealed bid oriented with direct proportion to the participant's reputation and review of the exchanged asset and asset service data tokens it avoids any price manipulation. This ensures a reputation-based adjustment of the asset pricing, creating a virtuous cycle in fidelity and incentivization.

– **Economic Analysis:** The proposed solution offers advantages to vehicle users, vehicle as well as Radar OEM. In our solution, we consider only limited vehicle OEM and Radar OEM, but it can be extended to multiple OEMs, which offer better price offerings and data collection. The Vehicle user can get a better price for their data and a discount on the data service offering. Vehicle OEMs can get a commission during each asset or asset service transaction. The Radar OEMs can get varied data for training their prediction service, offering an optimized signature service.

– **Legal Analysis:** The ERC721 smart contract implemented in our data mon-etization solution is evaluated from technical performance, but the legal con-straint is not considered. In the case of enterprise implementation, the require-ment for a legal audit [15] by either a legislative framework or regulation is to be performed in the future.

6 Experimental Evaluation

We evaluate the data monetization architecture implemented in the Substrate framework through embedded smart contract pallets and other networking,

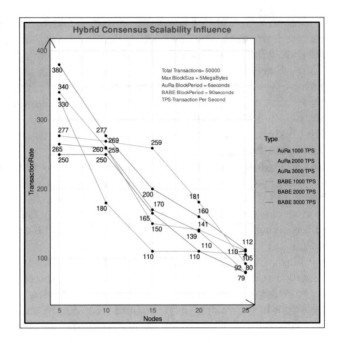

Fig. 3. Hybrid Consensus Scalability Influence

transaction processing, and consensus blockchain functionalities. We implement the Radar OEM and Automotive OEM middleware for Java-based data processing and off-chain interactions. The blockchain client and the middleware are transformed into a Docker container and deployed using Kubernetes on the TAS cloud. TAS is a public cloud hosting company based in Sophia Antipolis, France. The substrate network is tested with different configurations of 5, 10, 15, 20, and 25 validator nodes that participate in the variations of 1) AuRa with GRANDPA and 2) BABE with GRANDPA. A client which submits the transaction to the load balancer offloads the transaction to a node chosen based on the round-robin algorithm. For the test, the client submits the transaction of *CreateAsset*, which creates an Asset NFT ERC 721 token in the network. Transactions (Extrinsic) are submitted by the multithreaded client of 1000 threads signed by 2000 pre-generated accounts. The two-hybrid consensus algorithms of Block proposer: AuRa (Authority Round) or BABE (Blind Assignment for Blockchain Extension) and Block finalization based on chain level agreement: GRANDPA (GHOST-based Recursive Ancestor Deriving Prefix Agreement) are tested. The Substrate data monetization smart contract, the Middleware, and the cloud deployment implementations are publicly available at: https://github.com/scyrilnaves/article-datamonetisation.

6.1 Hybrid Consensus Analysis

In this section, we analyze the scalability of AuRa and BABE consensus and the GRANDPA finalization algorithm. The scalability results for AuRA and BABE are represented in Fig. 3. The results of AuRa are explained by the $O(n^2)$ message complexity which shows a decreased throughput in proportion to the number of nodes. In addition, the GRANDPA finalization message complexity bottleneck of $O(n^2)$ further augments the decrease in throughput. Still, the difference in this algorithm is that finalization is performed on the chain of blocks rather than individual blocks of AuRa. Also, another inference based on the result is that higher input Transaction per second (TPS) greater than the optimum 1000 increases the forks in the network as well, decreasing the throughput. This fork is explained by the processing and consensus-induced bottleneck affecting the liveness and consistency of chains but resolved by GRANDPA at additional computation.

BABE algorithm has an algorithm message complexity of $O(n)$ but suffers a decrease in scalability throughput due to the occurrence of forks in the system, given the susceptibility of the consensus algorithm to elect multiple validators for a single block height. This results in a larger number of secondary blocks rather than primaries which are attributed to the nature of the algorithm. Accompanied by the computation effort of the Verifiable Random Function, which affects the throughput of BABE while absent in the AuRa algorithm. We realize that the BFT algorithms of Aura and BABE have a scalability issue. Still, due to its hybrid nature with GRANDPA, it offers finalization and better protocol security. Aura consensus is more stable than the BABE protocol. BABE, along with GRANDPA protocol, suffers from consistency issues [11,12] as explained earlier but offers a higher degree of security in the form of impossibility to predict the next successive block proposer in the form of verifiable random function.

7 Conclusion

In this work, we analyzed the data monetization use case by creating a virtuous cycle of data flow incentivizing the Vehicle_OEM, Radar_OEM, and Vehicles who share the data. We exhibit the extensibility of the architecture to external systems and provide Radar Signature Services offering global fairness, interoperability, privacy by design, asset data reputation, and a secured network. We further evaluate the cloud-based implementation and understand the Substrate Hybrid Consensus from an application-based blockchain with the embedded smart contract, including its BFT family consensus scalability limitations. The performance throughput results for the above architecture are ideal for enterprise production implementation while also assuring the finalization of the chain, which is vital to construct any monetization scheme. As part of future improvements which can further enhance the data monetization protocol, we consider the following: 1) Auction: Auctions can be made more transparent by a commit reveal scheme where no time-bound constraints and hidden bids are eventually public when all the bids are received and committed. 2) Privacy: Differential Privacy [16] can be applied for more granular privacy control on the data ensuring more concealment than the masking techniques.

References

1. Sondergaard, P.: Gartner Symposium/ITxpo, October 2011. https://www.causeweb.org/cause/resources/library/r2493. Accessed 05 Apr 2023
2. Rossi, T.: Autonomous and ADAS test cars generate hundreds of TB of data daily. Tuxera (n.d.). https://www.tuxera.com/blog/autonomous-and-adas-test-cars-produce-over-11-tb-of-data-per-day/. Accessed 05 Apr 2023
3. McKinsey. (n.d.). Fueling growth through data monetization. McKinsey. https://www.mckinsey.com/capabilities/quantumblack/our-insights/fueling-growth-through-data-monetization. Accessed 04 Apr 2023
4. Parity Technologies. (n.d.). Substrate Technology. Parity Technologies. https://substrate.io/technology/. Accessed 07 Apr 2023
5. Ocean Protocol Foundation, Ocean Protocol: Tools for the Web3 Data Economy. https://oceanprotocol.com/tech-whitepaper.pdf. Accessed 24 Mar 2023
6. Bosch: Road Signature. https://www.bosch-mobility.com/en/solutions/automated-driving/road-signature/. Accessed 07 Apr 2023
7. Hardt, D.: RFC 6749: The OAuth 2.0 Authorization Framework. RFC Editor, 2012. https://www.rfc-editor.org/rfc/rfc6749. Accessed 12 Apr 2023
8. Dziembowski, S., Eckey, L., Faust, S.: Fairswap: how to fairly exchange digital goods. In: Proceedings of the 2018 ACM SIGSAC Conference on Computer and Communications Security, pp. 967–984 (2018). https://doi.org/10.1145/3243734.3243857
9. Banerjee, P., Ruj, S.: Blockchain Enabled Data Marketplace - Design and Challenges, November 2018. arXiv e-prints, arXiv:1811.11462
10. European Commission: 2018 reform of EU data protection rules. http://eur-lex.europa.eu/legal-content/EN/TXT/PDF/?uri=CELEX:32016R0679&from=EN. Accessed 4 June 2019
11. Stack Exchange: BABE GRANDPA Stalled. https://substrate.stackexchange.com/questions/214/recovering-from-stalled-finality-babe-grandpa. Accessed 17 Apr 2023
12. Wang, Y.: The adversary capabilities in practical byzantine fault tolerance. In: Roman, R., Zhou, J. (eds.) STM 2021. LNCS, vol. 13075, pp. 20–39. Springer, Cham (2021). https://doi.org/10.1007/978-3-030-91859-0_2
13. DIMO, Data Loss Horror, DIMO (2022). https://docs.dimo.zone/overview/introduction/what-is-dimo. Accessed 05 Apr 2023
14. Meijers, J., Putra, G.D., Kotsialou, G., Kanhere, S.S., Veneris, A.: Cost-effective blockchain-based IoT data marketplaces with a credit invariant. In: 2021 IEEE International Conference on Blockchain and Cryptocurrency (ICBC), 2021, pp. 1–9 (2021). https://doi.org/10.1109/ICBC51069.2021.9461127. Accessed 07 Nov 2022
15. Raskin, M.: The law and legality of smart contracts, 1 Georgetown Law Technology Review, vol. 304, 22 September 2016. SSRN. https://ssrn.com/abstract=2959166, https://doi.org/10.2139/ssrn.2842258
16. Desfontaines, D., Pejó, B.: SoK: differential privacies. In: Proceedings on Privacy Enhancing Technologies, vol. 2020, pp. 288–313 (2019). Accessed 18 Apr 2023

Blockchain and Sustainability in the Public Sector: A Risk Management Perspective

Fernando Escobar$^{(\boxtimes)}$ ⓘ, Henrique Santos ⓘ, and Teresa Pereira ⓘ

ALGORITMI Research Centre/LASI, University of Minho, Guimarães, Portugal
fernando.escobar.br@gmail.com

Abstract. Blockchain (BC) adoption in the public sector can potentially address sustainability issues. Thus, blockchain is seen as an enabler and a booster of sustainability and Sustainable Development Goals accomplishments. However, BC adoption in the public sector brings challenges concerning risk management. This work presents insights from 19 expert interviews organised as a risk management model for blockchain adoption and sustainability in the public sector, following the risk management process from the standard ISO 31000. The model describes nine identified risks, five proposed treatments, and five metrics for monitoring. This will support scholars and practitioners focusing on developing knowledge about adopting BC under the perspective of sustainability and the contribution to the risk management body of knowledge.

Keywords: blockchain · public sector · sustainability · risk management

1 Introduction

Beyond cryptocurrencies, blockchain (BC) can be adopted in diverse areas where a "trust" problem is observed [48], making it a promising technology to transform many activities [8] and contributing to the digital transformation in perhaps all industries [30]. One of the many relevant industries is the public sector, where BC's characteristics can promote fundamental changes [8,43].

As an emerging technology, blockchain has the potential to address issues of sustainability [14,26] and contribute to a more sustainable world [40]. In this vein, sustainability is the cornerstone of BC models within the United Nations' 2030 Agenda [18] where BC is seen as an enabler [32] and a booster [26] of sustainability and Sustainable Development Goals (SDGs) accomplishments.

Although it can improve sustainable development, BC can bring challenges concerning risk management (RM) [35,38]. Over the years, the interest of academics in RM associated with BC adoption has grown [21]. However, the focus has mostly occurred on the health/supply chain segments [16], lacking studies that generally assess risks, which can be applied to other domains, namely sustainability in the public sector [38]. This paper presents an approach to risk

J. M. Machado et al. (Eds.): BLOCKCHAIN 2023, LNNS 778, pp. 175–185, 2023.
https://doi.org/10.1007/978-3-031-45155-3_18

management based on insights from 19 experts to understand how BC can bring sustainability effects in the public sector.

The paper is structured as follows. Section 2 presents a background and related works. Section 3 describes the method adopted and Sect. 4 details the risk management model. The work ends with a discussion and conclusion in Sect. 5.

2 Background and Related Works

BC can play a relevant impact in sustainability and achieving the SDGs [32] as reported by diverse works. For instance, BC can contribute to SDGs achievements, being associated with other emerging technologies (e.g., Artificial Intelligence, Big Data, Internet of Things, and 5G communications) in several domains, including in the food supply chain [24,39,54]. Moreover, at the strategic level, the blockchain potentially handles the challenges associated with integrity, traceability, and transparency [50]. Concerning the impact, various papers have described BC's effect on all 17 SDGs [26,33] even under different perspectives, such as a supply chain, energy efficiency, and secure and reliable smart cities [40].

Regarding risks, the emergence of BC has inherent risks [49], and several works discuss some of them. There are four sources of risks in BC: systemic, governance, social welfare, and privacy [49]. From the point of view of the benefits, BC adoption minimises risk and improves security while risks from different sources (i.e., geographic, economic, political, technological, social, market, environmental, logistic, managerial, financial, legal/ethical, regulatory, and cultural) must be managed [34]. Also, BC is applied to study endogenous risk, including moral, information delivery, production organisation and procurement, and logistic risks [19]. On the other hand, adopting an RM perspective, the mean-variance approach can be applied to explore risks with air logistics in the blockchain [11].

3 Methods

To obtain a close expert view, interviews – the most widely used qualitative research approach in numerous disciplines [37] – were performed. An expert is a person with technical, process, and interpretative knowledge in his area with deep knowledge of specific experiences [6]. To gain insight into sustainability and BC in the public sector, experts were asked to what extent a public sector business process can benefit from BC adoption with sustainability effects, considering eight aspects (questions): behavioural incentives, process standardisation, compliance issues, supply chain traceability, resource management, circular economy, credit management, and SDG indicators. To deal with it were conducted 19 structured interviews with experts during January 2023. According to Glaser et al., at least ten interviews are needed to analyse the patterns adequately [20]; on the other hand, Saldaña argues for conducting between 20 and 30 interviews to gain a deep understanding of the phenomenon [45] – with 19, we are well above to Glaser et al., and close to Saldaña's minimum. Table 1 summarises the interviewees. As for ethical considerations, respondents were volunteers and agreed with the recording and were informed about the anonymity guarantee.

Table 1. Expert interviews by role and country.

Interviewee Role	BR	HK	GB	SP	DE	FR	US	PT	TOTAL
Public Sector's Process Executor	6							1	7
Blockchain Service Provider	2	1	1	1		1	1		7
Blockchain and/or Sustainability Researcher	3			1	1				5
# of interviews by country	11	1	1	2	1	1	1	1	
# of interviews									19

An interview protocol with five parts was developed to collect empirical data containing: (1) presenting the researcher and the objective of the interview; (2) presenting the research model; (3) demographic questions (i.e., role, country, and years of experience in the role); (4) questions – presenting each of the eight aspects shown above and the benefits of using BC, before performing each question; and (5) a final and open question about BC in the public sector or its potential impact on sustainability. For this paper, we analysed the concerns presented by interviewees in parts 4 and 5. The interviews were performed online by Microsoft Teams, with video recording and transcriptions, and carried out lasted, on average, 56 min each. The public sector process executor role, with an average of 12.2 years of experience, contributed with senior expertise in observing and analysing the potential of BC adoption. The BC service provider role, with an average expertise in BC of 5.5 years, brought the freshness of criticism and suggestions of design choices to respond to concerns (e.g., the need for sensors for input data integrity on the BC). Finally, with an average of 6.4 years of research, the researcher's role brought the theoretical view to contribute to the study.

4 Addressing Risk Management

To organise the insights from the interviews, this paper utilises the risk management process of ISO 31000 [25]. Table 2 consolidates the insights from interviewees, organising them by risk identification, risk treatment, and monitoring as a risk management model for BC adoption and sustainability in the public sector.

Table 2. RM model for BC adoption and sustainability in the public sector

Risk Identification	Risk Treatment	Monitoring
R1 - Data Integrity	T1 - Blockchain Design Choices	M1 - Number of Intermediaries
R2 - Data Privacy	T2 - Blockchain Governance	M2 - Resource Consumption
R3 - Organisational Culture	T3 - Privacy Techniques	M3 - Organisational Reputation
R4 - Compliance Issues	T4 - Standardisation	M4 - Total Process Cost
R5 - Environmental	T5 - Business Process Analysis and Transformation	M5 - Carbon Emission
R6 - Financial Freedom		
R7 - Performance		
R8 - Stakeholders		
R9 - Technical Issues		

4.1   Risk Identification

Risk identification aims to find, recognise, and describe risks that might help or prevent an organisation from achieving its objectives [25]. The following risks were identified through the interviews conducted with experts.

R1 - Data Integrity. Blockchain guarantees data integrity [12], being known as the truth machine [10]. However, there is a broad concern with risks associated with data collection (with the need for IoT and sensors – which are accurate and subject to external audit) and the interfaces with the "real world" (by oracles).

R2 - Data Privacy. BC achieves the goal of maintaining user anonymity and privacy [12], that mitigate, by default, privacy and security risks [42]. The main issue by interviewees is related to giving the users (holders) ownership over their data – design choices are primordial to address these issues [40,42], observing EU's General Data Protection Regulation (GDPR) compliance [22].

R3 - Organisational Culture. Organisational culture combined with structure and governance are aspects of an organisation that include strategy, values, cultural norms, authority and hierarchy relationships, and codes of ethics and conduct [41]. Some interviewees stated that governance aspects and organisational culture in the public sector are "natural barriers" that must be overcome to adopt BC and Decentralised Autonomous Organisation (DAO), with the potential to revolutionise public sector performance to support decision-making rebuilding the engagement of the people in the decision process, mainly at the local level, with eParticipation models. The main risk is related to the resistance that cultural barriers can offer to BC projects in the public sector.

R4 - Compliance Issues. Blockchain projects must comply with existing legal preconditions [46], besides corporate policies and codes of ethics influencing its adoption [14]. In general, interviewees confirmed that compliance aspects, even more in the public sector, must be observed, concluding that this risk compromises BC adoption. Also, most interviewees pointed to BC design choices to address (or mitigate) compliance issues, e.g., by adopting Zero Knowledge Proof [56]. Moreover, according to interviewees, the regulator should also be aware of the characteristics of the technology and its potential benefits to updating the legal framework, with sustainable concerns, on the way to a green blockchain.

R5 - Environmental. Blockchain can be analysed as an open system, considering the interfaces with the environment and the stakeholders in the blockchain ecosystem [31,44] also with sustainability concerns [27]. BC can address several environmental issues [55], mainly in supply chain [55], resource management [40,55], and tokenised credits from environmental projects [9,55]. The main risk in this category is related to energy consumption and carbon footprint.

R6 - Financial Freedom. Blockchain is linked with financial inclusion [28] as a socio-technical extension of the republican political vision [7]. Despite the advances that blockchain and the decentralised approach bring to individual freedoms, some respondents recognised the risk to financial freedom and privacy inherent in the perspective of Central Bank Digital Currency (CBDC).

R7 - Performance. Blockchain performance analysis includes the throughput, latency, and scalability [29]. The main risks identified are related to DAO, especially in a large community that can negatively affect decision-making, as voting often becomes a time and resource-intensive process; also, the delay in the massification and maturation of BC solutions can impact their viability.

R8 - Stakeholders. Stakeholder's engagement influences the adoption of BC [3]; also, blockchain can facilitate the management of dialogue between stakeholders [14]. Blockchain can help reduce the current information asymmetry between stakeholders, enabling more informed decision-making [36,55] and even decentralised [46]. The main risk gathered in the interviews relates to stakeholder identification in BC initiatives in the public sector and how to deal with them.

R9 - Technical Issues. Before implementation, specific criteria need to be considered for practitioners to make the right decisions, among them technical criteria [1]. The risks identified during the interviews are standardisation, interoperability, and integration. There is an intense debate about the need for standardisation rules – which may limit the evolution of technology, according to some – and the definition of interoperability criteria between the different BCs – which would broaden the reach of solutions, not limiting their evolvement.

4.2   Risk Treatment

Risk treatment aims to select and implement options to manage the risks identified [25]. The following treatments were identified through interviews conducted with experts and are related to at least one risk identified.

T1 - Blockchain Design Choices. Design choices address network typology, consensus mechanisms, and what to write on-chain and off-chain that can mitigate or solve several identified risks. In general, the interviewees indicated concerns about the network typology, highlighting that models permissioned or consortium would be more suitable for the public sector due to environmental sustainability issues and costs, in addition to governance and control.

T2 - Blockchain Governance. Several sustainable benefits will demand a new governance model for BC adoption [14,18], mainly based on permissioned BCs [5,28]. The interviewees pointed out the necessity of open governance based on principles of an open culture where the data of these BCs were open to public consultation, with transparency, accountability, and, consequently, incentive participation. Also, some highlighted the role of governance in legitimising changes and standardisation and the potential to mitigate financial freedom risk.

T3 - Privacy Techniques. Privacy is one of the top challenges concerning BC in the public sector [16,39]. Design choices (e.g., what to write on-chain) are primordial to address privacy issues [40,42], observing GDPR and Electronic Identification and Trust Services (eIDAS Regulation) compliance [22]. To anonymise the data and improve data privacy control, the interviewees indicated some encryption techniques, such as Zero Knowledge Proof [56] and Homomorphic encryption [47], in a pathway to a Self-Sovereign Identity.

T4 - Standardisation. International standardisation will have a positive effect in engendering an acceleration in BC applications [24]. Standards can help assure innovations and processes are interoperable, reliable, and secure [23], including adopting common data standards [51], eventually with a taxonomy [4]. The interviewees pointed out the need to develop and adopt standards that allow the integration of BC with other technologies (e.g., IoT and AI) and the standards for wallets to disseminate and massify solutions. The Education Credentials' use case of European Blockchain Service Infrastructure [17] is a good example of implementation with high compliance with the laws, regulations, and established standards, such as GDPR, eIDAS, European Self-Sovereign-Identity Framework (ESSIF), and W3C Verifiable Credentials and Presentations.

T5 - Business Process Analysis and Transformation. BC in the public sector when combined with Business Process Management offers excellent potential for realisation and response to some processes' weaknesses [53]. The adoption of new technology, as well as digital transformation projects [15], is an opportunity for public organisations to shape organisational behaviour with an impact on business processes, reimagining them [40,52] and the stakeholders' relationships [52], making them more standardised, transparent, and streamlined [24]. According to the interviewees, the analysis, design, and transformation of processes must observe the laws, regulations, and standards to endow the solution with the characteristic compliance by design, responding to various risks.

4.3   Monitoring

Monitoring aims to assure and improve processes' quality, effectiveness, and outcomes, including planning, gathering, and analysing information [25]. Some metrics were identified through interviews conducted with experts and are related to at least one risk identified and/or at least one treatment proposed.

M1 - Number of Intermediaries. Reducing the number of intermediaries is one of the benefits of BC adoption [36,55]. In addition to making systems more efficient, this reduction mitigates risks related to organisational culture, performance, and stakeholders, impacting design choices and governance.

M2 - Resource Consumption. Reducing resource consumption (e.g., paper, energy) positively impacts the systems' efficiency and mitigates environmental risk, influencing design choices.

M3 - Organisational Reputation. BC could increase confidence in the operations of public organisations [13], contributing to the organisation's reputation, increasing trust, and reducing the population's apathy, which mitigates risks related to the organisational culture and influences governance, favouring the adoption of solutions and their sustainability over time.

M4 - Total Process Cost. The total cost of the process is an objective measure that supports the process efficiency and with BC tends to be significantly reduced [2] for both the customer and the organisation. This measure is influenced by design choices and by the analysis and transformation of processes.

M5 - Carbon Emission. Carbon emission is a measure of energy demand that is impacted by design choices, potentially influencing environmental risks.

5   Discussion and Conclusion

Concerning SDGs and blockchain adoption, SDG16 - Peace, Justice, and Strong Institutions is the main block of the United Nations' Agenda 2030, mainly in the public sector, where BC can contribute to bringing transparency and traceability of the entire process, not just the result, building a strong democracy, being a way to undo the population's apathy toward the public sector and protect public assets. However, several interviewees highlight that BC is not a silver bullet and that the effect and benefits depend more on the application than the technology itself and the integration with other technologies. Furthermore, various respondents pointed out that many benefits of BC adoption could be achieved with other technologies, except for increased trust.

However, the adoption of BC in the public sector is not without risks. In this sense, the risk management perspective adopted in this work contributes to deepening the discussion, bringing insights from 19 experts, from which it was possible to identify risks and propose treatments and metrics for monitoring, discussing to what extent a public sector business process can benefit from BC adoption with sustainability effects, materialised in the risk management model. The risk matrix in Fig. 1. illustrates the correspondence between the identified risks and the respective treatments (characterised in grey) and indicates the metric when related, demonstrating the consistency of the proposed model – as can be seen, there is at least one related treatment for each identified risk.

	R1	R2	R3	R4	R5	R6	R7	R8	R9
T1					M2, M5		M1, M2, M4		
T2			M3				M1		
T3									
T4									
T5			M1				M1, M4	M1	

Fig. 1. Risk Matrix.

This work has theoretical and practical implications due to developing knowledge about adopting BC under the sustainability perspective and the contribution to the RM body of knowledge. But, it has limitations. First, the discussion is based on qualitative interviews (and more than half of the interviewees are from one country), which may contain bias despite being comprehensively planned. Second, the description of the RM model can limit the analysis, compromising the generalisation potential – empirical approaches may be used in further works.

Acknowledgement. This work has been supported by FCT - Fundação para a Ciência e Tecnologia within the R&D Units Project Scope: UIDB/00319/2020.

References

1. Aysan, A., Bergigui, F., Disli, M.: Using blockchain-enabled solutions as SDG accelerators in the international development space. Sustainability (Switzerland) **13**(7) (2021). https://doi.org/10.3390/su13074025
2. Babkin, E., Malyzhenkov, P., Pakhomova, K.: Blockchain as a tool for effective value co-creation strategy realization: an application to an insurance case. In: Forum/Posters/CIAO! DC@ EEWC (2019)
3. Balci, G., Surucu-Balci, E.: Blockchain adoption in the maritime supply chain: examining barriers and salient stakeholders in containerized international trade. Transp. Res. Part E Logist. Transp. Rev. **156**, 102539 (2021). https://doi.org/10.1016/J.TRE.2021.102539
4. Barclay, I., Cooper, M., Hackel, J., Perrin, P.: Tokenizing behavior change: a pathway for the sustainable development goals. Front. Blockchain **4** (2022). https://doi.org/10.3389/fbloc.2021.730101
5. Benítez-Martínez, F.L., Romero-Frías, E., Hurtado-Torres, M.V.: Neural blockchain technology for a new anticorruption token: towards a novel governance model, J. Inf. Technol. Polit. **20**(1), 1–18 (2023), https://doi.org/10.1080/19331681.2022.2027317
6. Bogner, A., Littig, B., Menz, W.: Introduction: expert interviews - an introduction to a new methodological debate. In: Interviewing Experts (2009)
7. Bychkova, O., Kosmarski, A.: Imagineering a new way of governing: the blockchain and res publica. **20**(1), 34–43 (2022). https://doi.org/10.1080/19331681.2022.2028695
8. Cagigas, D., Clifton, J., Diaz-Fuentes, D., Fernandez-Gutierrez, M.: Blockchain for public services: a systematic literature review. IEEE Access **9**, 13904–13921 (2021). https://doi.org/10.1109/ACCESS.2021.3052019
9. Calandra, D., Secinaro, S., Massaro, M., Dal Mas, F., Bagnoli, C.: The link between sustainable business models and blockchain: a multiple case study approach. Bus. Strategy Environ. (2022). https://doi.org/10.1002/BSE.3195
10. Casey, M.J., Vigna, P.: The Truth Machine: The Blockchain and the Future of Everything. St. Martin's Press, New York (2018)
11. Choi, T.M., Wen, X., Sun, X., Chung, S.H.: The mean-variance approach for global supply chain risk analysis with air logistics in the blockchain technology era. Transp. Res. Part E Logist. Transp. Rev. **127** (2019). https://doi.org/10.1016/j.tre.2019.05.007
12. Dai, F., Shi, Y., Meng, N., Wei, L., Ye, Z.: From Bitcoin to cybersecurity: a comparative study of blockchain application and security issues. In: 2017 4th International Conference on Systems and Informatics, ICSAI 2017, vol. 2018-January (2017). https://doi.org/10.1109/ICSAI.2017.8248427
13. De Filippi, P.: Blockchain Technology as an Instrument for Global Governance. Digital, Governance and Sovereignty Chair, pp. 1–16 (2021)
14. Di Vaio, A., Hassan, R., Palladino, R.: Blockchain technology and gender equality: a systematic literature review. Int. J. Inf. Manag. **68**, 102517 (2023). https://doi.org/10.1016/J.IJINFOMGT.2022.102517
15. Escobar, F., Almeida, W.H., Varajão, J.: Digital transformation success in the public sector: a systematic literature review of cases, processes, and success factors. Inf. Polity **28**(1), 61–81 (2023). https://doi.org/10.3233/IP-211518
16. Escobar, F., Santos, H., Pereira, T.: Blockchain in the public sector: an umbrella review of literature. In: Prieto, J., Benítez-Martínez, F.L., Ferretti, S., Arroyo

Guardeño, D., Nevado-Batalla, T.P. (eds.) Blockchain and Applications, 4th International Congress . BLOCKCHAIN 2022. LNNS, vol. 595. Springer, Cham (2023). https://doi.org/10.1007/978-3-031-21229-1_14

17. European Commission: EBSI Verifiable Credentials Playbook - EBSI Specifications (2022).   https://ec.europa.eu/digital-building-blocks/wikis/display/EBSIDOC/EBSI+Verifiable+Credentials+Playbook

18. Francisco Luis, B.M., Pedro Víctor, N.C.U., Valentín, M.M., Esteban, R.F.: Blockchain as a service: a holistic approach to traceability in the circular economy. In: Muthu, S.S. (ed.) Environmental Footprints and Eco-Design of Products and Processes, pp. 119–133 (2022). https://doi.org/10.1007/978-981-16-6301-7_6

19. Fu, Y., Zhu, J.: Big production enterprise supply chain endogenous risk management based on blockchain. IEEE Access **7**, 15310–15319 (2019). https://doi.org/10.1109/ACCESS.2019.2895327

20. Glaser, B.G., Strauss, A.L.: Discovery of grounded theory: strategies for qualitative research (2017). https://doi.org/10.4324/9780203793206

21. Happy, A., Chowdhury, M.M.H., Scerri, M., Hossain, M.A., Barua, Z.: Antecedents and consequences of blockchain adoption in supply chains: a systematic literature review. J. Enterp. Inf. Manag. **36**(2), 629–654 (2023)

22. Haque, A.B., Islam, A.K., Hyrynsalmi, S., Naqvi, B., Smolander, K.: GDPR compliant blockchains - a systematic literature review (2021)

23. Horner, J., Ryan, P.: Blockchain standards for sustainable development. J. ICT Stand. **7**(3), 225–247 (2019). https://doi.org/10.13052/jicts2245-800X.733

24. Hughes, L., Dwivedi, Y.K., Misra, S.K., Rana, N.P., Raghavan, V., Akella, V.: Blockchain research, practice and policy: applications, benefits, limitations, emerging research themes and research agenda. Int. J. Inf. Manag. **49**, 114–129 (2019)

25. International Organization for Standardization: ISO 31000: Risk Management: Principles and Guidelines. ISO (2009)

26. Jiang, S., Jakobsen, K., Bueie, J., Li, J., Haro, P.H.: A tertiary review on blockchain and sustainability with focus on sustainable development goals. IEEE Access **10**, 114975–115006 (2022)

27. Ketprapakorn, N., Kantabutra, S.: Toward an organizational theory of sustainability culture. Sustain. Prod. Consum. **32**, 638–654 (2022)

28. Kewell, B., Adams, R., Parry, G.: Blockchain for good? Strateg. Chang. **26**(5), 429–437 (2017). https://doi.org/10.1002/jsc.2143

29. Kuzlu, M., Pipattanasomporn, M., Gurses, L., Rahman, S.: Performance analysis of a Hyperledger fabric blockchain framework: throughput, latency and scalability. In: 2019 IEEE International Conference on Blockchain (Blockchain), pp. 536–540 (2019). https://doi.org/10.1109/Blockchain.2019.00003

30. Li, J., Greenwood, D., Kassem, M.: Blockchain in the built environment and construction industry: a systematic review, conceptual models and practical use cases. Autom. Constr. **102**, 288–307 (2019)

31. Mangla, S.K., Kazancoglu, Y., Ekinci, E., Liu, M., Özbiltekin, M., Sezer, M.D.: Using system dynamics to analyze the societal impacts of blockchain technology in milk supply chainsrefer. Transp. Res. Part E Logist. Transp. Rev. **149** (2021). https://doi.org/10.1016/j.tre.2021.102289

32. Medaglia, R., Damsgaard, J.: Blockchain and the united nations sustainable development goals: towards an agenda for IS research. In: Proceedings of the 24th Pacific Asia Conference on Information Systems: Information Systems (IS) for the Future, PACIS 2020 (2020)

33. Medaglia, R., Misuraca, G., Aquaro, V.: Digital government and the united nations' sustainable development goals: towards an analytical framework. In: DG.O2021: The 22nd Annual International Conference on Digital Government Research (DG.O'21), pp. 473–478. Association for Computing Machinery, July 2021

34. Min, H.: Blockchain technology for enhancing supply chain resilience. Bus. Horizons **62**(1) (2019). https://doi.org/10.1016/j.bushor.2018.08.012

35. Monrat, A.A., Schelén, O., Andersson, K.: A survey of blockchain from the perspectives of applications, challenges, and opportunities. IEEE Access **7**, 117134–117151 (2019). https://doi.org/10.1109/ACCESS.2019.2936094

36. Mora, H., Mendoza-Tello, J.C., Varela-Guzman, E.G., Szymanski, J.: Blockchain technologies to address smart city and society challenges. Comput. Hum. Behav. **122**(n/a), 1–19 (2021). https://doi.org/10.1016/j.chb.2021.106854

37. Myers, M.D., Newman, M.: The qualitative interview in IS research: examining the craft. Inf. Org. **17**(1), 2–26 (2007). https://doi.org/10.1016/j.infoandorg.2006.11.001

38. Nobanee, H., et al.: A bibliometric analysis of sustainability and risk management (2021). https://doi.org/10.3390/su13063277

39. Nurgazina, J., Pakdeetrakulwong, U., Moser, T., Reiner, G.: Distributed ledger technology applications in food supply chains: a review of challenges and future research directions (2021). https://doi.org/10.3390/su13084206

40. Parmentola, A., Petrillo, A., Tutore, I., De Felice, F.: Is blockchain able to enhance environmental sustainability? A systematic review and research agenda from the perspective of sustainable development goals (SDGs). Bus. Strategy Environ. **31**(1), 194–217 (2022). https://doi.org/10.1002/bse.2882

41. PMI: PMBOK® Guide - Seventh Edition. Project Management Institute (2021)

42. Pournaras, E.: Proof of witness presence: blockchain consensus for augmented democracy in smart cities. J. Parallel Distrib. Comput. **145** (2020). https://doi.org/10.1016/j.jpdc.2020.06.015

43. Rodríguez Bolívar, M.P., Scholl, H.J.: Mapping potential impact areas of blockchain use in the public sector. Inf. Polity Int. J. Gov. Democr. Inf. Age **24**(4), 359–378 (2019)

44. Rousseau, D.: General systems theory: its present and potential. Syst. Res. Behav. Sci. **32**(5) (2015). https://doi.org/10.1002/sres.2354

45. Saldaña, J.: The Coding Manual for Qualitative Researchers, no. 14. Sage, Thousand Oaks (2016)

46. Schulz, K.A., Gstrein, O.J., Zwitter, A.J.: Exploring the governance and implementation of sustainable development initiatives through blockchain technology. Futures **122** (2020). https://doi.org/10.1016/j.futures.2020.102611

47. Singh, P., Masud, M., Hossain, M.S., Kaur, A.: Blockchain and homomorphic encryption-based privacy-preserving data aggregation model in smart grid. Comput. Electr. Eng. **93**, 107209 (2021)

48. Tan, E., Mahula, S., Crompvoets, J.: Blockchain governance in the public sector: a conceptual framework for public management. Gov. Inf. Q. **39**(1), 101625 (2022). https://doi.org/10.1016/J.GIQ.2021.101625

49. Tasca, P.: Internet of value: a risky necessity. Front. Blockchain **3** (2020). https://doi.org/10.3389/fbloc.2020.00039

50. Teh, D., Khan, T., Corbitt, B., Ong, C.E.: Sustainability strategy and blockchain-enabled life cycle assessment: a focus on materials industry. Environ. Syst. Decis. **40**(4), 605–622 (2020). https://doi.org/10.1007/s10669-020-09761-4

51. Tsolakis, N., Niedenzu, D., Simonetto, M., Dora, M., Kumar, M.: Supply network design to address United nations sustainable development goals: a case study of blockchain implementation in Thai fish industry. J. Bus. Res. **131** (2021). https://doi.org/10.1016/j.jbusres.2020.08.003
52. de Villiers, C., Kuruppu, S., Dissanayake, D.: A (new) role for business - Promoting the United Nations' Sustainable Development Goals through the internet-of-things and blockchain technology. J. Bus. Res. **131**, 598–609 (2021)
53. Viriyasitavat, W., Xu, L.D., Bi, Z., Pungpapong, V.: Blockchain and internet of things for modern business process in digital economy - the state of the art. IEEE Trans. Comput. Soc. Syst. **6**(6), 1420–1432 (2019). https://doi.org/10.1109/TCSS.2019.2919325
54. Walshe, R., Casey, K., Kernan, J., Fitzpatrick, D.: AI and big data standardization: contributing to united nations sustainable development goals. J. ICT Standard. **8**(2) (2020). https://doi.org/10.13052/jicts2245-800X.821
55. World Economic Forum: Building Block(chain)s for a Better Planet. Technical report, September 2018
56. Yang, X., Li, W.: A zero-knowledge-proof-based digital identity management scheme in blockchain. Comput. Secur. **99**, 102050 (2020)

Blockchain-Based Federated Learning: Incentivizing Data Sharing and Penalizing Dishonest Behavior

Amir Jaberzadeh[1], Ajay Kumar Shrestha[2] (ORCID), Faijan Ahamad Khan[1],
Mohammed Afaan Shaikh[1], Bhargav Dave[1], and Jason Geng[1](✉)

[1] Bayes Solutions, 840 Apollo St, El Segundo, CA 90245, USA
`jason@bayes.global`
[2] Vancouver Island University, 900 Fifth St, Nanaimo, BC V9R 5S5, Canada
`ajay.shrestha@viu.ca`

Abstract. With the increasing importance of data sharing for collaboration and innovation, it is becoming more important to ensure that data is managed and shared in a secure and trustworthy manner. Data governance is a common approach to managing data, but it faces many challenges such as data silos, data consistency, privacy, security, and access control. To address these challenges, this paper proposes a comprehensive framework that integrates data trust in federated learning with InterPlanetary File System, blockchain, and smart contracts to facilitate secure and mutually beneficial data sharing while providing incentives, access control mechanisms, and penalizing any dishonest behavior. The experimental results demonstrate that the proposed model is effective in improving the accuracy of federated learning models while ensuring the security and fairness of the data-sharing process. The research paper also presents a decentralized federated learning platform that successfully trained a CNN model on the MNIST dataset using blockchain technology. The platform enables multiple workers to train the model simultaneously while maintaining data privacy and security. The decentralized architecture and use of blockchain technology allow for efficient communication and coordination between workers. This platform has the potential to facilitate decentralized machine learning and support privacy-preserving collaboration in various domains.

Keywords: Federated Learning · Blockchain · Data Trust

1 Introduction

In recent years, data sharing has become increasingly important for collaboration and innovation in various fields. The adoption of secure and trustworthy multi-center machine learning poses numerous challenges, including data sharing, training algorithms, storage, incentive mechanisms, and encryption. In this paper, we aim to tackle these challenges and propose a comprehensive solution for collaborative machine-learning applications. However, managing and sharing data in a secure and trustworthy manner poses several

© The Author(s), under exclusive license to Springer Nature Switzerland AG 2023
J. M. Machado et al. (Eds.): BLOCKCHAIN 2023, LNNS 778, pp. 186–195, 2023.
https://doi.org/10.1007/978-3-031-45155-3_19

challenges, such as data silos, privacy, security, access control, and data consistency. Data governance has been proposed as a common approach to managing data, but it still faces several challenges, and as a result, data trust has emerged as a nascent sub-area of data management [1].

This research paper proposes a trustworthy and robust framework for federated learning participants. Federated Learning (FL) is a privacy-preserving distributed Machine Learning (ML) paradigm [2]. The proposed comprehensive framework integrates data trust, InterPlanetary File System[1] (IPFS), blockchain, and smart contracts to establish a secure and mutually beneficial data-sharing distributed FL platform. The framework is designed to provide incentives, access control mechanisms, and penalties for any dishonest or malicious behavior—sharing bad data or non-compliance with protocols. The framework aims to foster trust among stakeholders, encourage data sharing for mutual benefit, and discourage actions that may compromise data security and accuracy. Our proposed approach is built on the use of smart contracts that enable monitoring of data sharing, access control, and compensation. To participate in the federated learning process, users must register and contribute their data while being required to provide a collateral deposit to deter dishonest behavior.

Our proposed framework prioritizes data privacy and security by utilizing the encryption-enabled InterPlanetary File System (IPFS) as a decentralized peer-to-peer file system to store and access data. By using IPFS, federated learning models can be trained on data that is stored on a distributed network of users' devices, reducing the need for centralized storage. The utilization of encryption enabled IPFS ensures that the user's data privacy is safeguarded throughout the learning process. The framework is designed to provide a fair and transparent approach to compensate users for their contributions while ensuring the privacy and security of their data.

The rest of the paper is organized as follows. Section 2 provides a succinct analysis of existing architectures and identifies their shortcomings. Our proposed model for the solution architecture is presented in Sect. 3. Section 4 contains the experimental results and discussion. Lastly, Sect. 5 concludes the paper by outlining future directions for further research and improvements to the proposed model.

2   Background and Related Works

The primary challenge in data governance is to dismantle data silos [3] and ensure data consistency, compatibility, privacy, security, access control, ownership, and rewards for sharing. It is, therefore, imperative to have data governance frameworks that can evolve with new technologies to address emerging challenges. FL is a learning technique where a central server connects with many clients (e.g., mobile phones, and pads) who keep their data private. Since communication between the central server and clients can be a bottleneck, decentralized federated learning (DFL) [4] is used to connect all clients with an undirected graph, which reduces communication costs and increases privacy by replacing server-client communication with peer-to-peer communication. DFL offers communication efficiency and fast convergence, and the advantages of FL are summarized in [5].

---

[1] https://ipfs.tech/.

Several variants of Federated Average (FedAvg) [2] exist with theoretical guarantees. In [6], the momentum method is used for local client training, while [7] proposes adaptive FedAvg with an adaptive learning rate. Lazy and quantized gradients are used in [8] to reduce communications, and in [9], the authors propose a Newton-type scheme. Decentralized (sub) gradient descents (DGD) are studied in [10–13], and DSGD is proposed in [14]. Asynchronous DSGD is analyzed in [15], and quantized DSGD is proposed in [16]. Decentralized FL is popular when edge devices do not trust central servers to protect their privacy [17]. Finally, the authors in [16] propose a novel FL framework without a central server for medical applications. The authors in [18] propose a secure architecture for privacy-preserving in smart healthcare using Blockchain and Federated Learning, where Blockchain-based IoT cloud platforms are used for security and privacy, and Federated Learning technology is adopted for scalable machine learning applications. In [19], authors propose a blockchain-based Federated Learning (FL) scheme for Internet of Vehicles (IoV), addressing security and privacy concerns by leveraging blockchain and a reputation-based incentive mechanism.

Compared to prior research on FedAvg, our paper proposes a decentralized framework to improve FL's resilience to node failures and privacy attacks. Unlike previous decentralized training approaches, our algorithm utilizes IPFS and efficient encryption methods to securely converge model training among many nodes.

3 System Model

3.1 Data Trust, Access Control and Incentive Method

Data trust ensures that data is available for data mining with increasing legal protections for privacy and sustains the underlying ownership of the data and digital rights, which is the primary focus of the data management field [20]. Our work emphasizes sharing data, transparency, control, and incentives for users in the federated learning setting. Specifically, it explores a particular type of technical platform, distributed ledgers with smart contracts. The smart contract is designed to oversee data sharing, compensation, and access control. Participants would be allowed to register and contribute their data to the federated learning process, and a collateral deposit would be required from each participant to discourage any dishonest behavior. The collateral deposit serves as a financial penalty for participants who fail to provide quality data or who intentionally provide misleading information. If a participant fails to provide accurate data or engages in any dishonest behavior, the deposit will be forfeited. The forfeited deposit will then be used to compensate other participants who have contributed accurate data to the federated learning process. Through the implementation of a smart contract, the total compensation for data sharing is updated and distributed to participants based on their contribution. The contract would also ensure that each participant can only register once, and that compensation can only be distributed when the total compensation amount is positive. The proposed smart contract system provides a reliable and secure framework for federated learning, user data, and blockchain integration. It offers a fair and transparent way of compensating participants for their contributions while ensuring the privacy and security of the data.

3.2   IPFS Storage

In addition to privacy concerns, data storage is another key challenge in federated learning. Traditional data storage approaches are not well-suited for federated learning, as they often require centralized storage of data. This centralized storage approach can increase the risk of data breaches and raises concerns around data ownership and control. To address this challenge, we have proposed using the InterPlanetary File System (IPFS), a peer-to-peer distributed file system that allows data to be stored and accessed in a decentralized manner. By using IPFS, federated learning models can be trained on data that is stored on users' devices, without the need for centralized storage.

3.3   Confidentiality and Privacy

In the context of IPFS, hashing is not sufficient to ensure the confidentiality and privacy of the stored data, as the content can still be accessed by anyone who has access to the network. To address this, both symmetric key encryption and asymmetric cryptography are applied [21], and we further use smart contracts for access control and to provide confidentiality and privacy to the data stored in IPFS. This approach ensures that even if an attacker gains access to the IPFS network, they would not be able to read the encrypted content without the secret. Although encryption may introduce some overhead in terms of performance and complexity, it is necessary for ensuring the security of data in IPFS.

We used the cryptography Python library to securely share and store machine learning models using IPFS. The library provides various cryptographic primitives and recipes for encryption, digital signatures, key derivation, hashing, and more, adhering to best security practices. The code initially connects to an IPFS daemon, loads a model from IPFS, generates an RSA key pair and an AES key, and encrypts the AES key with the public key using hybrid encryption. The actual data is encrypted using the symmetric key (AES), and the symmetric key is encrypted using an asymmetric key (RSA). This ensures that the data can only be decrypted by the intended recipient who possesses the corresponding private key. Our research further implements a method for encrypted model states to be fetched from a group of workers, decrypted with the AES key, and returned as decrypted model states. This allows multiple parties to share their model states securely. In addition, we have also implemented a method for pushing a model state to IPFS. The model state is encrypted using the AES key and the AES key is encrypted with the public key. This encryption mechanism allows the model state to be stored on the IPFS network in an encrypted form, ensuring that only authorized parties can access it. To optimize memory usage, our code maintains a list of model hashes that have been pushed to IPFS and clears the list once a specified number of models have been pushed. This optimization technique helps prevent system resources from being overwhelmed and causing performance issues. By clearing the list after a certain number of models have been pushed, the code ensures that memory usage remains within reasonable limits.

3.4   Decentralized Network Architecture

Our research proposes a blockchain-based architecture for federated learning, consisting of a smart contract and IPFS. The smart contract coordinates the FL task, distributing rewards and penalizing bad actors. The metric used for rewarding workers based on performance considers factors such as model accuracy, consistency, precision and recall on the unseen test dataset. Our proposed architecture contains essential information such as the participants' details, model evaluations submitted at the end of each round, and the reward to be distributed. The IPFS, on the other hand, stores the models trained by participants at the end of each round.

As shown in Fig. 1, blockchain-based federated learning involves two classes of actors: the requester and the workers. The requester initiates the FL task by deploying the smart contract, pushing the initial model to the IPFS, and specifying additional parameters such as the number of rounds and the reward to be distributed. The requester can push any model for any task to IPFS. On the other hand, workers participate in the FL task created by the requester. They train the model through a round-based system on their own data and earn rewards based on their performance.

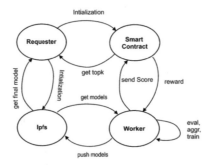

Fig. 1. Blockchain and IPFS-based federated learning.

Workflow. To begin the task, the requester deploys the smart contract and sets the number of training rounds (N) and the total reward (D) for workers. The requester also pushes the initial model to the IPFS-based model storage, which will be used as the basis for the trained models. Workers can then join the task by interacting with the smart contract, and once enough workers have joined, the requester triggers the training phase through the smart contract. During the training phase, workers train the model for N rounds on their local data, with each round beginning with workers retrieving the trained models of all other workers from the IPFS-based model storage to evaluate them on their local data. The scores are pushed to the smart contract, which aggregates them to obtain the Top K best-performing workers in the previous round. Rewards are distributed to the workers based on their performance. The trained models are then pushed to the IPFS-based model storage, and the process repeats for N rounds. The requester is not involved in any interaction during the training phase, but some operations such as score aggregation can be offloaded to the requester machine to save on computational resources and transaction costs. Once the training phase concludes, the requester can retrieve the final global model from the IPFS-based model storage and close the task by calling a function of the smart contract.

Smart Contracts. The smart contract contains various functions that enable the task requester to initialize, initiate, and oversee the FL task. It also allows workers to participate in the task, submit evaluations, and exit from it. Below is a brief overview of the different functions within the smart contract.

initializeTask. This function is called by the requester to initialize the FL task. It takes two parameters: the URI of the machine learning model and the number of rounds in the FL task. The function requires a deposit to be made in the smart contract.

startTask. This function is called by the requester to start the FL task. It changes the status of the task to "running".

joinTask. This function is called by the workers to join the FL task. It registers the worker in the smart contract and returns the URI of the machine learning model.

submitScore. This function is called by the workers to submit the score of their local model after each round's evaluation phase.

removeWorker. This function is called by workers to remove themselves from the task.

nextRound. This function is called by the requester to advance the FL task to the next round.

getSubmissions. This function is called by the requester to get the submissions from all workers for the current round.

submitRoundTopK. This function is used to get the top k rank of the workers who will be rewarded for their performance in a task or job. This information is used to distribute the rewards among the top-performing workers.

distributeRewards. This function is used to reward top-performing users in a round by splitting the total reward among them. The first users receive half of the total reward, while the remaining users receive a smaller share.

3.5 Aggregation/Averaging Method

As shown in Fig. 2, in federated learning, workers train the model on their local data, and their models are stored in IPFS-based model storage. The workers retrieve their own model and the models of other workers from the storage and append them to a dictionary. To improve the model's accuracy, the workers use an averaging function that takes all the stored models as input and returns the average model. The averaging is done by adding up all the models and dividing the result by the number of workers who contributed to their models.

Overall, this process allows for multiple workers to collaborate on model training without relying on centralized data storage. By averaging the models, the final global model can be improved by incorporating the knowledge of all workers in the task. By contrast, FedAvg in centralized fashions leads to massive communication between the central server and clients, which causes channel blocking. Furthermore, the central server is vulnerable to attack, compromising the privacy of the entire system.

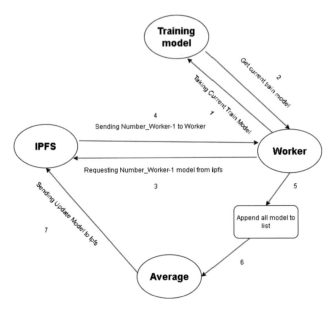

Fig. 2. Aggregation/Averaging methods in our DFL framework.

4   Experimental Results and Discussions

We evaluated the proposed platform using the MNIST dataset[2]. We used a simple feed-forward Neural Network (CNN) model with N layers to classify 0–9 handwritten numbers. The algorithms were developed using the PyTorch framework on the Ethereum blockchain. The simulation modeling was done on Ganache, a local Ethereum blockchain for testing. The training dataset comprised 60,000 images, while the test dataset consisted of 10,000 images. The training dataset was divided evenly between workers at the start of the training and each worker used the test dataset for their evaluations and scoring. Our implementation employed a decentralized client and server system and ran on a local machine. When the requester initiated the process, each worker trained the model taken from IPFS sequentially and securely saved the trained model to the IPFS until the maximum number of epochs was reached. Alternatively, workers could be spawned in parallel and train the model simultaneously on multiple devices. Here we tested our results with one machine that runs each step sequentially.

4.1   Performance Analysis

We first trained the classification model with our Decentralized Federated Learning framework and showed a convergence of above 95% accuracy under 90 epochs as shown in Fig. 3. We also studied other performance metrics like precision and recall and got 0.973 and 0.97 respectively which shows great performance in the classification task.

---

[2] https://pytorch.org/vision/stable/generated/torchvision.datasets.emnist.html#torchvision.datasets.emnist.

The total training time by 3 workers was 6525.46 s. Each worker takes about 36 min to converge on a Xeon CPU with 8 cores which is a comparable time to convergence in decentralized Federated Learning frameworks. We also compared the impact of double encryption on the convergence time. As shown in the left graph of Fig. 3, there is an additional 2 min and 34 s overhead for all three workers or 51 s for each worker. The communication cost for our double encryption and decryption process and secure key pair transfer protocol was only 2% of the time required for convergence with the same accuracy. The accuracy for each worker is plotted sequentially for each round of 3 epochs. As shown in Fig. 3, when the models start to train, all three workers start with low accuracies that improve within their first round (i.e., 3 epochs) and followed by the next worker's training.

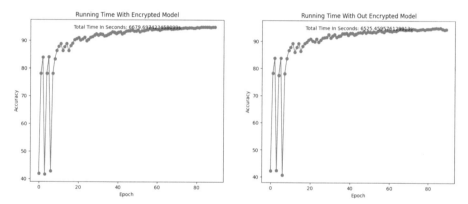

Fig. 3. Comparison of Running Time: Encrypted vs Unencrypted Models for 3-Worker Model.

4.2   Accuracy (Workers vs Epochs) Analysis

The graphs in Fig. 4 show the accuracy of a federated learning model trained over multiple epochs, with the left graph showing results for a model trained with 3 workers and the second graph showing results for a model trained with 5 workers. Accuracy is the percentage of correct classifications that a trained machine learning model achieves. By analyzing the graphs, we can see that both models reach an acceptable accuracy over a similar number of epochs.

This shows that dividing data between more workers does not have a negative impact on model convergence but can speed up the training process and scale up the training process. It can also reduce the required compute power for each worker which qualifies low-end devices to be used as compute nodes. In the left graph with 3 workers, the model's accuracy has a more stable pattern, which is due to having more data to train on for each worker. In a realistic case by increasing the training dataset, we can improve the stability of models with more workers. Overall, we can conclude that the decentralized federated learning model is performing well and improving over time, but the number of workers should be chosen as a ratio of the training dataset.

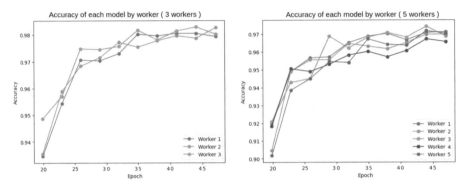

Fig. 4. Accuracy vs Epochs for 3-worker (left) and 5-worker (right) models.

5 Conclusion

We proposed a decentralized federated learning architecture that leverages blockchain, smart contracts and IPFS for secure and efficient training of a global model with decentralized data. Experimental results showed that our proposed framework achieved above 95% accuracy under 90 epochs with a comparable convergence time to centralized federated learning frameworks. We also compared the impact of double encryption on the convergence time and showed that it only resulted in a minimal overhead cost of 2%. Overall, our proposed approach addresses several challenges associated with managing and sharing data in a secure and trustworthy manner by providing a comprehensive framework that establishes trust among stakeholders, promotes data sharing that benefits all involved parties, and deters any actions that could compromise the security and accuracy of the shared data. Future research can explore the scalability and feasibility of the proposed model in a real-world scenario.

References

1. Delacroix, S., Lawrence, N.D.: Bottom-up data trusts: disturbing the 'one size fits all' approach to data governance. International Data Privacy Law **9**, 236–252 (2019)
2. Brendan McMahan, H., Moore, E., Ramage, D., Hampson, S., Agüera y Arcas, B.: Communication-efficient learning of deep networks from decentralized data. In: Proceedings of the 20th International Conference on Artificial Intelligence and Statistics, AISTATS 2017 (2017)
3. Seaman, D.: From isolation to integration: re-shaping the serials data silos. Serials: The Journal for the Serials Community **16**, 131–135 (2003)
4. Zinkevich, M.A., Weimer, M., Smola, A. Li, L.: Parallelized stochastic gradient descent. In: Advances in Neural Information Processing Systems 23: 24th Annual Conference on Neural Information Processing Systems 2010, NIPS 2010 (2010)
5. Sun, T., Li, D., Wang, B.: Decentralized federated averaging. IEEE Trans. Pattern Anal. Mach. Intell. (2022). https://doi.org/10.1109/TPAMI.2022.3196503
6. Hsu, T.-M.H., Qi, H., Brown, M.: Measuring the Effects of Non-Identical Data Distribution for Federated Visual Classification (2019)
7. Deng, Y., Kamani, M.M., Mahdavi, M.: Adaptive Personalized Federated Learning (2020)

8. Sun, J., Chen, T., Giannakis, G.B., Yang, Z.: Communication-efficient distributed learning via lazily aggregated quantized gradients. Adva. Neural Info. Proc. Sys. **32** (2019)
9. Li, T., et al.: FedDANE: a federated newton-type method. In: 2019 53rd Asilomar Conference on Signals, Systems, and Computers, pp. 1227–1231. IEEE (2019). https://doi.org/10.1109/IEEECONF44664.2019.9049023
10. Nedić, A., Ozdaglar, A.: Distributed subgradient methods for multi-agent optimization. IEEE Trans Automat Contr **54**, 48–61 (2009)
11. Chen, A.I., Ozdaglar, A.: A fast distributed proximal-gradient method. In: 2012 50th Annual Allerton Conference on Communication, Control, and Computing, Allerton 2012, pp. 601–608 (2012). https://doi.org/10.1109/Allerton.2012.6483273
12. Jakovetic, D., Xavier, J., Moura, J.M.F.: Fast distributed gradient methods. IEEE Trans. Automat. Contr. **59**, 1131–1146 (2014)
13. Matei, I., Baras, J.S.: Performance evaluation of the consensus-based distributed subgradient method under random communication topologies. IEEE J. Sel. Top. Sign. Proces. **5**, 754–771 (2011)
14. Lan, G., Lee, S., Zhou, Y.: Communication-efficient algorithms for decentralized and stochastic optimization. Math. Program. **180**, 237–284 (2020)
15. Lian, X., Zhang, W., Zhang, C., Liu, J.: Asynchronous decentralized parallel stochastic gradient descent. In: 35th International Conference on Machine Learning, ICML 2018, vol. 7, pp. 4745–4767 (2018)
16. Reisizadeh, A., Mokhtari, A., Hassani, H., Pedarsani, R.: Quantized decentralized consensus optimization. In: Proceedings of the IEEE Conference on Decision and Control, vol. 2018-Decem 5838–5843 (2018)
17. Sun, T., et al.: Non-ergodic convergence analysis of heavy-ball algorithms. In: 33rd AAAI Conference on Artificial Intelligence, AAAI 2019, 31st Innovative Applications of Artificial Intelligence Conference, IAAI 2019 and the 9th AAAI Symposium on Educational Advances in Artificial Intelligence, EAAI 2019 (2019). https://doi.org/10.1609/aaai.v33i01.33015033
18. Singh, S., Rathore, S., Alfarraj, O., Tolba, A., Yoon, B.: A framework for privacy-preservation of IoT healthcare data using federated Learning and blockchain technology. Futur. Gener. Comput. Syst. **129**, 380–388 (2022)
19. Wang, N., et al.: A blockchain based privacy-preserving federated learning scheme for Internet of Vehicles. Digital Comm. Netw. (2022). https://doi.org/10.1016/j.dcan.2022.05.020
20. Shrestha, A.K., Vassileva, J., Deters, R.: A Blockchain platform for user data sharing ensuring user control and incentives. Frontiers in Blockchain **3**, 48 (2020)
21. Shrestha, A.K.: Designing Incentives Enabled Decentralized User Data Sharing Framework. UofS Harvest (2021). https://hdl.handle.net/10388/13739

An Introduction to Arbitrary Message Passing

Jiasun Li[1] and Zhengxun Wu[2(✉)]

[1] George Mason University, Fairfax, VA 22030, USA
jli29@gmu.edu
[2] Fairfax, USA
wuzhengxun@outlook.com
https://sites.google.com/view/jiasunli

Abstract. We introduce an emerging arbitrary message passing (AMP) problem across multiple blockchains, which concerns transmitting general information from one blockchain to another. We develop a framework to highlight defining properties of AMP solutions and highlight that AMP protocols naturally generalize asset bridges and are special types of oracles.

Keywords: Arbitrary Message Passing · Blockchain · Interoperability

1 Introduction

Since Ethereum popularized smart contracts and enabled a large variety of decentralized applications (dApps), many new blockchains have emerged as smart contract platforms. As different blockchains do not have native access to each other's on-chain information (e.g. new state updates), there is an increasing demand for mechanisms to interconnect different blockchains and facilitate cross-chain communication. Many efforts also aim to enable "arbitrary message passing" (AMP), that is, to allow any information of a user's choice to be communicated across blockchains. AMP protocols tend to have wide applications. For example, the choice of which AMP protocol to use became one of the most heatedly debated items in the recent governance vote involved in the automated market maker Uniswap's expansion to Binance (See https://gov.uniswap.org/t/cross-chain-bridge-assessment-process/20148).

Despite AMP's potential value for the broader blockchain ecosystem and numerous projects being actively developed under the theme, the community still has limited knowledge about these emerging applications. This paper thus aims to fill this gap. Specifically, in this paper, we introduce the AMP problem and discuss its scope. The exercise allows us to distill a framework to analyze the mechanism, security (trust assumptions), and performance (cost/speed) of

---

Authors are listed alphabetically.

Z. Wu—Independent.

© The Author(s), under exclusive license to Springer Nature Switzerland AG 2023
J. M. Machado et al. (Eds.): BLOCKCHAIN 2023, LNNS 778, pp. 196–203, 2023.
https://doi.org/10.1007/978-3-031-45155-3_20

various AMP solutions. We envision the potential of arbitrary message passing in the broader blockchain/dApp ecosystem, pinpoint the risks they may bring along, and then enumerate open research questions.

The AMP problem is closely related to several themes in the literature. On the one hand, as ways to allow different blockchains to communicate with each other flexibly, AMP can be viewed as a generalization of cross-chain asset "bridges." [1] provide an overview of bridges from a blockchain scaling perspective — Their focus is mainly on bridges between layer 1 infrastructure and layer 2 applications, while AMP also additionally involves cross-domain (L1-to-L1) communications. [2] sample several cases of bridge hacking events — Their focus is mainly on bridges that support asset transfers, while arbitrary message passing also goes beyond asset transfers. Neither paper cover more recent projects that explicitly target the AMP problem. On the other hand, as concrete implementations of cross-chain communication, our analysis of AMP protocols also relates to the theoretical framework that formalizes cross-chain communication with surveys of a few early solutions [3], and the design philosophy for building interoperable blockchain systems [4].[1]

2 The AMP Problem

At a high level, AMP protocols aim to solve a specific cross-chain communication problem, so that a user can send a transaction that potentially contains arbitrary information or implements arbitrary business logic on a source chain (e.g. Ethereum) to trigger certain state updates on a destination chain (e.g. Solana). The state updates on the destination chain will contain information or reflect complicated business logic that corresponds to the initiating transaction on the source chain. Since the AMP space is fast-moving, there is currently no standardized terminology to synthesize the various concepts used in practice.[2] This lack of a common language hinders the comparison of existing solutions and further analysis of their respective properties. Therefore, in this section, we first define the AMP problem and discuss its scope.

Definition 1. *An arbitrary message passing (AMP) protocol allows a user to send a transaction on a source chain to trigger certain state updates on a destination chain. Both the initiating transaction on the source chain and the responding transaction on the destination chain may potentially contain arbitrary information and implement complicated business logic.*

An AMP protocol as defined by Definition 1 renders existing asset transfer bridges as special cases. The latter simply corresponds to Definition 1 in which the initiating transaction on the source chain is to lock certain funds in source-chain tokens while the corresponding transaction in the destination chain is

---

[1] Also see [5] and [6] for general discussions of blockchain interoperability.
[2] For example, Ethereum docs define "generalized message passing bridges" that share similarities with AMP protocols [Link], although the definition there additionally requires asset transfer functionality.

to mint the same amount of funds but in destination-chain tokens. Arbitrary message passing, as the name suggests, aims to go beyond simple asset transfers and instead implement a richer set of business logic.

That said, passing arbitrary messages creates new challenges that are not present among specific-purpose message-passing applications such as asset transfer bridges. One challenge comes from the difficulty in specifying in advance what kind of general messages may be passed. Because it is technologically feasible for validators to fake transactions and relay bogus calls to the destination chain, it is crucial to prove that an AMP call on the destination chain is indeed triggered by a genuine call on the source chain. For specific message-passing applications, the destination gateway may hard-code certain checks (e.g. an asset transfer bridge would only accept a relayed message if it contains a source-chain lock transaction with a valid signature) to mitigate this concern. However, for arbitrary message passing, it is difficult to hard-code such requirements in advance given the flexible nature of potential messages passed.[3] This concern thus requires additional security guarantees of the protocol.

Another unique challenge to AMP concerns error recovery and state management, as the architecture used in AMP-based dApps is more flexible and subtle compared to traditional ones. In traditional dApps, a smart contract receives tokens and executes business logic in an atomic fashion. AMP-based dApps, however, may receive tokens on the source chain and executes business logic on the destination chain; then in case the business logic fails to execute on the destination chain, how to handle tokens sent to the contract on the source chain becomes a concrete challenge unique to AMP protocols. It is thus important for AMP protocols to find a graceful way to handle such problems and avoid users losing tokens sent to the contract.

On the other hand, as ways to bring off-chain information on-chain, AMP can also be viewed as a specific type of "oracle" [7]. Since the "off-chain" information an AMP protocol brings on-chain is recorded (or at least is expected to be) on another blockchain, AMP protocols still enjoy specific features that a general oracle would not have, and these features may facilitate the design of AMP protocols. For example, AMP protocols typically do not have to worry about whether the source data is "correct," a problem at the center of almost all oracle problems, which is however guaranteed for AMP protocols by the integrity of the consensus process on the source chain.

Based on the reasoning above, we can interpret the AMP problem as a (proper) superset of the asset transfer bridging problem and a (proper) subset of the oracle problem.

---

[3] Therefore, as a practical implication, we note that any destination-chain contract that may be called by the destination-chain gateway should be aware of this potential risk. Such risks are also not purely theoretical. For example, Li.Finance, a bridge aggregator service, got hacked for $600K in 2022 Q1 [Link] since they allow arbitrary messages and didn't check for potential attacking messages encoded off chain.

3   An Analytical Framework

The AMP problem would be trivial if there were to exist a trusted third party who commits to constantly monitoring the source chain and sending transactions on the destination chain according to certain pre-specified logic. However, relying on a trusted third party destroys the very purpose of decentralized applications. The AMP problem aims to develop mechanisms to incentivize participation as well as to ensure good behaviors of a decentralized group of actors, and therefore simulate the role of the hypothetical centralized third party. In other words, AMP solves a specific consensus problem concerning the state of another blockchain.[4]

In the rest of our analysis of AMP, we assume that the source and destination chain both operate correctly, that is, they satisfy the basic properties of a blockchain. Then for a protocol to be considered securely implementing arbitrary message-passing, it needs to satisfy the following properties:

1. (Generality) The ability to pass messages across different chains should not depend on the content of the messages being delivered, nor on the applications invoked to initiate the message passing. For example, the protocol should be expected to be VM-agnostic — that is, it facilitates communication between blockchains of potentially different execution architectures.
2. (Safety) Each message executed on the destination chain is expected to have and only have one corresponding triggering message on the source chain. As a result, on the one hand, if there is no initiating message on the source chain, then no responding message can be faked on the destination chain; on the other hand, if a responding transaction on the destination chain has already been executed, then the original initiating transaction on the source chain cannot be reused to trigger another responding transaction on the destination chain. Furthermore, the relayed messages must be tamper-resistant, so that the responding transaction on the destination chain must follow pre-specified logic from the triggering transaction on the source chain.
3. (Liveness) A successful triggering transaction on the source chain must see a successful responding transaction on the destination chain within a reasonable amount of time. In other words, the liveness requirement requires that the responding transaction on the destination chain features censorship resistance so that it will not be suppressed following a successful initiation transaction on the source chain.

Besides the defining features listed above, an AMP protocol, different from a normal asset transfer protocol, may also desire some additional properties, including, say:

1. Atomicity, that is, to require both the triggering and responding transactions on the source and destination chains to either succeed or fail;[5]

---

[4] In this sense, we can also view asset transfer and oracle protocols as solving consensus problems on different targets. For asset transfer, it is about an asset-specific event such as locking or burning funds while for oracle, it is about any information (either on chain or in the physical world).

[5] For example, a Hashed Timelock Contract (HTLC) features atomicity.

2. Ordered delivery, that is, to require the sequence of transactions on the source chain to be preserved on the destination chain (i.e., if a user has two transactions TX1 and TX2 on the source chain, with TX1 executed before TX2, then on the destination chain the responding transaction to TX1 should also be executed before TX2).[6]

These properties may not be necessarily implemented in the protocol layer but rather delegated to specific applications.

To satisfy the above properties, many AMP protocols have emerged (including Axelar, Hyberlane, LayerZero, Nomad, Warmhole, and zkBridge, etc.). As we analyze these protocols, we find that they tend to share the following components:

1. Gateway: a smart contract that keeps track of messages to be passed. Gateways typically come in pairs, with one deployed on the source chain and the other on the destination chain. An AMP protocol monitors state changes within the source-chain gateway and triggers state changes within the destination-chain gateway.
2. Interface (sender): an application-level API built on top of the AMP protocol to send messages.
3. Interface (receiver): an application-level API built on top of the AMP protocol to receive messages.
4. Signer: an agent involved in the AMP protocol who ensures no passing of tampered/fake messages.
5. Relayer: an agent who passes messages from the source-chain gateway to the destination-chain gateway.
6. Slasher: an agent who checks messages to prevent fraud.

For exposition ease, we further group the six components into two modules: The first three components (gateways, sender interfaces, and receiver interfaces) consist of the *on-off ramps* module (of AMP protocols) while the last three components consist of the *intermediary* module (between the on-off ramps of the source and destinations chains). We should point out that in practice, these various components may have different names across protocols.

AMP protocols may also feature some optional components:

1. Gas services: Since AMP typically involves multiple types of tokens to pay gas on the source chain, gas on the destination chain, and protocol fees, emerging protocols often provide gas services that allow end users to pay all expenses in a single type of token (e.g., the source chain's native token). Such gas services, however, may bring along additional risks. For example, if an AMP protocol receives all payments in the source-chain token, then the protocol effectively shorts the destination-chain token and longs the source-chain token — while app users (or app builders if they pay for users) effectively long destination-chain tokens and short source-chain tokens. During periods with high price volatility, AMP protocols may incur non-negligible price risks.

---

[6] In practice, transaction nonces are often used to ensure ordered delivery.

2. Liquidity routing protocols: Such services are typically developed to support cross-chain swaps — say, to swap token A on the source chain to token B on the destination chain, which generally follows the process below: (1) send token A to the source chain's gateway, or if token A is not supported by a bridge, swap token A through an automated market maker (AMM) to a token I that *is* supported, and then send token I to the source chain's gateway, (2) the gateway locks or burns token A (or token I) on the source chain, (3) signers sign the Merkle root of the block that includes the above transaction, (4) relayers relay the message to the destination chain's gateway, (5) slashers verify the validity of the message on-chain (or off-chain slashers submit proofs during a challenge window in optimistic cases), (6) the gateway on the destination chain mints token B, or if token I is not token B, mints wrapped token I on the destination chain and then swaps token I to token B with an AMM.[7]

3.1 Extensibility

We further discuss how AMPs can be extended to perform add-on features, focusing on the two most used functionalities: (1) callContractWithToken, which is a contract deployed on the destination chain that conducts certain functionality upon receiving some tokens from the source chain; and (2) atomicity, which is a property that ensures the following fact: If the triggering transaction on the source chain is finalized, the triggered transaction on the destination chain can also be ultimately finalized.

Axelar's `callContractWithToken` function implements the callContractWithToken extension. Regarding atomicity, Axelar's design is also friendly to implementing it, as we propose two methods below:

1. Use application-specific monitors to check a signed transaction's inclusion on the destination chain, and if necessary use exponential backoffs to trigger its re-submission until it is included in the destination chain. This design keeps atomicity guarantees at the application level, which provides more flexibility and allows the underlying AMP to focus on ensuring message validity.
2. Enrich Axelar's software to allow transactions encode types such as "atomic," "no-wait," or "MEV-prevention." Then validators are required to ensure "atomic" transactions' execution on the destination chain, do not wait for finality on the source chain to submit "no-wait" transactions to the destination chain, and submit "MEV-prevention" transactions to say Flashbots' relay, etc.

Hyperlane recently added the `dispatchWithTokens` function to implement the callContractwithtoken functionality. Hyperlane's design also accommodates the implementation of atomicity, similar to that for Axelar as described earlier.

---

[7] Some protocols like Hyperlane have built-in liquidity routing modules, while others like Axelar delegate liquidity routing to third-party services.

It is tricky to implement atomicity on LayerZero since it requires coordination between oracles and relayers for failed transaction submissions. A similar assessment goes to Wormhole, for which there is not much programming flexibility on the intermediary layer.

4   Conclusion

This paper provides a preliminary look into the emerging application of cross-chain arbitrary message-passing (AMP) protocols. We develop a general framework of AMP protocols and sample several existing projects for illustration. We also analyze these protocols' respective mechanisms, security, and performances. That said, given the nascent nature of AMPs, there remain many open questions, which we conclude with.

First, what arbitrary messages to pass? The name of arbitrary message passing suggests the ability to pass "arbitrary" messages — but how arbitrary? This question is not merely about technological capability but rather about economic meaningfulness, as theoretically any arbitrary messages (up to a size limit) could be included as graffiti in source-chain transactions and be passed on to the destination chain (e.g. verifying the source and destination transactions' respective graffiti hashes). That said, if the arbitrary messages passed do not necessarily trigger transactions on the destination chain (but are only recorded on the destination chain), would it still be meaningful? How large is the set of meaningful messages that would benefit from arbitrary message passing?

Second, how large is the value of atomicity? We view a key feature of message passing to be the atomic coupling of source chain actions and destination chain actions. Without atomicity, one can always sequentially initiate a source-chain transaction first and then follow up with a target-chain transaction. Apparently, this atomic feature may improve user experience, facilitate time-sensitive cross-chain actions, and relax trust in destination-chain transaction execution (but as we discussed earlier, it may not mitigate trust concerns but rather migration trust of the execution of the destination-chain leg to trust of the message passing protocol). Is the list of these benefits exhaustive? For what applications are any aspects of these benefits the most salient?

Third, arbitrary message passing under limited smart contract support? Note that all the solutions so far discussed in practice require both the source and destination chains to be programmable to a certain level. This programmability requirement thus precludes solutions that bridge bitcoin or other blockchains with limited smart contract functionality. Given bitcoin's importance in the cryptocurrency ecosystem, however, how large is the value of message passing involving bitcoin, if any? If the value is large, then how to additionally develop practical message-passing solutions between bitcoin and smart contract platforms?

Finally, what applications should be built on AMP protocols? Candidates include bridge aggregators,[8] which tends to witness the highest volume among

---

[8] Notably examples include Swing, Bungee, Chainswap, Li.Finance, and BoringDAO, etc.

all bridges. However, it is often difficult to analyze the security of these bridges given that many are still closed-sourced.[9]

References

1. McCorry, P., Buckland, C., Yee, B., Song, D.: Sok: validating bridges as a scaling solution for blockchains, Cryptology ePrint Archive (2021)
2. Lee, S.-S., Murashkin, A., Derka, M., Gorzny, J.: Sok: not quite water under the bridge: review of cross-chain bridge hacks, arXiv preprint arXiv:2210.16209 (2022)
3. Zamyatin, A., et al.: SoK: communication across distributed ledgers. In: Borisov, N., Diaz, C. (eds.) FC 2021. LNCS, vol. 12675, pp. 3–36. Springer, Heidelberg (2021). https://doi.org/10.1007/978-3-662-64331-0_1
4. Hardjono, T., Lipton, A., Pentland, A.: Towards a design philosophy for interoperable blockchain systems, arXiv preprint arXiv:1805.05934 (2018)
5. Neulinger, A.: Towards a comparison framework for blockchain interoperability implementations. In: 2022 IEEE Crosschain Workshop (ICBC-CROSS), pp. 1–3. IEEE (2022)
6. Pillai, B., Biswas, K., Hóu, Z., Muthukkumarasamy, V.: Cross-blockchain technology: integration framework and security assumptions. IEEE Access **10**, 41 239–41 259 (2022)
7. Eskandari, S., Salehi, M., Gu, W.C., Clark, J.: Sok: oracles from the ground truth to market manipulation. In: Proceedings of the 3rd ACM Conference on Advances in Financial Technologies, pp. 127–141 (2021)

---

[9] Separately, although we found bridges to feature vastly different mechanisms, subtle trust assumptions, and varying security guarantees, they are typically hidden by existing bridge aggregators. Hence, they might expose users to the vulnerability of the least secure underlying bridge.

A Survey of Decentralized Storage and Decentralized Database in Blockchain-Based Proposed Systems: Potentials and Limitations

Muhammed Tmeizeh[1]([⊠])(ID), Carlos Rodríguez-Domínguez[2](ID), and María Visitación Hurtado-Torres[2](ID)

[1] Faculty of Engineering and Information Technology, Palestine Ahliya University, 1041 Bethlehem, Palestine
mohammedt@paluniv.edu.ps

[2] Department of Software Engineering, Higher Technical School of Computer and Telecommunications Engineering, University of Granada, 18071 Granada, Spain
{carlosrodriguez,mhurtado}@ugr.es

Abstract. Blockchain technology has geared towards many fields and utilities. Blockchain, as an efficient data storage, is one of the use cases that aims to provide a trustable and immutable data storage. This study provides a systematic review and evaluation of peer-reviewed publications aimed at pointing out the potential of blockchain as a data storage and/or queryable distributed database. The study presented herein shows that the undertaking of using blockchain technology as a storage or database technology is increasing exponentially due to the decentralization, immutability, and security that blockchain offers. The methodology used to conduct this study encompassed searching in several scientific databases and selecting related publications indexed in the Journal Citation Reports (JCR) ranking. The outcomes of the selected studies have been extracted, synthesized, and presented. As a conclusion of this analysis, we have detected that blockchain suffers from many challenging limitations, such as the absence of a well-defined data indexing structure, data retrieval methods and queries, significant response times, and storage capacity and performance limitations.

Keywords: Blockchain · Decentralized Database · Decentralized data storage

1 Introduction

Decentralized data storage might be one of the most prominent issues in the future of data storage since it provides security features such as immutability, integrity, and authenticity. Blockchain offers the characteristics of a decentralized data storage, also solving many potential security breaches [1]. Particularly, in blockchain, data is appended into blocks through transactions. Once a block is

© The Author(s), under exclusive license to Springer Nature Switzerland AG 2023
J. M. Machado et al. (Eds.): BLOCKCHAIN 2023, LNNS 778, pp. 204–213, 2023.
https://doi.org/10.1007/978-3-031-45155-3_21

appended to the chain, it is difficult to alter or compromise, making the saved data highly resistant to potential attacks. Furthermore, the distribution of data among participating nodes enhances data availability.

Furthermore, many studies have also tried to apply blockchain to develop database platforms. However, there are some open issues, most of them related to the conception of blockchain itself. For instance, the main goal of blockchain is to store information, but in contrast to a traditional database, blockchain stores the data as a ledger of transactions. In addition, databases are usually not designed to manage data blocks, but blocks are commonly considered the core of any blockchain system. In terms of throughput, traditional databases have a much higher capacity than blockchain, which can only serve a small number of transactions per second. Regarding the latency, for example, Bitcoin needs 10 min to confirm a new block. Moreover, the capacity of a blockchain is about hundreds of gigabytes (GB), and adding new nodes to the network worsens the case in terms of capacity, throughput, and latency [2].

Anyhow, despite the stated problems, many studies utilize blockchain as a storage or database. Most of those works aim to exploit the decentralization and inherited security features that blockchain can offer. On the other hand, understanding the security properties of blockchain plays the leading role in utilizing and enhancing the degree of trust that blockchain endows [3].

Some works have been done aiming to survey the storage capability of blockchain platforms, such as [29] but this paper aims to provide an overview of the current state of the art in blockchain-based solutions, highlight and classify them, and demonstrate the potential of blockchain as a data storage and/or database.

The main contributions of this paper to the blockchain research field are:

- Present the opportunities and challenges of integrating a blockchain with off-chain storage technologies.
- Evaluate the proposed models in terms of improving storage capacity.
- Study how the proposed models meets data security and privacy requirements.
- Obtain a picture of knowledge about the hybrid blockchain solutions that aim to utilize blockchain as a storage medium or distributed database.

The remainder of this paper is organized as follows: Sect. 2 introduces an overview of blockchain. Section 3 summarizes the article selection methodology we used to carry out this study. Section 4 introduces several types of blockchain storage applications. Section 5 shows blockchain and decentralized database studies and applications. Section 6 discusses the challenges for blockchain storage and hybrid blockchain database solutions. Finally, Sect. 7 concludes the paper and outlines some future work.

2   Overview of Blockchain

Bitcoin was proposed by Satoshi Nakamoto in 2008 and is the core platform of the existing blockchain. Bitcoins attract much attention from researchers and market. In general, all blockchain platforms contain a public ledger where confirmed transactions are stored in a chain of blocks [2]. Each confirmed block is appended to the end of the chain so that it grows continuously. Each block contains a hash of the previous block with a timestamp and transaction data, as shown in Fig. 1. Miners are responsible for adding new blocks to the chain through a process called mining. Once a miner adds a new block, the chain is updated on all nodes. Since the ledger is distributed among all participating nodes, many approaches called consensus algorithms are used to achieve a consensus state in the blockchain, e.g., Proof of Work (PoW), Proof of Stake (PoS).

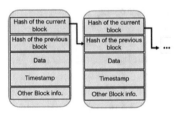

Fig. 1. Structure of a block

3 Research Methodology

This study has been chosen to analyze an overview of blockchain-based solutions for decentralized storage and databases by using a systematic literature review approach [4], with the following steps: Initially, the study is divided into two categories: Blockchain as storage and Blockchain as database. Then, to find relevant publications, the main keywords used are: "Blockchain AND (secure storage OR immutable storage OR store OR dynamic storage OR storing)", "Blockchain AND (decentralized OR distributed OR IPFS OR P2P)", "Blockchain AND (searchable OR search OR query OR analyzing)", "Blockchain AND (decentralized database OR distributed database OR database OR SQL)". Then, to create a shortlist of articles, only journal articles are considered; articles published before 2019 were excluded, and journals that are less than Q3 according to JCR were also excluded. Finally, abstracts are scanned to reduce the shortlist, and the selected articles are then subjected to a full-text search. The most relevant articles are opted out, and others are not considered. The search was conducted between December 2022 and March 2023. SpringerLink, Scopus, and Google Scholar were used.

4 Blockchain and Distributed Storage

Since blockchain is a promising solution for secure and tamper-proof data storage, many researchers have proposed solutions that use blockchain as a data store. In this section, 11 papers were reviewed.

Y. Chen et al. [5] introduced an e-health system that enables users to save and share their Electronic Medical Health Records (EHRs) in a secure and authorized manner. The proposed system stores encrypted EHRs in the Cloud using asymmetric encryption algorithms. Blockchain is used to index the files storing the EHRs, provide and store access control, and to store some textual private information, such as address, phone, and so on. Using Cloud storage adds some potential disadvantages to privacy concerns.

N. Alrebdi et al. [6] suggest a system to store Electronic Medical Records (EMRs) using distributed ledger technology (DLT), InterPlanetary File System (IPFS) and Cloud storage. The main idea is to save an encrypted copy of the medical file in a Cloud storage, while the access to this file will be directed from an smart contract of the blockchain. IPFS is used to generate the hash value for the given file and store this hash alongside with the patient ID in the blockchain. As illustrated, the blockchain itself does not store the patients' medical files.

S. Arslan et al. [7] modify the blockchain miners behaviours in order to send the processed data to nodes called storage nodes. In the study the video surveillance systems store its produced data in those nodes. The multimedia data split into smaller blocks named Group of Pictures (GOP) that compressed to some extent. Since compression process consumes the minor resources the study assumes that PoW can be avoided, due of using the private blockchain which assumes that nodes are trusted beforehand. Although, single point of failure should be considered, since data is saved without replicating it to multiple nodes.

L. Meng et al. [8] use a three-layer storage system: the user layer, the processing layer, and the blockchain layer. The file is divided into slices and encrypted using AES and RSA. The encrypted slices are then stored in IPFS, which is connected to a Hyperledger fabric that provides an index table of key-value data used to reach the data stored in IPFS. The key is the participant's ID, which is given a 160-bit hash value using the SHA1 algorithm, and the value is the hash of the encrypted file slices. The study provides experimental results for the use of allocation functions. Both Minimal Slice Number (MSN) policy and Minimal Node Number (MNN) policy are used, the study shows that MMN outperforms MSN in terms of utilizing IPFS space. In the study, although double encryption was used to enhance security, it is associated with high performance and latency.

V. Mani et al. [9] provide role-based access System for Electric Health Records (EHRs), it uses smart contracts over the Hyperledger Fabric to store the access permissions to EHRs of patients. The information of EHRs are stored in blockchain as hashed values, which will be used to reach the records in IPFS. The EHRs are stored as encrypted files inside the IPFS. Since Byzantine Fault Tolerance (BFT) consensus does not need to run mining process in Hyperledger Fabric to work, the authors used it to verify the patient preferences to preserve the patient privacy before sharing any record with others. Furthermore, Coutch DB is used as a state database. Although the proposed system offers security features, Various Types of EHRs such as videos and audios are not supported.

S. Vimal et al. [10] propose a study combines IPFS and blockchain, it aims to increase the efficiency of file storage and sharing inside IPFS using blockchain incentive coins named FileCoine. The authors suppose that any retrieval or storage process Will be paid by the user. Moreover, the study introduces many types of proof Services such as proof of storage, proof of replication and the proof of space-time, those services aim to proof that data still exists, proof that data is replicated with more than one copy, and lastly to proof that data is stored at the time that provider claims to. In addition, a blueprint of data ownership is proposed which is built on the top of Bitcoin blockchain.

Another proposed system by G. Gürsoy et al. [11] used the array mappers inside Smart contract of Ethereum to store pharmacogenomics data. The whole pharmacogenomics data is stored on-chain as key value form, where the key is the ID of the entered data and the value is the attributes of it such as the gene name. As illustrated the Smart contract serves as database. To retrieve any data the user or the caller will invoke the contract using its transaction ID. As the study provides efficient way to store the pharmacogenomics using smart contracts in private blockchain, some security and privacy concerns arise in case of using consortium or public blockchain. Cryptographic algorithms can be used to preserve security and privacy since the data will be transparent to all nodes.

F. Liu et al. [12] proposed work to save images using Distributed Image Storage Protocol inside IPFS. The image will be split into several fragments which will be then stored in IPFS nodes. Meanwhile, the blockchain is used to save the hashed value of the image. The study focused on designing a storage protocol which utilizes and reduces the space of storage in IPFS. Nevertheless, authentication and authorization over the stored images is needed.

O. Khalaf et al. [13] introduced an Optimized Dynamic Storage of Data (ODSD), it aimed to provide authorized, secure, ownership and traceable data storage using blockchain infrastructure. The broadcasted Wireless Sensor Network (WSN) data will be saved as a transaction inside the blocks of blockchain. Although the framework provides a new way to store WSN data it still has a potential to improve the privacy of the saved data since the data blocks are transparent to all participant nods of blockchain.

An additional paper by A. Sharma et al. [14] endeavors to enhance multi-tenant storage system using blockchain. The work offers on-chain and off-chain storage. The data stored in the on-chain is the metadata of the tenant data, while the tenant data is stored in off-chain database storage. A health care application is used by the authors as a case study to implement a multitenant software as a service SaaS application. Authenticity of off-chain files are verified using the on-chain hashed values. The off-chain database replication still one of the most important topics that require to be solved in hybrid blockchain solutions.

Cryptographic accumulator-based data storage scheme is designed by Y. Ren et al. [15], the study provides a new idea that aims to construct a scalable data storage for industrial internet of things (IIoT). It used the concept of sharding to split the blockchain network to multiple sub-chains running in parallel at the same time, hence the transactions will be divided into pieces that will be processed and verified together. The authors used local repairable code (LRC) sharded technology that provides good scalability and efficiency in reducing communication cost and data repairing among the used nodes in the sub chains. Bilinear accumulators are also used to enhance data security.

5 Blockchain and Database

The new existing solutions for blockchain and databases can be divided into two categories: The first is Blockchain enhanced with database capabilities, which aims to improve the performance and usability of blockchain to enable systematic data collection. The second category is the database refined with blockchain functions, which aims to ensure security and immutability of data. In this section, we have studied 7 works

5.1 Blockchain Refined with Database Features

G. Gürsoy et al. [22] introduce a new idea to store genomics inside a MultiChain private blockchain network using nested index of database, aiming to give the ownership, integrity, control and sharing of the data. The main challenge was storing a large scale-data on the blockchain network itself rather than using an associated storage, e.g. Cloud storage. The study offers a tool to enhance accessing and analyzing on-chain data. The nested indexing systems provides an identifier reference for each pushed transition, this reference is used to reach the intended data inside the transaction in a timely efficient way.

C. Lai et al.[23] designed a Merkle tree independent index structure that aims to provide query traceability over the data stored in blockchain. The data itself is stored on cloud, while tokens and indexes stored in blockchain to be used in searching and traceability. The system provides encryption, decryption, verification, and permission services using smart contracts. Testing is done on Hyperledger fabric. Furthermore, asymmetric encryption is used, certificate authority is provided to distribute public keys for the participant nodes.

K. Yue et al. [24] provide an overview for the approaches that can be used to migrate the block and transaction of the Bitcoin in order to perform Online Analytical Processing (OLAP) for the transactions, the SQL database used to perform the analytic over the migrated Bitcoin data.

In the study conducted D. Przytarski et al. [25] by the authors aim to provide a declarative query language for blockchain, they split the work into two categories. The first one is the front end which is the query syntax the second one is the back end which deal with. Many blocks are needed for queries that apply on expandable objects to ensure the state of data is compatible.

5.2 Database Refined with Blockchain Features

Z. Ge et al. [26] provide an analysis of performance for some hybrids database blockchain systems in terms of performance. The surveyed systems are BigchainDB [16], BigchainDB with parallel validation [16], BlockchainDB

[17], Veritas (Kafka) [21], and Veritas (Tendermint). The authors use Tendermint [18] to provide BFT, the Kafka [20] is used to provide Crush Fault Tolerance environment. Veritas with Tendermint and kafka were used to provide an equivalent way of blockchain platforms. The authors stored ledger using Sparse Merkle Tree [19] since it is more efficient to use it for verifying loges in terms of proof of existence, despite it needs more storage space. The study proposes a similar Veritas mechanism using Tendermint as a broadcasting service. The results showed that Veritas (Kafka) outperformed others in terms of number of performed transaction per second.

A study by M. Muzamma et al. [27] introduce an open-source system called ChainSQL that integrates database with blockchain. ChainSQL aims to solve the issue that distributed databases suffer from such as reliability. The actual data is stored in a MySQL database, whereas Ripple Blockchain is used to stores the transactions that will be used for the purpose of auditing over the stored data. In ChainSQL, transactions for updating data state are grouped into a proposal set. Then a consensus algorithm is used to confirm the proposal set.

Furthermore, T. Knez et al. [28] proposed a blockchain based transaction manager that used the blockchain alongside IPFS as a secure storage. IPFS is used to store the ontology data files while the blockchain stores the file identifiers. The system aims to track the changes done over the files, it lets users to query the data using SPARQL language. SPARQL endpoint connected to a local database using Apache Jena Triple Database. The local database is connected to blockchain to get the latest copy of ontology transitions to keep consistent.

6   Discussion

Table 1. Summary of Distributed Storage Solutions.

Ref.	Type of data	Capacity	Queryable	Blockchain Role	Off-Chain Storage
[5]	EHRs	Large	No	Indexing	Cloud Storage
[6]	EMRs	Large	No	Indexing	IPFS, Cloud Storage
[7]	Videos	Limited	No	Date Storage	–
[8]	Files	Large	No	Indexing	IPFS
[9]	EHRs	Limited	No	Indexing, Access Control	IPFS
[10]	Files	Limited	No	Incentives, Proof Services	IPFS
[11]	String	Small	Yes	Date Storage	–
[12]	Images	Large	No	Indexing	IPFS
[13]	Sensor Data	Small	No	Date Storage	–
[14]	EMRs	Large	No	Verify and Proof Services	Cloud Storage
[15]	IoT data	Limited	No	Sub-chains Networks	Bilinear accumulator

As illustrated from Table 1, most studies used off-chain to solve the capacity problem of blockchain, where the role of blockchain is an indexing system to reach the intended file inside the external storage. However, the studies that used blockchain as a direct data storage suffer from capacity limitation e.g. [7,11], and [13]. Furthermore, few studies provide extra layers of queries to retrieve the data such as [11]. Also, as we can figure out, blockchain technology shown the potential to be utilized with eHealth systems e.g. [5,6,9], and [14].

Many approaches use IPFS to store the data such as [6,9], and [10]. On the other hand, another approaches that aim to simulate or enhance scalability of blockchain such as [7] and [15]. Furthermore, some approaches try to use smart contracts or transactions of blockchain to store their data partially or totally such as [5,7], and [11]. As illustrated from the proposed works using off-chain has an added value of improving the capacity, but there in a need to modify some off-chain characteristics to grantee the data tamper proofing.

Table 2. Summary of Blockchain and Database Solutions.

Ref.	Article brief description	Article added value
[22]	Storing genomics in Multichain	Storing, accessing, and analyzing large-scale data on the blockchain
[23]	Blockchain index structure	Provides data traceability and encryption
[24]	Migrate Bitcoin data to a SQL DB	Perform OLAP over Bitcoin data
[25]	Store objects on the blockchain	provides a declarative query language
[26]	Analyse hybrid blockchain database	Increase transactions using Veritas(Kafka)
[27]	Integrates databases with blockchain	Enhance distributed database reliability
[28]	Blockchain based transaction manager for ontology data	Connect local ontology databases with blockchain to offer data traceability

Table 2 shows that the [22,23], and [25] studies have the same goal of storing data directly on the blockchain, but each work focuses on specific database features, for instance [22] focuses on accessibility, analysis, and indexing, [23] on data traceability, and [25] on providing a query language for expandable objects stored across multiple blocks. The [27] and [28] studies have a similar goal of integrating databases with blockchain to leverage blockchain features, while [27] focuses on traditional SQL databases. [28] works with ontologies database. In [22,23] and [25] it's possible to retrieve data from blocks, but it's not easy to run aggregation functions over the data, which was the reason for migrating data in [24]. In [26], the performance improvement was achieved by using Veritas with Tendermint, which improves blockchain broadcasting structure.

7   Conclusion

Decentralization in storage and databases provides notable features such as security, and availability. Blockchain has enormous potential when dealing with decentralization features. However, blockchain suffers from challenges that do not perfectly fit the requirements of decentralizing storage and databases. We conjecture that this survey can help readers gain an in-depth understanding of the challenges that face blockchain by summarizing, reviewing, and evaluating peer-reviewed publications published in highly ranked journals based on JCR.

One of the essential challenges is the lack of a well-defined index structure, which reduces response time and limits query capability. Throughput is also slow compared to traditional databases, limiting the ability to process a large number of transactions. In addition, blockchain growing size and adding new nodes exacerbates capacity and performance limitations. Moreover, the use of off-chain introduces new challenges such as data security, and availability.

Nevertheless, blockchain has a high potential to be used as a storage or database if it is modified either in its structure or by integrating it with a well-defined solution such as sub-chains, cloud or IPFS. Blockchain to store EMRs, images, videos, and WSN data has shown that there is a wide range of applications for blockchain-based systems. We conjecture that the future of Blockchain in this area can be done through many approaches, one of which is providing well-structured off-chain storage and/or enhancing it with database features such as a standard data model and indexing system. For future work, the authors aim to conduct a research paper proposing a blockchain-refined database solution.

Acknowledgements. This research was funded by the Spanish Ministry of Science and Innovation (State Research Agency), grant number PID2019-109644RB-I00, and Junta de Andalucía (Andalusian Regional Government), grant number B-TIC-320-UGR20.

References

1. Swan, M.: Blockchain: Blueprint for a New Economy. O'Reilly Media, Inc., Sebastopol (2015)
2. Nakamoto, S.: Bitcoin: a peer-to-peer electronic cash system (2009). https://bitcoin.org/bitcoin.pdf
3. Zhang, R., Xue, R., Liu, L.: Security and privacy on blockchain. ACM Comput. Surv. (CSUR) **52**, 1–34 (2019)
4. Kitchenham, B., Pearl Brereton, O., Budgen, D., Turner, M., Bailey, J., Linkman, S.: Systematic literature reviews in software engineering - a systematic literature review. Inf. Softw. Technol. **51**, 7–15 (2009)
5. Chen, Y., Ding, S., Xu, Z., Zheng, H., Yang, S.: Blockchain-based medical records secure storage and medical service framework. J. Med. Syst. **43**, 1–9 (2019)
6. Alrebdi, N., Alabdulatif, A., Iwendi, C., Lian, Z.: SVBE: searchable and verifiable blockchain-based electronic medical records system. Sci. Rep. **12**, 1–11 (2022)
7. Arslan, S., Goker, T.: Compress-store on Blockchain: a decentralized data processing and immutable storage for multimedia streaming. Clust. Comput. **25**, 1957–1968 (2022)

8. Meng, L., Sun, B.: Research on decentralized storage based on a blockchain. Sustainability **14**, 13060 (2022)
9. Mani, V., Manickam, P., Alotaibi, Y., Alghamdi, S., Khalaf, O.: Hyperledger healthchain: patient-centric IPFS-based storage of health records. Electronics **10**, 3003 (2021)
10. Vimal, S., Srivatsa, S.: A new cluster P2P file sharing system based on IPFS and blockchain technology. J. Ambient Intell. Humaniz. Comput. 1–7 (2019)
11. Gürsoy, G., Brannon, C., Gerstein, M.: Using Ethereum Blockchain to store and query pharmacogenomics data via smart contracts. BMC Med. Genomics **13**, 1–11 (2020)
12. Liu, F., et al.: A hybrid with distributed pooling Blockchain protocol for image storage. Sci. Rep. **12**, 3457 (2022)
13. Khalaf, O., Abdulsahib, G.: Optimized dynamic storage of data (ODSD) in IoT based on Blockchain for wireless sensor networks. Peer-to-Peer Netw. Appl. **14**, 2858–2873 (2021)
14. Sharma, A., Kaur, P.: Tamper-proof multitenant data storage using Blockchain. Peer-to-peer Netw. Appl. 1–19 (2022)
15. Ren, Y., et al.: Data storage mechanism of industrial IoT based on LRC sharding Blockchain. Sci. Rep. **13**, 2746 (2023)
16. McConaghy, T., et al.: Bigchaindb: a scalable Blockchain database. White Paper, BigChainDB (2016)
17. El-Hindi, M., Binnig, C., Arasu, A., Kossmann, D., Ramamurthy, R.: BlockchainDB: a shared database on Blockchains. Proc. VLDB Endow. **12**, 1597–1609 (2019)
18. Buchman, E.: Tendermint: byzantine fault tolerance in the age of Blockchains. University of Guelph (2016)
19. Dahlberg, R., Pulls, T., Peeters, R.: Efficient sparse Merkle trees. In: Brumley, B.B., Röning, J. (eds.) NordSec 2016. LNCS, vol. 10014, pp. 199–215. Springer, Cham (2016). https://doi.org/10.1007/978-3-319-47560-8_13
20. Kafka. https://kafka.apache.org/. Accessed 20 Mar 2023
21. Gehrke, J., et al.: Veritas: shared verifiable databases and tables in the cloud. In: CIDR (2019)
22. Gürsoy, G., Brannon, C., Ni, E., Wagner, S., Khanna, A., Gerstein, M.: Storing and analyzing a genome on a Blockchain. Genome Biol. **23**, 1–22 (2022)
23. Lai, C., Wang, Y., Wang, H., Zheng, D.: A blockchain-based traceability system with efficient search and query. Peer-to-Peer Netw. Appl. 1–15 (2022)
24. Yue, K., Chandrasekar, K., Gullapalli, H.: Storing and querying Blockchain using SQL databases. Inf. Syst. Educ. J. **17**, 24 (2019)
25. Przytarski, D., Stach, C., Gritti, C., Mitschang, B.: Query processing in blockchain systems: current state and future challenges. Futur. Internet **14**, 1 (2021)
26. Ge, Z., Loghin, D., Ooi, B., Ruan, P., Wang, T.: Hybrid Blockchain database systems: design and performance. Proc. VLDB Endow. **15**, 1092–1104 (2022)
27. Muzammal, M., Qu, Q., Nasrulin, B.: Renovating blockchain with distributed databases: an open source system. Futur. Gener. Comput. Syst. **90**, 105–117 (2019)
28. Knez, T., Gašperlin, D., Bajec, M., Žitnik, S.: Blockchain-based transaction manager for ontology databases. Informatica **33**, 343–364 (2022)
29. Zahed Benisi, N., Aminian, M., Javadi, B.: Blockchain-based decentralized storage networks: a survey. J. Netw. Comput. Appl. **162**, 102656 (2020)

A Blockchain-Based Framework Solution to Enhance the Remote Assessment Process in HEIs

Paulo Victor Dias[1] 🆔, Firmino Silva[1,2(✉)] 🆔, and António Godinho[1,2(✉)] 🆔

[1] ISLA – Research Center – Polytechnic Institute of Management and Technology, 4400-107 Vila Nova de Gaia, Portugal
{a22008554,p40588,p40030}@islagaia.pt
[2] COPELABS, Lusófona University, Campo Grande 376, 1749-024 Lisbon, Portugal

Abstract. Major improvements in eLearning system can result from the powerful benefits of blockchain technology. Much personally identifiable information is passed between the student, the school, and third-party providers as part of the educational process. This requires a system that combines information security and the ability to send data across an extensive network in a purely virtual way. Blockchain appears to be tailor-made to assist in safeguarding and protecting this new education model. The reliability of a distance assessment system deserves the scientific community's attention, which addresses the definition of assessment strategies and the adoption of technologies that promote a high degree of integrity of the distance assessment system. Enormous challenges are found and include the use of smart contracts, mitigation of plagiarism and fraud, copyright, authenticity, integrity, and immutability of academic assessment documents. The main goal of this research paper is to create a blockchain framework that enables the secure exchange of sensitive information in the remote assessment of a Higher Education Institution environment while maintaining the data's authenticity, immutability, and overall security, to prevent e-cheating.

Keywords: Remote Assessment · Blockchain · HEI · e-Learning · EVM · Remix

1 Introduction

Adopting digital technologies in education has not only transformed traditional teaching methods, enabling students to learn in new and innovative ways, but has also provided new approaches to the assessment process, particularly online assessment. Global crises like the recent COVID-19 pandemic enable online assessment approach as one of the most interesting potential applications. This alternative should be considered for these situations and adopted as equally applicable according to the assessment methodology followed in a course of any curricular unit of a Higher Education Institution (HEI). The challenges posed by this process are of various types, from the availability of different technologies and devices to the reliability of the process that can guarantee a high level of trust and mitigate or eliminate plagiarism and other types of fraud in an online assessment

J. M. Machado et al. (Eds.): BLOCKCHAIN 2023, LNNS 778, pp. 214–223, 2023.
https://doi.org/10.1007/978-3-031-45155-3_22

system [1]. In online activities like remote assessments and use of smart contracts [2] in academics, potential issues are often centered around the reliability and credibility of the information students provide. Additionally, concerns may be related to mitigating scientific paper plagiarism, ensuring copyright compliance, maintaining the authenticity and integrity of academic documents, and ensuring their immutability. The success of the remote assessment process depends heavily on the effective management of fraud risks, which require either mitigation or complete elimination.

In a nutshell, assessment systems can be considered a process of low reliability among the actors involved. The need for a mechanism that increases the confidence of the information transacted without depending on the initiative of the actors is necessary. Thus, blockchain [3] appears to be tailor-made to assist in safeguarding and protecting this new approach of e-Learning [4–6] as well as remote assessment process [7, 8]. The educational sector is the newest application of blockchain [9] and guarantees academic documents' authenticity, integrity, nonrepudiation, and immutability [10].

The main goal of this research paper is to present a blockchain framework that enables the secure exchange of sensitive information in the remote assessment in HEI environment while maintaining the data authenticity, immutability, and overall security, to prevent e-cheating. We will also evaluate in a controlled environment the applicability of the solution and its costs to create a more complete sense of the system's behavior. The main steps of the methodology we applied (Design Science Research [11]) are as follows: 1) Study the state of the art on remote assessment systems based on blockchain in HEI; 2) Define the requirements of the blockchain framework; 3) Build and test a prototype of a blockchain system to validate the studied concept.

The paper is structured as follows: the 2nd section resumes state of the art. A prototype development is detailed in the 3rd section. Results and discussion are in the 4th section, and the Conclusions are summarized in the last section.

2 State of the Art

Several studies point to major challenges that identify piracy, plagiarism, inappropriate use of copy and paste, e-cheating, and fraud as the main disadvantages of using remote assessment platforms [12, 13]. The disruption of the educational process in various educational institutions led by the Covid-19 pandemic and the outbreak of problems related to e-cheating and plagiarism that stemmed from it led [14] to an investigation with three central questions: "1) Is cheating inherent in distance learning? 2) What kind of cheating is used in distance learning? 3) What are the measures taken to tackle this problem?". The authors concluded that for the first question, studies are still not consensual, without a clear perception if e-Learning is better or worse than face-to-face learning. The types of e-cheating used are many and varied for the other two questions. Meanwhile, two approaches were found useful to control plagiarism: preventive and therapeutic, with the preventive approach having more effect in preventing plagiarism.

The research conducted by Kocdar et al. [13] aimed to study students' perceptions of cheating, plagiarism, and trust in e-assessments according to their experience. Results revealed no significant difference in students' perceptions and feelings towards cheating and plagiarism before the e-assessment experience. On the other hand, students' perceptions of cheating and plagiarism regarding their learning mode showed significant differences. Another finding was that distance education students had lower trust in e-assessment than in face-to-face courses.

The retrospective research of Levine and Pazdernik [15] yielded a four-prong anti-plagiarism program and its impact on the incidence of plagiarism in a post-Doctor program. The research showed that a combination of structured education modules related to education, presented some positive results. The Turnitin plagiarism detection software, implementation of policies and procedures, and support from the institution's writing center, significantly impacted mitigating these aspects. The prototype development requirements, raised from these studies, were taken into account to define the basis of work in the present research.

3   Prototype Development Approach

This section presents the development of the blockchain-based framework solution as a sequence of previous work [8], which aimed to study the state-of-the-art methods of distance learning in HEI, to identify the main guidelines used to mitigate e-cheating on the distance learning environment and to define the requirements for developing a framework to support remote assessments processes. The logical process flowchart and other additional information can be seen there. The present work builds a prototype supported by blockchain and will contribute to the improvement of information security, integrity, and authenticity in the remote assessment learning system in HEI.

The logical functioning of the framework runs as follows: lecturers and students share and access assessments through a combination of private and public keys. After completing each assessment, the submission is digitally signed, timestamped, and recorded on the blockchain. HEI is responsible for registering and revoking the permissions of teachers and students in the system and deploying the smart contracts in the blockchain. These smart contracts will make the interface between the stakeholders and the blockchain, automating the process in a transparent and auditable way. The blockchain works as a decentralized database for the student's progress in pursuing an academic diploma. This way, characteristics such as authenticity, integrity, and nonrepudiation are incorporated into the system to mitigate attempts of misconduct during online assessments. Although plagiarism detection is a very important issue during the remote evaluation process, the current proposal has the main focus on creating a more reliable infrastructure that allows better auditability of the information.

3.1   The Prototype Implementation

The programming language used for developing smart contracts was Solidity (https://soliditylang.org/) and it was also used an online Integrated Development Environment (IDE) named Remix (https://remix-project.org/) to simulate the deployment of smart contracts in the Ethereum Blockchain network. Figure 1 illustrates the class diagram of the smart contracts. The full code is displayed in the Github repository.

The Ethereum blockchain was chosen for three main reasons: credibility (of extreme importance for different sectors of the economy, companies of different sizes and academic projects), material availability (considerable amount of documentation available

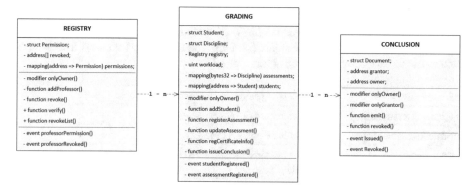

Fig. 1. Smart contracts class diagrams

to support a network), and cost and complexity to develop and test the framework (Remix IDE gives the possibility to run smart contracts on virtual machines with the simplicity of doing everything on a single webpage).

Smart Contract 1: REGISTRY. This smart contract is the first to be deployed in the system. It is responsible for registering the teacher's information who is allowed to interact in the network. Only the HEI can deploy this smart contract in the blockchain. Also, there are two restricted functions, *addProfessor* and *revoke*, where HEI is the owner. The last two functions are unrestricted. All four functions are described below (*r* states for restricted, and *ur* states for unrestricted):

addProfessor *(r)*	register a new professor in the network
revoke *(r)*	modify the status of any mapped participant in the network
verify *(ur)*	returns a Boolean value when checking a participant's status
revokeList *(ur)*	returns an array of Ethereum addresses of participants that had been revoked in the network

To add a teacher to the blockchain, the function *addProfessor* is called by HEI. This function requires the teacher's address in the blockchain and the expiring date as input parameters. If HEI needs to invalidate some of the participants' statuses in the network, the function *revoke* is called. To avoid any unauthorized processing of these functions, a modifier called *onlyOwner* is used. Lastly, two functions were created for verification purposes. These functions are unrestricted and can be called by anyone on the blockchain. The first one is *verify* which calls for the user's status in the network. The outputs will be false if the participant was not previously revoked and true if it was. Another way to output true is if the expiration date has expired. The second is *revokeList*, which returns an array list of all Ethereum addresses revoked in the network upon the call.

Smart Contract 2: GRADING. Although its deployment is restricted to the HEI, the constructor defines a teacher as the owner of the smart contract. It requires as inputs the number of assessments that a student must complete to conclude the studding unit, the address of the previously deployed REGISTRY, and the address of the teacher that will execute the functions of the new smart contract. There are four functions as described below (*r* states for restricted, *i* states for internal):

addStudent *(r)*	enrolls a new student in the network
registerAssessments *(r)*	allows to record the assessments
updateAssessments *(r)*	registers the student's progress
issueConclusion *(i)*	creates a new instance of smart contract after all assessments are concluded

The first main function *addStudent* registers students based on two parameters, their public address in the blockchain and identification in a Bytes32 format. Next, *registerAssessments* allows the designated teacher of the studding unit to record the identification and its assessment types. The third main function is *updateAssessment,* which the teacher calls to register each time a student progresses in fulfilling the assessments. Besides, this function automatically verifies whether the student has achieved the number of assessments necessary to conclude the studding unit. If this is true, the internal function *issueConclusion* is activated and executed, creating a new instance of smart contract called CONCLUSION. To avoid any unauthorized execution of these functions, a modifier called *onlyOwner* is also used in this smart contract. In this case, another requirement is added to the modifier. Besides only the owner being allowed to execute the functions, its permission must not be expired.

Smart Contract 3: CONCLUSION. This is responsible for generating the students' certificate of conclusion for the subject in which he was enrolled. It uses as inputs the students' address and the corresponding teacher's address of the studding unit in question. If needed, pedagogical supervisors can check the student's status by finding the CONCLUSION containing the student's address. This means that it can be verified if the certificate has been disabled or if the teacher was unauthorized for any reason. There are two main functions in this smart contract:

issue	emits the conclusion certificate
destroy	invokes *selfdestruct* function to correct mistakes in the system

Since this is an ending point in the student's progression for completing the studding unit, some tools must be available to correct mistaken updates when recording the inputs. An example of wrongdoing could be a teacher wrongly recording an assessment credit in GRADING, prompting an anticipated CONCLUSION issuance. Thus, the function *destroy* could be called. This function would disable the smart contract and clear its code and data in the most recent block in the chain.

4 Results and Discussion

This section presents the steps of an experiment conducted to validate our proposal. With the given experiment results, we also estimate the cost charged by Ethereum Virtual Machine to run our prototype. Evaluations regarding the Ethereum's throughput speed and transaction delay were not considered because of the nature of the system. Information update is mostly done on exams week, which happens a few times on an academic semester. The total number of transactions are low and not dependent on real-time update. A discussion about the results and possible alternatives is also presented.

4.1 Testing Experiment

When we open the Remix running on a Virtual Machine (VM), a set of accounts with individual addresses and an amount of ether and gas limit per address is pre-assembled. We selected an address that starts with 0x5B38 to represent the HEI in the system. Then, REGISTRY is deployed. The first step is to register a teacher. It requires the teacher's address, a hash value arising from the combination of the HEI and teacher addresses, and a Unix timestamp for the expiry date. This is the main step because the next smart contract is deployed only after the teacher's registration. Three other functions can be called in this smart contract: *revoke*, which receives as input the teacher's address to be withdraw from the network and can only be executed by the HEI; *revokeList*, which is a list that shows all the addresses that were revoked; and *verify*, that checks if the address referring to a teacher is registered in the network. All functions mentioned were tested, resulting in a status of *"true Transaction mined and execution succeed"*. The cost of gas to invoke each transaction was recorded as well.

Table 1. Gas cost of the prototype for each function.

REGISTRY		GRADING		CONCLUSION	
Function	Gas	Function	Gas	Function	Gas
constructor	569.231	constructor	1.395.271	constructor	278.133
addProfessor	93.450	addStudent	83.419		
revoke	75.506	registerAssessment	78.887		
verify	26.534	updateAssessment	86.247		
		issueConclusion	339.476		
		regCertificateInfo	84.066		

To deploy the next smart contract, GRADING, some parameters are requested in advance: the number of assessments required to complete the course, which in this example was chosen as three; the address of the deployed smart contract REGISTRY (0x0fC5); and the address of the teacher previously registered, that will be responsible for operating this new smart contract (0xAb84). With the contract deployed, several functions are available for the contract owner (teacher 0xAb84) to execute. All functions were executed and recorded. For the *addStudent* function, the teacher (0xAb84) requests GRADING (0x9D83) to execute with the inputs being: 0x7873 as the student's address and 0x5061 (Paulo Victor Dias) as the student's id initial Bytes32 address. The next transaction that registers an assessment for the course took as input: an id, in Bytes32, with the value 0x4173 (Assessment n1); and an expiration date, in Unix, 1671321599 (17.12.2022 at 23h59m59s). Then, the following transaction updates the student's progress toward the completion of the studding unit. It took as input: the id of the assessment (0x4173), the status of completion (true), and the address of the student that completed this assessment (0x7873). To verify if the subject was correctly assigned and follow the student's progress, function *verifyStudent* was also tested. Next, the *update-Assessment* function reaching the point where the number of assessments required to fulfill the studding unit was completed, automatically issuing a conclusion certificate through a third internal smart contract, CONCLUSION. The arguments passed to the conclusion certificate were: "0" → Issuer; "1" → Receiver; "2" → Date. The last transaction tested was the *regCertificateInfo* function. This function updates the conclusion certificate of the student with additional information if the teacher intends to. In our example, the inputs were as follows: a title for the certificate in Bytes32, 0x436f (Conclusion of Discipline n1); a description, also in Bytes32, 0x4176 (Avg grade: 18); and address for the issuer, 0xAb84; and the address of the student, 0x7873.

To estimate the cost of running our prototype in the Ethereum network, we considered the costs related to transactions and execution of contracts and functions. The scenario to approximate the gas cost necessary to deploy and execute our proposal is one teacher responsible for one studding unit, with twenty students, and three assessments of the same type, required to be fulfilled by each student. We also assumed that all students attended the classes and received a conclusion certificate. After the gas estimation, we also present the cost in US$. Table 1 compiles the gas cost for every transaction executed in the testing experiment. If we analyze the details of the table, it can be noticed that one cost stands out in matters of gas use, the constructor's deployment. There are a few reasons for that. One is the size of the inputs, which is more significant than in the other functions. Furthermore, there are internal variables that are initiated in non-volatile memory that depend on extensive instructions on the Ethereum VM. However, constructors are instantiated only once when the smart contract is deployed, making the cost a one-time paying fee. With the gas costs of Table 1 we can estimate how much gas is needed to run our prototype in the scenario abovementioned. $Cost_1 = 569.231 + 93.450 \times P$ expresses the amount of gas needed to deploy REGISTRY and register the permission for every eligible teacher. $Cost_2 = P \times (C \times (1.395.271 + (78.887 \times A_T)))$ represents the cost of deploying instances of GRADING for each teacher. In addition, it is also considering the cost of registering assessments' types for every REGISTRY. $Cost_3 = S \times 83.419$ calculate the gas needed to enroll new students to each studding unit. $Cost_4 = S \times A_N \times 86.247$ calculates the total cost of updating students' progress. $Cost_5 = S \times 339.476$ calculates the total gas consumed for issuing the conclusion certificate for each student. The letters in the equations are: P = number of teachers; C = number of studding units; A_T = types of assessments; A_N = number of assessments; S = number of students. The total gas cost was 15.769.559. From the calculations, we could infer relevant insights regarding what weighs more in gas consumption when running our prototype. For an average class with twenty students and three assessments of the same type (e.g., two midterm exams and one final exam) for the conclusion, we noticed that the execution of functions responsible for handling student's data, such as *updateAssessment* and *issueConclusion*, dictate the total cost. The explanation for this relies on the number of times these functions are called. Although they do not require much gas to be individually executed, their total cost grows proportionally since they are called several times per student during the semester.

To estimate the cost, we sum up all the gas needed to register the teacher, the students, and their achievements while they complete all assessments, and to create an instance of CONCLUSION for each student. We did not insert the costs of initially deploying the smart contracts REGISTRY and GRADING since it would tend to zero as these costs are shared by all students entering and leaving the HEI over time. So, the costs per classroom are 13.805.057 gas. According to the ETH/USD exchange rate of the day of the simulation, the total gas amount was US$25,189.

4.2 Discussion

Our solution was simulated in a scenario where the Ethereum blockchain still used the Proof-of-Work consensus protocol. Other types of blockchain networks are available concerning their architecture: private, hybrid, or public permissioned. If a private blockchain were used, three main attributes could be directly affected: network arrangement, cost and privacy. As all participants are known in private blockchains, the authentication and authorization mechanism could be done differently, exempting the need for a smart contract for validation. Also, attaching our solution to a public blockchain bounds us to its protocols. Alternatively, detaching the solution from a real cryptocurrency, which is charged for transacting in the Ethereum network, could lower the cost of implementation. Even though private blockchains need to take the infrastructure costs into account, it would be easier for the institutions to forecast the costs for the system in the long term. As the nature of the project is of recurrent updates of information for many users, together with a volatile price for transaction (gas price), this might offer a downside when in comparison. Private blockchains can decrease privacy issues in exchange for the centralization of information and less transparency. Since the core of the solution does not absorb much advantage from the decentralization and transparency attributes that public blockchains have, private blockchains can indeed bring improvements to the system. Having both public and private blockchains in a hybrid arrangement could work in a way that all transactions with sensible information are made in a private blockchain, and the metadata of these processes could be registered in a public blockchain. This solution brings more complexity to implementation, and raises the cost. It also points back to a less decentralized system as it would be needed a central authority to validate the information in the private network and update it to the public. Since the system deals with remote assessments, the information has a higher privacy issue, and the need to update the information on a public network, even in the form of metadata, does not bring much benefit to the solution as a whole. Public Permissioned blockchains provide an environment where a group of participants could be granted different permissions on the network (e.g., writing or read-only permissions). This arrangement could lessen the cost of executing the system because it would not need a complex validation system with cryptocurrencies or computational power. Hyperledger Fabric (HF) (https://www. hyperledger.org/use/fabric) could be a good example of a blockchain network to make this solution possible. Here, there is still an exchange between full decentralization and transparency. Still, given the purpose of the system, it may be considered for future improvements to test it in the HF and compare results.

Following, we discuss privacy and security matters that may derive from this work. In relation to privacy issues, consideration should be given to whether private information about stakeholders may be disclosed in any way. No grades or other personally identifiable information is uploaded when recording students' progress in the system. What is recorded is how many assessments were carried out until it publishes the completion of the subject that the student was taking. Furthermore, all students are identified by their public key, so even if that data goes on the blockchain, there is no way to correlate it with a student. Security affairs are another crucial aspect of the smooth functioning of the framework. For this, one must examine how each stakeholder participates in the network and how they can interfere maliciously. Analyzing the students' perspective, they cannot access any smart contract; this is the teacher's role. Thus, no student can intentionally harm or interfere in the system. Looking at the teacher' perspective, they have the authority to register the assessments and update students' information in the blockchain but cannot interfere in other studding units. A teacher can deliberately provide fake information about a specific student's progress but can be easily identified by auditing the transactions. Moreover, they hold themselves accountable for their transactions in the blockchain, making it difficult for them to misconduct. Finally, from the perspective of HEIs, they are responsible for registering the participants and deploying smart contracts but cannot interfere with updating evaluations. Therefore, they cannot surpass the teacher's authority in the process.

5   Conclusion

The COVID-19 outbreak brought new attention to a topic that was already building its way in relatively slow steps: remote education. The boom in online platforms to fulfill problems with in-person education also shed light on the vulnerabilities of the existing systems regarding personal data, security in assessments, and the discrepancy of access to solutions between HEIs. The framework proposed was conceived as a blockchain infrastructure and a set of smart contracts to record academic progress and issue course completion certificates in a transparent and automated manner. Covers a variety of tools, such as cryptographic pair keys, to support the goal of mitigating academic dishonesty in online evaluations. Looking from the cost perspective, the system shows limitations for a real implementation since running a framework that needs recurrent updates, depending on a volatile cryptocurrency, would make it difficult to HEIs to project the costs of using the system per semester. Even with the value of ETH simulated, the total cost calculated would become impractical for most parts of the HEIs. In essence, the developed framework met the desired proposal, fulfilling its purpose of delivering a safer network for improving data integrity, the participants' authenticity, and validating the assessments carried out. Future work will consider plagiarism detection tools to be activated automatically by a smart contract, accepting or rejecting the submission of an assessment depending on an acceptable percentage. Transforming the current system into a plugin to be used in LMS platforms such as Moodle could also be considered, since it could facilitate and increase the adhesion of the solution by HEIs. Also, migrating the system from the Ethereum blockchain to the Hyperledger Fabric blockchain may be more suitable for the nature of the project.

References

1. Holden, O.L., Norris, M.E., Kuhlmeier, V.A.: Academic integrity in online assessment: a research review. Frontiers in Education **6** (2021)
2. Ante, L.: Smart contracts on the blockchain – A bibliometric analysis and review. Telematics and Informatics (16 Oct 2020)
3. Pilkington, M.: Blockchain technology: principles and applications. In: Research Handbook on Digital Transformations, pp. 225–253. Elgaronline (2016)
4. Johnes, S.: The impact of blockchain technology on the eLearning industry. eLearning Industry (07 Nov 2021). [Online]. Available: https://elearningindustry.com/impact-of-blockchain-technology-on-the-elearning-industry. Accessed 23 Fev 2022
5. Heitz, C., Laboissiere, M., Sanghvi, S., Sarakatsannis, J.: Getting the next phase of remote learning right in higher education. McKinsey (23 Apr 2020). [Online]. Available: https://www.mckinsey.com/industries/education/our-insights/getting-the-next-phase-of-remote-learning-right-in-higher-education. Accessed 13 Jan 2022
6. Oganda, F.P., Lutfiani, N., Aini, Q., Rahardja, U., Faturahman, A.: Blockchain education smart courses of massive online open course using business model canvas. 2020 2nd International Conference on Cybernetics and Intelligent System (ICORIS), pp. 1–6 (2020)
7. Li, C., Guo, J., Zhang, G., Wang, Y., Sun, Y., Rongfang, B.: A Blockchain System for E-Learning Assessment and Certification. In: 2019 IEEE International Conference on Smart Internet of Things (SmartIoT), pp. 212–219 (9–11 Aug 2019)
8. Dias, P.V., Silva, F.: Blockchain and digital signature supporting remote assessment systems: a solution approach applied to higher education institutions scope. Adva. Tourism Technol. Sys. **23**, 05 (2023)
9. Steiu, M.-F.: Blockchain in education: Opportunities, applications, and challenges. First Monday **25**(9), 7 (2020). September
10. Morais, A.M.D., Lins, F.A.A.: Uso de Blockchain na Educação: Estado da arte e desafios em aberto. Revista Científica Multidisciplinar Núcleo do Conhecimento **22**(10), 78–100 (2020)
11. Hevner, A., Chatterjee, S.: Design Research in Information Systems: Theory and Practice, vol. 22, p. 320. Springer Science & Business Media (2010)
12. Arkorful, V., Abaidoo, N.: The role of e-learning, advantages and disadvantages of its adoption in higher education. Int. J. Instruct. Technol. Dista. Learn. 29–42 (Jan 2015)
13. Kocdar, S., Karadeniz, A., Peytcheva-Forsyth, R., Stoeva, V.: Cheating and plagiarism in e-assessment: Students' perspectives. Open Praxis **10**(3), 221–235 (2018)
14. Mohamed, M., Shaikha, A.: Whither plagiarism in distance learning academic assessment during COVID-19?. In: 2020 Sixth International Conference on e-Learning (econf), pp. 1–5 (2020)
15. Levine, J., Pazdernik, V.: Evaluation of a four-prong anti-plagiarism program and the incidence of plagiarism: a five-year retrospective study. Assess. Eval. High. Educ. **43**(7), 1094–1105 (2018)

Detecting Fraudulent Wallets in Ethereum Blockchain Combining Supervised and Unsupervised Techniques - Using Autoencoders and XGboost

Joao Crisostomo[1]($\boxtimes$), Victor Lobo[1,2], and Fernando Bacao[1]

[1] Nova IMS, Lisbon, Portugal
`jdiogo.rcosta@gmail.com`
[2] CINAV, Portuguese Naval Academy, Almada, Portugal

Abstract. The illicit activity in Blockchain reached an all-time high in 2021. In this work, we combined two machine learning techniques, Autoencoder (AE) and Extreme Gradient Boosting (XGBoost), to improve the performance of predicting illicit activity at the account level. The choice of autoencoding technique allows us to be able to detect new MOs (modus operandi) from fraudsters. With an Autoencoder trained only with healthy accounts, we are not misleading the model to focus on specific MOs as it happens with tree-based models. This allows us to introduce a dimension that could capture future fraudulent behaviours. Furthermore, the dataset was generated considering the real applicability of the model, i.e. it mimics what can realistically be obtained in a practical situation. With this approach, we were able to improve the state-of-the-art performance. In our test set the precision-recall AUC (area under the curve) of our final model increased by 12,07% when compared with our baseline.

Keywords: Anomaly detection · Ethereum · Blockchain · Autoencoder · XGBoost · Fraud Detection · Malicious Accounts · Machine Learning

1 Introduction

Ethereum, the second-largest blockchain network, is a decentralised platform that runs smart contracts: executable code without any possibility of downtime, censorship, or third-party interference. These characteristics make Ethereum a powerful tool for building decentralized applications (dApps), which can potentially disrupt a wide range of industries. Ethereum blockchain is the main player with smart contract-enabled coins whose adoption has been increasing since 2016 and has achieved the status of the most used type of blockchain technology in 2022.

However, like any complex system, this blockchain platform is vulnerable to fraudulent behaviours such as money laundering or illegal activities. In 2020 alone, the economic losses in blockchain-related security incidents were up to 17.9 billion dollars [1]. Plus, over the last five years, the total cryptocurrency value received by illicit addresses has been increasing, hitting a new all-time high in 2021 [1]. Due to this trend, researching machine learning applications to detect and avoid these threats, increasing the overall security of the Ethereum Blockchain Network and allowing a wide adoption is, more than ever, an urgency. According to the literature, the research so far has been mainly focused on using only a specific type of machine-learning approach, supervised or unsupervised techniques and always concerning a specific type of anomaly, such as fraudulent behaviour or smart contract exploitation. Furthermore, the datasets used don't take into consideration the constraints of gathering data in real time. Also, in some situations, the information available at the moment of decision could drift apart from those datasets.

Our research aims to provide a different perspective in detecting malicious accounts for the Ethereum Blockchain, tackling this problem with the combination of an Autoencoder and an XGboost. The Autoencoder was trained using only numerical features from healthy accounts. Then we added the RMSE between the autoencoder input and output to the dataset, calling this new feature - AE score. Afterwards, we trained an XGBoost model with all features from our ETL process plus the AE score. The ETL process to generate the dataset takes into account the real-time constraints and the data available at the moment of decision. The experiment results were promising, improving our precision-recall AUC metric by 12.07% from our baseline - XGboost standalone.

This paper starts with Sect. 2 - Related Work, where the basic concepts of the Ethereum blockchain are introduced and presents the developed research regarding machine learning methods to detect and mitigate malicious activity in the Ethereum blockchain. Then, in Sect. 3, we introduce the methodology used in this experiment, detailing the building blocks of our final model and the process of data gathering. Following this, Sect. 4 describes the experiment results and achievements. Finally, it presents a conclusion where we discuss the limitations of this research and further work.

2 Related Work

Blockchain technology has the potential to revolutionise a wide range of industries by providing a secure and transparent way to record and transfer value.

Ethereum blockchain was developed in 2015 by Vitalik Buterin [2]. Its cryptocurrency is called Ether, while the programs that run on Ethereum are called smart contracts (ERC-20) and allow complex decentralized DApps [3]. The EVM, Ethereum Virtual Machine, is the Ethereum computing infrastructure for Ethereum nodes. From the beginning, and until September 2022, the consensus mechanism was based on Proof of Work (PoW). However, nowadays, Ethereum uses a Proof of Stake (PoS) consensus mechanism [4]. The Ethereum network has two types of accounts, the EOA - Ethereum Owned Account and Smart Contract Account. In order to interact with this network, the user has to create an EOA type of account in which a private key to digitally sign transactions is generated. The Smart Contract Account is a self-executable code that can be invoked by another Smart Contract or EOA. Its transaction types can be fund transfers, contract deployment, or contract execution. The fund transfer is the traditional transfer type, where an EOF transfers funds to another EOA. A contract transfer type is deployed when a new contract is added to the blockchain without a 'to' address, and the data field is used to store the contract code. The contract execution type occurs when a smart contract is triggered to execute its code. In this case, the 'to' address field is the smart contract address [5].

Like any complex system, Ethereum has a number of challenges to overcome [6]. Scalability is one of the main concerns. While Ethereum processes 20 transactions per second, Visa can process 1700 transactions per second [7]. Another challenge is the high energy consumption due to the consensus mechanism. To address this, the consensus mechanism was updated last September from PoW to PoS [4]. The user experience of Ethereum's dApps is another topic of improvement. These Apps are complex and intimidating for non-technical users, which limits their adoption. Additionally, Ethereum and other blockchain platforms operate in a largely unregulated environment, which can create uncertainty and potential legal risks for developers and users. As blockchain technology becomes more widely adopted, regulatory frameworks will likely be developed to provide more clarity and guidance for stakeholders. For last, with any decentralised system, security is a major concern for Ethereum. Ethereum's community is constantly working to improve the platform's security, including identifying and fixing vulnerabilities in the protocol and encouraging best practices among developers.

Anomaly detection is one of the main issues within the security and to address it we can apply Machine Learning techniques. To do so, we first have to define the type of anomalies that can occur in the Ethereum blockchain network. It can be divided into three categories [8]. Smart contracts, a key feature of Ethereum, are self-executing contracts with the terms of the agreement between buyer and seller being directly written into lines of code. However, it can also be vulnerable to bugs and exploits that can result in the loss of funds or sensitive data [1]. We can consider these exploits as anomalies and train machine learning models to learn the patterns within the exploitation. Another category is hack attacks [9]. Ethereum and other blockchain platforms rely on a decentralised network of nodes to validate and record transactions. If malicious actors manage to gain control of a significant portion of the network, they could potentially carry out attacks such as double-spending or censorship. Finally, the most well-known issue when we are dealing with a medium to transfer value - financial fraud. Within this category, we have scammers, money laundering and stealing of funds or sensitive data. All in all, can be considered anomalies.

Investing in methods for detecting and mitigating malicious activity on the network is crucial to improve the security of the Ethereum platform. Using Artificial intelligence (AI) and data to research methods to detect and mitigate abnormal activity, such as identification and blocking of suspicious accounts or transactions, may help improve the network's overall security and resilience which in turn could increase the adoption and usage of this technology.

In this regard, *Farrugia et al.* [10] studied the detection of illicit wallets on the Ethereum blockchain, where a trained XGBoost model was used dataset built upon a set of statistical measures from raw transactions, considering both ETH and ERC-20 tokens (smart contracts). Using 10-fold cross-validation the authors achieved a ROC-AUC performance of 99,40%. Despite this result, the dataset was balanced, meaning that the ratio of fraudulent wallets is approximately 50%. This behaviour does not correspond to the reality where a block has on average 150 transactions and most of the time non of them are from or to an illicit EOA.

Moreover, *Kumar et al.* [11] applied tree-based machine-learning models to detect malicious accounts on the Ethereum Network. With a detection accuracy ROC-AUC of 96,54% for EOA, providing a benchmark for the performance achieved with supervised learning models to detect malicious accounts.

A set of supervised machine learning algorithms, to detect illicit accounts on Ethereum Blockchain, were proposed by Ostapowicz et al. [12] where the XGboost model proved to be the best in class with a precision of 78,03% and recall of 31,32%.

On the other hand, *Scicchitano et al.* [13] proposed an iterative learning process based on auto-encoders to detect anomalous behaviour in Ethereum Classic (ETC). Our idea to apply an autoencoder prior to a Gradient Boosting algorithm comes from this work.

Baek et al. [14] in turn, proposed a machine-learning method to detect money laundering at the Wallet level using only transactional data from wallets with high trading volumes.

A phishing scam framework was proposed by Wen *et al.* [18], testing a set of machine learning supervised techniques, where the AdaBoost algorithm was the champion Model.

Podgorelec et al. [14, 15] proposed a machine learning-based method to identify anomalous transactions during the signing process that occurs within the Ethereum blockchain implementation. With this approach, it is possible to use reinforcement learning to train and improve the model during execution.

Finally, a one-class graph deep learning framework was proposed by *Patel et al.* [16, 17] to detect anomaly transactions in the Ethereum blockchain.

All of these approaches focus only on one model solution with supervised techniques. Our goal is to improve these models by combining supervised techniques with unsupervised techniques. Specifically, we aim to apply an autoencoder to detect out-of-normal account behaviour as an input to our supervised tree-based machine-learning model.

3 Methodology

By combining an unsupervised with a supervised technique, our goal is to improve the benchmark performance achieved in the literature focusing on the financial fraud type of anomaly that occurs in the Ethereum blockchain. We will also work with this challenge as an imbalanced problem, as it is in a real environment. Therefore, the dataset was obtained following a perspective of using this development in a real use-case scenario, near real-time. To do so, we started from a set of 250 anomalous accounts with their fraud disclosed date, identified in [18] and validated by *etherscan.io*. Then, we have selected three months of transactions backwards starting from that disclosed date, as in Fig. 1.

Regarding the normal accounts, we selected the ones that have transactions in the same blocks as the abnormal transactions that were flagged. Selecting also three months of data ending in those same blocks, as demonstrated in Fig. 2.

This approach targeted 9000 unique accounts. After excluding the smart contract account types we reduce the sample to 4727 healthy accounts. For the abnormal accounts, after removing the smart contract account type, we got 99 unique accounts. This set has

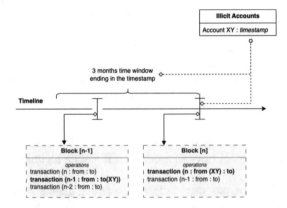

Fig. 1. Schematic of the time window application during data gathering.

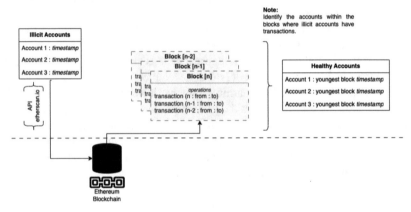

Fig. 2. Schematic of the data gathering approach.

accounts that not only were hacked but also that were newly created for the purpose of executing illicit activity.

Considering the same observation window, we then collected all the transactions for the identified accounts, reaching a total of 877 379 transactions [19]. With the transaction information for our set of accounts, we have generated the features by extracting information from the transaction structure data such as {max, min, mean, and total} eth value {sent, received}, and transaction velocities {sent, received} from transaction timestamp. We have also applied the same method for the smart contract info. All the categorical types of features were encoded using the *Catboost* encoder, where the encoding takes into account the target distribution [20]. In [19], we have all the generated features.

This approach generated 72 features with less than 95% of null values per account, though our experience tells us that all the extracted features may not be essential to our classifier. Therefore, during the modelling phase, we iteratively removed the less important features without degrading the performance.

In order to improve the sensitivity to new fraud MOs (modus operandi) we tried to combine two distinct machine learning approaches in the modelling phase - as presented in Fig. 3. First, we applied an auto-encoder to the healthy accounts, using only the numeric features. Then, choosing the model which provided the best results from the literature [12], the XGboost model, we fed this model with the output of the auto-encoder and the already generated features, to predict if an account is fraudulent or not.

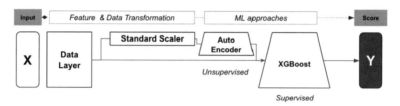

Fig. 3. Schematic of the proposed modelling approach used in this research.

The architecture of the autoencoder is composed of three dense layers with a dropout of 10% in the second layer. All these three layers use a Relu activation function. For the decoding part, it mirrors the encoder, with the output layer configured with a Sigmoid activation function, since we want to predict a value between 0 and 1.

This simulation was performed with Python language using Cloud AWS Sagemaker instance *ml.t3.medium.* The main libs used were *sci-kit-learn* version 1.0 and *Keras* version 2.2.4.3. The Notebooks used in these simulations and also the ETL steps are available in the git repository: *eth-ad-autoencoder-xgboost,* in [21].

4   Experimental Results

In order to have a feeling that our dataset will be able to discriminate our illicit accounts we proceed with a 2D space reduction of a non-linear transformation applied to the original feature space using t-distributed stochastic neighbour embedding (t-SNE) [20], Fig. 4. With this, we noted that several illicit accounts can be clustered together, highlighting that with the current feature space, we have information that helps in the prediction. However, we also have some isolated points. For the latter, we expect that the autoencoder helps in its identification by characterising it as outliers (RMSE significantly higher than a healthy account).

The autoencoder was trained with a subset of the numeric features (26 features) using healthy account information only. The performance metric used was RMSE which was then used as a new feature of the XGboost model. To avoid overfit we trained the autoencoder using a 3-fold cross-validation. With this methodology, we are adding a measure of how far from the expected values a given point (account) is - we can call this outlier distance. Finally, we train an XGboost Classifier with all the features plus the autoencoder score, to predict if a given EOA is an illicit account.

In the XGboost model, we used early stop with the validation dataset to avoid overfitting. For the hyperparameter tuning, we used the hyperopt python library [22].

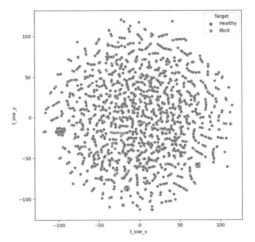

Fig. 4. Two-dimensional t-SNE scatter plot.

This Python library uses a form of Bayesian optimization for parameter tuning. The main parameters for optimization were: (i) learning rate, (ii) number of trees, (iii) max tree depth, (iv) reg_alpha (L1 regularization), and (v) reg_lambda (L2 regularization). Figure 5 (a) shows the performance measures for an XGBoost model trained without the autoencoder feature.

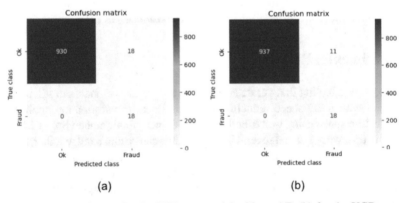

Fig. 5. Confusion matrix: (a) for the XGBoost model without AE, (b) for the XGBoost model with AE.

Besides the performance of ROC-AUC (98.96%), as presented in Table 1, the precision-recall metric (PR-AUC) shows us a poor performance. With this model, we have detected twice as much as the real fraud accounts, meaning a precision of 50%. Now by adding the AE score to the model, we have improved our precision-recall AUC metric by 12.07% in the test set. This improvement is a result of the decrease in false positives, as demonstrated in Fig. 5 (b). In both models, we were able to detect all the true positive samples.

The marginal increase in the ROC-AUC, due to the imbalance of our dataset, hides the real improvement in performance observed with the Precision/Recall-AUC.

Regarding the feature importance, when we delve into SHAP values feature importance - as in Fig. 6, some of the top features presented here - *total_trx_etherSent*, *sent_trx_avgValue*, *time_trx_avgBetweenReceivedTxn* - are also identified in [10].

Table 1. Performance summary of the two models.

Test Set	ROC AUC	PR AUC	Recall	Precision
Model without Auto-encoder	98.16%	50.00%	100.00%	50.00%
Model with Auto-encoder	98.86%	62.07%	100.00%	62.07%

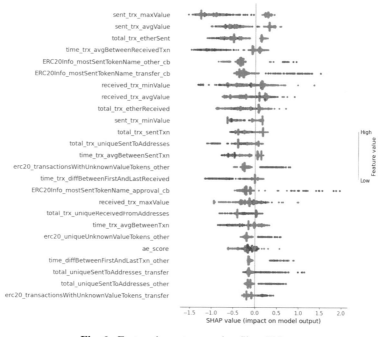

Fig. 6. Feature importance using Shap Values.

Furthermore, for high values of the following features: *ERC20Info_mostSentTokenName_other_cb*, *ERC20Info_mostSentTokenName_transfer_cb*, *erc20_uniqueUnknownValueTokens_other*, *total_uniqueSentToAddresses_other*, we have observed that it has a positive impact in the model, meaning that if an account interacts heavily with an unknown type of tokens, smart contracts, it's risk score is higher. The *"_cb"* stands for *Catboost* encoding. On average, the low values of the autoencoder score - ae_score-, impact the model negatively, towards the normal transactions, as we expected. For the high values, this is not so clear as we do not have a clear distribution of the positive impact on the model output. These dynamics can be translated as the following: autoencoder score gives more certainty to the model when the output is the class normal account.

With the performance values for ROC-AUC of 98,86% and F1 Score of 75% and to the best of our knowledge, we improved the state-of-the-art from the literature, where *Kumar et al.* [11] had a ROC-AUC performance of 96,54% and Ostapowicz et al. [12] achieved an F1 score of 44,70%.

5 Conclusions

We have presented a novel approach to detect abnormal behaviour regarding fraudulent accounts on the Ethereum blockchain where not only the chosen process to create the dataset is completely new and closer to the real environment, but also it combines two different machine learning approaches with which we were able to improve the benchmark in the detection of illicit activity on the Ethereum blockchain.

Despite the promising results of this experiment, we should test it with a larger dataset. Additionally, we could improve the autoencoder using genetic programming to search for the best combination of layers and enrich the dataset with new features that are presented in the Ethereum blockchain architecture. We also could try to split the dataset into one with only hacked accounts and another with new accounts created exclusively to perform illicit activity.

The proposed approach is a step forward in using several techniques to build a better overall model to detect abnormal behaviour in the Ethereum blockchain. However, several other approaches can be explored - such as graph neural networks or transformers-, and also, improving the feature space with new information from a third party or another layer of the blockchain should be tested.

Furthermore, it will be interesting to study the impact of using an autoencoder as an input feature to other supervised techniques applied to anomaly detection. If so we could generalize this approach into a broad spectrum of applications.

References

1. Yan, C., Zhang, C., Lu, Z., Wang, Z., Liu, Y., Liu, B.: Blockchain abnormal behavior awareness methods: a survey. Cybersecurity. **5**, 1–27 (2022)
2. Buterin, V.: A next-generation smart contract and decentralized application platform. finpedia.vn. Available: https://finpedia.vn/wp-content/uploads/2022/02/Ethereum_white_paper-a_next_generation_smart_contract_and_decentralized_application_platform-vitalik-buterin.pdf
3. Dannen, C.: Introducing Ethereum and Solidity (2017). https://doi.org/10.1007/978-1-4842-2535-6
4. Zhang, W., Anand, T.: Ethereum Architecture and Overview. In: Zhang, W., Anand, T. (eds.) Blockchain and Ethereum Smart Contract Solution Development: Dapp Programming with Solidity, pp. 209–244. Apress, Berkeley, CA (2022)
5. Transactions. In: ethereum.org [Internet]. [cited 11 Feb 2023]. Available: https://ethereum.org/en/developers/docs/transactions/

6. Website. Available: Active areas of Ethereum research. In: ethereum.org [Internet]. [cited 2 Jan 2023]. Available: https://ethereum.org/en/community/research/
7. Choo, K.-K.R., Dehghantanha, A., Parizi, R.M.: Blockchain Cybersecurity, Trust and Privacy. Springer Nature (2020)
8. Kamišalic, A., Kramberger, R., Fister, I.: Synergy of Blockchain Technology and Data Mining Techniques for Anomaly Detection. https://doi.org/10.2196/18623
9. Hassan, M.U., Rehmani, M.H., Chen, J.: Anomaly Detection in Blockchain Networks: A Comprehensive Survey. arXiv [cs.CR]. 2021. Available: http://arxiv.org/abs/2112.06089
10. Farrugia, S., Ellul, J., Azzopardi, G.: Detection of illicit accounts over the Ethereum blockchain. Expert Systems with Applications, 113318 (2020). https://doi.org/10.1016/j.eswa.2020.113318
11. Kumar, N., Singh, A., Handa, A., Shukla, S.K.: Detecting Malicious Accounts on the Ethereum Blockchain with Supervised Learning. Lecture Notes in Computer Science 94–109 (2020). https://doi.org/10.1007/978-3-030-49785-9_7
12. Ostapowicz, M., Żbikowski, K.: Detecting Fraudulent Accounts on Blockchain: A Supervised Approach. arXiv [cs.CR] (2019). https://doi.org/10.1016/j.eswa.2007.08.093
13. Scicchitano, F., Liguori, A., Guarascio, M., Ritacco, E., Manco, G.: Deep Autoencoder Ensembles for Anomaly Detection on Blockchain. Lecture Notes in Computer Science, 448–456 (2020). https://doi.org/10.1007/978-3-030-59491-6_43
14. Baek, H., Oh, J., Kim, C.Y., Lee, K.: A model for detecting cryptocurrency transactions with discernible purpose. In: 2019 Eleventh International Conference on Ubiquitous and Future Networks (ICUFN) (2019). https://doi.org/10.1109/icufn.2019.8806126
15. Podgorelec, B., Turkanović, M., Karakatič, S.: A Machine Learning-Based Method for Automated Blockchain Transaction Signing Including Personalized Anomaly Detection. Sensors 20 (2019). https://doi.org/10.3390/s20010147
16. Patel, V., Pan, L., Rajasegarar, S.: Graph Deep Learning Based Anomaly Detection in Ethereum Blockchain Network. Network and System Security 132–148 (2020). https://doi.org/10.1007/978-3-030-65745-1_8
17. Patel, V., Rajasegarar, S., Pan, L., Liu, J., Zhu, L.: EvAnGCN: evolving graph deep neural network based anomaly detection in Blockchain. Advanced Data Mining and Applications 444–456 (2022). https://doi.org/10.1007/978-3-031-22064-7_32
18. Website. Available: "Blacklisted/sanctioned Addresses." n.d. Accessed 11 February 2023. https://dune.com/harrydenley/BlacklistedSanctioned-Addresses
19. Website. Available: Crisóstomo, João. n.d. List_of_Raw_features.pdf at Main · Joaocrisostomo/eth-Ad-Autoencoder-Xgboost. Github. Accessed 11 February 2023. https://github.com/joaocrisostomo/eth-ad-autoencoder-xgboost
20. Dorogush, A.V., Ershov, V., Gulin, A.: CatBoost: gradient boosting with categorical features support. arXiv preprint arXiv:1810 11363
21. Website. Available: Crisóstomo, João. n.d. Eth-Ad-Autoencoder-Xgboost. Github. Accessed 11 February 2023. https://github.com/joaocrisostomo/eth-ad-autoencoder-xgboost
22. Website. Available: Hyperopt: Distributed Asynchronous Hyperparameter Optimization in Python. n.d. Github. Accessed 11 February 2023. https://github.com/hyperopt/hyperopt

TariScript: Bringing Dynamic Scripting to Mimblewimble

Cayle Sharrock$^{(\boxtimes)}$ and Schalk van Heerden

Tari Labs, 20 Woodlands Dr, Johannesburg 2091, South Africa
{cj,sw}@tari.com

Abstract. Mimblewimble is a cryptocurrency protocol with good privacy and scalability properties. A trade-off of Mimblewimble is the requirement that transactions are interactive between sender and receiver. TariScript is presented as an extension to Mimblewimble that adds scripting capabilities to the protocol. We describe the theoretical basis for TariScript and present the modifications required to make it secure. The trade-offs and use cases for TariScript are briefly covered.

Keywords: Blockchain · Mimblewimble · Scripting · TariScript

1 Introduction

This paper introduces TariScript, a dynamic scripting extension for Mimblewimble. Scripting unlocks many applications, including support for unilateral payments, covenants, atomic swaps, and side-chain pegs. We briefly review how vanilla Mimblewimble handles transactions to illustrate why, for example, unilateral payments are impossible. We touch on what other projects are doing to overcome these limitations. We then detail the theoretical framework for TariScript and conclude with a description of a concrete implementation of TariScript, its use-cases and limitations.

1.1 Mimblewimble

Mimblewimble is a scalable, confidential cryptocurrency protocol developed by the anonymous developer, Tom Jedusor [1].

Confidentiality comes from the use of Pedersen Commitments [2] to blind values, rather than publishing transaction values in the clear, as in Bitcoin [3], for example.

The key scalability advantage that Mimblewimble gains over Bitcoin is that it is only necessary to know the emission schedule, the current unspent transaction output (UTXO) set, and some housekeeping data in order to verify that the accounting is correct. In particular, data associated with spent transaction outputs can be discarded without compromising the security guarantees of the protocol [1].

© The Author(s), under exclusive license to Springer Nature Switzerland AG 2023
J. M. Machado et al. (Eds.): BLOCKCHAIN 2023, LNNS 778, pp. 234–243, 2023.
https://doi.org/10.1007/978-3-031-45155-3_24

The key to the protocol is the homomorphic balance over the expected supply of coins and the set of outputs in circulation. The details are given in [1] and an accessible explanation is available at [4], but we provide a simplified summary of the key points here for convenience.

Given a blinding factor k_i and value v_i using generators from a suitable elliptic curve G and H respectively, we denote the Pedersen commitment C_i as the combination, $C_i \equiv v_i \cdot H + k_i \cdot G$.

The emission of new coins follows a pre-determined schedule such that s_b are the number of new coins minted in block b. For a chain containing M unspent outputs after N blocks, the following balance must hold

$$\sum_i^M C_i = \sum_b^N s_b + \Delta_N \tag{1}$$

where $\Delta_b = \delta_b \cdot G$ is termed the total accumulated public excess after block b. In general, the sum of all the blinding factors, k_i must equal the total accumulated excess, i.e.,

$$\delta_b = \sum_i k_{ib} \ \forall i \in \text{UTXO set after block } b \tag{2}$$

There is a similar balance for every Mimblewimble transaction:

$$\sum_{j \in \text{outputs}} C_j - \sum_{i \in \text{inputs}} C_i \equiv \text{fee} \cdot H + \Delta \tag{3}$$

where $\Delta = \delta \cdot G$ and the excess is again the sum of the blinding factors, taking note of a sign change between outputs and inputs, i.e.,

$$\delta = \sum_{j \in \text{outputs}} k_j - \sum_{i \in \text{inputs}} k_i$$

The public excess for the transaction is collectively calculated by the parties in the transaction. A signature committing to the public excess, signed by each party is stored in the blockchain data. A running total of the total accumulated public excess is maintained by validators to verify the balance in Eq. (1) after each block.

When miners validate a block, they lump every transaction into effectively a single transaction and calculate the block-equivalent of Eq. (3):

$$\sum_{j \in \text{outputs}} C_j - \sum_{i \in \text{inputs}} C_i \equiv \sum_{m \in \text{txs}} (\text{fee}_m \cdot H + \Delta_m) \tag{4}$$

1.2   Tari

Tari [5,6] is a Mimblewimble-based proof-of-work blockchain. Tari aspires to provide a scalable, private, digital assets network and smart contract platform.

To achieve this, the Tari protocol requires several features unsupported by Mimblewimble, including unilateral (i.e. non-interactive) payments, covenants and side chain peg-in transactions [7].

A review of the prior art finds that some, not all, of these features may be possible to implement using "scriptless scripts" [8]. In particular, unilateral payments are not possible with scriptless scripts since they cannot avoid the interactive requirement of Mimblewimble transactions.

Unilateral payments were employed in LiteCoin extension blocks [9,10].

Beam [11], adds support for general smart contract applications via a virtual machine extension. Neither approach directly augments the core protocol with native scripting. We believe that Tari is unique in this approach.

In addition, we seek to devise a solution that unlocks all of Tari's desired use-cases with a single, generalised approach, hence TariScript.

2   TariScript

TariScript is a protocol extension that adds dynamic scripting capabilities to Mimblewimble. Some implementation details are omitted in the interests of brevity but a full description is available in the Tari specifications [12].

We begin with the observation that the balance equations (1) and (4) are necessary but not sufficient to secure Mimblewimble. Additional rules, including the inclusion of a range proof [13], proving that the value transferred is non-negative, and the excess signature, reduces the set of all possible transaction expressions to the subset of transactions considered secure under the Mimblewimble protocol.

TariScript adds another, strictly non-expansive, constraint to the spending rules of UTXOs. This means that TariScript only reduces the subset of valid transactions; never increases it. Therefore the security guarantees of Mimblewimble are not affected. While this argument does not constitute a security proof of TariScript, it is a strong intuitive argument that the approach is valid, with some caveats that we will cover shortly. A rigorous security proof for TariScript is an avenue for future work.

An outline of the TariScript specification is as follows:

1. Every UTXO must carry a script, α.
2. The script can accept arbitrary input parameters, θ, provided by the spender of the UTXO. The script is executed to produce a result, r, i.e., $\alpha(\theta) \to r$.
3. The result r must be singular (i.e. not an array) *and* must be a valid public key under G. If this is the case,
4. we assign the result, r to a new variable introduced in TariScript, the *script public key*, K_s,
5. The UTXO may be spent if, other requirements notwithstanding, the spender demonstrates knowledge of *both* the commitment blinding factor k (this is the vanilla Mimblewimble requirement), and the script private key, k_s, such that $K_s = k_s \cdot G$. The latter is demonstrated by signing the script and its input with k_s.

There is an implicit caveat that the script and its metadata be non-malleable (no parties can modify it once the sender has broadcast the transaction) and immutable (it cannot be removed once it is in the blockchain).

To address malleability, the script, α, is signed and the signature is included with the UTXO. Additionally, a signature signing the script and the input parameters are provided when spending the UTXO.

To address immutability, we first review a feature of vanilla Mimblewimble called cut-through [1]:

If Alice sends coins to Bob, and Bob sends the same coins to Charlie in the same block, miners, at their discretion, may omit Bob's UTXO entirely, and Eq. (4) will still hold because Bob's commitments simply cancel out. Note that, only the commitment is cut-through. Since kernels are never cut through, and fees are specified in the kernel, the fees paid by Bob are still tracked and the overall accounting still balances.

Cut-through poses a problem for TariScript since Alice is required to provide script with UTXO B. If B is cut-through, the script is lost and we violate the immutability constraint of TariScript. Therefore a new mechanism, the script offset, γ, is introduced to explicitly prevent cut-through in Tari blocks.

2.1   Preventing Cut-Through with the Script Offset

First, the sender chooses a new secret scalar, k_o, from the curve which we call the sender-offset private key. The sender calculates the corresponding sender-offset public key, $K_o = k_o \cdot G$ and includes this key with the UTXO metadata.

As per the TariScript rules, the recipient must know the script private key, k_s, corresponding to the script public key, $K_s = k_s \cdot G$, which results from the script execution, $\alpha(\theta) \to r = K_s$. In other words, to spend an output, a spender must provide input, θ, to the script such that it resolves to some K_s for which spender knows the private key.

The recipient proves knowledge of k_s, by signing the script and input data with k_s when spending the output.

Then, the script offset, γ, is calculated by summing all script private keys for every input in the transaction, and subtracting the sum of all sender-offset private keys in the transaction. That is,

$$\gamma = \sum_{i \in \text{ inputs}} k_{si} - \sum_{j \in \text{ outputs}} k_{oj} \tag{5}$$

The script offset is broadcast along with the usual Mimblewimble transaction data. Any third party can verify that the script offset is correct by ensuring that

$$\gamma \cdot G \equiv \sum_{i \in \text{ inputs}} K_{si} - \sum_{j \in \text{ outputs}} K_{oj} \tag{6}$$

When miners construct blocks, they do a similar aggregation for script offsets that they do for transaction excess:

$$\gamma_{\text{total}} = \sum_{m \in \text{txns}} \gamma_m \tag{7}$$

It is a simple exercise to prove that Eqs. (5) and (6) apply for a transaction block if γ is replaced by γ_{total}.

The block script offset, γ_{total}, is stored in the block header.

It naturally follows that the script offset γ means that no third party can change or remove any input or output from a transaction or the block, since doing so will invalidate the script offset balance equations, (6) or its block-level equivalent. By extension, the script offset also prevents cut-through. In the scenario above, if a miner cut out B from the transactions between Alice, Bob, and Charlie, the script offset calculation would differ by $k_{sB} - k_{oB}$ and the validation would fail.

To address malleability concerns, both the script and sender-offset public key must be signed by the sender-offset private key. This signature is included with the UTXO. In addition, every input contains a valid script signature, which signs the script, the input data and the script public key with the script private key.

3   Implications and Trade-Offs

3.1   Script Lock Key Generation

It may appear the burden for wallets has tripled since each UTXO owner has to remember three private keys: the spend key, k_i, the sender offset key k_o and the script key k_s for every transaction. In practice, this is not the case.

Spend keys are typically deterministically derived from a single seed phrase, for example, using hierarchical deterministic (HD) wallets [14].

On closer inspection of (6), the sender-offset private key does not actually need to be stored at all. Once the script offset, γ and script signature are calculated and broadcast, the offset is never required again, and can be discarded.

Script key management is application-dependent. For unilateral payments, the script key is a static key associated with the recipient's node or wallet, similar to an Ethereum address, and so only a single private key is required. For default payments, as shown in the examples, the recipient is free to provide any value they like, which can be discarded once the transaction is finalised. In other applications, the script key may derived from a suitable path in an HD wallet, as per the spend keys.

3.2   Blockchain Size

The most obvious drawback to TariScript is the effect it has on blockchain size. UTXOs are substantially larger, with the addition of the script, metadata signature, script signature, and a public key to every output. The increase depends on the script and its serialisation, but typically, UTXOs are 13%–40% larger than vanilla Mimblewimble. These can eventually be pruned but will neverthe-less increase storage and bandwidth requirements. TariScript inputs are two to four times larger than vanilla Mimblewimble inputs. The latter consist of a com-mitment and output features. In TariScript, each input includes a script, input data, the script signature, and an extra public key. In addition, every block header contains an extra field, the total script offset, that cannot be pruned away.

In terms of performance, TariScript introduces two additional signature veri-fication operations and an additional balance requiring a single curve operation. Script evaluation, $\alpha(\theta) \rightarrow r$, will depend on the implementation. Tari's imple-mentation, as discussed in the next section, is very efficient with strict upper bounds on script size and complexity. In practice, we find that the Tari network is limited by block space and network constraints during busy periods rather than transaction processing speed [15].

3.3   Chain Analysis

Another potential drawback of TariScript is the additional information that is handed to entities wishing to perform chain analysis. Having scripts attached to outputs will often clearly mark the purpose of that UTXO. Users may wish to re-spend outputs into standard, default UTXOs in a mixing transaction to disassociate TariScript funds from a particular script.

3.4   Horizon Attacks

The Mimblewimble protocol allows outputs, and consequently, the output scripts to be discarded (pruned) once they are spent.

In practice, nodes maintain a cache of full blocks before pruning outputs. The depth of the cache is termed the pruning horizon. The horizon plays a key role in simplifying node synchronisation and handling short chain forks that inevitably arise in proof-of-work chains. Nodes are able to verify that scripts are respected as long as chain re-organisations are not deeper than the pruning horizon.

In particular, when a new node joins the network, it will not be able to know whether the scripts attached to spent transactions older than the pruning horizon were faithfully executed to reach the current chain head.

The usual Mimblewimble guarantees remain, including the overall coin bal-ance, but there is now an avenue of attack for a malicious party: Force a chain re-organisation chain beyond the pruning horizon, and alter the script of a spent output for which the attacker knows the spend key; thus enabling spending of the output. This applies in the specific case of unilateral payments, where the attacker knows the spend key, but not the script key.

This is termed a "horizon attack". There are three ways to mitigate or prevent it:

1. After receiving a unilateral payment, the receiver can spend the output to her-self with a standard interactive payment. This prevents the horizon attack

completely. Wallets can be programmed to do this automatically and periodically, at the cost of an additional on-chain transaction, batching output spends to reduce on-chain costs.

2. Make the pruning horizon very deep, several weeks' worth, say, to the point that the cost of re-organising the chain for the horizon attack becomes uneconomical.
3. Run at least one node in archival node, meaning that UTXOs are never pruned. Only a single honest archival node across the entire network is required to be able to eventually bring all other nodes back into the correct consensus.

All three mitigations strategies can be run independently and in concert to reduce the risk of a horizon attack to one of theoretical concern.

4   Examples

4.1   Standard Mimblewimble Transactions

The simplest script that maintains the current Mimblewimble payment functionality is the identity script,

$$\alpha(\theta) \rightarrow \theta \tag{8}$$

Here the spender can provide an arbitrary public key as the script input and it will be interpreted as the script public key, K_s. In practice, a spender would use a nonce, $r = k_s$, with $K_s = k_s \cdot G$ and the UTXO commitment's blinding factor, k, which they alone know, to spend the output in a transaction.

4.2   Unilateral Payments

The strategy for leveraging TariScript for unilateral payments is as follows: Assume Alice wants to pay Bob at some static "address". Mimblewimble does not have addresses, *per sè*, but any public key, $K_{sb} \equiv k_{sb} \cdot G$ that Bob makes publicly available will suffice.

The problem can be reduced to one in which Alice is able to publish the transaction unilaterally, such that Bob can independently identify and claim the output without any communication from Alice, and such that Alice cannot spend the output herself.

The solution centers on Alice using Bob's public key as the script public key, K_s. Alice then combines her sender-offset key, k_{ob}, and Bob's public key, to derive a shared secret using a Diffie-Hellman key exchange [16]:

$$k_b = \mathrm{H}\big(k_{ob} \cdot K_{sb}\big) = \mathrm{H}\big(k_{sb} \cdot K_{ob}\big) \tag{9}$$

where H is a suitable hash function that produces a valid scalar under G. Alice also encrypts the value of the commitment with the shared secret and stores it at any convenient location in the transaction metadata.

This concludes part one of the problem.

To prevent Alice being able to spend the transaction, she provides a script, $\alpha(\cdot) \rightarrow K_{sb}$. That is, the script returns the script public key when a null input is provided. Since Alice does not know k_{sb}, she cannot sign the script when trying to spend the input. However, Bob can and thus part two of the problem is solved.

Once broadcast, any node can verify that the transaction is complete, verify the signature on the script, and verify the script offset.

Bob can scan all transactions looking for a script that matches $\alpha(\cdot) \rightarrow K_{sb}$. If so, he recovers the spend key using his private key (also the script private key in this case) and the sender-offset public key as per the third term in Eq. (9).

4.3   TariScript Script - A Concrete α Implementation

Tari [5] uses a simple Forth-like stack-based language [17], similar to Bitcoin script in its implementation of α. We use the term TariScript to refer to both the protocol modifications and the set of opcodes and execution rules that define the script language.

Scripts contain opcodes representing commands that are executed sequentially. The commands operate on a single data stack that initially contains the input data. The set of commands include mathematical and cryptographic operations, stack manipulation and conditional logic [18].

After the script completes, it is successful if and only if it has not aborted and there is exactly a single element remaining on the data stack. The script fails if the stack is empty, contains more than one element, or aborts early.

To prevent denial-of-service or resource exhaustion attacks, both the script and stack are limited in size. Any stack overflow results in script failure. In addition, there are no opcodes for loops or timing functions, guaranteeing that every script will terminate.

Example Script - Time-Locked Spending Conditions. Interesting and complex transactions can be constructed dynamically, and without needing to make any further changes to the Mimblewimble protocol.

As an example, Alice wants to send some Tari to Bob. But, if he doesn't spend it within a certain time frame (up till block 4000), then she wants to be able to spend it back to herself.

This type of transaction is impossible in vanilla Mimblewimble and, in fact, is also outside the reach of scriptless scripts, which have no concept of blockchain context, like the block height.

However, the TariScript script in Fig. 1 achieves the desired result.

The spender (Alice or Bob) provides their public key, some P_x, as input to the script. The first opcode, DUP, duplicates the top element of the stack, leaving two copies of the public key. Then, PushPubkey(P_b), pushes Bob's public key, P_b, onto the stack.

Subsequently, CheckHeight(4000) pushes the difference between the current block and block 4000 to the stack. GeZero replaces the top stack element with 1 if the value was positive, or 0 if it was negative.

```
Dup PushPubkey(P_b) CheckHeight(4000) GeZero IFTHEN
PushPubkey(P_a) OrVerify(2) Drop
ELSE EqualVerify ENDIF
```

Fig. 1. An example time-locked contract in TariScript. This script serialises to 84 bytes, of which 64 bytes are taken up by the two public key representations.

IFTHEN pops an element of the stack and executes the commands up until the ELSE opcode if the value is equal to one, and the commands between ELSE and ENDIF otherwise.

Let's assume the chain is at block 3990. All the commands up to EqualVerify are skipped. EqualVerify pops two items off the stack, which currently contains P_b,P_x,P_x and does nothing if they are equal, or aborts if they are not. Thus, the script will only continue if P_b == P_x.

The script is now complete, leaving the single value of P_x on the stack, which must be Bob's public key. This key is used as the script public key, K_s, as per the TariScript rules. Bob will have signed the script and its input with k_s as part of the valid transaction.

Alternatively, if the block height is 4000 or above, then the expression in line 2 from Fig. 1 is executed. First, PushPubkey(P_a) pushes Alice's public key onto the stack, leaving a data stack P_a,P_b,P_x,P_x.

Then OrVerify(n) pops n items off the stack. If the new top element is equal to *any* of the popped elements, the script continues, otherwise the script aborts. In this example then, both Alice and Bob's public keys are popped and the script will continue if, and only if, P_x matches one of them.

Assuming this is the case, the Drop opcode drops the superfluous key, leaving P_x, which must be one of Alice or Bob's public keys. As before the spender will also have provided a suitable signature using their private key.

5 Conclusion

TariScript provides a novel way of extending Mimblewimble with dynamic scripting abilities, while retaining its scaling and confidential properties. This is offset by larger outputs and blocks. The scripts are also subject to horizon attacks, which must be mitigated with a long pruning horizon or prevented by output sweeping and running at least one archival node. The benefits significantly outweigh the drawbacks, since this single extension enables multiple features needed to give Mimblewimble generalised smart contract capabilities, including unilateral payments, covenants and side-chain pegs.

Acknowledgements. The authors are indebted to David Burkett for reviewing several iterations of TariScript and for initially suggesting the basis for the script offset. We would also like to thank Hansie Odendaal, Michael Berry, Stanley Bondi and Phillip Robertson for critically reviewing initial drafts of TariScript, suggesting improvements and helping with the implementation in Tari.

References

1. Jedusor, T.: Mimblewimble (2016). https://docs.beam.mw/Mimblewimble.pdf
2. Pedersen, T.P.: Non-interactive and information-theoretic secure verifiable secret sharing. In: Feigenbaum, J. (ed.) CRYPTO 1991. LNCS, vol. 576, pp. 129–140. Springer, Heidelberg (1992). https://doi.org/10.1007/3-540-46766-1_9
3. Nakamoto, S. Bitcoin: a peer-to-peer electronic cash system (2008). https://bitcoin.org/bitcoin.pdf
4. Sharrock, C.: Mimblewimble transactions explained (2020). https://tlu.tarilabs.com/protocols/mimblewimble-transactions-explained
5. Tari: The protocol for digital assets. https://tari.com. Accessed 6 Apr 2023
6. The Tari RFCs. https://rfc.tari.com/. Accessed 6 Apr 2023
7. Back, A., et al.: Enabling blockchain innovations with pegged sidechains (2014). https://blockstream.com/sidechains.pdf
8. Gibson, A.: Schnorrless scriptless scripts (2020). https://reyify.com/blog/schnorrless-scriptless-scripts
9. Burkett, D.: LIP004. https://github.com/DavidBurkett/lips/blob/master/lip-0004.mediawiki. Accessed 29 Mar 2023
10. Burkett, D., Lee, C., Yang, A: LIP003. https://github.com/litecoin-project/lips/blob/master/lip-0003.mediawiki. Accessed 3 Apr 2023
11. One side payments (2019). https://github.com/BeamMW/beam/wiki/One-side-payments. Accessed 4 April 2023
12. Sharrock, C., Berry, B., van Heerden, S., Odendaal, H.: RFC-201: TariScript. https://rfc.tari.com/RFC-0201_TariScript.html. Accessed 6 Apr 2023
13. Bünz, B., Bootle J., Boneh, D., Poelstra, A., Wuille, P., Maxwell, G.: Bulletproofs: short proofs for confidential transactions and more. Cryptology ePrint Archive (2017). https://eprint.iacr.org/2017/1066.pdf
14. Wuille, P.: Hierarchical deterministic wallets. https://github.com/bitcoin/bips/blob/master/bip-0032.mediawiki. Accessed 17 May 2023
15. Odendaal, H.: Stress test of 2021/12/21 (console to mobile wallets). https://github.com/tari-project/tari-data-analysis/blob/master/reports/stress_tests/20211214-make-it-rain/Stress%20test%20of%2020211214%20-%20analysis.md. Accessed 17 May 2023
16. Merkle, R.: Secure communications over insecure channels. Commun. ACM **21**(4), 294–299 (1978)
17. Moore, C.: Programming a problem-oriented language (1970). http://forth.org/POL.pdf
18. Sharrock, C.: RFC-202: TariScript opcodes. https://rfc.tari.com/RFC-0202_TariScriptOpcodes.html. Accessed 6 Apr 2023

Smart Contract Analyzer. A Tool for Detecting Fraudulent Token Contracts

Adnan Imeri[⊠][iD], Thierry Grandjean, Oussema Gharsallaoui, Ismail Dinc, and Djamel Khadraoui

Luxembourg Institute of Science and Technology (LIST), 5, avenue des Hauts-Fourneaux, 4362 Esch-sur-Alzette, Luxembourg
adnan.imeri@list.lu
http://www.list.lu

Abstract. Smart contracts have demonstrated new ways to manage and trade digital assets, conduct financial transactions, and transform business processes. Several concepts have emerged to enable investors to own or trade digital assets. Trading platforms relying entirely on decentralized, known as decentralized exchanges, allow unrestricted financial transactions to exchange digital assets. Beyond the opportunities offered, using the decentralized environment remains complex to understand by most of its users, consequently giving adversaries opportunities to benefit from investors based on scamming schemes. The cryptocurrency market is damaged by malicious actors that aim to drain investor funds via scamming token smart contracts. This research paper initially highlights related problems with fraudulent token contracts. Further, it proposes a solution for identifying several fraudulent schemas in the crypto ecosystem via a dynamic algorithmic solution supported by the SC Analyzer tool based on real-time data.

Keywords: smart contract analysis · scamming · cryptocurrency · centralization · trapdoor · rug pull

1 Introduction

The distributed decentralized transaction environment presents a complex, fuzzy, and hard to understand by potential investors. Currently, transactions on the blockchain are carried out via Smart Contract (SC) operations (functions). SC functions tasks depend on how the SC has been designed. For example, SC has particular functions, which carry out operations for distributing ownership of digital assets, e.g., cryptocurrency, among many accounts (or investors). Different standards like ERC-20 or ERC-721 exist for designing Token Smart Contracts (TSC), offering general guidance on the main functionalities guidance on the main functionalities of the TSC (Sect. 2).

In recent years, digital asset investors have faced adversaries investing their funds in different crypto-related projects. According to [12,22], a billion USD of

J. M. Machado et al. (Eds.): BLOCKCHAIN 2023, LNNS 778, pp. 244–253, 2023.
https://doi.org/10.1007/978-3-031-45155-3_25

investor funds are lost (stolen) by malicious actors via decentralized exchanges. Scams such as rug pulls, honeypot/trapdoor, or centralized aspects of the TSC are common issues in decentralized finance (DeFi) [27]. That is because, in general, some smart contract (SC) developers with a good understanding of the financial operation in DeFi can easily create digital assets, e.g., tokens on the blockchain, and get them decentralized exchanges (DEX) without a code audit. SC is generally "smart" to the extent that they have been developed to be. The SC has several vulnerabilities that an adversary looking to gain advantage might exploit, e.g., "The DAO Attack"[1]. A sort of vulnerability is using a function of SC not to perform a specific task, such as transferring assets from owner to investor[2]. This kind of SC is known as a "honeypot/trapdoor"[3]. Its vulnerability stands in the trapping of investors once they intend to transfer (exchange) some digital assets from a specific TSC. This trap means that the users can own digital assets but cannot transfer ownership from their account, i.e., selling them in different exchanges. This vulnerability is supported by scamming schemes where the adversary parameterizes a TSC in a way to gain benefits from investor funds. For example, an attacker can pretend that the TSC contains sophisticated functionalities that yield attractive rewards, e.g., 50% and more for the investors who stake the coin in their wallets for a period of time. Once the user buys these tokens from DEX, e.g., Uniswap or Pancakeswap (trusted marketplaces), the TSC will block the purchased tokens by disabling selling or swapping operations. By denying these activities to the investors, the attacker will be the only one with the power to use the SC function to exchange or swap a specific amount of tokens, certainly at higher prices using DEX.

This paper aims to expose some scamming schemas in cryptocurrency marketplaces. Initially, we elaborate on problems related to "honeypot/trapdoor", "rug pull", and "centralization" aspects of the SC. Further, we present our scientific approach expressing an algorithmic solution to identify and spot these problems for investors. We developed a proof of concept implementation to support our solution and showed initial promising results.

The outline of this paper is as follows. Section 2 presents contextual information about concepts used in this research. In Sect. 3, we define the problem in detail. Section 4 presents related works. Section 5 shows the proposed solution. Further, in Sect. 6, the proof of concept (PoC) is presented, including results from the current state of our tool. Finally, Sect. 6 concludes this paper and gives future direction.

---

[1] Understanding The DAO Attack: https://www.coindesk.com/learn/understanding-the-dao-attack/.

[2] We refer to the term "investor" as any potential investor, any platform user, that owns, buys, or sells the digital asset.

[3] We use the term "trapdoor" as a synonym for the honeypot, as we believe it is far more descriptive.

2   Background: Blockchain and Digital Assets

Blockchain (BC) is a distributed decentralized database that allows storing append-only transaction data, gathered, cryptographically chained, and maintained by a consensus algorithm.

Smart Contracts (SC) are computer code deployed on BC and executed based on pre-defined parameters. SC are autonomous, self-executed programs to fulfill specific requirements [15].

Digital Assets (DA) - Present anything that has or presents value, and is digitally stored, identified, and explored. With the presence of technology, DAs are more often present and used by the financial industry, supply chain, multimedia companies, and many others to represent the value of specific assets.

Crypto Asset Standards: Different standards enable the creation of crypto assets (digital tokens). The purpose of such standards was to standardize the development and deployment of digital tokens on a blockchain platform; therefore, all tokens are the same. That enables the exchange of such tokens with other tokens or exchanging them for monetary value. The most known token standards are ERC-20, ERC-721, ERC-777, and many others [25].

Centralized Exchange (CEX): A CEX requires the creation of an account (KYC verified) on the CEX platform, e.g., Binance. The trader who wants to exchange DT from an asset needs to deposit the amount desired to exchange [16].

Decentralized Exchanges (DEX): Contrary to a CEX, in DEX, e.g., Uniswap, PancakeSwap, creating an account on any platform is not required [17,24]. Users use their wallets to interact with DEX and trade assets.

Liquidity: is the amount of Asset A (reserve A) and the amount of Asset B (reserve B) provided by the user called a liquidity provider. The liquidity provider got in return an amount of liquidity token representing the provider contribution.

Wallet: It's composed of two parts: a private and a public key. The private key should remain private (secret) and is used to sign transactions in the BC, whether the public key is publicly accessible and it's extracted from the private key using ECSDA reduced in size and hashed to be an address [26]. That address is what identifies the "wallet" in the BC.

3   Problem Definition: Digital Crypto Asset Trapdoor

The marketplaces for digital cryptocurrencies and other "crypto" assets contain tremendous amounts of money. Considering this, it becomes natural that strategies to profit from such a market are developed. We have identified at least three major problems enabling malicious users to benefit from crypto marketplaces. The first one is the trapdoor or honeypot SC. The second one is related to liquidity draining by SC owner known as rug pull. Third is the "centralization" aspect of the SC.

The honeypot or trap door is a scam SC deployed by malicious users to benefit from selling digital tokens. The SC typically presents a "regular" contract related to operating a business or DAO, with the goal of using Web3 to address a particular problem. That is usually done by representing ERC/BEP20[4] token. Once the investors, i.e., buyers of represented ERC/BEP20 tokens, invest in such digital assets, they are potentially trapped due to the impossibility of reselling or sweeping these Tokens for FIAT or other tokens. In such a scenario, we consider a trap door for the investors, as all monetary values are transferred to the malicious user. Similarly, in the case of rug pull, the malicious users, i.e., fraudulent developers teams behind the Token, aim to manipulate the Token price and basically steal user money. The strategy before initially pumping the prices, i.e., self-buying and allocating a large supply of tokens in their wallet (s), before pulling the entirely the liquidity. That leads to Token price dumping, making Token (digital assets) worthless at DEX. Another way of rig pulling is by limiting the selling options (linked to "centralization" aspects of SC) of Tokens for the users. According to [22], in 2021, an estimated $7.7 billion was stolen via the rug pull scam SC. Further, for the case of the "centralization" aspect of the SC, we refer to a specific set of SC functions that enables certain privileges for the owners of the SC. These functions are distinguished with keyword *onlyOwner*, the semantics of which exclusively enables the Owner of SC to execute. Besides being advantages to govern the aspects i.e., functions of SC, the same might be used by SC owners to encode hidden exploits that be used to manipulate liquidity (rug pulls), deny actions for some SC users (selling orders), enlist users to "black" lists, deny token sweeping, and other potential activities.

4   Related Works Studies

SC carries massive digital assets that can be exchanged for real money, i.e., FIAT. As a result, they are appealing to attackers who aim to exploit TSC through malicious schemes. SC must be well-designed, implemented, and tested before deployment. In this sense, many efforts have been made to enhance the capabilities of the SC. Several research proposes automatic checking of the SC before deploying. The researchers from [3] have shown formal verification of the SC to ensure the correct specification of functionalities. For example, a researcher from [13] shows a scientific approach for automatically detecting inconsistent behavior of ERC-20 SC. The research in [20,21] presents scientific aspects of formal verification of the SC that enables the correctness of the SC by design. In [18], authors use machine learning and fuzzy testing to assess the SC vulnerability. For being able to develop SC in the best manner, some best practices are expressed in [1,2]. These practices enable consideration of design principles to avoid major consequences in the execution of the SC. Furthermore, standards, notably Openzeppelin [10], provide clear design principles related to ERC-20, ERC-721, and many other design principles. Several tools already exist for analyzing SC vulnerabilities, notably Oyente

---

[4] BEP20: https://academy.binance.com/en/glossary/bep-20.

[4], Securify [5], and Smartcheck [6]. Some of them cover a large range of vulnerabilities, while others concentrate on a specific topic like EasyFlow [9] concentrating on overflow detection. In [8,11], it highlights SC vulnerabilities and presents a framework for SC analysis and testing. Research in [7] targets similar problems relying on deep learning scam detection framework but mainly concentrating on byte code. Similarly, in [14], deep analysis of the honeypot is performed manually. Solutions, as shown in [19], are limited in the scope of operation, considering limited DEX-es and basically consider only "honeypot", i.e., trapdoor-related problems relying on specific SC function and excluding real-time behavior of SC. However, most of the research and operational solution does not include creating a tool dedicated to analyzing the behavior and the state evolution of a deployed ERC-20 TSC to determine if the concerned TSC is malicious. Our approach performs a dynamic check on the TSC on a private node to avoid transaction execution on the Blockchain. That will allow us to 1) perform a dynamic check of the TSC to detect trapdoor or rug pull scams and 2) check all TSC functions related to owner governance, e.g., to detect any tax (fee) amount anomaly. Compared to the existing solutions, we consider our approach unique in simulating the actual, i.e., current state behavior of an ERC/BEP20 TSC.

5   The Proposed Solution for Detecting Fraudulent Token Contract

In this section, we present the scientific solution for detecting the malicious behavior of TSC to tackle (spot) the mentioned problem in Sect. 3. The proposed algorithmic solution comprises parameters that extract, measure and quantify information required for identification and spotting honeypot/trapdoor SC, rug pull, and "centralized" aspects of SC, supported by a tool called *SC Analyzer*. The algorithmic expression shown in pseudocode in 1 presents dedicated logic expressing sequential steps to perform checks over TSC. Initially, it requires from the user the *Token_SC_ID* (SC address) as input. The tool automatically identifies which BC framework, i.e., Ethereum (ETH) and Binance Smart Chain (BSC) the SC address belongs. In the backend, it requires *Web3* connection, i.e., BC nodes to interact in real-time with BC networks (ETH, BSC), as well as the total number of DEX-es (Total_Nr_DEX) towards which it performs TSC analysis. As a result, it provides a response for *Token_SC_ID* as (or not) honeypot/trapdoor, rug pull, and "centralization" TSC. The process continues by getting the *Application Binary Interface (ABI)* for the given SC, which further allows using *Web3* methods to interact with BC. It checks, for any DEX provided, the balance of the given TSC. This means for any available DEX, and it checks if the TSC has liquidity, meaning allows users to sweep tokens with other tokens. If the algorithm finds that the balance in all DEX is zero or *transfer method fails*, meaning that only the owner can "sell", it spots it as trapdoor TSC; otherwise, it shows that the TSC is not a trapdoor. Our simulator manages the algorithm for detecting transfer method failure or when the liquidity is zero liquidity. The transfer method tests are performed without buying or selling any token, instead via the buy-sell simulation. Right before starting the simulation, we take a snapshot of the current state of the blockchain and simulate the buy-sell in a private network, applying our algorithmic method. Such simulation enables it to avoid any loss of funds for the user. Further, all data received are stored in a separate database for traceability, TSC behavior, and other analysis.

Similarly, we check many data points to identify potential TSC rug pull cases. Following Algorithms 1, we apply parameters that are useful for identifying rug pulls and "centralized" aspects of TSC. As a result, it also yields a message associated with data for potential rug pull. It enlists all TSC functions that might impact the behavior of SC by the owner of the TSC, i.e., "centralization".

For being able to identify the potential rug pull, our algorithm introduces a specific section "Tokenomiks"[5] It collects information about the "economy" or "financial" mappings of the TSC. The set of information about "Liquidity Locked", "Total Supply", "Added Liquidity", "Buy Tax", "Sell Tax", "Reward", "Marketing Wallet" and "Reflection" that are presented under term "Tokenomiks". The algorithm checks if the TSC has locked liquidity. It checks if the selling tax is high or if the "yields" enormously high reward. In addition, our algorithms check that token holders have a high percentage, e.g., more than 15% of added liquidity. Certainly, that would enable the specific TSC owner or associated user to drain liquidity and make the token worthless. Certainly, we apply background checks to avoid false positives by double-verifying the token holder's address and transactions. For more graphical readings and to verify if the token's price has skyrocketed in a short time, we provide direct access to graphs associated with the TSC. In addition, we read from third-party APIs to check if the TSC developer team is KYC or DAX-ed. We combine all these parameters based on data retrieved via *"Web3 methods"* combined with BSC or ETH APIs to identify potential rug pull. Once the conditions are fulfilled from our algorithm, we flag TSC as a potential rug pull.

With respect to "centralization", our tool collects all functions that are labeled *onlyOwner*, including transaction execution history by these functions. Moreover, our solutions improve scam detection to a certain extent, as described following. Firstly, the basic check over the *onlyOwner*, as most of the tokens on BSC and ETH use the *Ownable* standard provided by Openzeppelin. Secondly, as a complement to the first step, we check keywords such as *"authorized"*, *"onlyMinter"*, *"onlyAuthorized"*. Third, we check occasional calls, i.e., function calls of the TSC by the owner. Finally, in the fourth step, we analyze each TSC function if ye have a require used with *"msg.sender"*. In most cases, the "msg.sender" function is used in a "required" instructions or in an "if" statement to whitelist wallet addresses. We store these functions in a specific database to perform further analysis. For identifying potential risks from such function, we refer to audits based on formal verification as shown in [23]. That enables users to read functions, meanings, and timestamps when the TSC owner uses them.

---

[5] Tokenomiks: https://coinmarketcap.com/alexandria/article/what-is-tokenomics.

Algorithm 1. Algorithmic expression for capturing rug pull and centralization SC

Require: Token_SC_ID, Web3_connection, Total_Nr_DEX
Ensure: Results: Token_SC_ID, is potential (or not) rug pull, centralization risks
1: $BC\_Framework \leftarrow Input(Token\_SC\_ID)$
2: $SC\_ABI \leftarrow ABI\_BC\_Framework$
3: $Intitailise : DEX\_List, SetDEX\_Set = 0, web3.BalanceOf(SC\_ABI)$
4: $Intitailise : Tokenomiks, DB$
5: **while** $(DEX\_List \geq 0)$ **do**
6:     CHECK:web3.BalanceOf in all DEX_List
7:     **if** $(web3.BalanceOf == 0 \lor transfer\_method == fails)$ **then**
8:         $Token\_SC\_ID \leftarrow trapdoor$
9:     **else**$(Token\_SC\_ID \leftarrow not\_trapdoor)$
10:         **if** $Locked\_liquidty == 0 \lor Token\_Holder \geq 15\%$ of $Added\_Liquditiy$
   or $Team \neq KYC \lor Team \neq DAX$ **then**
11:             $Token\_SC\_ID \leftarrow righpull$
12:         **end if**
13:     **end if**
14: **end while**
15: $DB \leftarrow generated\_data(Token\_SC\_ID), only\_OwnerFunctions$
16: END

5.1 Proof of Concept Implementation

Figure 1, depicts the *SC Analyzer* global architecture. It is composed of three main components: i) Graphical User Interface (GUI), ii) Data Access Layer (DAL), and iii) Data Layer Component (DLC). *SC Analyzer* is an interactive tool that allows users to interact with BC frameworks by providing specific information via GUI. For developing GUI we used *React.js*. For implementing algorithms for checking TSC characteristics, handling communication (including Web3 connection), and performing data analysis, we used the DAL layer, developed in *Node.js*. Furthermore, we use a particular database to store large amounts of data to log SC behavior and traceability and perform further analysis. For this component, we have chosen MongoDB. The DLC is composed of different BC layers, such as BSC and ETH. In total, the development is composed of approximately 1650 lines of code. The minimal technical specifications for deploying *SC Analyzer* are 8 GB RAM Memory, 2 Core CPU, and 150 GB Storage.

5.2 Initial Results

To test our approach, we performed the following steps. Initially, we performed human-based analyses to identify suspicion TSC and performed data collection over ETH and BSC frameworks. Over this data set, we perform tests of the TSC by proving their ID as an input to *SC Analyzer*. In approximately 368 tested TSCs from the suspicious cases, in both platforms, our tool has been able to

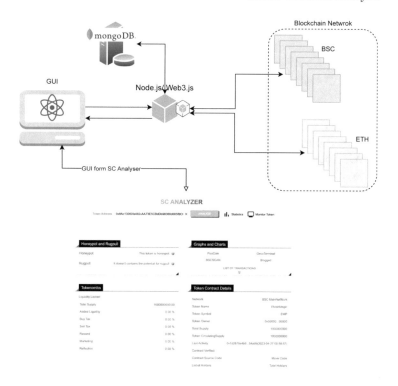

Fig. 1. The global architecture and GUI of SC Analyzer.

identify 93 cases of fraudulent TSC, i.e., trapdoor and rug pull. Around 120 TSC have been flagged with "centralization" issues.

For elaboration, an example, we have TSC identified via 0xddc561592 f4403179454 c97479925c2e49ccf1a2, which is identified and spotted as a trapdoor/honeypot. Further, our tool has been able to identify sophisticated rug pull cases. The TSC identified via "0x9Ac13060fa482cAA73E5CBdDb98360d6 65fBC603" presents rig pull (and trapdoor), as the malicious owner 0x82d51d79041f753e16f0d0ee03d9635 aa3b508da executes the *approveo* function via this transaction. This function allows adding tokens in the wallet *0xaC073D85F5c56ce851Fad5ca681298000bE352 6d* without emitting any event to avoid block explorer indexing. Then the wallet receiving tokens will sell the tokens received in the transaction as verified in this transaction hash.

6   Conclusion and Future Works

Despite their enormous potential and sophisticated technological solutions, on many occasions, the SCs haven't always been applied for good purposes. The decentralized environment presents a huge challenge for investors in understanding operational and technical levels. In this paper, we proposed a solution for some well-known problems that cause many investors to lose funds. Our scientific method permits initially extraction of TSC information, allowing investors to see the real-time characteristic related to TSC, such as "Smart Contract Details", "Tokenomiks", "Graphs", and "onlyOwner". That already presents a one-stop shop for every TSC. Further, it verifies BEP/ERC-20-based TSC potential risks for honeypot/trapdoor, rug pull, and "centralization" of TSC by checking different parameters.

In terms of limitation, the proposed tool is currently mainly applicable to ERC-20 TSC. Therefore, it does not cover the ERC721 or DAO-based SC. Furthermore, we aim for additional automation aspects, especially identifying new DEX-es and performing data collections, i.e., wallets, as we currently manually perform that task. In addition, to avoid any false positives, we manually checked the results. Particularly for the case of the rug pull as we check token distribution over wallets from the TSC.

Future works aim to train Machine Learning (ML) algorithms to perform analysis and to improve the automation process for automatic analysis of data sets (ETH and BSC). Further, we aim to extend our tool by enabling it comparison of standard TSC functions with the deployed SC. On the other hand, we aim to implement SC security practices to be verified dynamically, i.e., in real-time based on real data.

References

1. Step by Step Towards Creating a Safe Smart Contract: Lessons and Insights from a Cryptocurrency Lab. https://eprint.iacr.org/2015/460.pdf
2. Security checklists for Ethereum smart contract development: patterns and best practices. https://arxiv.org/pdf/2008.04761.pdf
3. Sun, T., Yu, W.: A formal verification framework for security issues of blockchain smart contracts. Electronics **9**(2), 255 (2020). https://www.mdpi.com/2079-9292/9/2/255#B20-electronics-09-00255
4. Luu, L., Chu, D.H., Olickel, H., Saxena, P., Hobor, A.: Making smart contracts smarter. In: Proceedings of the 2016 ACM SIGSAC Conference on Computer and Communications Security, pp. 254–269 (2016)
5. Tsankov, P., Dan, A., Drachsler-Cohen, D., Gervais, A., Bünzli, F., Vechev, M.: Securify: practical security analysis of smart contracts (2018). https://arxiv.org/pdf/1806.01143.pdf
6. Tikhomirov, S., Voskresenskaya, E., Ivanitskiy, I., Takhaviev, R., Marchenko, E., Alexandrov, Y.: SmartCheck: static analysis of ethereum smart contracts (2018). https://orbilu.uni.lu/bitstream/10993/35862/3/smartcheck-paper.pdf
7. Hu, H., Bai, Q., Xu, Y.: SCGguard: deep scam detection for ethereum smart contracts. In: IEEE Conference on Computer Communications Workshops, pp. 1–6 (2022)
8. Akca, S., Rajan, A., Peng, C.: SolAnalyser: a framework for analysing and testing smart contracts (2019). https://www.pure.ed.ac.uk/ws/files/118990635/SolAnalyser_A_Framework_AKCA_DOA29092019_AFV.pdf
9. Gao,J., Liu, H., Liu, C., Li, Q., Guan, Z., Che, Z.: EASYFLOW: keep ethereum away from overflow. https://arxiv.org/pdf/1811.03814.pdf

10. https://www.openzeppelin.com/
11. Ma, T.: Cybersecurity and ethereum security vulnerabilities analysis. Highlights Sci. Eng. Technol. **28**(34), 375–381 (2023)
12. Crypto Investors Lose $1 Million in Fake MetaMask Token Rug Pull Scam (2021). https://www.gadgets360.com/cryptocurrency/news/metamask-token-scam-usd-1million-lost-defi-cybercrime-dextools-2677970. Accssed 12 Apr 2023
13. Chen, T., et al.: TokenScope: automatically detecting inconsistent behaviors of cryptocurrency tokens in ethereum. In: Proceedings of the 2019 ACM SIGSAC, pp. 1503–1520 (2019)
14. Torres, C.F., Steichen, M., State, R.: The art of the scam: demystifying honeypots in ethereum smart contracts. arXiv preprint arXiv:1902.06976 (2019)
15. Imeri, A., Agoulmine, N., Khadraoui, D., Khadraoui, A.: Blockchain-based multi-party smart contract for service digitalization and automation. In: Arai, K. (ed.) FTC 2021. LNNS, vol. 359, pp. 793–806. Springer, Cham (2022). https://doi.org/10.1007/978-3-030-89880-9_59
16. Lehar, A., Parlour, C.A.: Decentralized exchanges (2021). https://doi.org/10.2139/ssrn.3905316
17. Aspris, A., Foley, S., Svec, J., Wang, L.: Decentralized exchanges: the "wild west" of cryptocurrency trading. Int. Rev. Financ. Anal. **77**, 101845 (2021). https://doi.org/10.1016/j.irfa.2021.101845
18. Liao, J.-W., Tsai, T.-T., He, C.-K., Tien, C.-W.: SoliAudit: smart contract vulnerability assessment based on machine learning and fuzz testing. In: 2019 Sixth International Conference on Internet of Things: Systems, Management and Security (IOTSMS), pp. 458–465 (2019). https://doi.org/10.1109/IOTSMS48152.2019.8939256
19. Honeypot Detector for BSC (2023). https://honeypot.is/. Accessed 3 Apr 2023
20. Imeri, A., Agoulmine, N., Khadraoui, D.: Smart contract modeling and verification techniques: a survey. In: 8th International Workshop on ADVANCEs in ICT Infrastructures and Services, pp. 1–8 (2020)
21. Almakhour, M., Sliman, L., Samhat, A.E., Mellouk, A.: Verification of smart contracts: a survey. Pervasive Mob. Comput. **67**, 101227 (2020)
22. Puggioni, V.: Crypto rug pulls: what is a rug pull in crypto (2022). https://cointelegraph.com/explained/crypto-rug-pulls-what-is-a-rug-pull-in-crypto-and-6-ways-to-spot-it
23. Smart Contract Audit - Web3 Security Leaderboard (2023). https://www.certik.com/products/smart-contract-audit. Accessed 12 Apr 2023
24. Uniswap whitepaper by Hayden Adams. https://hackmd.io/@HaydenAdams/HJ9jLsfTz. Accessed 26 Apr 2023
25. Token Standards—ethereum.org (2023). https://ethereum.org/en/developers/docs/standards/tokens/. Accessed 28 Apr 2023
26. Monrat, A.A., Schelén, O., Andersson, K.: A survey of blockchain from the perspectives of applications, challenges, and opportunities. IEEE Access **19**(7), 117134–51 (2019)
27. Harvey, C.R., Ramachandran, A., Santoro, J.: DeFi and the Future of Finance. Wiley, Hoboken (2021)

Blockchain Context Canvas: A Tool to Align Developers and Stakeholders

Alfredo Colenci Neto[1(✉)] ⓘ and Daniel Capaldo Amaral[2] ⓘ

[1] Technology Faculty of São Carlos - Fatec, São Carlos, SP, Brazil
alfredo.colenci@fatec.sp.gov.br
[2] University of São Paulo - USP, São Carlos, SP, Brazil
amaral@sc.usp.br

Abstract. Blockchain technology has had significant prominence in recent years for being a solution that integrates the various actors of a business scenario providing security in transactions, thereby reducing intermediaries and increasing efficiency in supply chains. Among the characteristics of these applications are the significant amounts of stakeholders, metadata, and business rules, as well as the knowledge gap between developers and business analysts. Herein, we propose and evaluate a tool called Blockchain Context Canvas that guides the discussion between users, analysts, and developers objectively, without users and analysts needing to discuss how the technology works, thus promoting greater efficiency and objectivity in the construction of the smart contract. The evaluation uses expert opinion. The tool was used in an application for the coffee supply chain and then evaluated by an expert developer experienced in blockchain solutions. The results indicate that the tool can serve as a roadmap for guiding real supply chain projects.

Keywords: blockchain technology · supply chain · blockchain context canvas

1 Introduction

Blockchain technology has stood out in recent years for being a solution that integrates the various actors in a business scenario, thereby effectively providing the guarantee of anonymity and traceability issues in a distributed environment.

Blockchain is essentially a distributed database system, which records transactional data in the form of a network of data blocks, verified by network computers and which, due to this logic, cannot be deleted by a single actor [1–4].

In this way, it enables transparency in the process, as participants can view the transactions related to them at any time and verify the suitability of documents through encrypted keys [5]. There are reports of use cases in a broad context of society, such as agriculture [6], sustainability [7], automation of insurance processes [8], sharing economy [9], among several other areas.

In a blockchain project, several aspects, including the context of the application, definition of actors and their responsibilities, assets, data structure, and connection with external applications, among others, must be considered. From a technical point of

J. M. Machado et al. (Eds.): BLOCKCHAIN 2023, LNNS 778, pp. 254–263, 2023.
https://doi.org/10.1007/978-3-031-45155-3_26

view, there is a need to define the consensus, endorsement policy, privacy and reliability issues, and associated service providers. These definitions culminate in the creation of a fundamental element of the network—the smart contract. It contains the business rules for carrying out transactions and, therefore, is the "heart" of the system. Furthermore, [10] shows that it cannot be changed after being published on the network.

It is noticed that the architecture of a blockchain application is not simple, and this is an obstacle that developers and users face during the design and implementation of systems [11]. Blockchain technology has a difficult understanding for business analysts, stakeholders, and users. Moreover, as it usually involves many stakeholders, it is difficult for the developer to synthetically understand everyone's problem, such that he can identify the best solutions in the organization of metadata and governance rules.

[12] analyzed 82 blockchain projects and showed that among the essential properties of this type of application is the interaction of three elements: different participants, the unit of value transacted and the necessary security filters. They must be combined to create a business model with this technology and they are what allow decentralized models, most open, not having a clear and visible monetization strategy [12]. However, despite enabling innovative business models, they make the task of specifying these different elements and their relationships complex [12], which makes it difficult to analyze and design this type of solution. [13] reinforces the statement. The improvements in transaction transparency and reliability are true value underlying blockchain technology, however, it will be necessary to develop frameworks to ensure the participation of all those involved.

One possible solution is to guide the discussion to the fundamental points that allow the developer, analyst, and product owners to reach a minimum consensus on the main aspects of the solution. Herein, we propose a software engineering canvas tool called Blockchain Context Canvas (BCC) to be used as a necessary support artifact for the design of a blockchain solution focused on the supply chain. This is a qualitative and exploratory research, which aims to conduct a bibliographic survey about blockchain technology and its elements, includes field interviews with actors for the mapping of the production process, and presents the case study of the application of the blockchain canvas. In an applied way, the article presents and validates the proposal through an application in a real scenario. The focus of the present study is the coffee production chain in a traceability context. However, it is intended that this study can be sufficiently generic to be used in several application areas.

2 Theoretical Reference

2.1 The Challenge in Analyzing Blockchain Solutions

[14] defines a blockchain as a distributed data structure, also called a ledger, in which all confirmed transactions are stored by data sets in units called blocks. Each block references the previous block in a chronological sequence, thus forming a chain, hence the name blockchain. Its structure allows different bodies to transact with each other on a computer network wherein there is no validation by a central institution. This is due to the combination of a peer-to-peer network, together with consensus mechanisms, cryptography, and market mechanisms.

Blockchain is essentially a distributed database system that records transactional data in the form of a network of data blocks, verified by the network computers and that, due to this logic, can no longer be deleted by a single actor [1–4]. It enables transparency in the process, as participants can view the transactions related to them at any time and verify the suitability of documents through encrypted keys [5].

[15] explains that blockchain provides a means for many different participants, regardless of their location, to record their transactions in a jointly shared digital ledger that exists as linked and synchronized copies on their respective computers. All transactions are timestamped at different intervals and linked to the previous event via a cryptographic hash. Instead of keeping separate records based on receipts and vouchers, participants can enter their transactions directly into the shared ledger. This entry, once confirmed by a consensus of the participants, is then cryptographically locked and impossible to alter.

One of the problems with blockchain applications is that they change the current paradigm of computing. Professionals in the market are used to thinking in terms of server–client applications, that is, they understand an information system as composed of centralized databases. Blockchain applications are distributed databases. This change has a considerable impact on not only the issue of permission but also the choice of metadata.

According to [14], the biggest problem of this centralized format is that it is a monopolistic, asymmetric, and opaque information system, which could result in trust problems, such as fraud, and corruption, as well as tampering and falsifying information.

Similarly, [16] says that traditionally, individual users who have limited computational power and local storage typically use a centralized server to facilitate collaboration. Such a service enables anywhere-available, real-time, and concurrent access to the database. However, it requires full trust in the central server, which is expected to correctly execute all requests. One immediate concern is that a malicious server can fool any client without being detected. Another issue arises from the client side: a client may manipulate the records arbitrarily for its own interest. A sophisticated mechanism must be adopted to prevent clients from issuing undesired updates.

Another challenge during the initial design of a project is the determination of the necessity to use a blockchain or not. Some studies, such as those of [17, 18], show that some teams face difficulties vis-à-vis addressing this question and the characteristics that should be considered for solution design.

The main elements of a blockchain application are the assets, actors, data and properties, channels, peers, ledger, and transactions. Some of these elements can be observed in a blockchain application. One difference, however, is with regard to content. Blockchain applications in supply chains usually involve numerous stakeholders. Considering that care needs to be taken to maintain a small amount of metadata, there is a conflict or trade-off that needs to be addressed by the analyst.

The result is that the process of analyzing and developing these solutions is not simple, as presented by [11]. The immutability of the blocks also brings the additional challenge of making changes to the data after they are put into production difficult often impossible.

In practice, this creates the need for precision in the application definition and numerous meetings so that everyone involved can develop a common understanding and alignment. This results in extensive durations of meetings and rework efforts. Moreover, misunderstandings can lead to the wrong network design, which is especially dangerous in chaotic blockchain applications because of their difficulty or even impossibility to be changed (in the case of open networks). Solutions used in tool analysis.

As it is a relatively new technology, no standard modeling technique exists for requirements engineering and organizational modeling within blockchain-oriented software (BOS) engineering.

[19] says that there is no standard notation available to design or model a BOS. A blockchain-based system could need a specialized notation to represent it. The lack of specialized notations can overcomplicate the adoption or migration to BOS, as the interaction between the blockchain and the application will not be properly specified.

Some authors propose the use of techniques already established in software engineering, such as Unified Modeling Language diagrams and, more specifically, use case, sequence, and activity diagrams. Some works have also used the entity relationship diagram and the i* framework as a way to solve the necessity for tools focused on the design of blockchain solutions.

Furthermore, in the literature, it is possible to find several papers focused on the use of business process modeling (BPM) to design blockchain-based solutions in supply chains [20–22].

Certainly, BPM supports the identification of the authors and flows within a supply chain, but it should be seen as an auxiliary tool, as it does not consider the technical issues of a blockchain project.

[23] proposed a methodology for designing a strategy to develop and validate the overall blockchain solution and integrate it in the business strategy with a focus on cost, key performance indicators, and strategy.

In fact, there is a paucity in terms of standard methodologies for designing a BOS to develop and validate blockchain solutions.

3 Method

The proposition and validation of design methods is not a simple task. [24] make a practical proposal, with an analogy to work in medicine. The primary goal in medicine is to propose a method (treatment or drug) that is administered to human patients to obtain the desired result. As in design, the validation process becomes a job of gathering evidence from controlled investigations of different types.

Similarly, [25] consider validation of design methods where ".... Becomes a process of building confidence in its usefulness with respect to a purpose" [25]. They propose the use of two variables: whether the method provides solutions correctly and effectively, and whether the method produces such solutions efficiently and acceptably. Combined, they would produce four types of validation, in quadrants, as follows: theoretical structural validity, theoretical performance validity, empirical structural validity, and empirical performance validity.

In the present study, the presentation of the method is the main objective, but we try to obtain a first level of validity evidence. We started by verifying the logical proposition by using the application in an example of a solution in the coffee supply chain and analyzing the logic, reliability of information, and absence of bias, according to [26], approving the method in the perception of the researchers involved in the proposition.

Finally, the results were presented to experienced developers of blockchain solutions to independently verify whether the result was consistent with expectations. We also took cognizance of their perception of the effectiveness of the method. As there is no empirical test under controlled conditions, it is assumed that this is a first evidence and that it is of the theoretical structural validity type, which is sufficient for other researchers to use and continue the evolution and testing of the method.

4   Blockchain Context Canvas Proposal

The bibliographic review of Sect. 2.1 showed the challenges. The main practical problems in the analysis lies in the knowledge gap combination due to the amount of stakeholders and difficulty in understanding the technology. On the one hand, the large number of stakeholders makes it difficult for developers and analysts to identify the essential elements, a fundamental issue to maintaining the simplicity and functionality of the blockchain network. Finally, the difficulty in understanding the technology leaves the stakeholders without knowing what to expect or how best to capitalize on the technology.

In addition to listing challenges, in the bibliographic review we identified all the elements necessary for the description of a blockchain-type application. Comparing challenges with elements, it was possible would be to identify essential elements that could make the discussion process more objective while avoiding a deviation regarding the functionality of the technology:

Project context: It is important during the initial discussions of the project to define the context within which the solution will operate. Notably, a production process with several actors may generate thousands of data per minute. It is common in blockchain projects for stakeholders to want to involve various aspects that are not part of the core business of the proposal. However, depending on the context that is intended to work, much of the data collected will be unnecessary. It is not the intention of a project of this magnitude to collect all the data. Thus, by defining the context, the application is limited to working only with what is essential.

Actors: You must define who the actors that will be part of the project are, as well as their roles and responsibilities, including whether there are conflicts of interest among them that could make the project unfeasible. In this step, you must define who will have the privilege of reading and writing the data.

Transactions: A blockchain is a ledger distributed over a network that records transactions (messages sent from one network node to another) executed among network participants. Transactions are understood to be the actions generated by the participants in the system. You must define which transactions represent the methods that can modify the assets within the blockchain ledger.

Assets: Blockchain assets are a type of digital asset that represents an object that will be identified on the blockchain network. In the production chain, an asset can have several formats and configurations, and its format must be mapped at each stage.

Data: As a way to define the structure of each asset, you should generate a description of the properties of each field belonging to the project context.

Core processes: In supply chain management, the flow of materials and services required in manufacturing a given product is managed, and it includes various intermediate storage and production cycles until the delivery to the final point of consumption.

Technical issues: Basic technical aspects should be considered at the onset of a blockchain-network-related project. These include project type (public, private, or hybrid), permissioning (permissionless or permisioned), and infrastructure on which the solutions will be implemented (cloud or on premise).

The model chosen to present these elements is a canvas, a method that became globally known due to the proposal of [27] when they created the Business Model Canvas (BMC) for modeling business plans. The BMC can be understood as a strategic planning tool that allows for the developing and sketching of new or existing business models easily through nine blocks.

Like the popular BMC blueprint, BCC uses building blocks that allow for the identification of key elements and the analysis of their relationships. Similar to BMC, BCC building blocks are designed to facilitate the framing and resolving of the most important challenges faced by these organizations in blockchain solutions in connection to the development of a smart contract.

The main advantages of this representation are ease of use, no need for lengthy training, low implementation cost, possibility of collaborative work between those involved, and fast and purpose-driven results of the blockchain project.

The final result was named BCC; Fig. 1 illustrates the artifact.

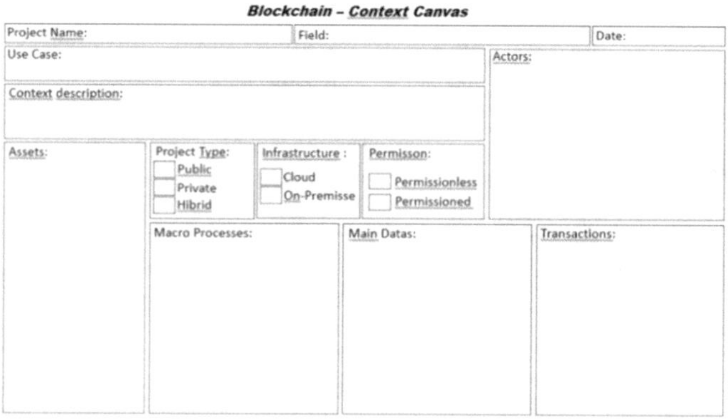

Fig. 1. Blockchain Context Canvas.

With the definitions presented above, one can even get the answers to questions pertaining to the real need to use a blockchain network or not: Is it necessary to have a shared database? Is it necessary to have multiple parties writing data? Are potential writers untrusted? Is disintermediation needed? [18].

As a means to validate the BCC proposal, the tool was applied to a real case focused on the coffee production chain.

5   Case Study

The proposed BCC was evaluated based on feasibility. In the operations management of a productive chain, a project for coffee traceability was used, wherein all the proposed steps were followed. This is a project to develop a prototype for coffee producers in the state of São Paulo, Brazil.

The first part, not reported here, dealt with data collection. A team of three people, including the two authors, collected data on the coffee supply chain in the interior of the State of São Paulo, Brazil, during 9 months). The result was the identification of actors and exchange flows of inputs and products along the outbound logistics of coffee beans from the rural producer to the industry.

The data obtained in the first stage, completed in January 2023, were used by another team, now with the participation of a developer specialized in blockchain-type applications. In this stage, which lasted 5 h, the team used the Context Canvas tool, its phases and artifacts, to describe an application capable of recording the traceability of coffee between the rural producer and the processing company, according to the data collected in field. Figure 2 shows the final result of the table generated with the developer.

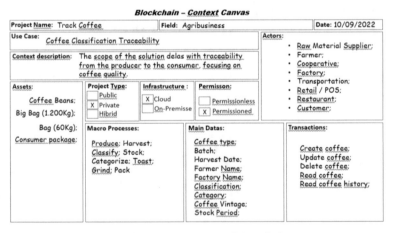

Fig. 2.   The context canvas of the solution.

After making the canvas between those involved, it was possible to create the project's smart contract, which took 2 h to complete. According to the developer, there was a significant gain compared to other projects that he participated. The smart contract was developed in Go language, and the CC-Tools library from the company GoLedger was used as support.

After creating the smart contract, it was possible to build the blockchain network in the AWS Cloud environment with three organizations. For this, the GoFabric tool was used to create the network on the Hyperledger Fabric platform. It is noteworthy that with the use of the canvas tool, the creation of the network took only three hours.

6 Expert Assessment

As a way to validate the proposed canvas, it was presented to an expert in building blockchain networks and smart contracts with a computer science background and four years of experience in blockchain projects.

"Without a doubt, one of the main difficulties we encounter when starting a blockchain project is the definition of the asset to be worked on. There is always a doubt about what will be identified and the main categories that make up the asset."

"I participated in a project where it took more than 30 days for those involved (stakeholders) to actually define what an airplane was, in the sense of what fields were needed in the formation of the asset. This becomes complex when the asset changes shape during the production chain."

"What I mostly use as a support tool in a blockchain project for supply chains is the mapping of the production process with the identification of the actors and their activities."

"Certainly, this proposed blockchain canvas would facilitate a lot the construction of the smart contract. I believe that with the making of this artifact, I can build the smart contract in a very short time using a library."

7 Final Considerations

The present article presented the first step on this research to propose a new methodology to initiate blockchain applications and its validation, as exemplified in the traceability project of the coffee production chain. It comprehensively presents the phases to be performed in order to facilitate the design stage and improve the value and quality of blockchain solutions.

We developed the main elements: phases and artifacts. We provide yeat one example of use applying them literature, as described in the article and exemplified in the case study. It was not possible to validate the results with the managers, but the synthesis of observations with the project team, in a non-structured interview, allowed the following observations about the case:

1 - The difficulty of discussing the theme with experts in the chain and even developers due to the complexity of the technology.

2 - The asset metadata could be identified, but it is difficult to determine if it would be sufficient, especially considering that once the network is started it cannot be changed.

3 - Despite the restrictions, it was possible to see that the use of the BCC, in this real project for coffee traceability, was of great help in the elaboration of the smart contract.

These results indicate some paths for the evolution of methodologies for designing applications with blockchain data. Among the aspects that can be improved, the following are observed: the importance of discussing ways to improve the initial stages of the software engineering process with the initiation of those involved with the technology and its potential and the need for techniques to validate the metadata and functions present in the contract.

This article contributes to knowledge with respect to the synthesis and presentation of the phases to be performed in order to facilitate the design stage of a blockchain application and, notwithstanding the focus on a specific application in the area of operations management, in this case, the coffee production process, it can serve as a reference to production engineering professionals for similar applications in supply chain studies.

As the next steps in this project, the application of the concept to four more real blockchain projects in different segments of operations management is underway, and it is expected to be published soon.

The authors would like to thanks the support of Monitora, a Marlabs company during the development of the research.

References

1. Sapra, R., Dhaliwal, P.: Blockchain: the perspective future of technology. Int. J. Healthc. Inf. Syst. Inform. **16**(2), 1–20 (2021). https://doi.org/10.4018/IJHISI.20210401.oa1
2. Kumar, A., Kumar, S.: A systematic review of the research on disruptive technology– Blockchain. In: 2020 5th International Conference on Communication and Electronics Systems (ICCES), pp. 900–905. IEEE (2020). https://doi.org/10.1109/ICCES48766.2020.091 38055
3. Weking, J., Mandalenakis, M., Hein, A., Hermes, S., Böhm, M., Krcmar, H.: The impact of blockchain technology on business models – a taxonomy and archetypal patterns. Electron Markets **30**, 285–305 (2020). https://doi.org/10.1007/s12525-019-00386-3
4. Swan, M.: Blockchain: Blueprint for a New Economy. O'Reilly Media Inc, Sebastopol, CA (2015)
5. Kimani, D., Adams, K., Attah-Boakye, R., Ullah, S., Frecknall-Hughes, J., Kim, J.: Blockchain, business and the fourth industrial revolution: whence, whither, wherefore and how? Technol. Forecast. Soc. Change **161**, 120254 (2020). https://doi.org/10.1016/j.techfore. 2020.120254
6. Jahanbin, P., Wingreen, S., Sharma, R.: Blockchain and IoT integration for trust improvement in agricultural supply chain. In: 27th European Conference on Information Systems (2019)
7. Brody, P., Pureswaran, V.: Device democracy: saving the future of the Internet of Things. IBM Institute for Business Value, Tech. Rep. (2014). http://www935.ibm.com/services/us/gbs/thoughtleadership/internetofthings/
8. Raikwar, M., Mazumdar, S., Ruj, S., Gupta, S.S., Chattopadhyay, A., Lam, K.Y.: A blockchain framework for insurance processes. In: 2018 9th IFIP International Conference on New Technologies, Mobility and Security (NTMS), pp. 1–4. IEEE (2018)
9. Fiorentino, S., Bartolucci, S.: Blockchain-based smart contracts as new governance tools for the sharing economy. Cities **117**, 103325 (2021)
10. Zheng, Z., et al.: An overview on smart contracts: challenges, advances and platforms. Future Gener. Comput. Syst. **105**, 475–491 (2020)

11. Jurgelaitis, M., Butkienė, R., Vaičiukynas, E., Drungilas, V., Čeponienė, L.: Modelling principles for blockchain-based implementation of business or scientific processes. In: CEUR Workshop Proceedings: IVUS 2019 International Conference on Information Technologies: Proceedings of the International Conference on Information Technologies, Kaunas, Lithuania, April 25, 2019, vol. 2470, pp. 43–47. CEUR-WS (2019)

12. Lage, O., Saiz-Santos, M., Zarzuelo, J.M.: Decentralized platform economy: emerging blockchain-based decentralized platform business models. Electron Markets **32**, 1707–1723 (2022). https://doi.org/10.1007/s12525-022-00586-4

13. Lee, J.Y.: A decentralized token economy: how blockchain and cryptocurrency can revolutionize business. Bus. Horiz. **62**(6), 773–784 (2019)

14. Tian, F.: An information system for food safety monitoring in supply chains based on HACCP, blockchain and internet of things. Doctoral thesis. WU Vienna University of Economics and Business (2018)

15. Appelbaum, D., Nehmer, R.: Designing and auditing accounting systems based on blockchain and distributed ledger principles. Feliciano School of Business 1–19 (2017)

16. Peng, Y., Du, M., Li, F., Cheng, R., Song, D.: FalconDB: blockchain-based collaborative database. In: Proceedings of the 2020 ACM SIGMOD International Conference on Management of Data, pp. 637–652 (2020)

17. Gatteschi, V., Lamberti, F., Demartini, C., Pranteda, C., Santamaria, V.: To blockchain or not to blockchain: that is the question. IT Prof. **20**(2), 62–74 (2018)

18. Wüst, K., Gervais, A.: Do you need a blockchain? In: 2018 Crypto Valley Conference on Blockchain Technology (CVCBT), pp. 45–54. IEEE (2018)

19. Porru, S., Pinna, A., Marchesi, M., Tonelli, R.: Blockchain-oriented software engineering: challenges and new directions. In: 2017 IEEE/ACM 39th International Conference on Software Engineering Companion (ICSE-C), pp. 169–171. IEEE (2017). https://doi.org/10.1109/ICSE-C.2017.142

20. Weber, I., Xu, X., Riveret, R., Governatori, G., Ponomarev, A., Mendling, J.: Untrusted business process monitoring and execution using blockchain. In: La Rosa, M., Loos, P., Pastor, O. (eds.) Business Process Management (BPM 2016). LNCS, vol. 9850, pp. 329–347. Springer, Cham (2016). https://doi.org/10.1007/978-3-319-45348-4_19

21. Caro, M.P., Ali, M.S., Vecchio, M., Giaffreda, R.: Blockchain-based traceability in agri-food supply chain management: a practical implementation. In: 2018 IoT Vertical and Topical Summit on Agriculture-Tuscany (IOT Tuscany), pp. 1–4. IEEE (2018)

22. Casado-Vara, R., González-Briones, A., Prieto, J., Corchado, J.M.: Smart contract for monitoring and control of logistics activities: pharmaceutical utilities case study. In: Graña, M., et al. (eds.) International Joint Conference SOCO'18-CISIS'18-ICEUTE'18. AISC, vol. 771, pp. 509–517. Springer, Cham (2019). https://doi.org/10.1007/978-3-319-94120-2_49

23. Perboli, G., Musso, S., Rosano, M.: Blockchain in logistics and supply chain: a lean approach for designing real-world use cases. IEEE Access **6**, 62018–62028 (2018)

24. Frey, D.D., Dym, C.L.: Validation of design methods: lessons from medicine. Res. Eng. Design **17**, 45–57 (2006). https://doi.org/10.1007/s00163-006-0016-4

25. Seepersad, C.C., Pedersen, K., Emblemsvåg, J., Bailey, R., Allen, J.K., Mistree, F.: The validation square: how does one verify and validate a design method. Decis. Making Eng. Des. 303–314 (2006)

26. Olewnik, A.T., Lewis, K.: On validating engineering design decision support tools. Concurrent Eng. **13**(2), 111–122 (2005)

27. Osterwalder, A., Pigneur, Y.: Business Model Generation: A Handbook for Visionaries, Game Changers, and Challengers, vol. 1. John Wiley & Sons (2010)

A Feature Model of Consensus Algorithms in Distributed Ledger Technology

Elena Baninemeh[1]([✉]) [iD], Slinger Jansen[1,2] [iD], and Bas Pronk[1]

[1] Information and Computer Science, Utrecht University, Utrecht, The Netherlands
{e.baninemeh,slinger.jansen,b.pronk}@uu.nl
[2] Lappeenranta University of Technology, Lappeenranta, Finland

Abstract. A distributed ledger is a database distributed across multiple systems, with each system holding a synchronized copy of the data. Distributed ledger technology has applications in various healthcare, finance, and cybersecurity domains. However, the intricacies of the features of consensus algorithms, which ensure consistency across different ledgers, remain challenging, as the relevant knowledge is scattered across a wide range of literature or in the form of tacit knowledge of software practitioners. This study presents a systematic data collection comprising an extensive literature review and a set of expert interviews to provide insights into designing and evaluating of consensus algorithms for web3 applications. The usability and usefulness of the extracted knowledge were evaluated by seven experienced practitioners in web3 development companies, resulting in an overview of 13 consensus algorithms, their features, and their impacts on quality models. With this comprehensive knowledge, web3 developers can expedite evaluating, selecting, and implementing consensus algorithms for distributed ledgers.

Keywords: consensus algorithm · algorithm selection · distributed ledger

1 Introduction

Distributed ledger technology (DLT) has emerged as a potential alternative to traditional centralized data management systems. Unlike centralized systems, DLT allows data to be stored and maintained among multiple peers in a network, without relying on a central authority [15]. DLT achieves this through the use of a consensus algorithm that establishes a shared state of the ledger among all network participants. Consensus algorithms are designed to address the challenges of maintaining a distributed ledger, such as ensuring data integrity and preventing malicious attacks [8].

DLT has been applied to various domains, including supply chain management, healthcare, finance, and more [17]. However, DLT is still an emerging technology that faces several significant challenges, including concerns about its

© The Author(s), under exclusive license to Springer Nature Switzerland AG 2023
J. M. Machado et al. (Eds.): BLOCKCHAIN 2023, LNNS 778, pp. 264–275, 2023.
https://doi.org/10.1007/978-3-031-45155-3_27

security and scalability [16]. Designing the ledger is a crucial challenge many new DLT projects face, involving making numerous decisions during the design process. The designer must make choices regarding the consensus algorithm, transaction validation mechanism, data storage structure, and access control policies, among other things. These decisions significantly impact the system's security, scalability, and efficiency, which can affect the project's success. Therefore, careful consideration and extensive research must be carried out during the design process to ensure that the DLT project can meet its objectives and deliver optimal performance [20].

Consensus algorithms are a critical component of DLT, as they ensure the consistency of distributed ledgers among network nodes [13]. Due to the wide range of threats that can affect the system, consensus algorithms come in different forms and designs. For instance, large public and cryptocurrency ledgers typically use a proof-of-work algorithm, which requires nodes to solve a complex mathematical puzzle before adding new data. On the other hand, smaller private blockchains often use distributed system consensus algorithms, such as PAXOS and RAFT, which rely on agreement protocols rather than computational puzzles to ensure data consistency [8, 13].

In this study, we proposed a systematic approach for collecting data on consensus algorithms to support web3 developers in selecting, creating, and employing consensus algorithms. Our study involved conducting a literature review and interviewing experts to evaluate the usefulness and usability of the extracted knowledge. The study identified 13 consensus algorithms and their features, providing valuable insights into designing and evaluating consensus algorithms.

In Sect. 2 we present the research challenge of capturing knowledge around features of consensus algorithms and propose to do so through literature study and expert interviews. Subsequently, we report on the creation of a feature model in Sect. 3 and distinguish between boolean and non-boolean features to provide a deeper understanding of feature models of consensus algorithms. In Sect. 4, we discuss how the feature model contributes to the state of the art around consensus algorithms, and we argue that, while consensus algorithms are important, their selection generally fully depends on the DLT platform that is selected first. We conclude and summarize our study in Sect. 5.

2   Research Approach

This study's main research question is, *"How can knowledge be captured regarding consensus algorithms to support web3 development companies with evaluating, designing, and implementing consensus algorithms?"*. It addresses the challenge of selecting an appropriate consensus algorithm for a distributed ledger technology (DLT). This is due to a large number of available alternatives, each with a wide range of features, and the inherent trade-offs between security, scalability, and decentralization, known as the consensus algorithm trilemma. The research project combines multiple research methods, including a literature study and expert interviews, to create an artifact that supports web3 development companies in evaluating, designing, and implementing consensus algorithms. The literature study identifies the role of DLT and consensus algorithms, extracting alternative consensus algorithms and their features and extracting feature models for consensus algorithms. The expert interviews aim to gather data and evaluate the completeness and usefulness of the preliminary design of the artifact, which will be evaluated in case studies. Finally, the research project uses Myers and Newman guidelines to conduct a series of qualitative semi-structured interviews with experts selected based on their expertise and experience. Table 1 shows the experts participating in this research. Seven domain experts, including Blockchain developers and Consensus algorithms experts from different organizations, have participated in the research to assist us with answering the research questions. Before reaching out to potential domain experts, a role description was created to accurately identify their areas of expertise and ensure that the right target group was approached. Subsequently, we sent emails to the chosen experts, providing them with the role description and details regarding our research topic. It is important to note that the selection of experts was carried out in a pragmatic and convenient manner, based on the expertise and experience they had indicated on their LinkedIn profiles. We employed a set of evaluation criteria, such as "Years of experience", "Expertise", "Skills", "Education", and "Level of expertise", to guide the selection process. The semi-structured interviews were conducted with experts, and each interview had a duration of 45 to 60 min. To minimize any preconceived notions, we employed a set of open-ended questions to extract as much information as possible from the experts. The interviews were conducted virtually using platforms like Skype and Zoom. Prior consent was obtained from the interviewees to record the interviews, which were later transcribed for analysis. The knowledge obtained from each interview was regularly shared and validated in subsequent interviews to ensure the incremental acquisition of accurate information. Finally, our findings and interpretations were presented to the interview participants for their final approval.

Table 1. The interview participants were experts in consensus algorithm design. Due to the specialized nature of this expertise, the response rates were low, but the quality of the interviews was high.

Occupation	Company	Years of Experience
Co-Founder	Lisk	5
Consensus Researcher	Humanode	6
Blockchain developer	Gimly Blockchain projects	4
Blockchain developer	dappdevelopment.com	6
Founder	Emerging Horizons	3
Co-Founder	WBNoDe	8
Consensus algorithm developer	Hyperledger Fabric	4

Concensus Algorithms - The literature on benchmarking the consensus algorithms for blockchains includes several studies, such as [7,11,13]. While two of these studies propose a Boolean decision tree, they have limitations, such as a restricted set of alternatives and features. The survey presented by Fu et al. [13] lacks robustness as it offers limited features and alternatives. So while these studies laid an excellent foundation for this study, we decided to dive deeper into the features that consensus algorithms provide and evaluate these with practitioners to provide an actionable set of knowledge about the features of consensus algorithms. Researchers, consensus algorithm designers, and consensus algorithm implementers can form better technology selection decisions with such knowledge.

Based on the literature study, we collected the different consensus algorithms as alternatives and their features that define a consensus algorithm. These features and alternatives are required for consensus algorithm selection. This wide variety of features and different algorithms has led to classifying consensus algorithm selection as a Multi-Criteria Decision-Making (MCDM) problem [5]. The full overview of all consensus algorithms and their sources can be found in Table 2.

From the consensus algorithms identified in the literature, many were considered by the interviewees to be either unused or unfamiliar to them. The responses regarding the number of significant alternatives can be found in Table 2.

The selection of consensus algorithms is closely linked to the choice of distributed ledger platforms, which greatly influences the type of consensus algorithms considered. As a result, lesser-known algorithms, such as proof-of-play, which are not currently employed by any major platforms, are generally not considered due to a lack of trust in their reliability.

Table 2. These tables compare some of the consensus algorithms mentioned in the literature (left) to those confirmed as relevant in interviews (right).

	Coverage	Alsunaidi & Alhaidari (2019)	Pathiajani, Kshirsagar & Pachghare (2019)	Wang et al. (2019)	Cachin., Vukolic (2017)	Ambili et al. (2017)	Xiao et al. (2019)	Bouraga (2021)	Zhang & lee (2019)	Kim & Nguyen (2018)
Ripple	66,67%			x	x	x	x		x	x
Pow	55,56%	x	x	x					x	x
Pos	55,56%	x		x			x		x	x
PBFT	33,33%	x					x		x	
Proof-of-luck	33,33%	x		x						x
Del. PoS	33,33%	x		x					x	
Tendermint	33,33%				x	x	x			
Iroha	33,33%		x		x					x
Chain	33,33%		x		x					x
Stellar	33,33%		x		x					x
Raft	33,33%	x	x							x
Proof-of-burn	22,22%			x						x
proof-of-stake-velocity	22,22%			x						x
Proof-of-activity	22,22%	x		x						
Hyperledger Fabric	22,22%		x		x					
R3 Corda	22,22%		x		x					
Sawtooth lake	22,22%				x					x

Interview Results	Coverage	Interview 1	Interview 2	Interview 3	Interview 4	Interview 5	Interview 6	Interview 7
Ripple	14,29%							x
Pow	100,00%	x	x	x	x	x	x	x
Pos	100,00%	x	x	x	x	x	x	x
PBFT	100,00%	x	x	x	x	x	x	x
Proof-of-luck	14,29%							x
Del. PoS	14,29%	x						
Tendermint	14,29%				x			
Raft	14,29%		x					
Hyperledger Fabric	14,29%		x					
Proof-of-elapsed-time	14,29%		x					
Polkadot	14,29%		x					
proof-of-authority	28,57%			x		x		
broof-of-burn	14,29%					x		

3   Feature Model

We have used data from domain experts and literature studies for identifying consensus algorithms. Each feature is assigned a Boolean or non-Boolean data type. Consensus algorithm Boolean features fall into three categories: design, security, and performance, with trade-offs among them. PBFT algorithms have higher throughput than PoW algorithms [13].

Table 3. This table displays Boolean Features, consensus algorithms, and mapping. 1s indicate supported consensus algorithm features while 0s signify a lack of support or insufficient evidence based on documentation analysis [6].

FA			Pow	Pos	PBFT	Proof-of-Authority	Del. PoS	Raft	Proof-of-elapsed-time
	Incentive	85,71%	1	1	1	1	1	0	1
Is there a reward for contributing to consistency or continuity?	Reward	42,86%	1	1	0	0	0	0	1
Is there a punishment for not contributing to consistency or continuity?	Punishment	14,29%	0	0	0	0	1	0	0
	TEE dependency	14,29%	0	0	0	0	0	0	1
Trusted execution enviorment (TEE) required for operation	TEE dependency	14,29%	0	0	0	0	0	0	1
	Data type	100,00%	1	1	1	1	1	1	1
Transactions are stored in the form of blocks. The blocks refer to their predecessor	Blockchain	100,00%	1	1	1	1	1	1	1
Transactions are stored as transactions and refer to previous transactions	Dag	0,00%	0	0	0	0	0	0	0
Transactions are stored as both blocks and transactions	Mixed	0,00%	0	0	0	0	0	0	0
	Fault tolerance	100,00%	1	1	1	1	1	1	1
Protocol can only gaurantee to work with crashing nodes	Crash	14,29%	0	0	0	0	0	1	0
Protocol is gauranteed to work with crashing and adversarial nodes	Byzantine	85,71%	1	1	1	1	1	0	1
	Permission model	100,00%	1	1	1	1	1	1	1
permission needed to acces network	Permissioned	57,14%	0	0	1	1	0	1	1
no permission needed to acces chain	Permissionless	42,86%	1	1	0	0	1	0	0
	Consensus finality	100,00%	1	1	1	1	1	1	1
The aggred values in the algorithm are not always correct	Probablistic	42,86%	1	0	0	0	1	0	1
The aggred values in the algorithm are always correct	Deterministic	71,43%	0	1	1	1	0	1	1

Design features of consensus algorithms refer to the structure and framework of the algorithm and include incentives, consensus finality, candidate formation, candidate configuration, leader selection, committee formation, and committee configuration. Incentives are the motivations for participating nodes to engage in the mining process and contribute to the consensus algorithm. These incentives can include rewards, such as cryptocurrency tokens, to encourage nodes to perform computational work and secure the network [19]. Consensus finality refers to the level of certainty or irreversibility that the consensus algorithm can achieve [21]. It determines when a transaction or block is considered finalized and cannot be altered or reversed. Candidate formation pertains to the criteria that nodes must meet to be eligible to participate in the consensus-building process. This includes factors such as node reputation, stakeholding, or computational power. Candidate configuration relates to how the group of participating nodes evolves during the consensus-building process. It involves determining which nodes are eligible to become candidates for adding new entries to the blockchain. Leader selection involves the process of choosing a node or a group of nodes responsible for proposing and validating new blocks or transactions [18]. This mechanism can vary depending on the consensus algorithm, with different approaches such as round-robin selection, random selection, or election-based selection. Committee formation refers to the process of selecting a subset of nodes responsible for validating and verifying new entries in the blockchain. These committees ensure the accuracy and integrity of the consensus process. Committee configuration focuses on how the composition of the committee evolves over time. This may include adding or removing nodes based on certain criteria or adjusting the committee size for improved scalability [9].

We conducted qualitative semi-structured interviews with experts to explore their knowledge regarding consensus algorithm features. The design features of consensus algorithms, except for two new features, Trusted Execution Environment (TEE) and file structure, have remained unchanged [3]. TEE relates to how proof is obtained to mine a block, while file structure pertains to the inherent structure of the consensus algorithm and is critical in determining leader selection, candidate formation, and committee formation.

Performance Features - This category of features focuses on how efficiently a consensus algorithm operates. The most important features in this category are throughput and latency. Throughput measures the number of new data entries a consensus algorithm can process per second, while latency measures the time for any data entry to be verified. Scalability is another characteristic in this category, indicating whether a consensus algorithm can function effectively when faced with many transactions or nodes. The extent to which a consensus algorithm can scale differs significantly among different algorithms. Lastly, fault tolerance is a feature that describes whether a consensus algorithm can tolerate Byzantine or crash faults.

One interviewee proposed a feature regarding the upgradability of consensus, stating that in some systems, a set of transactions can change the whole system and synchronize the entire network independently with the consensus at every node. The same developer referred to Substrate as a network that employs such upgradability methods. Another expert emphasized that sustainability is critical for companies building a ledger. They stated that some parties would not build their application on a proof-of-work network as it is not sustainable enough. This concern has a significant impact on the decision-making process, as developers with sustainability in mind tend to choose an algorithm other than proof-of-work, given the significant amount of energy it requires to operate [14].

Security Features - The security features of a consensus algorithm pertain to the degree of protection it provides against various attack vectors and threats. The primary attack vectors include a 51% attack, a Sybil attack, a denial of service (DoS) attack, and an eclipse attack. A 51% attack occurs when a malicious entity controls more than 50% of the network's computing power and can modify the ledger according to their will [4]. On the other hand, a Sybil attack is an attack in which the attacker creates fake identities to gain influence within the network. In a DoS attack, the attacker disrupts the ledger service by making it unavailable to other nodes. Lastly, an eclipse attack is a type of attack where the attacker takes over other nodes and forces them to only communicate with other malicious nodes. During our interviews, we observed that most participants were not familiar with the eclipse attack, and even those who were aware of it did not consider it significant. As a result, we did not include it in the feature model. Additional security features include authentication, non-repudiation, censorship resistance, and adversarial tolerance. Authentication requires nodes to authenticate themselves before participating in consensus. Censorship resistance ensures that no one can censor the data transmitted across the network [11]. Non-repudiation guarantees that no one can deny making a data entry in the ledger. Adversarial tolerance indicates the maximum percentage of adversarial nodes that a consensus algorithm can withstand (Fig. 1).

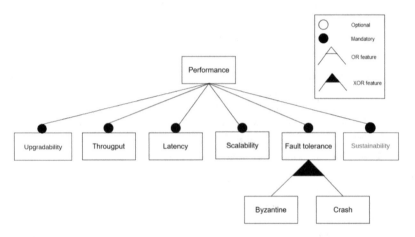

Fig. 1. Feature model of consensus algorithm performance features. Features that were proposed in the interviews but did not find support in literature or interviews are colored in red, whereas features that were added during the expert interviews with support in literature are shown in green.

Non-Boolean Algorithm Features - The experts identified five non-Boolean consensus algorithm features, being "Popularity in the market", "Maturity of the company", "Developer Resources (People)", "Sustainability", and "Scalability". The assigned values for these features are based on a 3-point Likert scale (High, Medium, and Low) and are used to evaluate a specific consensus algorithm.

Features of Consensus Algorithms - Data was extracted from various sources, including web pages, white papers, scientific papers, documentation, forum discussions, books, videos, and dissertations, to develop the initial list of consensus algorithm features. The initial list comprised 43 Boolean and five non-Boolean features. Subsequently, seven domain experts were involved in the research phase to refine the list of potential consensus algorithm features [7].

The Boolean and non-Boolean feature alternative mappings can be found in Table 3 and Table 4, respectively [6]. The former table contains binary codes for each Boolean feature indicating whether it is present. The yellow column shows the percentage of alternative algorithms with a particular feature. Table 4 assigns a score to each non-Boolean feature based on parameters such as transaction latency, block confirmation time, transaction throughput, the maximum number of nodes, energy consumption, number of validating nodes, permission model, and popularity. The parameters were carefully selected to ensure their relevance and accuracy in evaluating the features. Specifically, the performance score was calculated using transaction latency and block confirmation time as parameters. Transaction latency is the time taken to confirm a single transaction and is measured in seconds. The maximum time to ensure a transaction is used to calculate the score for this parameter. Block confirmation time, on the other hand, is the time required to confirm a single block and is categorized into three discrete values: "low", "middle", and "high". These values are used in the calculation of the performance score. The consensus algorithms' scalability was calculated using transaction throughput and the maximum number of nodes. Transaction throughput represents the number of transactions the network can process per second and is expressed in maximum transactions per second. The maximum number of nodes is the maximum number of nodes that a consensus algorithm can accommodate before its performance is significantly impacted. This parameter is measured in terms of the total number of nodes. To measure sustainability, energy consumption was used as a parameter. Energy consumption refers to the amount of energy consumed by a ledger of a given alternative type and is categorized into three discrete values: "low", "middle", and "high". Decentralization is measured using two parameters: the number of validating nodes in the network and the permission model. The number of validating nodes represents the total number of nodes that can participate in the consensus-finding process and is expressed in the number of validating nodes. The permission model determines whether a consensus algorithm operates in a permissioned or permissionless network. In a permissioned network, validating nodes require permission to participate in the consensus process, whereas in a permissionless network, there are no such restrictions. The popularity of a consensus algorithm is determined by the number of platforms that use it [4].

Conflicting views on attack vectors were resolved by prioritizing studies that provide clear reasoning for their conclusions. These algorithms are equipped with a mechanism that safeguards against particular attacks. This information has been identified in several surveys that have explored various consensus algorithms. However, some studies have reached different conclusions regarding attack vectors. For instance, [2] and [4] diverge views on whether proof-of-stake defends against double-spending attacks. In cases where conflicting information exists, studies that provide clear reasoning for their conclusions on attack vectors are preferred over those that assert that a particular algorithm offers protection against a given attack.

Table 4. The mapping among non-Boolean features and selected alternatives.

Non Boolean Features		PoW	PoS	Pbft	PoA	Raft	PoET
Performance		**Low**	**Med**	**High**	**High**	**High**	**Med**
Time it takes to confirm a transaction	Latency	>100s	<100s	<10s	<3s	<10s	<124s
Time it takes to confirm a block	Block conformation time	Low	High	High	High	High	High
Score for performance	Score	2	4	6	6	9	4
Scalability		**High**	**High**	**Low**	**High**	**Low**	**Med**
Amount of possible nodes the network can handle	Max nodes	1000000	100000	16	3000	20	100
Amount of transaction p/s	Throughput	<100	<1000	<2000	<300	>10000	<100
Score for Scalability	Score	6	6	2	5	3	4
Sustainability		**Low**	**Med**	**High**	**High**	**High**	**High**
Energy consumption per node	Energy consumption	High	Med	Low	Low	Low	Low
Decentralization		**High**	**High**	**Low**	**Low**	**Low**	**Med**
Amount of nodes participating with consensus	Validating nodes	100000	10000	16	12	20	100
Do nodes need permission to become validators?	Permission(-less, -ed, both)	-less	-less	both	-ed	-ed	-ed
Score for decentralization	Score	6	6	3	2	2	4
Popularity		**High**	**High**	**Med**	**Med**	**Low**	**Low**
The amount of platforms that use this alternative	supporting platforms	11	10	4	4	1	1

4   Discussion

Blockchain consensus algorithm selection is a crucial aspect that requires careful consideration, as the technology's affordances and consequences have long-term effects on the blockchain application. Changing consensus algorithms can be costly and complex, as evident in Ethereum's transition to a proof-of-stake consensus algorithm [10]. Hence, selecting an algorithm that best fits the current and future requirements of the blockchain project is essential. Several decision-support models have been proposed for selecting the appropriate consensus algorithm [7,12], and modular blockchain platforms have become popular due to their flexibility in selecting consensus algorithms [1].

In this research project, we interviewed engineers and consensus algorithm designers, which introduces a validity concern. While engineers focus on the consequences of selecting a platform based on the consensus algorithm, algorithm designers are more concerned with the principles of the algorithm itself. Designers have more freedom to consider the adaptability of the algorithm, while engineers typically work within the given framework. These interviews revealed that consensus algorithm selection is not a significant concern for most distributed ledger application developers during the building process. This significantly impacts the natural validity of the research project. Initially, our search focused on consensus algorithms and platform developers, but finding and convincing them to participate was challenging. This difficulty can be attributed to the need for more highly specialized employees in companies located primarily in the United States. As a result, we expanded our search to include blockchain application developers and consultants. While this introduces the previously mentioned validity concern, these additional experts provided valuable insights. They had extensive experience with various blockchain application requirements, which often influenced the choice of consensus algorithm features. Consequently, we could still deduce the essential features in consensus algorithm selection. Another validity concern emerged during the interviews when experts proposed features related to platform selection rather than consensus algorithm selection. This indicated confusion among some experts regarding which features are attributed to consensus algorithms and which belong to other layers of a blockchain application. We conducted a second literature review after the interviews to address this concern. Each proposed feature was carefully examined to determine its relevance to the algorithm layer or other layers of the blockchain application. As a result, two features suggested by interviewees were excluded.

5   Conclusion

In this work, we outline the features that distributed ledger technology designers need to consider with regard to consensus algorithms. While these decisions are not made regularly, we show they have significant consequences for the platform.

We find consensus algorithms are not frequently evaluated, designed, and implemented. While many others have offered decision support systems for consensus algorithm selection, we propose, as our scientific contribution, that more in-depth analysis and reporting of consensus algorithms is necessary. Subsequently, we use the feature modeling language to elicit the relevant features for consensus algorithms to provide a complete overview of what consensus algorithms have to offer than was available in the literature. The feature models in Sect. 3 are useful for scientists working in consensus algorithms and practitioners evaluating, designing, and implementing consensus algorithms.

Our practical contribution is the list of consensus algorithms that are supplied in Tables 2, and the features identified in some of the more common consensus algorithms in Tables 3, and 4. Experts evaluated and verified these tables to ensure our data was complete and correct. Using our conceptual model, we unearthed six of the most common consensus algorithms and mapped 48 distinguishing, unique features of these algorithms for use by industry practitioners. Practitioners designing distributed ledger technologies can use these models in the future to understand the trade-offs of the architectural decisions they will be making.

The findings of this research project open up several potential research for future work. To enhance the precision of the decision model, it is crucial to conduct a dedicated performance study that comprehensively compares the performances of a wide range of consensus algorithms. Currently, there is a scarcity of published studies comparing consensus algorithm performance, which can be attributed to the challenges associated with comparing performance due to the multitude of factors involved that extend beyond consensus alone. Existing studies that do compare performances often focus on a limited subset of algorithms in diverse environments. Although a combination of these studies has allowed us to deduce performance differences between algorithms, a dedicated performance study would offer a more precise and accurate assessment of the performance characteristics of each consensus algorithm. Such a study would provide valuable insights for informed decision-making in selecting the most suitable consensus algorithm for specific distributed ledger applications.

References

1. Alchemy: Modular vs. monolithic blockchains (2022). Accessed 5 Oct 2022
2. Alsunaidi, S.J., Alhaidari, F.A.: A survey of consensus algorithms for blockchain technology. In: 2019 International Conference on Computer and Information Sciences (ICCIS), pp. 1–6. IEEE (2019)
3. Ampel, B., Patton, M., Chen, H.: Performance modeling of hyperledger sawtooth blockchain. In: 2019 IEEE International Conference on Intelligence and Security Informatics (ISI), pp. 59–61. IEEE (2019)
4. Bamakan, M.H., Motavali, A., Bondarti, A.B.: A survey of blockchain consensus algorithms performance evaluation criteria. Expert Syst. Appl. **154**, 113385 (2020)
5. Baninemeh, E., Farshidi, S., Jansen, S.: A decision model for decentralized autonomous organization platform selection: three industry case studies. Blockchain: Res. Appl. **4**, 100127 (2023)
6. Baninemeh, E., Jansen, S., Pronk, B.: A feature model of consensus algorithms in distributed ledger technology. https://bit.ly/42TYrb8
7. Bouraga, S.: A taxonomy of blockchain consensus protocols: a survey and classification framework. Expert Syst. Appl. **168**, 114384 (2021)
8. Cachin, C., Vukolić, M.: Blockchain consensus protocols in the wild. In: 31 International Symposium on Distributed Computing (2017)
9. Chaudhry, N., Yousaf, M.M.: Consensus algorithms in blockchain: comparative analysis, challenges and opportunities. In: 2018 12th International Conference on Open Source Systems and Technologies (ICOSST), pp. 54–63. IEEE (2018)
10. Farshidi, S., Jansen, S., España, S., Verkleij, J.: Decision support for blockchain platform selection: three industry case studies. IEEE Trans. Eng. Manage. **67**(4), 1109–1128 (2020)
11. Ferdous, M.S., Chowdhury, M.J.M., Hoque, M.A., Colman, A.: Blockchain consensuses algorithms: a survey. arXiv preprint arXiv:2001.07091 (2020)
12. Filatovas, E., Marcozzi, M., Paulavičius, R.: A MCDM-based framework for blockchain consensus protocol selection. Expert Syst. Appl. **204**, 117609 (2022)
13. Fu, X., Wang, H., Shi, P.: A survey of blockchain consensus algorithms: mechanism, design and applications. Sci. China IS **64**(2), 1–15 (2021)
14. Jones, B.A., Goodkind, A.L., Berrens, R.P.: Economic estimation of bitcoin mining's climate damages demonstrates closer resemblance to digital crude than digital gold. Sci. Rep. **12**(1), 1–10 (2022)
15. Kannengiesserer, N., Lins, S., Dehling, T., Sunyaev, A.: Trade-offs between distributed ledger technology characteristics. ACM Comput. Surv. (CSUR) **53**(2), 1–37 (2020)
16. Monrat, A.A., Schelén, O., Andersson, K.: A survey of blockchain from the perspectives of applications, challenges, and opportunities. IEEE Access **7**, 134–151 (2019)
17. Ølnes, S., Ubacht, J., Janssen, M.: Blockchain in government: benefits and implications of distributed ledger technology for information sharing (2017)

18. Ongaro, D., Ousterhout, J.: In search of an understandable consensus algorithm. In: 2014 {USENIX} Annual Technical Conference ({USENIX}{ATC} 2014), pp. 305–319 (2014)
19. Singh, A., Kumar, G., Saha, R., Conti, M., Alazab, M., Thomas, R.: A survey and taxonomy of consensus protocols for blockchains. J. Syst. Archit. **127**, 102503 (2022)
20. Suciu, G., Nădrag, C., Istrate, C., Vulpe, A., Ditu, M.C., Subea, O.: Comparative analysis of distributed ledger technologies. In: 2018 Global Wireless Summit (GWS), pp. 370–373. IEEE (2018)
21. Yadav, A.K., Singh, K., Amin, A.H., Almutairi, L., Alsenani, T.R., Ahmadian, A.: A comparative study on consensus mechanism with security threats and future scopes: blockchain. Comput. Commun. **201**, 102–115 (2023)

How Decentralized Autonomous Organizations are Affecting Real Estate

Gonçalo Cruz⬥, Tiago Guimarães⁽✉⁾ ⬥, Manuel Filipe Santos⬥,
and José Machado⬥

Algoritmi Reasarch Center, School of Engineering, University of Minho, Azurém Campus,
4800-05 Guimarães, Portugal
tsg@dsi.uminho.pt

Abstract. Blockchain technology is innovating the real estate industry, providing greater accessibility, and streamlining the process for investors, buyers, and sellers. Three approaches that utilize blockchain technology for real estate are: virtual real estate, real estate tokenization, and fractional investment. Virtual real estate enables users to purchase digital properties in virtual worlds and monetize their investment. Real estate tokenization creates digital tokens that represent ownership of real estate assets, while fractional investment enables users to share ownership of properties by purchasing tokens that represent a percentage of the asset. Decentralized Autonomous Organizations (DAOs) and Non-fungible tokens (NFTs) can offer a new way for clients to access exclusive real estate properties and services. In this research we suggest a Real Estate membership by buying properties or NFTs. In a hypothetical use case, clients can enjoy special advantages such as priority access to off-market properties, individualized counseling, and revenue-sharing options. However, there are still challenges to overcome, such as regulation, mass adoption, and lower gas fees that can be analyzed deeper on future research.

Keywords: Blockchain · Non-fungible Tokens · Decentralized Finance · Decentralized Autonomous Organizations · Real Estate · Tokenization · Metaverse

1 Introduction

In recent years, non-fungible tokens (NFTs) and decentralized finance (DeFi) have emerged as significant areas of innovation and research. These technologies have the power to change the way we interact with digital assets [1, 2]. In this article we will discuss our current understanding and research into the application of NFTs in the DeFi market and explore how they can change present trading practices in Real Estate. We will start by describing the fundamental aspects of Blockchain technology, NFTs, and DeFi. We will then review the existing literature on the use of NFTs and DeFi, and how these technologies affect the Real Estate Market, summarizing the key findings and arguments from this literature. Finally, we will conclude by discussing the current state of knowledge on this topic and identifying any gaps or areas for future research. We are also going to explore How Defi and NFTs can create privileges among Real Estate Clients.

2 Methodology

This chapter outlines the methodological considerations for the study's development. While there may be some limitations that arise during the research process, they can only be included after the completion of all research work [3]. In this research process, we first started by defining a problem [4]. The problem was "How Defi and NFTs can create privileges among Real Estate Clients." To address this problem, a methodological approach focused on formulating hypotheses was chosen, with the aim of finding hypothetical use cases that combine the concepts related to the problem. We also interviewed a Real Estate expert to add more content to the article and to validate the suggestions made by us and the future work aswell.

3 Blockchain

The concept of blockchain technology has garnered significant attention in recent years, with its potential applications ranging from financial transactions to supply chain management, among many others. The main features of a blockchain are being a decentralized, distributed, digital ledger that is used to record transactions across many computers so that the record cannot be altered without the alteration of all subsequent blocks and the consensus of the network [5]. One of the key features of a blockchain is its distributed nature. Rather than being stored on a single central server, the ledger is distributed across the entire network of computers participating in the blockchain (Peer-to-Peer). This makes the system more resistant to tampering, as there is no single point of failure that can be exploited [2] Blockchain interoperability, also name for cross-chain technology, is the movement of data and assets between several blockchain networks. It makes it possible for users to take use of the distinctive features and functions that each network has to offer by transferring digital assets, smart contracts, and other data between diverse blockchain platforms [6].

3.1 Non Fungible Tokens (NFTs)

NFTs are unique digital assets stored on a blockchain, representing ownership of a specific item [7]. They are different from fungible tokens because they cannot be exchanged for other assets. NFTs are useful for tracking ownership and authenticity of unique items like fine art, real estate, and virtual goods [8]. NFTs offer a higher potential for diversification than DeFi tokens and cryptocurrencies, making them an attractive option for risk-averse investors. NFTs can act as a loan, and function as financial funds for stablecoins in the DeFi market. OpenSea is the largest NFT marketplace where users can purchase, exchange, and trade exclusive digital assets. Digital art and collectibles are popular categories of NFTs [9].

3.2 Decentralized Finance (DeFi)

DeFi (decentralized finance) is a rapidly growing sector that utilizes blockchain technology to provide decentralized and autonomous financial services and products [2]. This approach challenges traditional financial institutions by allowing anyone with internet access to access and use financial services securely and transparently through decentralized protocols [10]. Some examples of financial instruments and services that can be created for the DeFi market include stablecoins, loans, and prediction markets. DeFi offers financial inclusion to those who might not have access to traditional financial services, providing a new, decentralized, and transparent way of conducting financial transactions [11].

3.3 Decentralized Autonomous Organization (DAO)

As indicated by the name, it is an organization that is autonomous, which means a system governed and managed by the people in that same system. DAOs are decentralized because they are managed by a set of rules that are written in smart contracts and carried out on a blockchain, as opposed to traditional organizations, which are normally run and governed by a central authority [12]. As a result, decision-making and administration are decentralized, and all organization members are actively involved in the governance process. Increased transparency, security, and efficiency are just a few benefits that DAOs have over conventional organizations because of their decentralized nature. DAOs might also challenge established organizational structures and provide new possibilities for decentralized cooperation and decision-making [13]. Some circumstances in which DAOs can be used are donations to charity, where any member is free to choose how and where to use their funds. Freelancer networks allow contractors to create organizations that pool their resources to pay for office space and software subscriptions. And Ventures and Grants; a venture fund that combines funding and decides which projects to support where repaid funds may then be divided among DAO participants. There are several DAO membership models. Membership can influence the voting process and other important DAO components [14].

Token-Based Membership
These governance tokens may often be exchanged on a decentralized exchange without a permit. Others need to be won by demonstrating liquidity or some form of "proof-of-work." In any case, the right to vote is available merely by having the token. Usually employed to control large-scale decentralized protocols or the coins themselves.

Share-Based Membership
Share-based DAOs are still relatively open but have additional permissions. Anyone interested in joining the DAO may do so by submitting a proposal and often by providing some kind of tribute, like coins or labor. Shares stand for ownership and direct voting. A member's proportionate share of the Treasury is theirs to always keep. Usually refers to smaller, more human-focused organizations like charities, worker cooperatives, and investing clubs. Additionally, it can control protocols and tokens.

Reputation-Based Membership

Reputation serves as evidence of involvement and provides DAO voting rights. Reputation-based DAOs do not provide contributors with ownership rights, in contrast to token- or share-based membership. DAO members must acquire a reputation by involvement; it cannot be purchased, transferred, or delegated. Prospective members can freely make proposals to join the DAO and request to acquire reputation and tokens as payment in return for their efforts, since on-chain voting is permissionless. Commonly used for decentralized protocol creation and governance, but also works well for a wide range of organizations, including charities, worker cooperatives, investment clubs, among others [15].

3.4 NFTsxDefi

NFTs and DeFi have promising potential for various use cases beyond their current applications. NFTs are now being utilized in DeFi as loan collateral and payment instruments. As DeFi continues to evolve, NFTs may be integrated into derivatives markets and fractional shares purchasing. Despite being new technologies, the future possibilities of NFTs and DeFi are innovative and exciting [9, 16].

3.5 NFTs and Real Estate

The usage NFTs and blockchain technology have created new prospects in the real estate sector. **Real estate tokenization**, **virtual real estate**, and **fractional investing** are three significant applications of NFTs that provide special advantages to investors and property owners. This essay examines these aspects, the businesses that provide them, the difficulties and chances NFTs in the real estate industry bring [17].

Real Estate Tokenization

Blockchain technology is used in real estate tokenization to produce digital tokens that reflect ownership of real estate properties. This makes it possible to purchase and sell without the use of intermediaries and offers the chance to use tokens as collateral for borrowing and lending. To save expenses and speed up transactions, a well-known platform called Propy automates the real estate buying process. In fact, they have already handled the sale of real estate NFTs in Mississippi and Florida in 2022 [18, 19].

Virtual Real Estate (Metaverse)

Virtual Real Estate allows users to purchase digital properties in the Metaverse, represented by NFTs on a blockchain, and commercialize them in digital markets. Decentraland is a leading example of this concept, where users can buy virtual land called LAND and develop buildings to sell or rent to other users. They also have their own cryptocurrency called Mana and a decentralized governance system led by users [20].

Fractional Investment

With fractional investments, users can buy tokens that represent a portion of an asset. This improves the real estate industry by making it easier to split properties into tokens. RealT is an example of a platform that does this. Together with managing rent and property, RealT also divides profits among investors according to the quantity of tokens each one holds [21].

4 Real Estate DAO

After deep research on all the topics associated with this article, and with the content of the interview with the Real Estate professional, we found that an interesting combination was making a Real Estate Decentralized Autonomous Organization for private investors. In this scenario, the goal was to create an exclusivity community of Real Estate investors that want to benefit from special investments and perks on a Real Estate agency. This is supposed to be a premium community which means that the access is going to be limited and with a high cost to enter it.

4.1 Access

One of the key features of the Real Estate DAO is its decentralized and open nature, which allows for broad and inclusive participation from anyone interested in real estate investments. To ensure this accessibility, the DAO offers two primary methods of access: purchasing a real estate asset that grants an NFT or acquiring that same NFT through a secondary marketplace from an investor who wants to sell it.

Real Estate Asset Purchase: Purchasing a real estate asset of a high value that provides an NFT is the first way to gain access to the Real Estate DAO. In this case, an interested party would purchase a real estate asset from the Real Agency or a vendor that is linked with it in exchange for the property and an NFT that signifies ownership of the purchased item and another that is the DAO entrance token. The DAO would then be accessible to the bearer of this NFT, enabling them to take part in the investment pool and cast votes on DAO proposals. Because of the underlying value of the property itself, the price of the real estate asset would probably be higher than the cost of the NFT alone.

NFT Purchase from a Secondary Market: The second option to enter the Real Estate DAO is by purchasing an NFT from a secondary market like Opensea or rarible. In this situation, the holder of an NFT designating ownership of the DAO pass may choose to sell it to a third party. The purchaser of the NFT would then have access to the DAO in the same manner as if they had purchased a real estate asset directly from the Real Estate Agency or a related vendor. The inherent tier of the NFT, the possibility that its value would rise in the future, and other market factors would all have an influence on the price in addition to supply and demand in the market.

4.2 Vote

The Real Estate DAO uses a decentralized governance model that puts a focus on fairness and equality in making decisions, and the voting mechanism is a crucial part of the strategy. Every member shares the power to impact the direction of the organization using a token-based voting system, with one vote granted for each NFT which represents a real estate tier. No matter their amount of investment (Tier) or activity in the DAO, this guarantees that every investor has the same opportunity to vote on proposals.

Voting Process: All NFT holders are eligible to vote when a proposal is put up for a vote and can also make proposals within the DAO. They just use the token(s) that correspond to their NFT(s) to cast their vote, either in support of the proposal or against it. Investors can vote or alter their vote throughout the voting session, which normally lasts for a predetermined amount of time. Depending on the number of votes cast in support of or against the plan, it is either accepted or rejected at the conclusion of the voting session.

4.3 Pool

The establishment of an investment pool is another crucial component of the Real Estate DAO. The money of the investors will be pooled for the purpose of investing in bigger possibilities that might not be available to individual investors. Investors will be able to recommend how the funds should be distributed and cast votes for or against investments, and the DAO will administer the pool. The investors will have the power to make suggestions on what to do with the money from the pool and vote on those same decisions. A way to acquire money is putting a % on each property sale within the community to this pool, or also put a monthly amount of money to the pool.

4.4 Tiers and Benefits

As an example, we can categorize membership NFTs into three types: **Tier1, Tier2,** and **Tier3**. Each type of NFT offers different benefits, from access to exclusive property listings and priority access, to new listings, free property inspections and invitations to VIP events and conferences. Let's take a closer look at what each type of NFT offers (Table 1).

Table 1. Type of Tiers

Type of Membership	Perks
Tier 1 (Lowest Level of Investment)	• First priority for property viewings • Access to exclusive events hosted by the real estate agency (e.g., parties, dinners, tours)
Tier 2 (Medium Level of Investment)	• Tier1 perks • Free consultation with real estate experts • Early access to newly launched properties
Tier 3 (Highest Level of Investment)	• Tier 2 perks • Private real estate advisor • Exclusive access to pre-construction projects

4.5 SWOT Analysis

Real estate and blockchain in general, are complex and dynamic industries that are constantly evolving. To gain a better understanding of the market and identify opportunities for growth, it can be helpful to conduct a SWOT analysis. This analysis will focus on the industry's internal strengths and weaknesses, as well as external opportunities and threats. Due to this analysis, we can gain valuable insights into the current state of the Real Estate Market and its potential in bringing together the Decentralized Autonomous Concept.

Strengths

- **Customer loyalty**: Exclusive benefits to members can create a strong bond between them and the Real estate Agency.
- **Revenue stream**: Revenue can be generated from the sale of NFTs and the membership fees.
- **Exclusive access to properties**: Exclusive property listings and priority access to new listings can attract customers who seek for early access properties.

Weaknesses

- **Technological barriers**: Customers may not have the technical skills/knowledge to use these technologies, which could become a barrier to access the associated benefits.
- **Uncertainty**: Since these technologies are recent and have volatile values, there is some uncertainty around their long-term value and mass adoption.
- **High cost of entry**: The NFTs cost may be a problem for "regular customers" since this is a premium membership, prices tend to be higher than normal.

Table 2 SWOT Analysis.

Opportunities

- **Integration with blockchain**: blockchain technology can provide more security and transparency than a normal central server, enhancing the credibility of the real estate agency.
- **Brand recognition**: Exclusivity and possible demand can contribute to a positive brand recognition and reputation.
- **Partnership opportunities**: new partnerships with other businesses (e.g., car dealerships, travel companies) to offer additional benefits to members.

Threats

- **Competition**: Other real estate agencies may develop similar premium membership models, which could increase competition.
- **Market downturn**: A downturn in the real estate market could reduce demand for premium real estate services and membership.
- **Regulatory issues**: There may be regulatory issues surrounding the use of NFTs, which could impact the viability of the business model.

To overcome the weaknesses and Threats identified in this SWOT Analysis, some measures should need to be taken into consideration:

- Technological barriers: Provide user-friendly interfaces, clear instructions, and comprehensive customer support to help customers with limited technical skills or knowledge.
- Uncertainty: Diversify your offerings beyond NFTs to reduce dependence on their long-term value and mass adoption.

- High cost of entry: Introduce flexible pricing options or membership tiers to make your premium services more affordable and inclusive.
- Competition: Differentiate your premium membership model by offering unique and valuable benefits, while also establishing a strong brand presence and engaging in effective marketing strategies.
- Market downturn: Diversify your services beyond premium real estate and explore alternative revenue streams to mitigate the impact of a market downturn.
- Regulatory issues: Stay updated with regulations surrounding NFTs and real estate, collaborate with legal professionals, and engage in industry advocacy to address any potential regulatory issues.

How Can We Bind Real Estate Assets Into Digital Tokens/NFTs?
Establishing asset verification mechanisms: One of the key challenges is ensuring that the digital token/NFT accurately represents the real-world asset it is associated with. Robust verification mechanisms need to be implemented to validate the ownership and authenticity of the real estate asset. This may involve conducting thorough due diligence, utilizing third-party audits, or leveraging technologies like blockchain to create immutable records of ownership and transactions.

Ensuring legal and regulatory compliance: Various legal and regulatory systems apply to real estate transactions. When tying real estate assets to digital tokens/NFTs, it is crucial to understand and follow these restrictions. To safeguard against fraudulent actions and legal conflicts, it is crucial to engage legal specialists to assure compliance and handle any potential legal difficulties.

How Can We Prevent Double or Multiple Bindings?
The establishment of safe and tamper-proof mechanisms for tokenization and asset management is essential to preventing duplicate or multiple bindings of real-world assets. Blockchain-based smart contracts can enforce the rules regulating the transfer and ownership of digital tokens/NFTs while also ensuring transparency. By doing this, it is made sure that once a token is linked to a piece of real estate, it cannot be stolen or improperly moved.

Using unique IDs and metadata: Each real estate asset can have a unique identification issued to it that is connected to the matching digital token/NFT. This identification may be referenced in the metadata or included inside the token. Utilizing unique IDs makes it simpler to track and confirm who owns certain assets, lowering the possibility of duplicate or multiple bindings.

5 Conclusion

After carefully analyzing the results of our research with the help of the SWOT analysis and the interview with a Real Estate professional that both validated and added content to this article, we conclude that although this article presents a possible solution where exclusivity on the real estate market with the use of Decentralized Finance and Non fungible tokens can be created, this is just a hypothetical use case with no data that supports it at this moment and there is still a long path in these new technologies and challenges to overcome. The biggest challenges to overcome to make Blockchain technologies more accessible and trustable to everyone are regulation, mass adoption in different sectors like Banking, Energy, Social etc., and lower gas fees to make transactions easier and attractive. When we talk about Real Estate Market, we have a more consistent and mature sector, but that still needs to rethink how to adapt to customer needs and to the evolution of technologies like Blockchain, Artificial Intelligence and Augmented Reality. For future research we would like to suggest analyzing the possibilities of integrating blockchain in Real State more in-depth and we the approach that we suggest is to interview both Real Estate Investors and more professionals on the area in a larger scale and then do a Data Analysis of the content and present it to a higher entity on real estate or to an agency that is interested on developing the idea. Another topic is the cost of employing these technologies, so the calculations need to be executed to evaluate the financial opportunities and challenges on implementing it. And to conclude, the current policies around these sectors need to be analyzed and created new ones if necessary.

Acknowledgements. This work has been supported by FCT – Fundação para a Ciência e Tecnologia within the R&D Units Project Scope: UIDB/00319/2020.

References

1. Böhme, R., Christin, N., Edelman, B., Moore, T.: Bitcoin: economics, technology, and governance. J. Econ. Perspect. **29**(2), 213–238 (2015). https://doi.org/10.1257/jep.29.2.213
2. Buterin, V.: A next generation smart contract & decentralized application platform (2014)
3. Barañano, A.M.: Métodos e técnicas de investigação em gestão: manual de apoio à realização de trabalhos de investigação (2004)
4. Quivy, R., Van Campenhoudt, L., Santos, R.: Manual de investigação em ciências sociais (1992)
5. Nakamoto, S.: Bitcoin: a peer-to-peer electronic cash system (2008). https://bitcoin.org/
6. Buterin, V.: Chain interoperability (2016). http://andolfatto.blogspot.ca/2015/02/fedcoin-on-desirability-of-government.html
7. Ranaldo, A.: Non-fungible tokens (2022). https://ssrn.com/abstract=4274986
8. Wang, Q., Li, R., Wang, Q., Chen, S.: Non-fungible token (NFT): overview, evaluation, opportunities and challenges (2021). http://arxiv.org/abs/2105.07447
9. Yan, T., Kelly, K.: The Year Ahead for NFTs (2022)
10. Zimmermann, R.: Decentralized finance (Defi) high potential with new rules of the game (2021). https://morethandigital.info/en/decentralized-finance-defi-high-potential-with-new-rules-of-the-game/
11. Karim, S., Lucey, B.M., Naeem, M.A., Uddin, G.S.: Examining the interrelatedness of NFTs, DeFi tokens and cryptocurrencies. Finan. Res. Lett. **47** (2022). https://doi.org/10.1016/j.frl.2022.102696
12. Wang, S., Ding, W., Li, J., Yuan, Y., Ouyang, L., Wang, F.Y.: Decentralized autonomous organizations: concept, model, and applications. IEEE Trans. Comput. Soc. Syst. **6**(5), 870–878 (2019). https://doi.org/10.1109/TCSS.2019.2938190
13. Wright, A.: The rise of decentralized autonomous organizations: opportunities and challenges (2021). https://letstalkbitcoin.com/is-bitcoin-overpaying-for-false-security
14. Chohan, U.W.: The decentralized autonomous organization and governance issues (2017)

15. Ethereum: Decentralized autonomous organizations (DAOs) (2023). https://ethereum.org/en/dao/
16. Balakrishnan, A., Yeakley, J.: The year ahead for DeFi (2022)
17. Moringiello, J.M., et al.: Blockchain real estate and NFTS. https://ssrn.com/abstract=4180876
18. Tan, E.: NFT-linked house sells for $650K in Propy's first US sale (2022). https://www.coindesk.com/business/2022/02/11/nft-linked-house-sells-for-650k-in-propys-first-us-sale/
19. Propy: Florida – the home of the first US real estate NFT (2023). https://propy.com/browse/first-us-real-estate-nft/
20. Ordano, E., Meilich, A., Jardi, Y., Araoz, M.: Decentraland a blockchain-based virtual world (2023)
21. Jackobson, R.: Real T legally compliant ownership of tokenized real estate (2019)

A Thematic Analysis to Determine the Future Design of Mobile Cryptocurrency Wallet

Richard$^{(\boxtimes)}$ (iD), Muhammad Ammar Marsuki, and Gading Aryo Pamungkas

School of Information Systems, Bina Nusantara University, West Jakarta, Indonesia
`richard-slc@binus.edu`

Abstract. The growth of cryptocurrency has increased the usage of mobile cryptocurrency wallet applications. Even though the cryptocurrency wallet is an essential tool in managing access to crypto assets, the application's usability still needs to improve in many aspects. This research aims to determine the essential factors for future mobile cryptocurrency wallet design. The user reviews of several popular crypto wallets are collected from several application stores. A machine learning approach is used to classify the review category based on their relevancy to user experience. The result shows numerous problems still exist even with the most adopted mobile crypto wallet. Hence, we propose a recommendation of features to solve the problem and help preserve the credibility of the mobile wallet.

Keywords: Mobile Cryptocurrency Wallet · User Experience · Thematic Analysis · Cryptocurrency Wallet Design

1 Introduction

As cryptocurrency (crypto) continues to gain popularity over the years and is expected to reach a total net worth of USD 48.27 billion in 2030, mobile crypto wallets have become an integral part of the lives of more than 80 million users in 2022 [1]. A mobile crypto wallet is a powerful tool that enables users to manage their digital assets, conduct transactions, and access real-time information on their investments through their mobile devices. The user experience of mobile crypto wallets can vary from one application to another, impacting the user's satisfaction with the overall product [2]. Therefore, it is essential to determine the required features to increase the user experience of the future design of mobile crypto wallets.

There are various mobile crypto wallet options in the current market. Nonetheless, some applications have a significantly larger user base than others. Maulana and Wang [3] conducted studies regarding the connection between user loyalty and user satisfaction in the Bitcoin exchange business and discovered that user satisfaction substantially impacts user loyalty. According to Rehman et al. [4], enhancing user satisfaction through improved user experience and trust increases cryptocurrency usage.

Based on the studies, highlighting the users' preferred features in mobile cryptocurrency wallets may lead to a better design in expanding the application's capabilities. The knowledge of users' preferred features is generated by analyzing reviews from the

J. M. Machado et al. (Eds.): BLOCKCHAIN 2023, LNNS 778, pp. 286–296, 2023.
https://doi.org/10.1007/978-3-031-45155-3_29

most popular cryptocurrency wallet. Since the number of reviews might be massive, a machine-learning approach is used to classify the reviews relevant to the wallet's usability (see Fig. 1). This research aims to discover what issues regarding UX exist in mobile crypto by analyzing the review segments of the application. The analysis findings will later indicate what features a mobile crypto wallet requires to increase the adoption of the applications.

2 Background and Related Works

2.1 Crypto Wallet Definition

Cryptocurrency is the well-known use case of blockchain in the financial and technology sector [5]. A cryptocurrency (crypto) wallet is software/hardware connected to a blockchain network to manage crypto assets. The crypto wallet can be categorized into hot and cold wallets. The difference between a hot and cold wallet is based on the offline/online mode in managing the crypto asset. The hot wallet is connected to the internet, which allows us to manage the crypto-asset in real-time. The cold wallet is a physical medium to store the key offline, leading to better security [6]. The easily accessible features of hot wallets have significantly contributed to the high number of adoptions of current crypto wallet usage [7]. A hot wallet typically comes in three forms: desktop, mobile application, and browser. Based on the popularity of mobile crypto wallets, this research focuses on the issue of usability in a mobile crypto wallet. The review is retrieved from the two leading operating systems, iOS and Android.

2.2 Related Works in User Experience (UX) in Crypto Wallet

A recent study has stated that UX research on blockchain-related technology, such as cryptocurrency, has not caught up to the current state of the blockchain [8]. The number of problems that occurred to users' misconception of features provided by the crypto wallet that leads to some sort of financial loss from the user side is solid evidence to base the statement before [7, 9]. Krombholz et al. surveyed to investigate UX on Bitcoin. The result shows that most users still do not understand the features available in the Bitcoin network, especially regarding security and privacy, often endangering their anonymity [10].

The misunderstanding might be due to the lack of good usability from a desktop and mobile-based crypto wallet in performing a fundamental task [5, 11]. The user tends to receive guidance in the form of highly technical language that is hard to understand and has insufficient information regarding what steps to follow to identify the problems and how to solve them [11, 12].

With trust as the primary motivation amongst most crypto wallet participants, these usability issues directly impact the credibility of crypto wallets, affecting a significantly low usage rate compared to the high adoption number [7, 13]. A better choice of UI that delivers rich user-centered information and mitigation systems for financial losses related problems is recommended for future wallet design to address general and domain-specific issues [14–16].

3   Methodology

We developed a systematic approach to analyzing the data needed for investigating the features preferred by the users from a mobile cryptocurrency wallet. We gather the data using the scraping method to acquire reviews from App Stores. The collected data then proceeded to the cleaning process, where we applied various pre-processing techniques to filter only the relevant data needed for this research. We employ the K-Fold validation procedure to ensure the sampled data's reliability and validity, which will affect the analysis's accuracy. Finally, we used statistical techniques called thematic analysis. This last step aims to identify trends, patterns, and potential issues in the mobile cryptocurrency wallet.

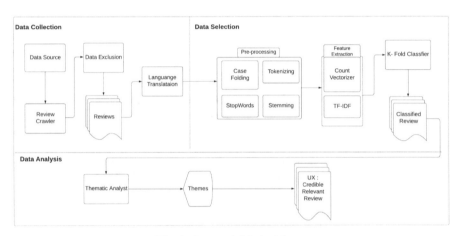

Fig. 1.   Research Methodology.

3.1   Data Collection

Data Source.   With a wide variety of mobile crypto wallets, we narrowed down the options included in this research. The top five mobile wallets are chosen based on their popularity, user ratings, and the number of reviews on the Google Play Store and Apple App Store. *Trust Wallet, Blockchain Wallet, MetaMask Wallet, Coinbase Wallet,* and *Coinomi Wallet* were chosen as the review object. The user reviews of these wallets are retrieved from Google Play Store and Apple App Store. This selection guarantees that the data collected are comprehensive and representative and allows us to compare user sentiment across different platforms and operating systems.

Review Crawler.   Since each wallet receives numerous amount of reviews, the manual gathering method is deemed to be impractical. The self-made crawler is built to collect reviews from the data source. Around 35,806 reviews were gathered from the App Stores and Google Play. The selected reviews include the text reviews, the rating score, the date, and the application version from when the review was uploaded.

Data Exclusion. Reviews containing less than four words are excluded to preserve the dataset's quality. The exclusion is taken because it might be uninformative, which could reduce overall data accuracy [17, 18]. After filtering out the noisy data, around 27,934 reviews are preserved in the review collection.

3.2 Data Selection

This research uses a machine learning approach to classify the review category. The preliminary processing techniques filter out irrelevant information and standardize the text data. Afterward, the text data is transformed into numerical vectors that machine learning algorithms can understand. The pre-processed data will be split for training the model and testing the set with a ratio of 80:20.

Pre-Processing. The pre-processing step holds a crucial role that involves transforming raw text data into a format suitable for the analysis. Several techniques are included in the transformation procedure, starting with converting the text into a uniform case (case folding) for consistency. The uniformed data then breaks down into individual words (tokenizing), the basis for removing stopwords. It cast out words that do not carry significant meaning to this analysis. Lastly, every word will be reduced to its root form to improve accuracy by reducing variation (stemming).

Feature Extraction. Four feature extraction methods were applied in this stage to assist the analysis of more extensive user reviews of the mobile crypto wallet. We identified the frequency of specific words and phrases in the text body using a count vectorizer. After identification, a frequency-inverse document frequency (TF-ID) was applied to determine the importance of each word and phrase. Sentiment analysis measures the overall scores of reviews by calculating the positive, negative, and neutral words, ranging from -1 (extremely negative) to 1 (extremely positive). At the end of this stage, the data set will be divided into training subsets and testing subsets for validating the accuracy of feature extraction models.

Table 1. Example of classified reviews.

Classification	Review Text	Explanation
Related to Cryptocurrency	The fees charged is unexpected, I deposited 50 USD and ended up only getting roughly 36 in my wallet	High fee transaction pricing lead to slight dissatisfaction from user
UX in general	Using the latest 1.10.2 crashes almost none, but still hangs when opening the app sometimes	Focus on how the application behaves—nothing to do with cryptocurrency
Irrelevant to UX	Hopefully this will take me to the moon!	Unrelated to the application or the cryptocurrency

Training Set. From 27,934 reviews, we take 1,000 random samples and categorize them according to their relation to UX. Relation in this context signifies a review relevant to distinctive features of mobile cryptocurrency wallets and information from existing studies. This process will divide the reviews into three categories: related to Cryptocurrency, UX in general, or Irrelevant to UX. Table 1 contains samples of the three types of reviews encountered and the reasons why they are classified into each category.

Machine Learning Model. After Finalizing the training set, we selected K-Fold validation as our machine-learning model. The combination of pre-processing, sentiments score, and our random samples had an F1 score (metric for evaluating machine learning that assesses the precision of a model) of 0.74 for the relevant UX reviews. A binary classifier that works by chance has an AUC-ROC (area under the ROC curve) value of 0.5, but a perfect classifier has a value of 1. After performing 10-fold cross-validation, our classifier achieved a mean AUC value of 0.84. Hosmer et al. [20] claim that this demonstrates outstanding discriminating ability.

3.3   Data Analysis

Thematic Analysis. The thematic analysis is chosen for its capability to detect and identify data to provide interpretations and generate patterns [19]. Following the identification process, the reviews were then coded in batches. We searched for themes within the codes, defining and naming each theme to reflect its content. The themes were refined as they progressed through the analysis process to ensure they accurately reflected the data content. The analysis identified four themes: domain-specific, security and privacy, misconception, and trust. The analysis then narrowed the identified reviews to focus on the most relevant reviews for each theme, resulting in 5,466 reviews. Table 2 shows the exact number of reviews narrowed down from each wallet.

Table 2. The number of classified and analyzed reviews per wallet and platform.

	Application Review Crawling	Automatically Classified Review	Analyzed Reviews
Trust Wallet	16,130	4,016	1,884
BlockChain	3,958	2,761	1,163
Metamask	3,794	1,498	613
Coinbase	2,360	2,581	1,401
Coinomi	1,692	850	405
Score Total	27,934	11,706	5,466

4 Result

The UX-related reviews must qualify for the requirement to ensure accuracy and reliability. The reviews need to be longer than three words to preserve clarity, and relevant keywords about cryptocurrencies, user experience, and mobile wallets must be present in the reviews to indicate their relevancy. We exclude unrelated reviews that did not meet the requirement to prevent false conclusions. The findings we obtain help us recognize the preferred features and responses from the issues conveyed to be implemented in the application.

4.1 Thematic Analysis Result

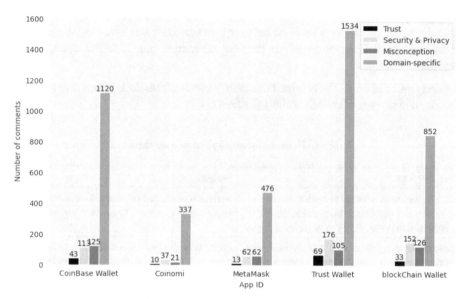

Fig. 2. Statistics of the identified theme.

Four themes were identified from the thematic analysis, with issues regarding domain-specific being the most common, followed by security and privacy, misconception, and trust.

Domain-Specific. This theme is identified based on the issues that are exclusive only to the mobile crypto wallet (Table 3).

Table 3. Findings in domain-specific theme.

Review Text	Insight
Allows most if not all coins. Allows multiple coin of the same wallet which is something Ive had a problem with in other apps	Wallets supporting multiple cryptocurrencies appeal more to the reviewer than other wallets that do not possess them
[mobile wallet name] did not spend effort in designing its interface. Its easily copiable by phishings. My account was hacked and as a result I lost $600!	Poorly designed UI that caused financial loss to the reviewer

We found out that users highly preferred mobile wallets that allow them to store multiple currencies. Mobile wallets that have already implemented this feature usually have numerous positive reviews. Awful user interface design is also frequently issued in the reviews. It is shown to reduce the user experience and, in severe cases, lead to monetary loss.

Security and Privacy. We obtained this from reviews that address issues regarding the security and privacy of mobile wallets (Table 4).

Table 4. Findings in security and privacy theme.

Review Text	Insight
Very secure imo, with all the features they gave you like password, biometrics, BIP39 passphrase and the ability to combined those	Variation of security options offered by the wallet leads to a more secure feeling for users
Never shared my password with anyone yet a random party was able to login to my wallet? Contacting the customer support only to be replied by a bot, got no choice other than deleting my account now	Poor security systems and customer support can cause the user to abandon the wallet

We know from the reviews that various security options are necessary, especially for second-factor authentication. Also, customer support is essential to aid users with sensitive personal information problems.

Misconception. This theme brings up the issue of drawbacks caused by user misconceptions (Table 5).

Table 5. Findings in misconception theme.

Review Text	Insight
Never really have any issues beside that it seems that my balance are really buggy and not accurate after transfering Idk why	Reviewer complained about the inaccurately displayed balance after transaction without knowing the reason behind it
Seeing all these negatives comment shows me that alot of people really don't understand how crypto works. It takes times for the balance to be sync with the blockchain. And for the high ETH transaction fee, its not the wallet fault just check this out [url to article about transaction fee]. Seems that nobody wants to take the time to learn how this stuff works	Reviewer stated that most of the user base still lacks common knowledge about cryptocurrency

Although some issues related to misconception might be caused by developer incompetence, we learned that the lack of common knowledge from the user's end on how cryptocurrency works usually causes the main problem.

Trust. This theme is acquired from identifying reviews that expressed user confidence in the mobile crypto wallet (Table 6).

Table 6. Findings in trust theme.

Review Text	Insight
A trusted wallet that puts you in control of your earnings	Users prefer more freedom in how to control their financial assets
Its really frustrating for me to trust this wallet when there are so many scams with people claiming to be their customer support!	The presence of proper customer support highly affects user trust

Currently, some mobile wallets indirectly dictate how customers manage their wallets. We learned that giving users as much freedom as possible is essential to gain users' trust. The importance of customer support is re-emphasized again in this theme as well.

4.2 Mandatory Features Recommended for Future Design

We propose implementing several features based on the insight we found in the thematic analysis result. The result that we expected after implementing the features mentioned above is a significant increase in positive reviews written over negative reviews for the application to help both existing and prospective users decide whether to adopt the mobile wallet.

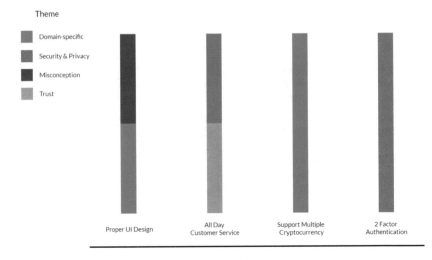

Fig. 3. The required features for a future mobile crypto wallet.

5   Conclusion

Four themes were identified from the data classification: domain-specific, security & privacy, misconception, and trust. Based on thematic analysis, research findings suggest several features for mobile crypto wallets to be implemented in the future. The features include a proper UI design, all-day customer service, support of multiple cryptocurrencies, and 2-factor authentication. Since the domain-specific theme has been found as the most common theme identified compared to others (see Fig. 2), we conclude that a proper UI design and a wallet that supports multiple cryptocurrencies are the most important features to be included in the future design of mobile crypto wallets. The other two features mentioned in Fig. 3 are also highly recommended to be implemented to build trust between the users and the developer, giving the wallet better credibility over competitors.

References

1. Crypto Wallet Market Growth & Trends. Grand View Research (2022). https://www.grandviewresearch.com/press-release/global-crypto-wallet-market. Accessed 31 Mar 2023
2. Badran, O., Al-Haddad, S.: The impact of software user experience on customer satisfaction. J. Manag. Inf. Decis. Sci. **21**(1), 1–20 (2018)
3. Maulana, D., Wang, G.: Analysis of factors affecting user loyalty on bitcoin exchange. In: 2019 7th International Conference on Cyber and IT Service Management (CITSM), pp. 1–5 (2019). https://doi.org/10.1109/CITSM47753.2019.8965350
4. ur Rehman, M.H., Salah, K., Damiani, E., Svetinovic, D.: Trust in blockchain cryptocurrency ecosystem. IEEE Trans. Eng. Manag. **67**(4), 1196–1212 (2019). https://doi.org/10.1109/TEM.2019.2948861

5. Moniruzzaman, M., Chowdhury, F., Ferdous, M.S.: Examining usability issues in blockchain-based cryptocurrency wallets. In: Bhuiyan, T., Rahman, M.M., Ali, M.A. (eds.) Cyber Security and Computer Science (ICONCS 2020). LNICST, vol. 325, pp 631–643. Springer, Cham (2020). https://doi.org/10.1007/978-3-030-52856-0_50

6. Ramirez, D.: Crypto hot wallet vs. cold wallet: what's the difference. https://www.nerdwallet.com/article/investing/hot-wallet-vs-cold-wallet. Accessed 28 Mar 2023

7. Albayati, H., Kim, S.K., Rho, J.J.: A study on the use of cryptocurrency wallets from a user experience perspective. Hum. Behav. Emerg. Technol. 3(5), 720–738 (2021). https://doi.org/10.1002/hbe2.313

8. Jang, H., Han, S.H.: User experience framework for understanding user experience in blockchain services. Int. J. Hum. Comput. Stud. **158**, 102733 (2022). https://doi.org/10.1016/j.ijhcs.2021.102733

9. Voskobojnikov, A., Wiese, O., Mehrabi Koushki, M., Roth, V., Beznosov, K.: The u in crypto stands for usable: an empirical study of user experience with mobile cryptocurrency wallets. In: Proceedings of the 2021 CHI Conference on Human Factors in Computing Systems (CHI 2021), pp. 1–14, Association for Computing Machinery, New York, NY, USA (2021). https://doi.org/10.1145/3411764.3445407

10. Krombholz, K., Judmayer, A., Gusenbauer, M., Weippl, E.: The other side of the coin: user experiences with bitcoin security and privacy. In: Grossklags, J., Preneel, B. (eds.) Financial Cryptography and Data Security (FC 2016). LNCS, vol. 9603, pp. 555–580. Springer, Heidelberg (2017).https://doi.org/10.1007/978-3-662-54970-4_33

11. Kazerani, A., Rosati, D., Lesser, B.: Determining the usability of bitcoin for beginners using change tip and coinbase. In: Proceedings of the 35th ACM International Conference on the Design of Communication (SIGDOC 2017). Association for Computing Machinery, New York, NY, USA (2017). https://doi.org/10.1145/3121113.3121125

12. Eskandari, S., Barrera, D., Stobert, E., Clark, J.: A first look at the usability of bitcoin key management (2015). https://doi.org/10.14722/usec.2015.23015

13. Ku-Mahamud, K.R., Omar, M., Bakar, N.A.A., Muraina, I.D.: Awareness, trust, and adoption of blockchain technology and cryptocurrency among blockchain communities in Malaysia. Int. J. Adv. Sci. Eng. Inf. Technol. 9(4), 1217–1222 (2019). https://doi.org/10.18517/ijaseit.9.4.6280

14. Chen, Y.P., Ko, J.C.: CryptoAR wallet: a blockchain cryptocurrency wallet application that uses augmented reality for on-chain user data display. In: Proceedings of the 21st International Conference on Human-Computer Interaction with Mobile Devices and Services, pp. 1–5 (2019).https://doi.org/10.1145/3338286.3344386

15. Sas, C., Khairuddin, I.E.: Design for trust: an exploration of the challenges and opportunities of bitcoin users. In: Proceedings of the 2017 CHI Conference on Human Factors in Computing Systems, pp. 6499–6510 (2017)

16. Jang, H., Han, S.H., Kim, J.H.: User perspectives on blockchain technology: user-centered evaluation and design strategies for dapps. IEEE Access **8**, 226213–226223 (2020). https://doi.org/10.1109/ACCESS.2020.3042822

17. Arora, A., Malhotra, R., Saxena, N., Sharma, S.: Wallet app credibility analysis based on app content and user reaction. In: 2018 5th International Symposium on Emerging Trends and Technologies in Libraries and Information Services (ETTLIS), pp. 263–268. IEEE (2018). https://doi.org/10.1109/ETTLIS.2018.8485230

18. Tavakoli, M., Zhao, L., Heydari, A., Nenadić, G.: Extracting useful software development information from mobile application reviews: a survey of intelligent mining techniques and tools. Expert Syst. Appl. **113**, 186–199 (2018). https://doi.org/10.1016/j.eswa.2018.05.037

19. Alhojailan, M.I., Ibrahim, M.: Thematic analysis: a critical review of its process and evaluation. West East J. Soc. Sci. **1**(1), 39–47 (2012)
20. Hosmer Jr, D.W., Lemeshow, S., Sturdivant, R.X.: Applied logistic regression, vol. 398. John Wiley & Sons (2013)

Blockchain Technology as a Means to Achieve Prior Transparency in Arbitration Processes. A Social Challenge

Javier Parra-Domínguez🄳, Yeray Mezquita(✉)🄳, Carlos Álvarez🄳,
and Fernando de la Prieta🄳

BISITE Research Group, University of Salamanca, Salamanca, Spain
{javierparra,yeraymm,carlos_alvarez,fer}@usal.es

Abstract. Any arbitration procedure must be transparent in order to guarantee that all parties have access to the information and supporting documentation they need to make informed decisions. This manuscript proposes a framework for transparency that makes use of blockchain technology to improve the fairness and integrity of any arbitration system. To comprehend the decisions made, the technological requirements and difficulties experienced by distributed platforms based on blockchain technology are explained. The proposed solution not only satisfies the specified technological needs, such as scalability and resilience but also fosters transparency by guaranteeing the verifiability of source data and the preservation of users' privacy. This system tries to get rid of corruption and safeguard consumer interests by employing a public network via which smart contracts can be executed. By expanding the scale of the network used, this strategy not only increases resilience but also improves the accountability and openness of the arbitration process.

Keywords: Arbitration · Blockchain · Transparency

1 The Need for Transparency

When we talk about transparency in society, it is mainly about providing clear, understandable information that does not give rise to any doubts [7]. More and more societies are determined to obtain information in this knowledge age. All citizens need to know information and get it from reliable sources to have sufficient knowledge and participate in integrating social progress that includes everyone. In an era of apparent progress in knowledge, it is a duty to be able to incorporate this information into new technologies, and that is why, from a social point of view, the integrating role of developments such as those based on the blockchain must be valued.

From the above, the importance of transparency as a social value can be deduced to the point of being mandatory for any organisation, either because

J. M. Machado et al. (Eds.): BLOCKCHAIN 2023, LNNS 778, pp. 297–306, 2023.
https://doi.org/10.1007/978-3-031-45155-3_30

it is concerned with corruption [11,12] or because of the need to incorporate value into the company's marketing [2]. If we focus on the social importance of transparency, we must advance in conceiving it as a collective and not an individual asset. The reflection and its social significance are so profound that transparency can be mixed with concepts such as ethics due to the need to achieve transparency as a form of moral action or what is understood as good behaviour [17,24]. Acting well is a value for individuals and organisations, and the morality is a conglomerate of values [13,23]. If we always work well in organisations, we can understand that there is a moral culture.

Currently, different studies have focused on the application of transparency criteria to the arbitration process itself, as can be seen in the previous points. However, there are not many articles that focus on providing transparency prior to the arbitration process in order to strengthen the consistency of arbitrators' decisions. In this paper, we propose a transparency framework before the arbitration process so that it can benefit more from the prior application of specific technologies.

At this point, it is essential to note the article's motivation. In the current framework of arbitration systems, arbitrators are often not provided with the most objective information possible, which can result in a process that needs to be more accurate. Therefore, we motivate the paper by indicating a scheme based on blockchain technology that allows for greater objectivity, as the information received by an arbitrator before reaching a decision is more precise and more confirmable.

The current work's purpose focuses on concretely making use of blockchain technology in the case studio of any arbitration system. In Sect. 3, we will introduce the tangible connection between blockchain, arbitration and transparency, incorporating the specific framework developed as a case study. The technological framework of the application will be presented in Sect. 4, concluding the manuscript with Sect. 5.

2   Arbitration Systems

Arbitration is understood as an alternative means of dispute resolution, being, in most cases, directly enforceable, as would be a judgment handed down by any state court or tribunal [22]. Specifically, arbitration systems facilitate processes of understanding between parties, allowing them to resolve their disputes through a process that is usually quicker than going to court.

A prominent figure in the arbitration process is the arbitrator, who must have been well chosen [5]. With time and specialised knowledge, even more than ordinary judges, resolving the dispute in compliance with the law, with greater attention to the allegations and expert reports, in less time and with the same impartiality as a judge or a court [4]. It is important to note that arbitration proceedings also involve costs and fees [26]. These payments can also be reduced if this process is preceded by incorporating technology that makes it more efficient in financial terms [6,20].

2.1   Arbitration and Transparency

If we refer to a process of transparency and link it to arbitration also as a process, we find the need to incorporate transparency itself into arbitration [21]. It is essential to bear this in mind because it already begins to consider the importance of new technologies in these processes, as they allow for greater transparency in the process [8,18]. We must also highlight the link of concern that exists at the international level.

One of our ambitions is to support the social challenge in blockchain technology and work on improving equity, inclusion and fairness. With the incorporation of blockchain technology, inclusion extends, for example, to older people, who are often concerned about the security of processes and feel insecure when there is no evident trust in the system being developed. In improving the quality of life, it is essential to consider the challenge blockchain technology poses for people; this is the case for older people. Any new technology, especially blockchain technology, requires new concepts for users to understand: distributed systems, key management, wallets and transactions. And as long as users need to understand the merit of blockchain technology, its potential benefit will be recovered for the time being. This is the case of the commercialisation of energy services, where the company that develops the commercial process is different from the large company that will subsequently provide the services and whose interests may be opposed and only verifiable by an arbitrator in a future dispute if it has the tools provided by technologies such as the blockchain.

Working for the general interest before the arbitration process is what we intend to do in this work, as indicated in the previous section. This interest is pursued with a clear objective: to prevent disputes from arising thanks to the clarity of contracts by incorporating blockchain technology [10] and, of course, based on the transparency of the laws and the case law that interprets them.

2.2   The Case of Arbitration in the Spanish Energy Market

Our motivation for this study is limited to the Spanish market. To understand the importance of arbitration today, it is essential to highlight all the information in Spain regarding all the most important bodies and laws involved.

In Spain, the law regulating arbitration is Law 60/2003 of 23 December 2003. This law was a step forward for the development of the arbitration system as a method for resolving disputes between commercial contracts. However, the primary legislation regulating arbitration in Spain is:

– Convention on the Recognition and Enforcement of Foreign Arbitral Awards 1958 (New York Convention).
– European Convention on International Commercial Arbitration 1961 (Geneva Convention).
– The Arbitration Act 60/2003 itself, the application of which is limited to Spanish territory[1].

---

[1] https://www.cnmc.es/ambitos-de-actuacion/energia last accessed 13/04/2023.

3   Blockchain, Transparency and Arbitration

Blockchain technology allows the creation and management of a distributed ledger among a network of nodes in which information is stored and known by all participants. Using a consensus protocol, the data stored in the ledger is tamper-proof, because no one would be able to modify it without the rest of the network noticing. That mechanism makes the peers of the network authorized owners of the data and transparency auditors. Thanks to the data generated being stored in the distributed ledger, it is offered transparency and immutability of the data to this kind of systems. Blockchain technology also provides a mechanism to deploy and run self-enforcing code, called smart contracts. A smart contract is a programmable software that allows for the inclusion of multiple clauses to facilitate agreements among various parties and automate straightforward processes [15].

In the use case proposed for study in this article, arbitration processes in the electrical industry, the use of blockchain technology provides the transparency necessary for the arbitrator to carry out his task of mediating between the parties in a more secure and unbiased manner. The process will be governed by smart contracts deployed in the blockchain, which will be in charge of the management of the flow of information generated within the platform, see Fig. 1.

Data transparency is critical when an arbitration process is needed to be carried out. So the use of blockchain technology can provide this kind of systems with the mechanisms needed to achieve true transparency between the parties interested. However, transparency being one of the main characteristics of blockchain technology, it also becomes one of its main challenges to be implemented in real use cases. Due to the application of laws such as the General Data Protection Regulation (GDPR) [1], it is essential that in services such as this, where a considerable number of actors are involved in maintaining the stored data, the privacy of users is not violated and the data is only accessed by those actors to whom permission has been granted (such as the company and the arbitration company for the use case proposed in this work) [3]. In the next section, Sect. 4, it will be explained in detail the technical requirements to allow the implementation of this system in the real-world industry.

4   Technical Framework

In this section, we are going to describe the technical requirements that the proposed system needs to be carried out for its correct implementation. In this sense, the technical requirements must be aligned with the challenges faced by distributed platforms based on blockchain technology. The literature has identified [14–16]: (i) that the systems are scalable and resilient, (ii) that the smart contracts comply with current legal regulations, (iii) that the information entered can be verifiable at source, and (iv) that the privacy of sensitive data generated by the users can be preserved.

Regarding the challenges of scalability and resilience, in the literature, the tendency is clear to use permissioned networks formed by the confederation of

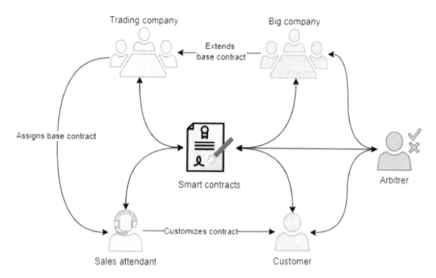

Fig. 1. Actors interaction diagram. In this diagram it is shown how the *Big company* extends the base contract with the products they offer, along with the minimum requirements a customer must comply with. Then the *Trading Company* assigns the base contract to one of their *Sales Attendant*, which will be in charge of finding customers for the *Big Company*. When the *Sales Attendant* finds a *Customer* it personalizes the requirements of the contract so it can sign it. Then, the *Customer* will receive the products offered in exchange of the payment signed. In case something goes wrong, or with the payment or with the services, it is called the arbitration process and an *Arbiter* will be in charge of the dispute between the *Customer* and the *Big Company*.

companies. However, there is a problem in the proposed system, and that is that the interests of end consumers would not be represented in the blockchain network, being large companies the only ones with power in this aspect. Therefore, the use of a permissioned blockchain network to manage and store the smart contracts for this platform should be ruled out unless a consumer syndicate is part of the network as a trusted intermediary, something that is not exempt from the corruption that could occur. In this sense, considering the scalability requirements of the proposed system, it is worth noting that while permissioned blockchains are often recommended in the literature due to their simpler consensus mechanisms and relaxed security assumptions, public blockchains like Ethereum and TRON offer scalable solutions. These platforms utilize consensus algorithms such as Proof of Stake (PoS) and Delegated Proof of Stake (DPoS), which can handle larger transaction volumes. Additionally, public blockchains provide greater network resilience by leveraging the size of the network. Furthermore, platforms like Tron or Ethereum offer minimal transaction fees, making them easily affordable for companies and users, thus providing a cost-effective solution for implementing and mass use of smart contracts on a public network.

On the other hand, when we talk about automatic interactions between trustless parties, we need to make use of smart contracts deployed within the blockchain platform. These smart contracts are self-enforcing and are composed of a set of clauses which parties agrees to comply with when using it. In the use case scenario proposed in this paper, a big company extends a base smart contract to the trading company. The trading company allows their sales attendants to seek customers and customize the base smart contract for each of them in an individual way. When a customer finds acceptable a customized contract, it signs it, as well as the big company that extended the base smart contract in the first place. When a payment it is overdue, the arbitration process starts and an arbiter is assigned to the cause. The challenge in this part of the technological framework, is to make the smart contracts legal compliant [16], which means that it is needed a way for both parts to break the contract, and a way for it to be paused while it is in a dispute process. In the context of smart contracts and blockchain technology, enforcing revocable and pausing requirements involves implementing specific technical mechanisms. Let's delve deeper into these technical details.

- Revocable Contracts: To enable revocable contracts, a common approach is to include a revocation function within the smart contract code. This function allows one or both parties to invoke it, triggering the termination or cancellation of the contract. When invoked, the revocation function typically executes certain predefined actions, such as refunding funds or transferring assets back to the relevant parties. It's crucial to ensure that the revocation process is securely implemented and properly authenticated to prevent unauthorized revocations.
- Pausing Contracts: Implementing a pause functionality in a smart contract requires additional technical considerations. One common method is to include a pause switch or flag within the contract's code. When the pause switch is activated, it halts the execution of the contract, effectively pausing any ongoing transactions or state changes. This feature is particularly useful during dispute resolution processes, allowing parties to freeze the contract's operations until the dispute is resolved. To ensure security and prevent abuse, proper access controls and authorization mechanisms should be implemented to govern who can activate or deactivate the pause switch.
- Arbitration and Dispute Resolution: In scenarios where an arbiter or mediator is involved in contract disputes, additional technical components need to be integrated. These may include the ability to appoint an arbiter, establish communication channels between the parties and the arbiter, and implement a mechanism for the arbiter's decision to be enforced within the smart contract. This can be achieved through a combination of off-chain communication and on-chain verification, ensuring that the arbiter's ruling is properly reflected and executed in the contract's logic.

Following the path marked by regulation, it is absolutely needed to find a way to preserve the anonymity of the customers and preserve the privacy of their actions within the platform. In this sense, the solution proposed for this scenario is the use of Zero-Knowledge Proofs (ZKPs) and ring signatures. Thanks to the use of ZKPs, along with a shuffle mechanism with the payments similar to the one proposed in Monero [25], the information stored in the blockchain is encrypted and only the ones with the keys are able to understand it, while the rest of the participants of the network are able to validate it without knowing what it is being validated.

Integrating the use of ZKPs in an Ethereum-based smart contract involves leveraging zk-SNARKs (Zero-Knowledge Succinct Non-Interactive Arguments of Knowledge) technology [9]. With zk-SNARKs, it is possible to prove the validity of certain statements without revealing the underlying data. In the context of a smart contract, this means that sensitive information can be kept private while still providing cryptographic proofs of its correctness. To achieve this, the smart contract would be designed to accept ZKP-based proofs from users, verifying the validity of their inputs without exposing the actual data. This way, the blockchain can maintain its decentralized nature while preserving privacy and confidentiality for sensitive transactions and computations. It is worth noting that integrating ZKPs in Ethereum requires the use of precompiled contracts or trusted execution environments for efficient ZKP verification [19], considering the computational complexity associated with zk-SNARKs.

To create true anonymity while complying with the Know your Customer (KYC) regulations, there should be an additional step in which the big company and the customer exchanges their keys when signing the smart contract, so they are the only ones in the network knowing what it is being done. In this regard, it is used a ring signature algorithm, which shuffles different addresses to mask the origin of a transaction [15], so for each transaction, exist a ring of signatures with the same probability of being their creator, while only the owner know which transactions are created by or meant for them.

The mentioned cryptographic key distribution process for a sample customer and a contract involves several steps to ensure secure and confidential communication. Here is a detailed, step-by-step description of a typical sample process:

1. Preparing the Smart Contract: First, the smart contract is prepared with the necessary functionality to handle ZKPs. This includes integrating cryptographic operations and data structures required for generating and verifying proofs.
2. Generating Keys: Both the customer and the big company need to generate their cryptographic key pairs. This typically involves generating a public-private key pair using a suitable cryptographic algorithm like RSA, ECC, or EdDSA.
3. Customer and Company Key Exchange: The customer and the big company securely exchange their public keys. This can be accomplished through secure channels such as encrypted email, secure messaging platforms, or secure file sharing protocols. This step ensures that each party possesses the necessary public key of the other party.
4. Contract Signing: The customer and the big company proceed with signing the smart contract, indicating their agreement to engage in the ZKP process. During the signing process, the customer's private key and the big company's

private key are used to digitally sign the contract, ensuring non-repudiation and integrity.

5. Initiating ZKP Protocol: Once the contract is signed, the ZKP protocol can be initiated. The customer generates a ZKP proof to demonstrate a specific statement without revealing any sensitive information. This typically involves using zk-SNARKs or other ZKP techniques to create a succinct proof of knowledge.

6. Proof Verification: The big company, acting as the verifier, receives the ZKP proof generated by the customer. Using the customer's public key obtained during the key exchange, the big company verifies the proof's validity without needing to know the actual sensitive data involved.

7. Interaction and Feedback: In some cases, the big company might provide feedback or request additional proofs from the customer to satisfy specific requirements or conditions outlined in the smart contract. The customer can generate and provide additional ZKP proofs as necessary.

8. Completion and Result Validation: Once all required ZKP proofs are provided and validated, the ZKP process is considered complete. The smart contract can proceed to execute further actions or trigger subsequent steps based on the verified results.

By following these steps, the cryptographic key distribution process ensures that the customer and the big company can securely communicate and engage in the ZKP process while maintaining the confidentiality and privacy of sensitive information.

5   Conclusions

This manuscript emphasizes the value of openness in society and the necessity of viewing it as a social good rather than an individual one. Organizations must include transparency in their policies and ethical guidelines since it is viewed as a moral deed. Corporate governance, communication, management, procedures, and marketing are the five primary levels of transparency in organizations that we have identified based on the literature.

In this work, we highlighted the need of transparency in arbitration systems, especially when powerful corporations employ unfair tactics against vulnerable clients. In order to enhance the arbitration process and provide technological support to arbitrators, thereby improving their decision-making, we introduced a blockchain-based framework for transparency. Our research reveals that the utilization of blockchain technology can significantly enhance the integrity and transparency of arbitration proceedings. A public blockchain network, whichever that make use of a POS or a DPOS consensus algorithm, is suggested to handle scalability and resilience and make sure that end users' interests are represented in the network. Legal compliance and the importance of smart contracts as a platform component are underlined. The Spanish legal system requires a means for ending contracts and halting them during disagreements. Privacy is frequently emphasized as being essential and important. According to the literature, ring signatures, and ZKPs can be employed to preserve the platform users' privacy and anonymity.

Acknowledgements. This research has been supported by the project "COordinated intelligent Services for Adaptive Smart areaS (COSASS), Reference: PID2021-123673OB-C33, financed by MCIN /AEI /10.13039/501100011033/FEDER, UE.

References

1. 2018 reform of EU data protection rules. https://ec.europa.eu/commission/sites/beta-political/files/data-protection-factsheet-changes_en.pdf
2. Baldassarre, F., Campo, R.: Sustainability as a marketing tool: to be or to appear to be? Bus. Horiz. **59**(4), 421–429 (2016)
3. Bertino, E., Kundu, A., Sura, Z.: Data transparency with blockchain and AI ethics. J. Data Inf. Qual. (JDIQ) **11**(4), 1–8 (2019)
4. Bingham, J.: Reasons and reasons for reasons: differences between a court judgment and an arbitration award. Arbitr. Int. **4**(2), 141–154 (1988)
5. Bloom, D.E., Cavanagh, C.L.: An analysis of the selection of arbitrators (1986)
6. Chang, S.E., Chen, Y.C., Lu, M.F.: Supply chain re-engineering using blockchain technology: a case of smart contract based tracking process. Technol. Forecast. Soc. Chang. **144**, 1–11 (2019)
7. Han, B.C.: The Transparency Society. Stanford University Press (2015)
8. Hasan, H.R., Salah, K.: Blockchain-based proof of delivery of physical assets with single and multiple transporters. IEEE Access **6**, 46781–46793 (2018)
9. Jake Frankfield, E.R.: Zk-SNARK: definition, how it's used in cryptocurrency, and history (2021). https://www.investopedia.com/terms/z/zksnark.asp. Accessed 17 May 2023
10. Janssen, M., Weerakkody, V., Ismagilova, E., Sivarajah, U., Irani, Z.: A framework for analysing blockchain technology adoption: integrating institutional, market and technical factors. Int. J. Inf. Manag. **50**, 302–309 (2020)
11. Kolstad, I., Wiig, A.: Is transparency the key to reducing corruption in resource-rich countries? World Dev. **37**(3), 521–532 (2009)
12. Lindstedt, C., Naurin, D.: Transparency is not enough: making transparency effective in reducing corruption. Int. Polit. Sci. Rev. **31**(3), 301–322 (2010)
13. Luo, Y.: An organizational perspective of corruption1. Manag. Organ. Rev. **1**(1), 119–154 (2005)
14. Mezquita, Y., Casado, R., Gonzalez-Briones, A., Prieto, J., Corchado, J.M., AETiC, A.: Blockchain technology in IoT systems: review of the challenges. Ann. Emerg. Technol. Comput. (AETiC) (2019). Print ISSN 2516-0281
15. Mezquita, Y., Gil-González, A.B., Martín del Rey, A., Prieto, J., Corchado, J.M.: Towards a blockchain-based peer-to-peer energy marketplace. Energies **15**(9), 3046 (2022)
16. Mezquita, Y., Valdeolmillos, D., González-Briones, A., Prieto, J., Corchado, J.M.: Legal aspects and emerging risks in the use of smart contracts based on blockchain. In: Uden, L., Ting, I.-H., Corchado, J.M. (eds.) KMO 2019. CCIS, vol. 1027, pp. 525–535. Springer, Cham (2019). https://doi.org/10.1007/978-3-030-21451-7_45
17. Parris, D.L., Dapko, J.L., Arnold, R.W., Arnold, D.: Exploring transparency: a new framework for responsible business management. Manag. Decis. **54**, 222–247 (2016)

18. Poorooye, A., Feehily, R.: Confidentiality and transparency in international commercial arbitration: finding the right balance. Harv. Negot. L. Rev. **22**, 275 (2016)
19. Reitwiessner, C.: EIP-196: precompiled contracts for addition and scalar multiplication on the elliptic curve alt_bn128 (2017). https://eips.ethereum.org/EIPS/eip-196. Accessed 17 May 2023
20. Saygili, M., Mert, I.E., Tokdemir, O.B.: A decentralized structure to reduce and resolve construction disputes in a hybrid blockchain network. Autom. Constr. **134**, 104056 (2022)
21. Shirlow, E., Caron, D.D.: The multiple forms of transparency in international investment arbitration: their implications, and their limits. Oxford Handbook of International Arbitration (Oxford University Press, 2020), ANU College of Law Research Paper (19.27) (2019)
22. Slaughter, A.M.: A global community of courts. Harv. Int. LJ **44**, 191 (2003)
23. Swartout, S., Boykin, S., Dixon, M., Ivanov, S.: Low morale in organizations: a symptom of deadly management diseases. Int. J. Organ. Innov. **8**(1), 17–23 (2015)
24. Turilli, M., Floridi, L.: The ethics of information transparency. Ethics Inf. Technol. **11**(2), 105–112 (2009)
25. Van Saberhagen, N.: Cryptonote v 2.0 (2013)
26. Ware, S.J.: Paying the price of process: judicial regulation of consumer arbitration agreements. J. Disp. Resol. 89 (2001)

Probabilistic Optimization of Optimistic Finality for the Waterfall Consensus Protocol

Sergii Grybniak[1] (ID), Yevhen Leonchyk[2] (ID), Igor Mazurok[2] (ID), Alisa Vorokhta[2](✉) (ID), Oleksandr Nashyvan[1] (ID), and Ruslan Shanin[2] (ID)

[1] Odesa Polytechnic National University, Shevchenko Av. 1, Odesa 65044, Ukraine
`sergii.grybniak@ieee.org`
[2] Odesa I. I. Mechnykov National University, Dvoryans'ka St 2, Odesa 65082, Ukraine
`alisa-vorokhta@stud.onu.edu.ua`

Abstract. Blockchain is a distributed ledger technology that provides an immutable record and store of transactions. Today, one of the key challenges facing blockchain technology is the time required to finalize transactions. Mass adoption of payment systems and the development of enterprise-class decentralized systems have created a demand for a significant acceleration of finalization time in blockchains' networks, to facilitate fast and efficient transactions while maintaining security and performance. This article discusses the Waterfall platform, which is based on a Directed Acyclic Graph (DAG) architecture. Waterfall implements a two-level consensus protocol combining Ethereum's approach with a new algorithm that provides single-slot finality. However, the optimistic consensus involves a security trade-off that requires the maintenance of network scalability and performance. The proposed protocol modifications aim to minimize the time of transaction finality by obtaining an optimal level of blockchain Coordinators' support for slot finalization, building a simulation model for testing the modifications, and mitigating the problem of non-relayed transactions. The outcomes of this study will be incorporated into the Waterfall platform software, to enhance its dependability, efficiency, and security.

Keywords: Blockchain · Distributed Ledger Technology · Consensus Protocol · Transaction Finality · Positive Voting · Non-Relayed Transactions

1 Introduction

Blockchain technology is a secure decentralized system of digital record-keeping [1, 2]. System security mainly depends on the nodes' ability to finalize transactions through a consensus protocol. Finalization is the process of confirming transactions and adding them to the blockchain ledger, ensuring transaction validity and irreversibility [3]. The consensus mechanism determines how finalization is achieved. As blockchain technology continues to evolve, new consensus methods may be developed to improve the finalization process.

The Bitcoin network [4], the most well-known blockchain in the world, has a probabilistic finality concept – the more blocks added to the blockchain ledger after a transaction, the less likely the transaction is to be reversed. For example, it is believed that

J. M. Machado et al. (Eds.): BLOCKCHAIN 2023, LNNS 778, pp. 307–316, 2023.
https://doi.org/10.1007/978-3-031-45155-3_31

the probability of a transaction being reversed after six added blocks is less than 0.1% [5].

On the Ethereum 2.0 blockchain, the finalized status of transactions is determined based on epochs [6]. Currently, the finalizing process takes about 2–3 epochs or 13–18 min. The reasoning behind this time duration was to strike a balance between decentralization, security, overall network load, customer expectations, etc. It is deemed appropriate for many use cases in various fields, including the decentralized finance (DeFi) industry [7]. However, in some business scenarios where many transactions are made, and for relatively small amounts, an average waiting time of approximately 15 min may not be acceptable. Note that there are 32 slots per epoch in Ethereum. Therefore, there are compelling arguments in favor of single-slot finality instead of epochs.

One possible solution to this problem is to develop an optimistic consensus protocol that ensures the specified probabilistic finality defined by a set of system parameters. This approach can significantly speed up the finalization of slots, and subsequently blocks themselves. However, to maintain network scalability and performance, a security trade-off must be made.

On the Waterfall platform [8], a two-level consensus protocol is implemented combining Ethereum's approach with a new algorithm that provides single-slot finalization, so that users have a choice. For example, figuratively speaking, if a user buys a cup of coffee, the deal may be promptly guided by the optimistic protocol, but more valuable arrangements should be made after waiting for 2–3 epochs.

The time required to finalize transactions also depends on how fast transactions are added to blocks from the pool, particularly their speed of propagation over the network and block occupancy. On the Waterfall platform, transactions are processed by Validators responsible for maintaining the integrity of Shard networks with DAG architecture [9]. As a result of the Shard working, the produced blocks of transactions are transmitted to network Coordinators for their linearization (ordering) and further finalization. However, there may be cases when Validators do not send received transactions to other Validators, leading to an uneven distribution of transactions between them. In addition, Coordinators may misbehave for various reasons, causing consensus delays.

The goal of our research is to enhance the consensus protocol "Waterfall: Gozalandia" [10] by minimizing the time of transaction finality. This will be accomplished through the completion of the following objectives:

- obtain the optimal level of blockchain Coordinators' support for slot finalization;
- build a simulation model for testing proposed protocol modifications;
- mitigate the problem of non-relayed transactions.

The study utilized various mathematical and statistical analysis techniques, as well as simulation modeling experiments in Python. The obtained outcomes will be incorporated into Waterfall platform software, which will enhance its dependability, efficiency, and security.

2   Literature Overview

The Ethereum consensus protocol has gained significant attention due to its recent transition to a new Proof-of-Stake (PoS) model [11], using a positive absolute supermajority rule that requires the support of two-thirds of total network stakeholders. This rule is intended to increase the security and decentralization of the blockchain network by ensuring that decisions are made with the agreement of a significant portion of network nodes. Despite the fact that Ethereum, as with any decentralized system, is not absolutely immune to attacks (e.g. [12, 13]), such an approach reveals itself as an effective governance mechanism in blockchain systems [14] and is constantly improving, including by cryptography and economic leverages.

The issue of how to finalize a single slot in PoS protocols is less researched in the literature than in other types of consensus protocols [15, 16]. The "Waterfall: Gozalandia" [10] presents an algorithm (so-called optimistic consensus) for single-slot finality, provided that epoch finalization is achieved in a timely manner. The main goal of this work is to acquire the optimal level of support from blockchain Coordinators accelerating the optimistic finalization of slots. In doing so, we consider Binomial distribution sampling that has been used in various fields to predict positive voting. Such an issue is well-studied in the case of two-alternative voting systems [18]. However, there are some distinctions in our case, since If a Coordinator has not positively voted, then it is considered faulty until it votes during the epoch.

In addition, efficient transaction dissemination is one of the crucial issues for transaction finality and is actively discussed in the blockchain research community. New relay protocols (e.g. [18, 19]) and incentivizing methods (e.g. [20–22]) are constantly emerging to enhance existing solutions and adapt them to new blockchain systems in accordance with their distinctive features. The Waterfall protocol also demands special modifications in the first place because of its DAG architecture, and significant increases in the number of Validators over the platform's evolution.

3   General Platform Design

Waterfall is a decentralized network in which there is a set of independent Shards built on the blockDAG principle. There is also a separate Coordinating network, which is to finalize the sequences of transactions in Shards. A set of coordinated peer-to-peer independent software processes, called Workers, is created to implement such an architecture, consisting of two information-related components (see Fig. 1).

The first component, called the Validator, operates in the DAG network of a specific Shard and is responsible for creating and validating blocks there. The second component, called the Coordinator, works in a common Coordinating network and is responsible for linearization and intershard interactions. From a technical point of view, Workers are deployed on many physical nodes (servers), one or more on each server. Workers running on the same server have a common transaction pool, a common network state, and an archive. This reduces the cost of network deployment. Due to the large number of Workers, this technical solution does not negatively affect the degree of decentralization of the system.

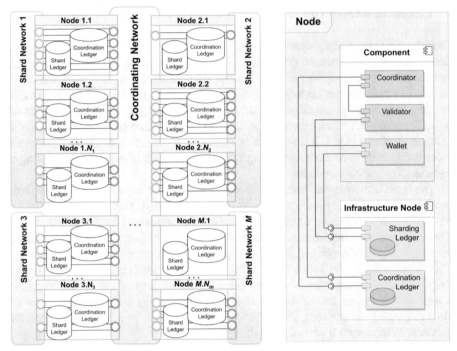

Fig. 1. Coordinating and Shard networks (the left panel) and the structure of a node (the right panel).

During the operation of the Coordination Network, all blockchain Coordinators are divided into 32 disjoint sets, according to the slots in which they work. In each slot, one of the Coordinators has the exclusive right to propose a block containing a list of DAG blocks with transactions to be finalized, and other Coordinators from this slot or subsequent slots can vote in support of this block. For the purposes of this paper, it is enough to know that slot results and, hence, the corresponding set of DAG blocks, are optimistically finalized if they have gained support in k – not necessarily consecutive – slots. In this work, we research a rule according to which a slot will be considered positively voted.

Some Coordinators for some reason may not send their vote for accepting the results of the slot. Such faulty participants make up a certain share f of the total number and are randomly distributed over the epoch slots. While in reality validator failures can occur at any time and for a fairly short period of time, we will only consider failures that occurred at the time of voting. Thus, each Coordinator can validate once per epoch, and the share of such Validators is limited by the value of f. The PoS consensus that provides epoch finalization requires the supermajority rule $f < 1/3$.

4   Probabilistic Slot Optimization

4.1   Slot Supporting Threshold

The number of votes supporting slot results can be considered as a random variable having the Binomial distribution with parameters n Coordinators per slot and success probability $p = 1 - f$. Figure 2 depicts probabilities of positive vote numbers as an example, with $n = 64$ and $p = 2/3$. Let us set the threshold value at which the slot will be considered as supporting the solution, $t = 1/2$. At the same time, the probability that honest (not faulty) Coordinators Y will collect a sufficient number of votes is equal to

$$P(Y > t \cdot n) = 1 - F_p(t \cdot n) = 0.9957, \qquad (1)$$

where F_p is the Binomial cumulative distribution function. This probability will increase as n increases. For example, if the number of Coordinators per slot is doubled $n = 128$, it will exceed 0.9999. On the other hand, the threshold $t = 1/2$ will not allow multiple solutions to be supported in a slot, which would lead to an unresolvable contradiction in the system. However, an increase in the value of t seems to be inappropriate, e.g. the probability of honest voters reaching a 'supermajority' (with $t = 2/3$) is equal to 0.5235, which is on average 16.75 slots per epoch.

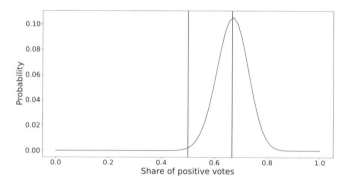

Fig. 2. The Binomial probability density function with $n = 64$ and $p = 2/3$.

From a practical point of view, the probability of reaching the majority by colluding malicious Coordinators within one slot is also of interest. As the most unfavorable case, we can assume that all the faulty participants ($f < 1/3$) are in collusion. Obviously, this is a complementary event to the abovementioned event, if we neglect the case when both groups (faulty and honest) get the same number of votes.

4.2   Simulating and Probabilistic Optimization

The considerations discussed above are applicable to the case when the share of all faulty network Coordinators f remains constant over slots. However, in practice, in the case of non-repetitive sampling (32 disjoint sets), the value of f may vary, although only slightly, for a sufficiently large total number of voters. For testing, a voting model was built under the condition $f = 0.333$, which imitates the work of the Coordinating Network with a different number of n voters per slot for one million epochs. The finding presented in Table 1 is in line with theoretical results.

Table 1. Average values of positively voted slots.

$n =$	64	128	256	2,048	8,192
$t = 1/2$	31.885	31.998	31.999	32.0	32.0
$t = 2/3$	16.932	15.898	16.575	16.356	16.801

Figure 3 illustrates the general case for a fixed number of Coordinators $n = 256$ when the parameters f and t vary over a wide range. For other values of n, the graph shape remains the same, but as the number of voters increases, the jump becomes sharper. The simulation results confirm the expediency of choosing the majority rule with $t = 1/2$.

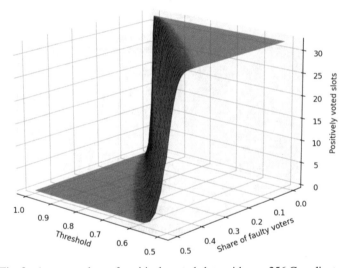

Fig. 3. Average values of positively voted slots with $n = 256$ Coordinators.

Let's assume that all the faulty Coordinators are in collusion and make a concerted effort to finalize a slot that is not going to be supported by the supermajority in the voting of an entire epoch. The attack begins when the leader of the slot is the Coordinator from among the conspirators that publishes its proposal for finalization. To validate this proposal (optimistic finalization), it is necessary that the conspirators are later able to gain control over the needed number of slots k (as a system parameter) faster than honest nodes that will support the "correct" competing proposal.

Here, as above, different numbers of Coordinators per slot acting during one million epochs were simulated. The shares of epochs when the group of conspirators with $f =$

1/3 was able to make a faulty optimistic consensus are presented in Table 2. With other values of n considered in Table 1, there was no faulty optimistic consensus. In all cases, the average numbers of optimistically finalized blocks in the Coordinating network were 20–21 per epoch, primarily due to faulty leaders (conspirators) producing faulty blocks that cannot be accepted by honest Coordinators. Therefore, one can recommend $k = 2$ for the Waterfall platform, since it will allow blocks to quickly reach optimistic finalization with a high confidence value.

Table 2. Shares of epochs with a faulty optimistic consensus.

$n =$	64	128
$k = 1$	0.029812	0.000439
$k = 2$	0.000314	–
$k = 3$	0.000002	–

5 Non-Relayed Transactions

Validators can withhold transactions for a variety of reasons. For example, Validators do not send transactions to others to include them in produced blocks on their own (so-called selfish mining), or Validators may be overwhelmed with incoming transactions (especially in high-throughput blockchain systems), causing them to drop some transactions to prioritize others. Non-relayed transactions can have significant consequences for the overall health of the network. In a supply chain management case, such transactions could result in delayed shipments or lost products; in a healthcare system, they could lead to delayed or inaccurate medical records, potentially endangering patient health, etc. As a whole, the problem of non-relayed transactions can have significant consequences for the security and efficiency of a blockchain network. Therefore, it is crucial for Validators to relay transactions promptly and efficiently, to ensure the integrity of the Shard network.

There are several approaches to prevent or at least to significantly reduce the appearance of non-related transactions in the Waterfall network:

- Due to the DAG structure, transactions are split into disjoint sets that must be processed by corresponding subnetworks without overloading specific nodes, even at peak times [23]. Thus, popular nodes receiving huge numbers of transactions will share with others those transactions that can be published only in other subnetworks.
- In the first stage, all nodes are located on cloud services and malicious owners cannot alter their software. Consequently, all Validators unconditionally follow the protocol [24] which was modified taking into account subnetworks, and transactions are propagated to all network nodes quickly and efficiently. Later, having a more significant number of Validators reduces the risk of non-related transactions caused by selfish mining or other malicious behavior.

- Burning the base transaction fee also reduces the attractiveness of selfish mining, but does not entirely eliminate it, since Validators get to keep tips for transaction publishing within the Waterfall tokenomics model [25].

One important problem with blockchain is the distribution of transactions and the replication of their pool. On some platforms, it is even considered good practice to not relay transactions from clients to other nodes. The transaction is "held" so that when creating a block, the node can include more transactions and receive a larger fee.

In Waterfall tokenomics, there is no significant direct economic interest for nodes to "hold" transactions in this way since fees are burned. Naturally, a node can get tips for hosted transactions, but the amount is not significant enough to justify breaking the protocol.

At the same time, it may be desirable to reduce synchronization traffic by transmitting the pool. Together, these two factors can stimulate changes in the software, which will lead to an increase in the publishing time of transactions received through faulty nodes that violate the protocol.

To eliminate the described problem in the Waterfall network, a mechanism of economic incentives is proposed, to reward compliance with the transaction distribution protocol. To accomplish this, each transaction transmitted from wallet to node is accompanied by an indication of the public key of the node through which the transaction is sent. Each wallet can transmit a transaction through several Entry Nodes at once. At the same time, the nonce of the transaction shows that it is the same transaction. However, different addresses of Entry Nodes make it possible to reward not only the producer of the block but also the Entry Node.

In this case, the reward is distributed as follows: $P = kP_i + (1 - k)P_b$, where P is the reward for publishing the transaction, P_i is the share of the reward of the block creator, P_b is the share of the reward of the Entry Node as the transaction provider. Currently $k = 0.5$, but the value of the distribution coefficient will be changed based on the results of the test network.

In addition to incentivizing compliance with the transaction distribution protocol, the described technical moment creates an additional point of responsibility if the receiving node signs the received transactions. The node that receives the transaction from the wallet checks the correctness of its address and, in case of an error and deliberate distortion, rejects the transaction.

In addition, a Validator's monitoring tool could be added to enhance the current transaction dissimilation protocol. It will detect transactions pending for a long time, and well-behaving Validators will primarily include them in produced blocks.

6   Conclusion

Using methods of mathematical and statistical analysis, and after conducting a series of simulation experiments, we managed to reduce the transaction finalization time significantly and give early probabilistic estimates of the optimistic finalization of transaction blocks in Shards. The proposed solutions make it possible to obtain the optimal level of decision support in the Coordinating Network. The problem of speed of transaction distribution and the presence of non-retransmitted transactions was mitigated by methods of economic incentives. Further work is aimed at designing a network monitoring tool for accurately estimating the likelihood of positive voting based on the results of the epoch, which can inform decision-making without additional assumptions on the value of f.

Currently, work is underway to implement the solutions obtained in the Waterfall platform software. The main elements of the protocol have been implemented, and load experiments are being conducted. Testing was carried out on t3.small and t3.medium AWS Servers with 2 cores CPU and 2 or 4 GB RAM respectively. According to preliminary data, this significantly increased the reliability, efficiency, and safety of decisions.

References

1. Sherman, A.T., Javani, F., Zhang, H., Golaszewski, E.: On the origins and variations of blockchain technologies. IEEE Secur. Priv. **17**(1), 72–77 (2019)
2. Bhutta, M.N.M., et al.: A survey on blockchain technology: evolution architecture and security. IEEE Access **9**, 61048–61073 (2021)
3. Anceaume, E., Pozzo, A., Rieutord, T., Tucci-Piergiovanni, S.: On finality in blockchains. arXiv:2012.10172v1 (2020). https://arxiv.org/pdf/2012.10172.pdf
4. Nakamoto, S.: Bitcoin: a peer-to-peer electronic cash system (2009)
5. Bitcoin Wiki: Confirmation. https://en.bitcoin.it/wiki/Confirmation. Accessed 9 Mar 2023
6. Ethereum.org: Proof-of-stake (PoS) (2023). https://ethereum.org/en/developers/docs/consensus-mechanisms/pos. Accessed 9 Mar 2023
7. Werner, S.M., Perez, D., Gudgeon, L., Klages-Mundt, A., Harz, D., Knottenbelt, W.J.: SoK: decentralized finance (DeFi). arXiv preprint arXiv:2101.08778 (2021)
8. Waterfall: a Highly Scalable EVM-based Smart Contract Platform. https://waterfall.foundation. Accessed 9 Mar 2023
9. Grybniak, S., Dmytryshyn, D., Leonchyk, Y., Mazurok, I., Nashyvan, O., Shanin, R.: Waterfall: a scalable distributed ledger technology. In: IEEE 1st GET Blockchain Forum, California, United States (2022). In press
10. Grybniak, S.S., Leonchyk, Y.Y., Mazurok, I.Y., Nashyvan, O.S., Shanin, R.V.: Waterfall: Gozalandia. Distributed protocol with fast finality and proven safety and liveness. IET Blockchain 1–12, 465–472 (2023)
11. Lau, K.: Ethereum 2.0. An Introduction (2020)
12. Schwarz-Schilling, C., Neu, J., Monnot, B., Asgaonkar, A., Tas, E.N., Tse, D.: Three attacks on proof-of-stake ethereum. In: Eyal, I., Garay, J. (eds.) Financial Cryptography and Data Security (FC 2022). LNCS, vol. 13411, pp. 560–576. Springer, Cham (2022). https://doi.org/10.1007/978-3-031-18283-9_28
13. D'Amato, F., Neu, J., Tas, E.N., Tse, D.: No more attacks on proof-of-stake ethereum? arXiv preprint arXiv:2209.03255 (2022)
14. Ethereum.org: Ethereum PoS Attack and Defense (2022). https://ethereum.org/da/developers/docs/consensus-mechanisms/pos/attack-and-defense. Accessed 9 Mar 2023
15. Buterin, V.: Paths toward single-slot finality. https://notes.ethereum.org/@vbuterin/single_slot_finality#Paths-toward-single-slot-finality . Accessed 9 Mar 2023
16. D'Amato, F., Zanolini, L.: A simple single slot finality protocol for ethereum. arXiv preprint arXiv:2302.12745 (2023)
17. Mayfield, P.: Understanding binomial confidence intervals (1999). http://1989-6580.el-alt.com/binomial_confidence_interval.htm. Accessed 9 Mar 2023

18. Han, Y., Li, C., Li, P., Wu, M., Zhou, D., Long, F.: Shrec: bandwidth-efficient transaction relay in high-throughput blockchain systems. In: Proceedings of the 11th ACM Symposium on Cloud Computing, pp. 238–252 (2020)
19. Naumenko, G., Maxwell, G., Wuille, P., Fedorova, A., Beschastnikh, I.: Erlay: efficient transaction relay for bitcoin. In: Proceedings of the 2019 ACM SIGSAC Conference on Computer and Communications Security, pp. 817–831 (2019)
20. Wang, X., Chen, Y., Zhang, Q.: Incentivizing cooperative relay in UTXO-based blockchain network. Comput. Netw. **185**, 107631 (2021)
21. Zhang, J., Huang, Y.: TF: A blockchain system with incentivized transaction forwarding. In: IEEE 42nd International Conference on Distributed Computing Systems (ICDCS), Bologna, Italy, pp. 213–223 (2022)
22. Babaioff, M., Dobzinski, S., Oren, S., Zohar, A.: On bitcoin and red balloons. In: The 13th ACM Conference on Electronic Commerce, pp. 56–73 (2012)
23. Antonenko, O., Grybniak, S., Guzey, D., Nashyvan, O., Shanin, R: Subnetworks in BlockDAG. In: IEEE 1st GET Blockchain Forum, California, United States (2022)
24. Ethereum.org: Ethereum Wire Protocol (2022). https://github.com/ethereum/devp2p/blob/master/caps/eth.md. Accessed 9 Mar 2023
25. Grybniak, S., Leonchyk, Y., Masalskyi, R., Mazurok, I., Nashyvan, O.: Waterfall: Salto Collazo. Tokenomics. In: 2022 IEEE International Conference on Blockchain, Smart Healthcare and Emerging Technologies (SmartBlock4Health), Bucharest, Romania, pp. 1–6 (2022)

The Role of Blockchain Technology in Ensuring Security and Immutability of Open Data in Healthcare

Tiago Guimarães$^{(\boxtimes)}$ ⓘ, Ricardo Duarte ⓘ, João Cunha ⓘ, Pedro Silva ⓘ, and Manuel Filipe Santos ⓘ

Algoritmi Reasarch Center, School of Engineering, University of Minho, Azurém Campus, 4800-05 Guimarães, Portugal
tsg@dsi.uminho.pt

Abstract. Clinical information is highly confidential due to its sensitive nature. Implementing health information systems has raised concerns regarding interoperability, privacy, and security. The storage and retrieval of this information also present the same problems. Therefore, any effort to introduce healthcare information systems must ensure patient data's safety, privacy, integrity, and immutability. Blockchain technology and the openEHR open data model have emerged to address these concerns, providing a solution that guarantees data security, interoperability between systems, and the accuracy of stored data queries. Two different architectures were developed and subjected to several performance tests to enhance security and immutability in open data models implemented in healthcare institutions. The results were analysed to determine which architecture provides more value to a healthcare institution. Subsequently, a discussion was held to draw appropriate conclusions.

Keywords: Blockchain · Blockchain in Healthcare · OpenEHR · Benchmarking

1 Introduction

The health sector is a sector that has unique requirements, and trusting the data from its activities is essential for its operations [1, 2]. With the growing amount of data generated, some problems arise, including unauthorised sharing, invasions, and theft of confidential data. These propensities lead to people's suspicions and doubts about the trust veracity of these institutions [2, 3]. It is essential to consider alternative approaches, like blockchain technology, to address these issues. Given its nature and characteristics, it offers a solution to the needs demanded by the sector [4, 5]. Allied with this technology comes the open data structure, openEHR. It enables reliable structuring, management, storage, and patient data integration across healthcare organisations. The main idea is to standardise concepts related to health used in databases or Electronic Health Record (EHR) systems in a set of libraries called archetypes [6].

© The Author(s), under exclusive license to Springer Nature Switzerland AG 2023
J. M. Machado et al. (Eds.): BLOCKCHAIN 2023, LNNS 778, pp. 317–327, 2023.
https://doi.org/10.1007/978-3-031-45155-3_32

The present work is divided into several sections, starting with a brief introduction and then a literature review about OpenEHR, blockchain and blockchain in healthcare. Section five is discussed why to use openEHR with blockchain. The following section presents the developed architectures, which will serve as the basis for the tests performed in section seven. Finally, section eight is presented a discussion of the results obtained and the conclusions, as well as future work tracing.

2   What is OpenEHR?

OpenEHR is an open-source, vendor-neutral standard for storing, retrieving, and exchanging electronic health records (EHRs). It was first developed in 2000 by a consortium of international health informatics experts to create a shared model for EHR systems that could be used across different healthcare settings and countries. The OpenEHR approach is based on archetypes and flexible and reusable templates for capturing clinical information. These archetypes can be combined and modified to create specific EHR instances that meet the needs of different healthcare organisations and individual patients [7].

OpenEHR is designed to overcome the limitations of traditional EHR systems, which are often siloed, proprietary, and difficult to customise. Using a standardised model for clinical data, OpenEHR enables interoperability between different EHR systems and between EHRs and other health information systems, such as decision support tools and population health management platforms. OpenEHR also supports open APIs and web services, allowing seamless integration with other healthcare applications and data sources [8–10].

3   Blockchain

Blockchain technology is a distributed and secure database that enables the recording of transactions in a reliable and immutable manner. Each transaction is validated by a network of peers, making data manipulation or falsification harder [11, 12].

According to Swan (2015) [11], blockchain can be defined as a "distributed digital ledger that uses cryptography to maintain secure and verifiable records." The ledger is composed of blocks, which contain information about recent transactions. Each block is connected to the previous block and is verified by a distributed network of computers.

According to Tapscott and Tapscott (2016) [12], blockchain is "a new technological platform that can help improve the way we record and manage data." Through blockchain, data exchange can be done without intermediaries or trusted third parties, allowing for a more secure and efficient transfer of value.

Blockchain is primarily known as the technology behind cryptocurrencies, but its potential for use extends far beyond that, including data management, supply chains, and electronic voting [11, 12].

4   Blockchain in the Healthcare Sector

Blockchain is an emerging technology that has gained prominence in various fields, including healthcare. It is a promising technology that can transform healthcare by providing a secure and efficient platform to store and share confidential medical information [13, 14]. According to de Hasselgren et al. (2019) [2], blockchain can provide an immutable and transparent record of health information, making sharing medical information more straightforward and secure.

Another significant benefit of using blockchain technology in the healthcare industry is data security. The blockchain can store medical data in an encrypted and decentralised manner, reducing the risk of data breaches and ensuring patient privacy. Furthermore, blockchain can help improve the efficiency of exchanging medical information between different healthcare providers and reduce medical errors [15, 16].

According to some authors, blockchain can help improve the transparency and efficiency of clinical trials and ensure the authenticity of collected data. Furthermore, blockchain can create immutable records of genomes and health data, which can be used to develop personalised treatments based on each patient's genetic characteristics [17, 18]. In addition to the above, several ongoing initiatives to use blockchain in healthcare include blockchain-based electronic medical record systems, health data sharing platforms, blockchain-based patient identity management systems, and drug tracking [19, 20].

Blockchain technology can potentially transform the healthcare industry, but its adoption still faces many challenges. It is necessary to work with stakeholders to overcome these obstacles and ensure that blockchain technology is safe, efficient, and properly regulated for use in healthcare [13, 21].

5   Why OpenEHR with Blockchain in Healthcare?

OpenEHR is a standard for electronic health record (EHR) systems that allows for creation of interoperable and vendor-neutral health data. Blockchain technology can enhance the security, privacy, and interoperability of OpenEHR by providing a decentralised and tamper-resistant platform for storing and sharing health data. By combining OpenEHR with blockchain, healthcare providers can create a secure and transparent environment for managing patient health data, improving patient outcomes, and reducing costs. Furthermore, blockchain technology can also support the creation of a patient-centric healthcare system, where patients have control over their health data and can choose who can access it [7, 13, 22].

In order to understand the Strengths, Weaknesses, Opportunities, and Threats (SWOT) in the combined use of openEHR with blockchain in healthcare, a SWOT analysis was conducted based on the publication of various authors.

The **strengths** highlighted by the authors are that OpenEHR provides a structured data model based on standards that can improve interoperability and data integration in healthcare [23]. Blockchain can enhance the security and privacy of healthcare data, ensuring that only authorised individuals can access sensitive information. The combination of OpenEHR with blockchain can provide an open, decentralised, and trustworthy solution for managing health data [22], and the combination of OpenEHR with blockchain technology can enhance the security of electronic health records, protecting patient data from breaches and unauthorised access [9].

The **weakness**es highlighted by the authors are that implementing OpenEHR and blockchain in healthcare may be expensive and require a high investment in information technology (IT) infrastructure and resources [24]. Implementing OpenEHR requires a certain level of technical expertise and may be complex for end users [7]. The use of blockchain has yet to be widely adopted in the healthcare industry, still facing significant challenges for its implementation, which may limit its effectiveness in certain areas [25].

The **opportunities** highlighted by the authors include the integration of OpenEHR with blockchain can provide a technical foundation for decentralised healthcare applications, such as digital health wallets and health tracking apps [19, 20]. Blockchain-based solutions can increase public trust in digital health systems and promote wider adoption of digital health technologies [2]. The growing demand for health data security and privacy protection may drive the adoption of more advanced technologies, such as OpenEHR with blockchain [26].

The **threats** highlighted by the authors include that data privacy may be compromised if blockchain technology is not implemented correctly or privacy policies are insufficient [25]. Resistance to change by healthcare professionals and patients may limit the adoption of digital health technologies, including solutions based on blockchain and openEHR [18], and the adoption of emerging technologies such as OpenEHR with blockchain may face regulatory and compliance challenges [25].

Combining OpenEHR with blockchain technology can offer numerous benefits to the healthcare industry, including the security of electronic medical records, the protection of patient data against violations and unauthorised access, patient control over their health data, and improvements in clinical research and population health. While adopting this technology may require a sophisticated IT infrastructure and changes in organisational culture, the growing demand for health data security and privacy protection can drive its implementation in the healthcare industry [9, 27].

6 Architecture

By integrating Blockchain technology with the open data model, openEHR, it is possible to ensure data interoperability and standardisation of all EHRs while maintaining data integrity, privacy, and security. In our case study, two scenarios were considered: Scenario 1, as represented in Fig. 1, and Scenario 2, as represented in Fig. 2.

Analysing the flow of Scenario 1, as depicted in the following figure, begins with the insertion of information and data about patients. These inputs are then transformed into EHRs and sequentially processed and analysed based on the specifications modelled in the open data model, openEHR, at the APIS layer through the gateway. Hyperledger Caliper was utilised to perform a series of insertions in the blockchain. Subsequently, the blockchain stores all the transactions in a secure, immutable, and private manner for future consultation. Finally, Hyperledger Caliper evaluates the transactions in terms of success, speed, the maximum, minimum, and average time to send and receive a response, and the average number of transactions processed per second. These metrics are presented on an HTML page.

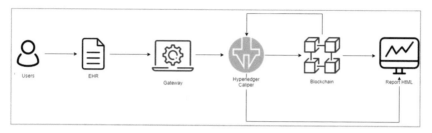

Fig. 1. First Scenario Architecture

Initiating the analysis of Scenario 2, as illustrated in the figure below, the flow begins similarly to the scenario described above. It commences with inserting clinical information for patients into the systems used by the institution. These inputs are then transformed into EHRs and sequentially processed and analysed based on the specifications modelled in the open data model, openEHR, at the APIS layer through the gateway. Continuing with the flow, a fork is observed. Following the upper arrow in the flow, the data is stored in the hospital's database, where it can be further processed and transformed as per the needs of the hospital's stakeholders.

On the other hand, the lower path involves creating a hash block for each object through the encoding process. An object ID is associated with the block to identify the person to whom the data pertains. The main objective of this technique is to validate if there has been any intrusion or alteration to the data, thus providing enhanced security. If any changes are made to the data in the object, the md5 will be updated, and the updated block will be stored on the blockchain. Proceeding with the flow, Hyperledger Caliper triggers the entries of the hashed objects into the blockchain. However, in this case, the constitution of the object changes, and it becomes just an ID and the hash. Next, the blockchain stores the transactions and plays a crucial role in verifying whether the recorded hash matches the updated hash. Finally, Hyperledger Caliper utilises metrics to measure performance, which are presented on an HTML page.

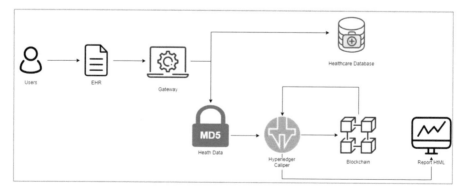

Fig. 2. Second Scenario Architecture

7   Results and Discussion

In this chapter, a visual demonstration of the network's performance obtained in the various stress tests will be carried out. Thus, it will be possible to measure and understand whether the objective of increasing the security and immutability of implementing open data models in a hospital environment has been achieved. This income statement performed two tests, one for the first and the other for the second. At each test, two types of graphs are shown. One evaluates the performance of the network in the gradual submission of people. The other considers the amount of memory that the network uses in total.

7.1   Insertion of Data into the Blockchain with 20000 Records, Scenario 1 and Scenario 2

Four insertions of 20000 people were performed on the blockchain for this test. Initially, the container was restarted. As the insertions were performed, the processing capacity of the network decreased, gradually increasing the average latency. This phenomenon is explained by the increasing amount of data stored in the blockchain. It is noteworthy that the processing speed was 5 TPS. Adding this to the physical capacity of the machine used for testing, the processing time increased considerably. A positive aspect that goes against speed stability corresponds to the absence of failures.

   In this test, it is possible to see a growing increase in processing time and memory usage in both scenarios. As mentioned, the amount of data entered is twenty thousand in each of the four iterations, thus pushing the blockchain's capacity to the limit. As the iterations were carried out, the network performance naturally decreased, which caused an increase in processing time. The same occurred with memory usage, where it is possible to conclude that as the volume of data increases, the memory used by the system also increases, which will cause a slower system.

   Upon analysing the two scenarios, as evidenced in Table 1, it is possible to deduce that Scenario 2 stands out with shorter processing times and memory usage in each iteration performed. Moreover, this scenario adds a layer of data security and better aligns with the internal requirements of the hospital. Considering the minimal impact on the existing implementation in the hospital setting, Scenario 2 is identified as the optimal choice. It is less resource-intensive, provides the desired benefits, and avoids disruptions. Further tests were conducted for this scenario, which will be presented in the subsequent chapters.

Table 1. Analyse between two Scenarios

		1st Interaction	2nd Interaction	3rd Interaction	4th Interaction
Process Time	Test – Scenario 1	583,74	688,3	768,45	1035,24
	Test – Scenario 2	60,33	312,37	442,89	763,4
Memory Usage	Test – Scenario 1	1679,55	2025,88	2231,38	2566,96
	Test – Scenario 2	529,39	752,26	1289,7	1631,93

7.2 Additional Test for Scenario 2

For the second test of scenario 2, a different configuration was performed, where the network defined the send rate automatically. Initially, the containers were cleared and then 1000, 5000, 10000, and 20000 people were inserted into the network. With the change made, a considerable variation in latency was observed compared to the previous tests. The achieved latency times show an improved, constant, and consistent speed. This occurs because the change allowed the system to choose the rate it could support.

Additionally, it is possible to observe that the sending speed decreases with each insertion. This evolution is expected as the network's capacity decreases as more data is inserted. It is possible to observe that memory increases progressively. There is more significant variation in memory because each iteration involves a different number of insertions, which justifies that memory is inconsistent and growing.

For the third test of scenario 2, the same configuration as test 2 was performed, where the network defined the send rate automatically. Initially, the containers were cleared, and then four insertions of 20000 people were performed each time in the network. With the changes made, considerable alterations were observed regarding latency times. The latency times increased and presented an improved, constant, and consistent speed because the change allowed the system to choose the rate it could support. Additionally, it is possible to observe that the sending speed decreased as the 20000 people were inserted. This happens because the network's capacity decreases with the continuous insertion of data, causing an overload of the network.

Regarding memory, it is possible to observe it increase progressively. In this case, memory increased consistently because the number of insertions was constant, meaning that the amount of processing by peers and orderers was practically the same, and the only thing that increased was the size of the databases. This increase is considered regular, constant, and growing because the number of records also increases. The results for both test it is visible in the following Table 2.

Table 2. Additional tests for Scenario 2

		1st Interaction	2nd Interaction	3rd Interaction	4th Interaction
Insertions	Scenario 2 – Test 2	1000	5000	10000	20000
	Scenario 2 – Test 3	20000	20000	20000	20000
Process Time	Scenario 2 – Test 2	3,87	3,9	4,7	4,79
	Scenario 2 – Test 3	5,16	5,4	5,14	6,38
TPS	Scenario 2 – Test 2	17,6	17,5	14,6	14,5
	Scenario 2 – Test 3	13,4	12,9	13,6	11,3
Memory Usage	Scenario 2 – Test 2	689,25	997	1139,45	1314,39
	Scenario 2 – Test 3	1566,44	1909,07	2267,47	2323,48

8   Discussion and Conclusion

Every solution requires a thorough and truthful evaluation, and engaging in an honest discussion about its efficacy is vital. In order to do so, it is helpful to utilise technology assessment methodologies like a SWOT analysis.

A SWOT analysis is a tool commonly used by organisations for managing their strategies and objectives. This tool assesses internal and external factors affecting the organisation, with SW factors being internal and OT factors being external attributes. By performing this analysis, one can determine the feasibility of implementing a given solution. The SWOT analysis yields qualitative and quantitative metrics, which can serve as indicators of the technology acceptance model as perceived by practitioners.

Strengths:

- Structured clinical data.
- Interoperability between all systems.
- Availability of data for better development of Business Intelligence and decision support systems.
- Reduction of obsolete data and poorly inserted records without clinical value and clean information analysis.

Weakness:

- Scalability and storage and processing capacity.
- Application with a real-time data update.
- Dynamics of OpenEHR structures and their versioning.

- Efficiency regarding processing time and resources.
- Relationship between OpenEHR structures and Multidimensional structures that support analytical processes considering analysis axes that cut across several patients or certain variables/characteristics of the same patient.

Opportunities:

- Standardisation of clinical records on a large scale.
- Training and integration of clinical modelling of these structures in training healthcare professionals and/or higher healthcare courses.
- Centralised data for analysis.

Threats:

- User resistance to adopting a new system by healthcare professionals.
- Negative aspects of the environment with the potential to compromise the proposed solution.
- External development of more efficient solutions.
- The emergence of a new standard with better conditions.

Integrating these technologies and methodologies in the health area is important for evolution in terms of speed of patient care and quality of service, without forgetting the importance of interoperability that OpenEHR provides for better communication between services and, consequently, BI integration.

OpenEHR models provide a disruptive way of storing data within healthcare and are primarily focused on standardising and managing clinical data, including capturing and storing structured and unstructured health information.

It is not explicitly designed for data analysis or business intelligence purposes, so current BI solutions must adapt and adjust how data is consulted efficiently.

According to my research, this integration between OpenEHR and BI has yet to be adequately investigated, so a gap in this area can be used for research. Future research work will be the realisation of artefacts to adapt a BI system in a generic way to the OpenEHR model and study the creation of a layer of Extraction Transformation and Load (ETL) that efficiently analyse data from the models.

Acknowledgements. This work has been supported by FCT – Fundação para a Ciência e Tecnologia within the R&D Units Project Scope: UIDB/00319/2020.

References

1. Onik, M.M.H., Aich, S., Yang, J., Kim, C.-S., Kim, H.-C.: Blockchain in Healthcare: Challenges and Solutions. Elsevier Inc. (2019). https://doi.org/10.1016/b978-0-12-818146-1.000 08-8
2. Hasselgren, A., Kralevska, K., Gligoroski, D., Pedersen, S.A., Faxvaag, A.: Blockchain in healthcare and health sciences—a scoping review. Int. J. Med. Inform. **134**, 104040 (2019). https://doi.org/10.1016/j.ijmedinf.2019.104040
3. Gamal, A., Barakat, S., Rezk, A.: Standardised electronic health record data modeling and persistence: a comparative review. J. Biomed. Inform. **114**, 103670 (2021). https://doi.org/10.1016/j.jbi.2020.103670

4. Hölbl, M., Kompara, M., Kamišalić, A., Zlatolas, L.N.: A systematic review of the use of blockchain in healthcare. Symmetry (Basel) **10**(10) (2018). https://doi.org/10.3390/sym101 00470
5. Prokofieva, M., Miah, S.J.: Blockchain in healthcare. Australas. J. Inf. Syst. **23**, 1–22 (2019). https://doi.org/10.3127/ajis.v23i0.2203
6. Ribeiro, T., Oliveira, S., Portela, C., Santos, M.: Clinical workflows based on OpenEHR using BPM. In: Proceedings of the 5th International Conference on Information Communication Technologies Ageing Well e-Health, ICT4AWE 2019, no. Ict4awe, pp. 352–358 (2019). https://doi.org/10.5220/0007878203520358
7. What is openEHR? https://www.openehr.org/about/what_is_openehr
8. Beale, T., Heard, S.: Architecture Overview, OpenEHR. Foundation (2007)
9. Cunha, J., Duarte, R., Guimarães, T., Santos, M.F.: Permissioned blockchain approach using open data in healthcare. Procedia Comput. Sci. **210**, 242–247 (2022)
10. Wulff, A., Haarbrandt, B., Tute, E., Marschollek, M., Beerbaum, P., Jack, T.: An interoperable clinical decision-support system for early detection of SIRS in pediatric intensive care using openEHR. Artif. Intell. Med. **89**, 10–23 (2018)
11. Swan, M.: Blockchain: Blueprint for a New Economy. O'Reilly Media, Inc. (2015)
12. Tapscott, D., Tapscott, A.: Blockchain Revolution: How the Technology Behind Bitcoin is Changing Money, Business, and the World. Penguin (2016)
13. Kuo, T.-T., Kim, H.-E., Ohno-Machado, L.: Blockchain distributed ledger technologies for biomedical and health care applications. J. Am. Med. Inform. Assoc. **24**(6), 1211–1220 (2017)
14. Mettler, M.: Blockchain technology in healthcare: the revolution starts here. In: 2016 IEEE 18th International Conference on e-Health Networking, Applications and Services (Healthcom), pp. 1–3 (2016)
15. Zhang, P., Schmidt, D.C., White, J., Lenz, G.: Blockchain technology use cases in healthcare. In: Advances in Computers, vol. 111, pp. 1–41 (2018)
16. Drescher, D.: Blockchain Basics: A Non-technical Introduction in 25 Steps, p. 255. Apress (2017)
17. Cunha, J., Duarte, R., Guimarães, T., Quintas, C., Santos, M.F.: Blockchain analytics in healthcare: an Overview. Procedia Comput. Sci. **201**, 708–713 (2022). https://doi.org/10. 1016/j.procs.2022.03.095
18. Kumar, T., Ramani, V., Ahmad, I., Braeken, A., Harjula, E., Ylianttila, M.: Blockchain utilisation in healthcare: key requirements and challenges. In: 2018 IEEE 20th International Conference on e-Health Networking, Applications and Services (Healthcom), pp. 1–7 (2018). https://doi.org/10.1109/HealthCom.2018.8531136
19. Ben Fekih, R., Lahami, M.: Application of blockchain technology in healthcare: a comprehensive study. In: Jmaiel, M., Mokhtari, M., Abdulrazak, B., Aloulou, H., Kallel, S. (eds.) ICOST 2020. LNCS, vol. 12157, pp. 268–276. Springer, Cham (2020). https://doi.org/10. 1007/978-3-030-51517-1_23
20. Khezr, S., Moniruzzaman, M., Yassine, A., Benlamri, R.: Blockchain technology in healthcare: a comprehensive review and directions for future research. Appl. Sci. **9**(9), 1736 (2019)
21. Agbo, C.C., Mahmoud, Q.H., Eklund, J.M.: Blockchain technology in healthcare: a systematic review. Healthcare **7**(2) (2019). https://doi.org/10.3390/healthcare7020056
22. Azaria, A., Ekblaw, A., Vieira, T., Lippman, A.: Medrec: using blockchain for medical data access and permission management. In: 2016 2nd International Conference on Open and Big Data (OBD), pp. 25–30 (2016)
23. Kalra, D., Beale, T., Heard, S.: The openEHR foundation. Stud. Health Technol. Inform. **115**, 153–173 (2005)
24. Benchoufi, M., Porcher, R., Ravaud, P.: Blockchain protocols in clinical trials: transparency and traceability of consent. F1000Research, vol. 6, p. 66 (2018)

25. Soltanisehat, L., Alizadeh, R., Hao, H., Choo, K.-K.R.: Technical, temporal, and spatial research challenges and opportunities in blockchain-based healthcare: a systematic literature review. IEEE Trans. Eng. Manag. (2020)
26. Angraal, S., Krumholz, H.M., Schulz, W.L.: Blockchain technology: applications in health care. Circ. Cardiovasc. Qual. Outcomes **10**(9), 1–4 (2017). https://doi.org/10.1161/CIRCOU TCOMES.117.003800
27. Dagher, G.G., Mohler, J., Milojkovic, M., Marella, P.B.: Ancile: privacy-preserving framework for access control and interoperability of electronic health records using blockchain technology. Sustain. Cities Soc. **39**, 283–297 (2018)

Managing Write Access Without Token Fees in Leaderless DAG-Based Ledgers

Darcy Camargo, Luigi Vigneri[✉], and Andrew Cullen

IOTA Foundation, Pappelallee 78/79, 10437 Berlin, Germany
{darcy.camargo,luigi.vigneri,andrew.cullen}@iota.org

Abstract. A significant portion of research on distributed ledgers has focused on circumventing the limitations of leader-based blockchains mainly in terms of scalability, decentralization and power consumption. Leaderless architectures based on directed acyclic graphs (DAGs) avoid many of these limitations altogether, but their increased flexibility and performance comes at the cost of increased design complexity, so their potential has remained largely unexplored. Management of write access to these ledgers presents a major challenge because ledger updates may be made in parallel, hence transactions cannot simply be serialised and prioritised according to token fees paid to validators. In this work, we propose an access control scheme for leaderless DAG ledgers based on access credits, a quantity generated by tokens to guarantee access, instead of paying fees in the base token. We outline a general model for this new approach and provide simulation results highlighting the performance of this solution in terms of credits consumed and delay guarantees according to varying traffic conditions and heterogeneous user behavior.

Keywords: Leaderless distributed ledgers · Dual-token economy · Priority-based write access · DAG-based ledgers

1 Introduction

Blockchains have sparked a revolution in the way information is shared in a trustless way. One of the main criticisms, though, is still related to performance: Bitcoin and Ethereum, the two most relevant projects by market cap as of 2023, are only able to process a few transactions per second [3], creating competition between users to obtain writing permission to the blockchain. Such a limited writing space is shared through auction-like mechanisms to discriminate which transactions deserve to be added to the ledger. As transactions compete for the limited writing available, often this system leads to large fees [11].

Various attempts have tried to make DLT projects more scalable, notably lightning networks, sharding and Layer 2 solutions [13]. Furthermore, more recently, there has been an increasing interest in directed acyclic graph (DAG) ledgers, which generalize the original chain structure introduced by the blockchain: in fact, when blocks are created at a high rate compared to their

J. M. Machado et al. (Eds.): BLOCKCHAIN 2023, LNNS 778, pp. 328–337, 2023.
https://doi.org/10.1007/978-3-031-45155-3_33

propagation time, many competing (or even conflicting) blocks are created leading to frequent bifurcations; DAG-based approaches allow to include blocks not only to the main chain, but also to these bifurcations using additional references [10, 12]. Since transactions can be written and processed in a parallel way, i.e., no total ordering artificially enforcing a pause between subsequent blocks, DAG-based ledgers promise improved throughput and scalability. A number of DAG-based distributed ledger technologies (DLTs) already provide strong performance for consensus and communications layers, such as Honeybadger [9], Hashgraph [2], Aleph [6] and more recently Narwhal [5].

Standard blockchains and leader-based DLTs are built on the dichotomy between the *user* that wants to issue transactions or other state-changing data and a *block issuer* (leader) responsible for creating the blocks that will actually include these data in the ledger. This standard model couples the consensus and access elements of the protocol in the block issuance, creating competition among block issuers to provide ledger access as a service to the base users. In these protocols users propose fees for their data and the block issuers select which data to include to maximise their profits. Such tokenomics schemes are known for being effective but carry a variety of drawbacks: exclusion of low-scale operations, value extraction from users, fee-bidding wars [11], market manipulation, unpredictable pricing and uncertainty of inclusion, to name but a few [4]. Some DLT protocols have attempted to develop zero-fee systems to varied degrees of success [1, 7, 10]. Among these projects, the DAG-based protocols have shown more promise due to the option of decoupling access and consensus rights, like in Prism [1] or in the IOTA Tangle [10].

In this work we propose a novel scheme for managing write access to leaderless DAG-based DLTs through *Access Credit*, a quantity that is passively generated based on tokens held and contributions to the protocol (e.g., being a validator). This Access Credit can be *consumed* to create new blocks, buy name services or interact with smart contracts. The key advantages of the proposed access control scheme are as follows.

- **Zero token fees**: we require Access Credit, instead of tokens, to create new blocks, whose cost is proportional to its computation and storage requirements as well as the global demand for write access.
- **Leaderless**: contrary to most existing access control schemes, our proposed solution does not rely on leaders or rounds; this greatly improves resilience against censorship and value extraction by powerful block creators.
- **Parallel ledger updates**: with a leaderless DAG-based ledger we enable parallel execution and writing, as blocks can reference multiple past blocks concurrently.

Furthermore, we validate our approach through Python simulations that show the effectiveness of our solution: as we will see, the parallel execution facilitated by the DAG ledger introduces additional complexity to keep ledger consistency across all network participants. We highlight to the reader that we present our proposal in a general manner, that is we do not refer to any existing solution such that the principles described in this paper can be applied to any leaderless DAG-based DLT, and as such, the analysis may lack implementation-specific details.

The rest of the paper is organized as follows: the system model is introduced in Sect. 2; then, in Sect. 3, we present our access control policy; after that, Sect. 4 presents our Python simulations in both single- and multi-node environment. Finally, we conclude our paper in Sect. 5.

2   System Model

Actors. We categorize the actors of a leaderless DLT as follows.

- **Accounts:** actors capable of holding tokens and Access Credit. As such, accounts are capable of gaining write access to the ledger by creating blocks. Please note that an account-based DLT is not necessarily mutually exclusive with the UTXO model; in fact, an account can be thought as an identity registered in the ledger to whom one of more UTXOs are associated.
- **Nodes:** the physical machines able to peer with each other to keep local versions of the ledger up-to-date through block processing and forwarding.

Remark: it is important to note that accounts being block creators do not necessarily act as *validators* as in blockchains, where block creators gather transactions from a shared mempool. In fact, such a shared mempool is not possible in a DAG-based ledger because blocks are written to the ledger in parallel. Our work focuses on management of write access for accounts, so although these accounts are block issuers, we assume that their motive is to write to their own data to the ledger rather than considering them as intermediaries for base users.

Blocks. A block is the fundamental data structure of DLTs carrying value transactions, data, smart contract executions, or any other information that may alter the ledger state. Blocks must also include a cryptographic relation with past blocks, the issuer's signature and fields to manage the consensus protocol (e.g., timestamps). In blockchain technologies, the cryptographic relation is the hash of the block that the issuer believes to be the last included in the ledger. On the other hand, DAG technologies have more malleability: in fact, as multiple blocks may be referenced, the simple act of issuing a block can be used as a statement about trust in numerous blocks. In this paper we define an associated "cost" per block, which we call *work*. Work is measured by a protocol gadget as a part of the node software, and it represents the computational load on the node while processing the block and applying the necessary state changes as well as the resource consumption in terms of bandwidth and storage. As an example, size (in bytes) of the block is one components of the work calculation.

Access Credit. The ledger keeps track of computing and storing both Access Credit and token balances. Access Credit is used to gain write access to the ledger. We refer to the amount of spent Access Credits as the *credit consumed*.

This quantity needs to be part of the block and it must be signed by the associated account so that the value cannot be altered. The credit consumed is then used to determine the priority of the block when there is competition to gain write access, as we shall explain in the following section describing our access control. Credits are generated when tokens are moved to a new address through blocks, smart contracts or other means. The amount generated follows the amount of tokens and the time spent in such an address[1]:

$$\text{AccessCredit} = \text{TokensMoved} \times \text{TimeHeld} - \text{CreditConsumed}. \qquad (1)$$

When the value in Eq. (1) is positive, there is a surplus of credits that will be given to a declared account, in an act we call *allotting credits*. Each protocol has its time mechanisms, and the only requirement in the term *TimeHeld* from Eq. (1) is that it is objective so each node agree on the credits held by each account. This property is trivial for UTXO-based ledgers, but can also be induced in any other protocol by using appropriate timestamping. To create sustainable economic incentives, we expect Access Credits to have an active market, where users can sell their spare access boosting network adoption and usage.

3   Credit-Based Access Control

Accounts are managed through wallets or light nodes to create blocks. Since accounts do not receive nor process the blocks produced in the network, they ignore the current congestion level and are enable to properly set the amount of Access Credit to consume. For this reason, they must either set up their own nodes or use third-party free or paid services to set a reasonable amount of Access Credit consumed and to forward the block to the rest of the DLT network. Upon request, nodes send information related to the real-time congestion level of the network, namely an estimation of the amount of Access Credit needed to successfully schedule a block. Then, the account can set the credit consumed of the newly created block depending on its preferences, similarly to the way priority is set when gas fees are paid in Ethereum [8]. Nodes can be thought as gateways that play a fundamental role in the congestion management of the entire architecture. In this work, we assume that consumed Access Credit is a quantity larger or equal to 0: while bounds are useful for spam protection (in the case of a lower bound) and to avoid overspending resources (in the case of an upper bound), we leave the study of this optimization as a future work.

Unlike standard blockchains where block producers try to extract value by selecting the most profitable blocks, in our approach the rules are defined at protocol level and each node participates without the possibility of extracting value. Our access control chooses which blocks get gossiped in the peer-to-peer network, where the Access Credit is consumed instead of being redistributed, hence nodes have no incentives to deviate from the protocol.

---

[1] To counteract a large accumulation of Access Credit that could undermine the decentralization of the network, we suggest to couple the Access Credit to a decay function. For the sake of simplicity, we do not consider this decay function here.

In the following, we present the main components of our proposal, namely the enqueueing phase, the scheduling mechanism and the policy to drop blocks during congestion.

Enqueuing. As blocks get gossiped, receiving nodes verify the correctness of their content (verification procedure varies depending on the specific implementation). For valid blocks, the protocol calculates the *Priority Score* as follows:

Consider a block B and the tuple (c_B, w_B) where c_B is the Access Credit consumed and w_B the work of block B. We refer to Priority Score S_B *the ratio between the Access Credit consumed and the work of block B, i.e., $S_B = \frac{c_B}{w_B}$.*

Once the Priority Score is computed, the block is enqueued into the *scheduling buffer*, which gathers all blocks not yet scheduled (more details in the Scheduling subsection). In our proposal, this buffer is a *priority queue* sorted by Priority Score. The insert of a new block in the buffer has linear complexity.

Scheduling. The scheduling policy is a mechanism that selects which blocks must be forwarded in the DLT network. We consider a scheduler that works in a service loop, where every τ units of time it selects the blocks with the largest Priority Score in the scheduling buffer such that the work units of the selected blocks are smaller or equal than m work units. In this scenario, the enforced network throughput limit is m/τ work units per second. When a block is chosen to be scheduled, it is forwarded to neighbouring nodes where it can be enqueued in their buffer if they have not yet received it, after which the block undergoes the same scheduling process in each new node. We do not assume any specific gossip protocol: flooding, i.e., forwarding indiscriminately to all neighbours, is a popular choice in DLTs, but this can be optimised to save network bandwidth.

Block Drop. Scheduling buffers have a limited size. In this work, when the buffers get full, the protocol will drop the block with the lowest Priority Score, removing it from the buffer. Additionally, to limit the effectiveness of long-range attacks, we also drop blocks whose timestamps become older than a certain threshold compared to the node's local clock.

4   Simulations

In this section, we present a performance analysis of our credit-based access control. We perform both simulations on a single node to collect metrics related to cost of new blocks and to the time spent in the scheduling buffer (Sect. 4.2) and on a *multi-node setting* to verify ledger consistency and analyse the rate of discarded blocks (Sect. 4.3).

4.1   Simulation Setup

In our setup, we consider 1000 accounts, i.e., block issuers. The token holdings belonging to those issuers are drawn from a Power Law distribution of the form $p(x) = 10/x^2$. The amount of tokens per issuer does not vary over the course of the simulations. Furthermore, each user gets 1 credit/second for every 10 tokens held: for example, a user with 25 tokens obtains 2.5 credit/second.

Blocks are generated according to a non-homogenerous Poisson Process, with alternating congested and uncongested periods of 3 min each. We define a congested period as the time interval where the sum of the block generation rate over all accounts is larger than scheduling rate. The simulation is run for one hour, that is 10 congested periods and 10 uncongested periods. Additionally, we impose a maximum scheduling rate of 100 blocks per second (for the sake of simplicity, all blocks have the same size). In our simulations, the number of blocks issued by an account is proportional to its token holdings. Moreover, we define four types of block issuers according to the way the block cost is set:

- **Impatient:** These accounts consume all of their Access Credits each time they issue a block. They do not respond in any way to the credit consumption they observe in the buffer.
- **Greedy:** These accounts look at the highest amount of Access Credit consumed in the scheduling buffer and consume 1 more Access credit than this. If this greedy policy dictates that they would need to consume more than they have, they simply do not issue anything until the price goes down or they have generated enough Access Credit.
- **Gambler:** These nodes consume the amount of Access Credit of one of the top 20 blocks in the priority queue, chosen randomly.
- **Opportunistic:** These nodes consume 0 Access Credit, regardless of what is seen in the scheduling buffer.

Finally, we assume the buffer having a maximum capacity of 500 blocks. Blocks are removed from the buffer when the buffer reaches its maximum capacity or a block spends 30 seconds in the buffer without being scheduled.

Potential changes in the parameters used in the multi-node simulator will be explicitly mentioned in Sect. 4.3.

4.2   Single-Node Simulator

Impatient Strategy. In this set of simulations, all accounts follow the *impatient* consumption strategy. In Fig. 1a, we plot the cost of a scheduled block and the sojourn time of the same block as the simulation time advances. As a reference, we also add the traffic load over time, which alternates congested and uncongested periods. When accounts act as *impatient* users, we realize that the cost of a block increases during less congested periods, while – during congestion – the cost of a block stabilizes at less than 30 credits with peaks up to 150 credits; conversely, in uncongested periods the credits spent is at least the double. This is because users tend to overspend when using this consumption strategy:

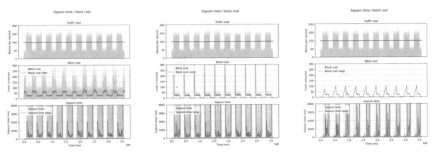

(a) Accounts are *impatient*.   (b) Accounts are *greedy*.   (c) Accounts are *gambler*.

Fig. 1. Traffic load (top figure), credits consumed (middle) and sojourn time (bottom) per block over time. Red line indicates the scheduling rate in the traffic load plot, and the moving average in the other plots.

during congestion, accounts have less time to accumulate Access Credit; the plot basically shows how much one can accumulate since the latest block it has issued.

The sojourn time, defined as the time a block spends in the scheduling buffer (remember, this is a single-node simulator, so the sojourn time is the time spent in a single buffer), is very low when the network is uncongested but experiences large oscillations during congestion: in particular, after the transition to congestion, the mean sojourn time spikes to around 2 seconds and then keeps oscillating between 0.5 and 1 second; a non-negligible number of blocks experience a much larger sojourn time as it can be seen by the blue line in Fig. 1a.

Greedy Strategy. Here, we show the same set of plots, but when all accounts act as *greedy*. A greedy consumption strategy optimizes the inefficiencies of the impatient one, which tends to overspend unnecessarily. The cost of a block, from Fig. 1b, is now very low (close to 0) with little traffic; however, the transition to a congested network creates a very steep increase in the cost of a scheduled block: for a short period of time, the average consumed Access Credits is larger than 300, then suddenly decreasing around 30. This strategy can be compared with first price auctions, carrying their intrinsic drawbacks as well. While several recent approaches have been trying to mitigate the fluctuations in the block cost and to improve the user experience [8,11], we stress that finding an optimal credit consumption policy is out of the scope of this paper.

Similarly, it is possible to see frequent oscillations in the sojourn times with spikes (i) at the beginning and (ii) at the end of the congested period: (i) the increased traffic load alters the dynamics of the system, lowering the rank in the priority queue for blocks not yet scheduled, and we notice that oscillations are visible throughout the entire congested period; (ii) additionally, when congestion ends, a lot of blocks sitting in the buffer for long (but not yet dropped) have the opportunity to be scheduled experiencing a large delay, witnessed by the spike at the end of each high-traffic period.

Gambler Strategy. In these simulations all accounts follow the *gambler* strategy. In Fig. 1c, we see that the spikes in the Credits consumed are largely reduced compared to the *greedy* scenario: we cannot see accounts consuming more than 100 credits. However, the cost of scheduled blocks stabilizes at a price only marginally lower than before.

4.3 Multi-node Simulator

In this section, we use a multi-node simulator which emulates a complete DAG-based DLT protocol, i.e., each node maintains a copy of the DAG, uses a selection algorithm to choose where attach new blocks and checks the validity of all arriving blocks. A number of specific DAG-based protocol details are included which our proposal do not necessarily depend on, but this allows us to at least provide preliminary results for integrating this approach into a working protocol. Each node also operates an account for issuing blocks, so we use the terms node and account interchangeably in this section. The same consumption policies are tested as for the single-node simulator, but we use a smaller network and shorter simulation times to facilitate detailed presentation of each node's outcome.

The simulations consist of 20 nodes connected in a random 4-regular graph topology, i.e., 4 neighbours each. The communication delays between nodes are uniformly distributed between 50 ms and 150 ms. The scheduling rate is 25 blocks per second. We use the same token distribution as in the single-node simulator. We initially consider a mix of greedy and opportunistic nodes.

We slightly modify the block generation process in this set of simulations: here, blocks are generated according to a separate Poisson process for each node and added to the node's local mempool from which they can create blocks. For the first minute of each simulation, blocks are generated at 50% of the scheduling rate, then for the following two minutes, the rate gets to 150% compared to scheduling rate, and for the final minute, it decreases to the 50% again. This traffic pattern simply seeks to show one cycle of increase in demand and then subsiding of demand.

Finally, we introduce the concept of *block confirmation*: let $CW_B \in \mathbb{N}^+$ be the *cumulative weight* of block B, which indicates how many times B has been referenced directly or indirectly by other blocks; then, a node locally considers block B as confirmed if $CW_B \geq 100$. Furthermore, a block is confirmed when all nodes have marked the block as confirmed. Confirmation rate is the rate at which blocks become confirmed. Cumulative weight in a DAG is analogous to the depth of a block in a blockchain which is often used for confirmation. Additionally, a block is *disseminated* when all nodes have seen the block. Dissemination rate is the rate at which blocks become disseminated.

Remark: "Scaled" plots are scaled by the node's fair share of the scheduler throughput which is proportional to their token holdings, so a scaled rate of 1 means they are getting 100% of their fair share. In plots showing metrics for all nodes, the thickness of the trace corresponding to each node is proportional to the token holdings of that node.

Experimental Results. We begin by considering the dissemination rates, as seen in Fig. 2a. Here, the greedy nodes are able to issue more than their fair share because the opportunistic nodes are opting not to consume any Credits and hence the greedy nodes get priority from the scheduler by consuming more.

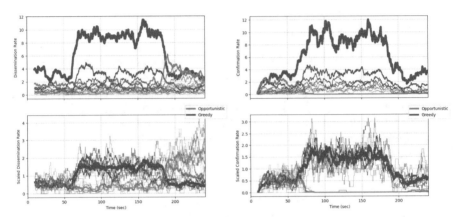

(a) Dissemination rates and scaled dissemination rates.

(b) Confirmation rates and scaled confirmation rates.

Fig. 2. Multi-node simulations.

Figure 2b illustrates the confirmation rates corresponding to the same simulation. These traces follow a very similar trajectory to the dissemination rates, but we notice that even when congestion dies down, the confirmation rates of the opportunistic nodes do not recover immediately as the dissemination rates did. In fact, many old delayed blocks from the congested period are stuck in the buffers of nodes across the network and as these old blocks begin to be forwarded when the congestion goes away, they are not selected by other nodes to attach to, so their cumulative weight does not grow and they do not become confirmed.

These multi-node simulator results only present a very limited scenario with basic credit consumption policies, but the results show promise for providing effective access control. However, they also begin to show some of the complexities of integrating this approach into complete DAG-based DLT protocols. Further studies need to be carried out for specific DAG implementations to fully understand the implications of our approach.

5   Conclusion

We proposed a credit-based access control mechanism for leaderless DAG-based DLTs. Our solution solves the problem of regulating write access without the need for token fees or serialisation of ledger updates into blocks by validators. The proposal is based on *Access Credits*, which are naturally generated for holding the base token. State-changing data must consume these credits to be included in the ledger, creating a utility loop where rewards are given in Access Credits.

Our simulations show that the consumed credits remain stable over time, even with large jumps of demand for ledger write access. Additionally, we showed that write access can be effectively regulated across multiple nodes in a peer-to-peer network in some simple scenarios. Leaderless DAG-based ledgers present enormous potential for advances in the DLT field, and this work will provide a foundation for similar schemes seeking to manage write access.

References

1. Bagaria, V., Kannan, S., Tse, D., Fanti, G., Viswanath, P.: Prism: deconstructing the blockchain to approach physical limits. In: Proceedings of ACM SIGSAC Conference on Computer and Communications Security, CCS 2019, pp. 585–602. Association for Computing Machinery, New York (2019)
2. Baird, L., Harmon, M., Madsen, P.: Hedera: a public hashgraph network and governing council (2020). Accessed Feb 2023
3. Bonneau, J., Miller, A., Clark, J., Narayanan, A., Kroll, J.A., Felten, E.W.: SoK: research perspectives and challenges for bitcoin and cryptocurrencies. In: 2015 IEEE Symposium on Security and Privacy, pp. 104–121. IEEE (2015)
4. Daian, P., et al.: Flash boys 2.0: frontrunning in decentralized exchanges, miner extractable value, and consensus instability. In: 2020 IEEE Symposium on Security and Privacy (SP), pp. 910–927. IEEE (2020)
5. Danezis, G., Kokoris-Kogias, L., Sonnino, A., Spiegelman, A.: Narwhal and tusk: a DAG-based mempool and efficient BFT consensus. In: Proceedings of European Conference on Computer Systems, EuroSys 2022, pp. 34–50. Association for Computing Machinery, New York (2022)
6. Gągol, A., Leśniak, D., Straszak, D., Świętek, M.: Aleph: efficient atomic broadcast in asynchronous networks with Byzantine nodes. In: Proceedings of the 1st ACM Conference on Advances in Financial Technologies, pp. 214–228 (2019)
7. LeMahieu, C.: Nano whitepaper (2022). Accessed Feb 2023
8. Leonardos, S., Monnot, B., Reijsbergen, D., Skoulakis, E., Piliouras, G.: Dynamical analysis of the EIP-1559 Ethereum fee market. In: Proceedings of the 3rd ACM Conference on Advances in Financial Technologies, AFT 2021, pp. 114–126. Association for Computing Machinery, New York (2021)
9. Miller, A., Xia, Y., Croman, K., Shi, E., Song, D.: The honey badger of BFT protocols. In: Proceedings of ACM SIGSAC Conference on Computer and Communications Security, CCS 2016, pp. 31–42. Association for Computing Machinery (2016)
10. Müller, S., Penzkofer, A., Polyanskii, N., Theis, J., Sanders, W., Moog, H.: Tangle 2.0 leaderless Nakamoto consensus on the heaviest DAG. IEEE Access **10**, 105807–105842 (2022)
11. Roughgarden, T., Syrgkanis, V., Tardos, E.: The price of anarchy in auctions. J. Artif. Int. Res. **59**(1), 59–101 (2017)
12. Sompolinsky, Y., Lewenberg, Y., Zohar, A.: Spectre: a fast and scalable cryptocurrency protocol. Cryptology ePrint Archive, Paper 2016/1159 (2016). Accessed Feb 2023
13. Zhou, Q., Huang, H., Zheng, Z., Bian, J.: Solutions to scalability of blockchain: a survey. IEEE Access **8**, 16440–16455 (2020)

BFT Testing Framework for Flow Blockchain

Yahya Hassanzadeh-Nazarabadi[(✉)], Misha Rybalov, and Khalil Claybon

Flow, Vancouver, Canada
{yahya,misha.rybalov,khalil.claybon}@dapperlabs.com

Abstract. This paper introduces *BFTune*, an innovative open-source testing framework for comprehensive functional testing against coordinated active attacks in the Flow blockchain. The proposed *BFTune* is the first testing framework of its kind, offering functional coverage, protocol realism, and programmable attacker models for the Flow blockchain, addressing a gap in existing blockchain testing tools. We detail the design and implementation of *BFTune*, present real-world BFT testing scenarios, and provide an extensive experimental assessment of its scalability and runtime overhead.

Keywords: Blockchain · Flow · Byzantine Fault Tolerance · Testing framework · Protocol realism

1 Introduction

Traditional blockchain systems rely on full nodes to keep the blockchain network operational [1,2]. Flow blockchain [3,4], conversely, decouples full nodes into five specific roles, which operate simultaneously for better throughput, performance, and scalability. The five specific node roles are Access Nodes, Collection Nodes, Consensus Nodes, Execution Nodes, and Verification Nodes. Access Nodes are the gateway between the blockchain and external applications, while Collection Nodes verify and group transactions into collections. Consensus Nodes form blocks by agreeing on the contents and order of collections, while Execution Nodes execute blocks, and Verification Nodes verify the execution result.

In contrast to existing blockchains, the consensus protocol in Flow only forms blocks and does not execute them, ensuring the blockchain's integrity and security. Once a block is finalized, its contents cannot be altered or reversed. The Execution Nodes execute a block, and the individual execution results are verified through Verification Nodes. A block is considered sealed if the majority of the Execution Nodes generate harmonious execution results and a certain threshold of the Verification Nodes verify and approve the block's execution result. Sealing a block signifies the immutability of its execution state transition, whereas finalizing a block entails the immutability of its content.

Byzantine Fault Tolerant (BFT): Flow blockchain is a decentralized Proof-of-Stake (PoS) [5] system that relies on a minimum number of participating nodes

© The Author(s), under exclusive license to Springer Nature Switzerland AG 2023
J. M. Machado et al. (Eds.): BLOCKCHAIN 2023, LNNS 778, pp. 338–347, 2023.
https://doi.org/10.1007/978-3-031-45155-3_34

to ensure honest behavior and strict adherence to the protocols associated with their respective roles. For the sake of safety [3], Flow necessitates that more than $\frac{2}{3}$ of the stake from the majority of roles is controlled by honest (i.e., non-Byzantine) participants for each role individually. As such, the Flow blockchain is designed to be resilient to Byzantine faults, much like other decentralized and permissionless blockchain systems, as long as the honest stake threshold per role is upheld.

Byzantine faults refer to failures, errors, or malfunctions resulting from nodes acting arbitrarily, inconsistently, or maliciously, which may threaten the blockchain system's safety [6]. Some examples of these faults include Sybil attacks [7], network partition attacks [8], and denial-of-service attacks [9]. Flow blockchain undergoes research, auditing, analysis, and development to address Byzantine behavior within protocol-defined limits. Validating its BFT implementation requires unit testing, integration testing, and functional testing. Unit testing verifies individual components' functionality, detecting defects early. Integration testing assesses component interactions, uncovering issues not found during unit testing. Functional testing evaluates BFT feature performance in real-world scenarios, ensuring security and reliability, and minimizing vulnerability risks in real-world implementations.

Original Contributions: In this paper, we introduce *BFTune*, an innovative open-source testing framework designed for comprehensive functional testing with complete protocol realism against coordinated active attacks throughout the entire transaction lifecycle in the Flow blockchain. We observe that no existing blockchain testing frameworks provide a combination of functional coverage, protocol realism, and a programmable attacker model [10–14].

The original contributions of this paper include the following: (1) We propose *BFTune*[1], the first open-source testing framework for functional Byzantine Fault Tolerant (BFT) testing on the Flow blockchain, incorporating support for programmable attacker models and protocol realism. (2) We implemented several real-world BFT testing scenarios showcasing how *BFTune* enables developers to assess and enhance the security and reliability of proposed protocols within the Flow blockchain ecosystem. (3) We conduct an extensive experimental assessment of *BFTune*, showcasing its scalability and runtime overhead.

2   Related Works

To the best of our knowledge, no open-source testing framework currently exists for functional testing against coordinated active attacks throughout the entire transaction life cycle in blockchain systems. BFTSim [10], recognized as the first simulator for BFT consensus protocols like PBFT [15], is limited to testing non-malicious Byzantine faults. These faults, not intentionally induced by malicious parties, can result from software bugs or communication issues. Similarly, Twins [12] serves as a unit testing framework for consensus protocols, aiding in

---

[1] https://github.com/onflow/bftune

identifying Byzantine behaviors in simulated environments. It creates duplicate copies of a node, sharing the same identity and network credentials, referred to as "twins".

Tool [11] is a highly scalable BFT simulator equipped with an attacker module that features several pre-defined attack vectors, such as network partitioning. However, unlike our proposed *BFTune*, Tool does not replicate real network protocols. Instead, it represents network characteristics through a high-level model in which message delays are determined by the variables drawn from a stochastic distribution.

Tyr [13] is a functional testing framework designed to detect consensus failure bugs in blockchains, specifically those that compromise the safety or liveness of the consensus process. Equipped with a dedicated detector for each type of consensus failure bug, Tyr monitors and analyzes node behavior in real-time during testing and identifies the nodes responsible for causing failure bugs.

Fluffy [14] is a differential integration testing framework designed to identify transaction execution bugs in the Ethereum [2] blockchain by comparing execution results across two distinct implementations. However, utilizing differential testing methods like Fluffy cannot guarantee the detection of all bugs. Every implementation involved in the differential testing process may be susceptible to the same bug, yielding identical results.

Table 1 compares our proposed blockchain testing framework, *BFTune*, and existing testing frameworks and simulators. The table emphasizes three essential features a reliable blockchain testing tool should possess: functional coverage, protocol realism, and attacker model support. *Functional coverage* refers to whether the tool offers functional simulation or functional testing to evaluate the blockchain's behavior comprehensively and ensure it performs as expected in real-world usage. *Protocol realism* indicates whether the tool conducts testing or simulation on the entire blockchain protocol stack without any abstraction. In other words, the tool should be capable of testing (or simulating) the original blockchain node code implementation without requiring any simplification or abstraction. Lastly, a dependable testing tool must support an *attacker model* to enable users to test (or simulate) arbitrary attack vectors. The tool should provide users with the means to create and test scenarios in which the blockchain is subjected to various attacks, verifying its robustness. As the table demonstrates, our proposed *BFTune* outshines existing blockchain testing tools by supporting functional coverage, protocol realism, and attacker model altogether.

Table 1. Comparison of Blockchain Testing Tools and Simulators

Name	Type	Functional Coverage	Protocol Realism	Attacker Model
BFTSim [10]	Simulator	✓	✗	Limited (consensus)
Tool [11]	Simulator	✓	✗	✓
Twins [12]	Testing Framework	✗	✓	Limited (consensus)
Tyr [13]	Testing Framework	✓	✓	✗
Fluffy [14]	Testing Framework	✗	✓	✗
BFTune (ours)	**Testing Framework**	✓	✓	✓

3 *BFTune* BFT Testing Framework

3.1 Overview

Communication between nodes is essential for the correct operation of distributed systems. Nodes depend on their message exchange capability to perform tasks and make decisions. The messages a node sends reflect its behavior, while the messages it receives shape its understanding of the system. An attacker can manipulate a node's behavior and system perception by controlling that node's message-sending and receiving capabilities. The attacker controls the node's behavior and influences its system perception by dictating which messages to send and selectively censoring the received ones.

In general, attacks on distributed systems based on message exchange can be modeled by a subset of colluding nodes, called *corrupt nodes*, deviating from expected protocols and exhibiting unexpected behavior. Other nodes perceive this behavior through messages received from those corrupt nodes. Such attacks can be abstractly modeled by a centralized *attack orchestrator* controlling the set of corrupt nodes, dictating their behavior, and manipulating their system perspective. The attack orchestrator selectively manipulates the sent and received messages, causing corrupt nodes to misbehave or provide false information, leading to orchestrated attacks. *BFTune*, distinct from other testing frameworks for blockchains, employs such a centralized attacker orchestrator to control corrupt nodes' message flow as a fundamental design principle.

In essence, *BFTune* allows Flow blockchain protocol developers to implement attack vectors using a predefined *Attack Orchestrator* interface, which is compatible with *BFTune*. Developers determine the total node count and the fraction of corrupt nodes. *BFTune* executes the BFT test by initiating the specified number of honest and corrupt nodes, with the *Attack Orchestrator* controlling the corrupt nodes' message flow during testing. Based on the implemented attack, the *Attack Orchestrator* inspects incoming and outgoing messages of the corrupt nodes under its control and decides to discard, corrupt, or pass messages through. Discarding prevents outgoing messages from being sent or incoming messages from being delivered. Corrupting involves manipulating outgoing messages, while passing allows messages to continue unaltered.

The sole difference between honest and corrupt nodes in *BFTune* is an additional component in corrupt nodes, allowing the *Attack Orchestrator* to control message exchange decisions. Apart from this, both node types share similar implementations and follow the same Flow blockchain protocol. This enables realistic testing with minimal abstractions for attack vector implementation. Developers focus on attack implementation without modifying the Flow protocol, simplifying the process and maintaining consistency with the main network. The *Attack Orchestrator* remains idle, passing messages unless relevant to the attack vector. Implementing and testing an attack in *BFTune* can be simplified by translating the attack logic into the logic of controlling messages that are emitted by or intended for corrupt nodes. This can be done by passing through, discarding, or corrupting messages.

3.2   Architecture

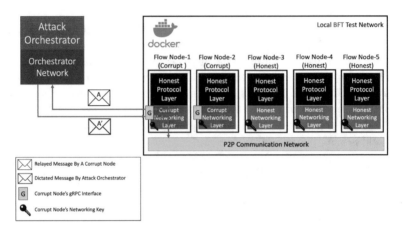

Fig. 1. An example of *BFTune* architecture overview in a local test network with two corrupt nodes and three honest nodes. The example illustrates the corruption path of an outgoing message originated by the (corrupt) Node 1.

Figure 1 depicts the *BFTune* architecture, comprising three core components: the *Attack Orchestrator*, the *Orchestrator Network*, and the *Local BFT Test Network*. The Local BFT Test Network is a collection of Flow node containers that run in a Docker environment. This network is named Local BFT because it is the only setup that allows the containers to run in either mode of honest or corrupt. For security reasons and to prevent the leakage of BFT testing code to production environments like the Flow blockchain main network, corrupt nodes are only composable in the Local BFT Test Network. Running them outside this network is safeguarded by a fail-safe approach.

Honest nodes adhere to the Flow protocol as designed, while the *Attack Orchestrator* manages corrupt nodes. The *Attack Orchestrator* is designed as an interface type that outlines a range of attack vectors for testing within *BFTune*. It directs predefined attack scenarios by controlling the incoming and outgoing messages of corrupt nodes. While the logic employed by each *Attack Orchestrator* may differ based on the specific attack scenario under examination, the communication protocol between the *Attack Orchestrator* and corrupt nodes remains consistent across all attack vectors in *BFTune*. This consistency is maintained within the *Orchestrator Network*, which is vital for accurate testing since reliable and efficient communication with no message loss between the *Attack Orchestrator* and corrupt nodes is crucial. The Orchestrator Network's primary responsibility is facilitating dependable message exchange between the *Attack Orchestrator* and corrupt nodes by establishing and maintaining gRPC streaming communication channels [16] from the *Attack Orchestrator* to individual corrupt nodes.

In *BFTune*, a Flow node is structured with a two-layer architecture that includes the protocol and networking layers. The protocol layer encompasses all blockchain services, such as transaction processing, block creation, execution, and verification. The networking layer ensures secure and reliable message distribution among Flow nodes. To guarantee protocol realism and comprehensive testing of the entire protocol stack of the Flow blockchain without protocol-level abstractions, both honest and corrupt nodes in *BFTune* operate with the same protocol layer as used in the Flow blockchain main network. The only distinction between a corrupt node and an honest node lies in the modified networking layer of the corrupt node (Fig. 1). This altered networking layer relays the corrupt node's outgoing messages to the *Attack Orchestrator* rather than distributing them within the Flow peer-to-peer (P2P) communication network. Similarly, the corrupt networking layer conveys incoming messages to the *Attack Orchestrator* instead of passing them to the node's protocol layer. Upon receipt of a relayed message from a corrupt node, the *Attack Orchestrator* executes the preprogrammed attack on the message to determine whether to pass through the message, discard, or corrupt it.

When opting to pass through an incoming message, the *Attack Orchestrator* directs the networking layer of the corresponding corrupt node to deliver the incoming message to the node's protocol layer. Similarly, the *Attack Orchestrator* allows an outgoing message to proceed by instructing the networking layer of the relaying corrupt node to distribute it within the Flow P2P communication network to its intended recipients. Discarding a relayed message entails the *Attack Orchestrator* disregarding the message and not providing any instructions to the relaying node, resulting in the message never being disseminated in the Flow P2P communication network. In contrast, corrupting a message requires altering its content before allowing it to pass through. The *Attack Orchestrator* is only capable of corrupting outgoing relayed messages. To accomplish this, the *Attack Orchestrator* commands the relaying corrupt node to transmit a modified version of the relayed message within the Flow P2P communication network. Additionally, the *Attack Orchestrator* can instruct a group of corrupt nodes to distribute self-crafted messages within the P2P communication network.

Figure 1 illustrates how *BFTune* corrupts outgoing messages. Like the honest networking layers, the corrupt networking layers possess the networking key of their corresponding corrupt nodes. This enables corrupt networking layers to sign the dictated corrupt messages of the *Attack Orchestrator* using the networking key of the corrupt node before dispatching them in the P2P communication network of the Flow blockchain. As a result, the corrupt messages that the *Attack Orchestrator* dictates appear to come from corrupt nodes to other Flow nodes. In other words, the *Attack Orchestrator* changes the behavior of corrupt nodes that the honest nodes perceive by dictating outgoing messages to them. In the scenario depicted in Fig. 1, the honest protocol layer on the corrupt node 1 generates message *A*, which it intends to send to a subset of other Flow nodes via the P2P communication network. Instead of disseminating message *A* within the P2P communication network, the corrupt networking layer of node 1 relays it to the *Attack Orchestrator*. Using its implemented attack vector logic, the *Attack Orchestrator* chooses to corrupt message A and dictates a tampered version, represented as message A'. When the corrupt networking layer of node 1 receives the dictated message A', it signs it with the networking key of node 1 and forwards it to the intended recipients, as instructed by the content of the dictated message. As far as the recipients are concerned, message A' appears to have originated from node 1.

It should be noted that the *Attack Orchestrator* can only decide whether to pass incoming messages through or discard them. Tampering with the content of incoming messages compromises their cryptographic integrity. Re-signing them by the original sender is infeasible because the receiving corrupt networking layer lacks the original sender's signing networking key. However, if the original sender is also a corrupt node, corrupting an incoming message is equivalent to corrupting the outgoing messages of the corrupt sender. Moreover, the *Attack Orchestrator* has no control over the messages sent by honest nodes to corrupt nodes except for discarding or passing them through.

3.3   Sample BFT Testing Scenarios

The following is a list of sample attack scenarios that have already been implemented in *BFTune* testing framework. Due to space constraints, the attack scenarios are summarized to include only the essential details. While some advanced attack scenarios, such as consensus double voting or execution fork, have not been included in this list, we plan to explore them in our future works.

Pass-Through Test [17]: The Pass-Through Test is a basic test used to assess the overall health of the BFT testing framework rather than evaluating a BFT scenario. In this test, the *Attack Orchestrator* implements a pass-through function where all incoming and outgoing messages from corrupt nodes are passed through, and the *Attack Orchestrator* records these messages. The test succeeds if the expected types and the number of messages are passed through, indicating no bugs in *BFTune* or its integration with corrupt nodes. The Pass-Through Test can also be used as a regression test to evaluate the performance of integrating *BFTune* into the Flow network.

Message Validation Test [18]: The Flow blockchain allows various message types to be exchanged between nodes, but nodes are restricted from emitting certain message types depending on their role. Sending disallowed message types by a node constitutes a Byzantine fault and undermines the safety of the blockchain protocols [3]. To ensure this does not happen, the P2P communication network of the Flow blockchain implements several message validation mechanisms to validate the received messages before propagating them further in the network. To test the functionality of this message validation system, a message validation test is conducted using an *Attack Orchestrator* that instructs corrupt nodes to emit disallowed message types. The test's success is determined by whether or not the protocol layer of the honest nodes receives the disallowed messages from the corrupt nodes. If no protocol layer of an honest node receives the dictated messages, the test is a success. On the other hand, if a protocol layer of an honest node receives any disallowed message, it indicates a potential flaw in the message validation pipeline and represents a vulnerability in the system.

Signature Requirement Test [19]: In the Flow blockchain, valid signatures from the networking signing key of the original sender are mandatory for every gossiped message. If a message violates this requirement, honest nodes drop it without processing or disseminating it further in the network. To evaluate this signature requirement functionality, the Signature Requirement Test is performed by an *Attack Orchestrator* that enforces a no-signature policy on a subset of corrupt nodes. The corrupt nodes disseminate the dictated messages from the *Attack Orchestrator* within the Flow network without signing them. The test's success is determined by whether or not the protocol layer of the honest nodes receives corrupt messages that violate the signature policy. If the protocol layer of an honest node does not receive any dictated message that violates the signature requirement, the test is considered successful. However, if any honest node receives such messages, it indicates a potential flaw in the message validation pipeline and results in a test failure.

4   Experimental Results

To evaluate the performance overhead of executing tests on *BFTune*, we take a *happy path* [20] functional test of the Flow blockchain and compare its performance on both the *BFTune* and a standard setup in the Docker environment. The *happy path* test initializes a functional Flow blockchain network and assesses transaction submission, block generation, execution, and verification health. We employ the Pass-Through scenario [17] (refer to Sect. 3) to conduct the *happy path* test on the *BFTune*. The experiment's base setup includes the minimum number of nodes required to initialize and operate a Flow blockchain network, specifically, 4 Consensus Nodes, 1 Access Node, 2 Collection Nodes, 2 Execution Nodes, and 1 Verification Node. In separate trials, we scale the number of Execution Nodes from 2 to 5 and the number of Verification Nodes from 1 to 5. It is important to note that the test runs a full Flow blockchain within a Docker environment, with each Flow node operating in its real-world configuration, devoid of abstraction. In addition, within the *BFTune* setup, the Execution and Verification nodes function in corrupt mode, relaying their messages by the *Attack Orchestrator*. We measure the entire test suite run time for each case, incorporating network bootstrapping, nodes' container startup, test scenario execution, and network and container teardown. We also calculate the average execution time of the test scenario independently, excluding network bootstrapping, startup, and shutdown.

Figures 2 and 3 display the average execution time of the *happy path* test suite and the individual *happy path* test itself when scaling the number of Verification and Execution Nodes. Each data point corresponds to the average of 10 test runs. The *Attack Orchestrator* in the Pass-Through test operates using a blocking and sequential approach, meaning it queues the relayed messages from corrupt nodes and processes them sequentially. As the number of corrupt nodes increases, the sequential message-processing behavior of the *Attack Orchestrator* leads to an increase in execution time. This outcome justifies the upward trend observed in the test execution time on *BFTune*. The sequential *Attack Orchestrator* for the Pass-Through test is deliberately selected for experimentation, as it exemplifies the maximum execution overhead for tests in *BFTune*. As demonstrated by these figures, compared to the standard setup (i.e., Honest), the functional tests' execution overhead for the Flow blockchain in *BFTune* is limited to a 50% increase in the individual test scenario execution time and a 15% increase in the overall test suite execution time. Given that *BFTune* is the first functional BFT testing framework for the Flow blockchain, this represents a reasonable execution time overhead to incur when testing attack scenarios.

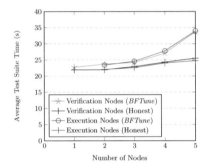

Fig. 2. Average *happy path* test *suite* execution time as scaling one role.

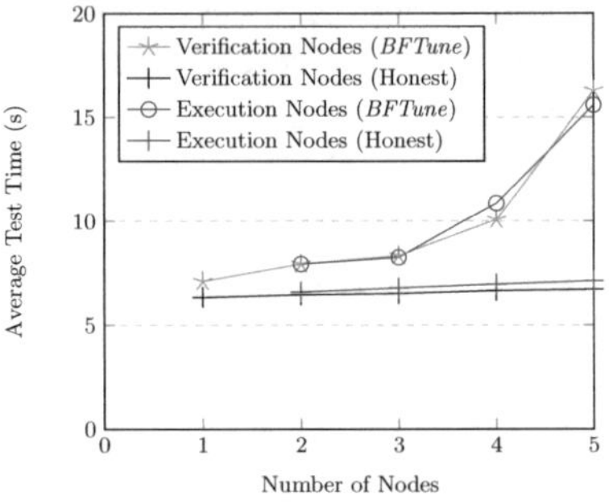

Fig. 3. Average *happy path* test execution time as scaling one role.

5   Conclusion

In this paper, we presented *BFTune*, a novel open-source testing framework designed for functional Byzantine Fault Tolerant (BFT) testing on the Flow blockchain, offering support for programmable attacker models and protocol realism. *BFTune* addresses the limitations of existing testing frameworks and simulators that lack the combination of functional coverage, protocol realism, and attacker model support altogether. We showcased several real-world BFT testing scenarios, demonstrating how *BFTune* enables developers to assess and improve the security and reliability of proposed protocols within the Flow blockchain ecosystem. Furthermore, we conducted an extensive experimental assessment of *BFTune*, illustrating its scalability and runtime overhead. We anticipate that *BFTune* will contribute to the ongoing development and refinement of the Flow blockchain and other blockchain systems seeking robust BFT testing solutions. As part of our future work, we plan to implement more advanced attacks on *BFTune*, including consensus double voting and execution fork scenarios.

References

1. Nakamoto, S.: Bitcoin: a peer-to-peer electronic cash system. Decentralized Bus. Rev. (2008)
2. Wood, G., et al.: Ethereum: a secure decentralised generalised transaction ledger. Ethereum Project Yellow Paper (2014)
3. Hentschel, A., Hassanzadeh-Nazarabadi, Y., Seraj, R., Shirley, D., Lafrance, L.: Flow: separating consensus and compute–block formation and execution. arXiv preprint arXiv:2002.07403 (2020)
4. Hentschel, A., Shirley, D., Lafrance, L., Zamski, M.: Flow: separating consensus and compute–execution verification. arXiv preprint arXiv:1909.05832 (2019)
5. King, S., Nadal, S.: PPCoin: peer-to-peer crypto-currency with proof-of-stake. Self-published paper (2012)
6. Lamport, L., Shostak, R., Pease, M.: The Byzantine generals problem. In: Concurrency: The Works of Leslie Lamport, pp. 203–226 (2019)
7. Douceur, J.R.: The Sybil attack. In: Druschel, P., Kaashoek, F., Rowstron, A. (eds.) IPTPS 2002. LNCS, vol. 2429, pp. 251–260. Springer, Heidelberg (2002). https://doi.org/10.1007/3-540-45748-8_24
8. Kuperberg, M.: Towards an analysis of network partitioning prevention for distributed ledgers and blockchains. In: 2020 DAPPS. IEEE (2020)
9. Saad, M., et al.: Exploring the attack surface of blockchain: a comprehensive survey. IEEE Commun. Surv. Tutor. **22**, 1977–2008 (2020)
10. Singh, A., Das, T., Maniatis, P., Druschel, P., Roscoe, T.: BFT protocols under fire. In: NSDI (2008)
11. Wang, P.-L., Chao, T.-W., Wu, C.-C., Hsiao, H.-C.: Tool: an efficient and flexible simulator for byzantine fault-tolerant protocols. In: DSN. IEEE (2022)
12. Bano, S., et al.: Twins: BFT systems made robust. arXiv preprint arXiv:2004.10617 (2020)
13. Chen, Y., Ma, F., Zhou, Y., Jiang, Y., Chen, T., Sun, J.: Tyr: finding consensus failure bugs in blockchain system with behaviour divergent model. In: Symposium on Security and Privacy. IEEE (2022)
14. Yang, Y., Kim, T., Chun, B.-G.: Finding consensus bugs in Ethereum via multi-transaction differential fuzzing. In: OSDI (2021)
15. Castro, M., Liskov, B., et al.: Practical Byzantine fault tolerance. In: OsDI (1999)
16. Google LLC. GRPC: a high-performance, open-source universal RPC framework (2015). https://grpc.io/docs/what-is-grpc/introduction/. Accessed 8 Mar 2023
17. Flow go integration tests - pass through. https://github.com/onflow/flow-go/tree/master/integration/tests/bft/passthrough. Accessed 8 Mar 2023
18. Flow go integration tests - message validation. https://github.com/onflow/flow-go/tree/master/integration/tests/bft/topicvalidator. Accessed 8 Mar 2023
19. Flow go integration tests - signature requirement. https://github.com/onflow/flow-go/tree/master/integration/tests/bft/gossipsub/signature. Accessed 8 Mar 2023
20. Flow go integration tests - happy path testing. https://github.com/onflow/flow-go/blob/master/integration/tests/common. Accessed 8 Mar 2023

Implementing a Blockchain-Powered Metadata Catalog in Data Mesh Architecture

Anton Dolhopolov$^{(\boxtimes)}$, Arnaud Castelltort, and Anne Laurent

LIRMM, Univ. Montpellier, CNRS, Montpellier, France
{anton.dolhopolov,arnaud.castelltort,anne.laurent}@lirmm.fr

Abstract. This paper explores the implementation of a blockchain-powered metadata catalog in a data mesh architecture. The metadata catalog serves as a critical component in managing data at scale, allowing for efficient discovery, access, and governance. By integrating blockchain technology, the metadata catalog can provide federated control, immutability, and transparency in managing metadata across a distributed network of data domains. This paper discusses the benefits of using blockchain technology in the metadata catalog and provides a proof-of-concept implementation of a blockchain-powered metadata catalog in a data mesh architecture using HyperLedger Fabric. The paper also highlights some challenges and potential solutions for adopting this approach, including scalability, interoperability, and governance concerns. Overall, this paper presents a novel approach for implementing a secure and federated metadata catalog in data mesh architecture that can improve the efficiency, reliability, and transparency of data management.

Keywords: Blockchain · Data Mesh · Metadata Catalog · Data Governance

1   Introduction

Data mesh [4] is one of the most recent grounded theoretical developments in the data management field. Contrary to the previous approaches such as data warehouses [11] or data lakes [13], this new paradigm is inspired by the microservices architecture [18]. It advocates the logical decentralization of the data platform. It builds upon the 4 core principles: distributed domains data ownership, data-as-a-product, self-serve data platform, and federated computational (data) governance. In data mesh, the decentralization of the analytical platform architecture components aims to enable organizational scaling and close the gap between data generation and data analyses.

Metadata management is another important element of an efficient data platform. The research shows that it facilitates discoverability, accessibility, interoperability, semantic comprehension, and utilization of the data [9,19]. However,

J. M. Machado et al. (Eds.): BLOCKCHAIN 2023, LNNS 778, pp. 348–360, 2023.
https://doi.org/10.1007/978-3-031-45155-3_35

most of the theoretical models and practical implementations focus on centralized metadata catalog architectures. Therefore, such catalogs may fall short with the original purpose of building the decentralized system like data mesh.

On the other side, blockchain technology envisions the decentralization of information system actors at its core. The recent works show that it can be beneficially applied for cross-organizational collaboration [12] and metadata management [5,14]. The attractiveness of this technology includes such properties as a historical, secure, and immutable ledger, unified data control in form of smart contracts, and use of the consensus algorithms for providing data distribution.

In this paper, we claim that blockchain technology can profit the institutions that want to adapt the data mesh paradigm. First, we describe the challenges of data governance in the mesh. Then, we draw upon the previous theoretical work of metadata catalog distribution by providing a proof-of-concept implementation of the catalog based on the open-source Hyperledger Fabric platform. The paper concludes with discussions and further research plans.

2   Challenges of Data Governance in Data Mesh

According to [16], data governance "... is a collection of information-related processes, roles, policies, standards, and metrics oriented to maximize the effectiveness of deriving business value from data". Thus, data mesh should offer a system where technical engineers, business users, legal representatives, and other participants can easily define business processes, regulatory policies, data access roles, and other elements that will shape the platform's inner operations.

The centralized governance imposes the rules and standards to follow by each team which enables interoperability. But it has organizational scaling issues as it has been shown for data warehouse and data lake systems [15]. At the same time, the decentralized governance leaves a lot of freedom for each domain team which brings the risk of building incompatible data products, duplicating the development efforts, missing compliance procedures, etc.

In [4], the author of the data mesh concept, Zhamak Deghani, highlights the importance of federated computational data governance as the need to "maintain a dynamic equilibrium between domain autonomy and global interoperability". It means a striking balance of global policies (centralization) and local freedoms (decentralization) is required.

Federated governance systems should provide the tools that help to collaborate and ease the definition and access to data schemes, semantic knowledge, and lineage, but also to automatically enforce the security, transparency, and legal policies. However, the main challenge surfaces when constructing such a system is based on already available tools or existing research.

In Sect. 1 we mentioned that in the research metadata management system often performs the functions of the data governance system. Nonetheless, the majority of scientific systems such as GOODS [8], AUDAL [20], DAMMS [21], Child et al. [3], Cherradi et al. [2], and industrial products as DataHub Project[1]

---

[1] https://datahubproject.io/.

and Apache Atlas[2] implement only the centralized approach. It means that the metadata is collected within the central repository, analyzed and linked in post-hoc fashion, and often made available to users via web portals.

Some recent works [7,10] attempt to implement federated data governance by using semantic web technologies, but these are still immature systems.

Therefore, there is an open research need for designing and developing federated metadata management systems.

3   Introducing Blockchain-Powered Metadata Catalog

Section 2 describes the challenges associated with building efficient data governance in data mesh. In this section, we briefly describe 3 types of metadata systems that can be implemented by companies. Afterward, we proceed with a description of how Hyperledger Fabric can be used for developing an effective federated catalog that satisfies security, traceability, transparency, immutability, and interoperability demands.

3.1   Metadata Catalog Types

The theoretical work of authors defines 3 types of metadata catalogs in [6].

In Type I all metadata records are kept in a central repository. To enable the data access control, it presents the *visibility* and *access* functions that are associated with each user. The initial metadata is generated upon data product release or update and pushed to the repository that is available to the participants of the mesh (potentially through the web portal interface).

In Type II the repository of records is distributed. Instead of pushing the metadata about data products into a central repository, each domain of the mesh hosts an instance of the metadata catalog. The system is also composed of a data synchronization algorithm and standardized data access policies.

Type III catalog is seen as a complete system's decentralization with no single repository or standardized procedures. Each domain (or a subset of domains) is free to implement the storage and governing policies however it likes, but it is still required to provide the metadata exchange, discovery, and query interfaces.

3.2   Using Hyperledger Fabric for Metadata Management

HyperLedger Fabric (HLF) platform [1] is an open-source project hosted by the Linux Foundation. Its modular architecture allows using various computational components depending on the system requirements: pluggable consensus protocol, identity providers, and transaction endorsement policy. If there are additional security demands, it allows the exploitation of network segmentation via channels and private data exchange algorithms.

We describe how Hyperledger Fabric fits the Type II catalog thereafter.

---

[2] https://atlas.apache.org/.

Chaincode and Platform Governance. A smart contract is a universal way of enforcing some pre-determined, beforehand agreed procedure over the given asset during the parties' interaction. HLF defines a notion of a *chaincode* that is assembled from one or more smart contracts and policies. It is the smallest software module deployed as a Docker container. Chaincode policies determine how the underlying contracts should be executed. For instance, a policy can define which nodes should perform an endorsement of the proposed transaction.

On the other hand, chaincode itself can be seen as a transparent and interoperable governing tool for interacting with the ledger. It guarantees a standardized way of modifying the information and performs an ongoing metadata integrity verification which fits the implementation requirements of the Type II catalog.

Upon the chaincode execution, it is also possible to emit events. Therefore, the catalog users (or simply nodes) can listen to the specific types of ledger updates and discover the newly published information.

Another distinctive characteristic of the HLF is that developers can define smart contracts with general-purpose programming languages like Java, Go, or JavaScript/TypeScript (NodeJS-based contracts). This is contrary to such platforms as Ethereum which makes the technology adaption curve easier in general. In some way, it also offers interoperability capabilities since it is possible to develop a number of chaincode contracts using any supported language. The only requirement for using such a contract is the agreement of the network majority.

Overall, platform control is done through the policies management mechanisms. Policies of the HLF (as a part of a single chaincode package or of the global channel configuration) represent how members come to the agreement of accepting or rejecting changes to the network, channel, or smart contract.

Private Communication. Permissioned (private) blockchain provides more advantages for building the Type II metadata catalog compared to the permissionless (public) one. For example, domain or user identification allows to design more secure systems with access control management. For more details, we refer the reader to [6] which outlines the benefits of using a permissioned blockchain.

In its nature, HLF is a permissioned blockchain platform that runs the private network of uniquely identified components by using the **M**embership **S**ervice **P**roviders (MSP). We may imagine the MSP as a certificate authority extension that establishes the identity of each element of the network - organizations, nodes, applications, and policies. The whole network is structured as a number of non-overlapping communication *channels*. To give an analogy, the channel resembles a subnet of the Classless Inter-Domain Routing network[3].

Private data exchange is another privacy-preserving technique that lets to share secret data only with intended parties. The channel segregation method limits the participating parties from access to the whole ledger. By contrast, the private exchange mechanism lets to monitor that data sharing has taken place without the data itself, for instance, when secretly passing the data product access credentials and recording it on the ledger.

---

[3] https://datatracker.ietf.org/doc/html/rfc1519.

Ledger and Consensus Algorithm. When implementing the metadata management system with the help of blockchain technologies, the ledger takes the role of the underlying metadata records storage medium. Its properties such as distribution, immutability (with historical information as a side effect), and cryptographic signatures suit well the required needs.

Data indexing improves the search performance in a big pile of records which is often the case for immutable ledgers. In HLF the equivalent indexing functionality is done via the ledger state database. This database keeps only the latest modification over the given asset identified by a unique key. It is implemented using either the key-value (LevelDB) or key-document (CouchDB) stores.

For maintaining the ledger in the synchronized state and to make the configuration upgrades possible in the network, HLF provides several options for implementing a consensus algorithm. For the extreme cases when the blockchain users operate in a trustless environment, it is possible to use the Byzantine Fault Tolerance protocol. It might be well suited for the cross-organizational (meta)data exchange initiatives (like in healthcare or finance) when security and network poisoning fault-tolerance are of high importance.

By contrast, if the environment is partially or fully trusted (like within the same organization), it might be sensible to implement a more performant Crash Fault Tolerant consensus algorithm. In fact, due to the modular architecture, it is possible to use distinct protocols for ledger or configuration updates.

In this section, we discussed 3 metadata catalog types: centralized, distributed, and decentralized. We also presented the advantages of employing a blockchain platform for building a distributed catalog. Next, we present our proof-of-concept catalog model and the implementation based on the Hyperledger Fabric.

4   Metadata Catalog Model and Implementation

Our research contribution is three-fold. First, we derive our model from the works on the federated data exchange [12] and provenance metadata management [5] by proposing novel asset structures to be used in the metadata catalog.

Second, we define a catalog architecture integrated with data mesh products and a smart contract that consists of several functions for using the ledger as the metadata information store.

Third, we provide 4 scenarios of how the proposed model and contract would be used for managing the data products metadata in the context of a data mesh.

4.1   Ledger Asset Structure

At the baseline, the metadata should describe the **D**ata **P**roduct (DP) from different perspectives to facilitate its discovery, addressing, understanding, and manipulation. The standard functional classification includes operational (location, size), technical (format, type, schema), and business (notions, context, process lifecycle) metadata.

Our proposed catalog ledger structure is comprised of the following assets:

– DataProductAsset - describes the main information such as name, location
– MetadataAsset - describes the DP's metadata such as owner, state, lineage
– ConsumptionRequestAsset - defines a request to use the published DP
– ConsumptionResponseAsset - defines a response to the open DP's request
– ConsumptionResponseUpdateAsset - defines an update to the response

These assets (with detailed definitions available in Appendix A) enable the data mesh governance in the following aspects.

The Lineage field allows products discovery and risk assessments. By traversing and studying the dependency graph users can project the product failure consequences, or they may decide to consume more coarse or refined products.

The Schema and SampleDataLocation fields help with DP understanding and interoperability. Schema may contain semantics, while samples represent the underlying data that is open to everyone. SampleDataLocation helps with the development of more refined products without administrative approval delays.

The asset ownership and integrity verification is implemented through the cryptographic hashes mechanism. The hash matching during the data product processing is essential for mitigating fraudulent activities.

Consumption-related assets (Request-Response-Update) are used for access control and usage tracking. It helps to implement security and regulation compliance measures in the first place.

4.2   Chaincode Operations and Catalog Architecture

As mentioned in Sect. 3.2, we consider the smart contract machinery, and the chaincode in particular, as an enforcement tool of the pre-agreed governance.

Our proposed chaincode contracts allow the user to define and record in the metadata catalog new assets with various types of information:

– RegisterDataProduct function records metadata regarding new product in the data mesh by creating DataProductAsset and MetadataAsset objects
– RequestDataConsumption function records information about data product access request by creating ConsumptionRequestAsset object
– ResolveConsumptionRequest function records the response to the data product access request by creating ConsumptionResponseAsset object
– UpdateDataProduct function records a new asset with updated metadata information and old asset references by creating new DataProductAsset and MetadataAsset objects
– UpdateConsumptionResponse function records a response update information by creating ConsumptionResponseUpdateAsset object (optional, but only relevant in case of a previous UpdateDataProduct function call)
– GetAllAssets function returns all available assets in the ledger

Classical Hyperledger Fabric network architecture includes a communication channel, blockchain nodes, ledger storage, and chaincode. The applications (or data products) and **C**ertificate **A**uthorities (CAs) are considered outside of the network but they are still communicating with nodes.

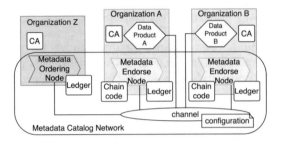

Fig. 1. Proof-of-concept metadata catalog architecture

We provide our proof-of-concept catalog architecture in Fig. 1. We use two organizations to demonstrate the data product registration and consumption interactions. It is a simplified plot since it does not contain all HLF platform elements (like Membership Service Provider or Private Data Store). However, it represents the most important parts of the system such as metadata nodes that own the chaincode and/or the ledger.

The endorsement node is used for validating the proposed transactions with new assets to be appended in the ledger. In the default policy setup, the majority of endorsing nodes is required meaning that both Organization A and Organization B should approve the transaction for accepting it. Organization Z holds only an ordering node that does not affect the approval or refusal decision, but it is solely responsible for ordering a number of transactions into a block that will be added to the ledger. It guarantees that blocks are identical across nodes.

In an example scenario, before connecting to the running metadata catalog network, a new organization would have to setup a CA infrastructure. This infrastructure is wholly owned by the organization and is necessary to generate identities for its domains, data products, feedback loops, etc. After intermediating the request to join the catalog, the MSP would assign the organization roles and rights based on the network configuration policies. The assigned roles and rights define how and who this organization can interact with.

4.3   Use-Cases for Distributed Metadata Catalog in Data Mesh

In this subsection, we present 4 use-cases for using the proposed distributed catalog in the context of a data mesh platform of the video-service company.

In the basic case, a **D**ata **P**roduct **O**wner from organization A (DPO-A) wants to register a new product in the catalog. For doing this, one has to use the RegisterDataProduct function by passing the DataProductAsset structure (a JSON that corresponds to the type fields) and some other parameters (name, description, consumed products). When the new asset transaction is validated, it will be recorded into the block and synchronized across the different nodes, and the new data product will become available to other domains. The sequence diagram of the whole process is demonstrated in Fig. 2 which also shows the final ledger and PromoAd Costs product state.

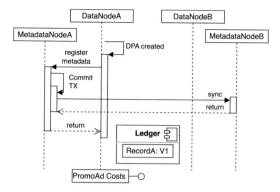

Fig. 2. Process of a metadata asset registration

In the second case that is shown in Fig. 3, a DPO from organization B (DPO-B) wants to consume a new **D**ata **P**roduct from organization A (DP-A) that was created before. For doing it, one has to use the RequestDataConsumption function by passing the desired product information (product A in this case) and the access rights. When the transaction is accepted it is reflected on the ledger and the request becomes available to the DPO-A for review. To approve or deny the request, DPO-A has to use the ResolveConsumptionRequest function by passing the request ID and the rights to be granted. If the response was positive (e.g. "read" right was granted), DPO-B can setup the product consumption, and the lineage relationship is established between DP-A and DP-B.

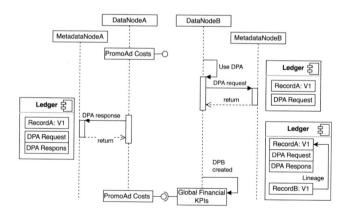

Fig. 3. Process of requesting a metadata asset and responding to the request

In the third case, we have a settled product consumption chain where the end users can see the movie promotional list. Then, a product owner from organization A decides to make an update of the existing DP-A. When the update is recorded, the versioning relationship arises since the ledger will contain a new asset pointing to the old DP-A asset. As soon as the update is propagated across the network, Node B is notified of the update and it can automatically stop the DP-B if there is a breaking change of DP-A (as shown in Fig. 4).

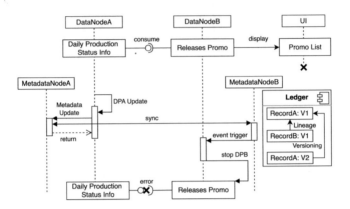

Fig. 4. Process of metadata asset update and the update event handling

In the last case shown in Fig. 5, a DPO-A also decides to update its data product consumption responses. Such action is optional but it is useful in the sense that it brings automatic forward compatibility of the previously approved requests. When the response update is recorded and propagated, Node B is notified and proceeds with testing the new setup of the data product from organization A. If the consumption attempt is successful, then it can re-publish the updated version of the product B metadata reflecting the recent changes of the upstream dependency graph and re-establishing the data flow.

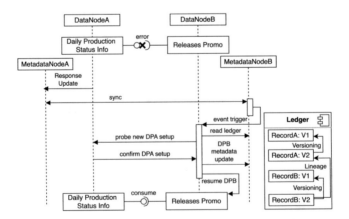

Fig. 5. Process of response asset update and automatic consumption recovering

5   Contribution Discussion

This section shows the advantages of the proposed model and its implementation, as well as missing requirements and potential solutions.

5.1   Model Pros and Cons

Sawadogo et al. [19] outlined 6 main requirements for the metadata systems in the data lake context: data indexing (DI), data versioning (DV), polymorphic data (PD), semantic enrichment (SE), link generation (LG), and usage tracking (UT). Nonetheless, these features are still relevant and necessary for constructing efficient metadata systems in the data mesh.

Dehghani [4] did not state any need for a metadata management system per se, but rather attributed all the mentioned properties as part of the data product design itself. Although, we find it conceptually easier to attribute these to a metadata catalog along with other federated governance requirements, including, but not limited to, standards and policies as computational blocks (PCB), system evaluation feedback loops (FL), and governance leverage points (LP) [17].

Since data mesh builds upon the micro-services architectures ideas, we also consider the necessary properties of being independently deployable (ID) and automatically testable (AT). A conclusive summary is displayed in Table 1.

Table 1. Supported metadata management system requirements

LP	FL	AT	LG	SE	PD	DI	DV	UT	PCB	ID
✗	✗	✗	*	*	*	✓	✓	✓	✓	✓

✓– supported ✗– unsupported ∗ – partially supported

5.2   Potential Model Improvements

Data product polymorphism (or polyglot data) can not be represented at once as a single asset definition with multiple dimensions (stream, batch, report, etc.). But it is possible to register multiple product metadata assets related to each output dimension. Semantic enrichment and link generation are not implemented automatically in the system. The user still has to pass such information as consumed DPs, schema, and description. It can be improved by extending the model for multiple product dimensions and by introducing artificial intelligence tools for automatic metadata extraction, such as data similarity and schema linking.

Automated testing is a big challenge that developers still face when making upgrades in any large distributed system. Apparently, it is also inherent to our proposed catalog model and it is difficult to point to any viable solution. Still, HLF has *channeling* support that can be used for testing any platform upgrades without affecting the main catalog network.

Feedback loops and leverage points are other important but challenging elements with no evident solution. Donella Meadows, the author of these concepts, mentions the use of global incentives that may benefit the implementation of feedback loops and leverage points. Introducing such incentives component easily aligns with the nature of blockchain platforms in the first place. However, we think that it is also required to have well-integrated federated governance and self-serve infrastructure platforms which is not the case today.

6 Conclusions and Further Research

Overall, this paper presented a novel approach for implementing metadata catalog in data mesh architecture. At first, we inspected the challenges of federated governance and the limits of existing systems. Second, we described the benefits of using Hyperledger Fabric. Third, we proposed a new, blockchain-based metadata model, distributed system architecture, and demonstrated 4 relevant metadata catalog use cases for enabling computational governance in data mesh.

Our further work will focus on extending the proposed model and running experiments for estimating the model performance throughput. We will also investigate graph-based distributed ledger technologies for improving catalog functionalities and make a comparative study of the construction process of a Type III metadata catalog on top of property graphs.

A Metadata Catalog Assets

```
type DataProductAsset struct {
    Name                 string    'json :"Name" '
    DataHash             string    'json :"DataHash" '
    DataLocation         string    'json :"DataLocation" '
    SampleDataLocation   string    'json :"SampleDataLocation" '
    Schema               string    'json :"Schema" '
    Version              string    'json :"Version" '    }
type MetadataAsset struct {
    ID            string            'json :"ID" '
    IsDeleted     bool              'json :"IsDeleted" '
    CreateTime    time . Time       'json :"CreateTime" '
    Description   string            'json :"Description" '
    Name          string            'json :"Name" '
    Owner         string            'json :"Owner" '
    OldAssetId    string            'json :"OldMetadata" '
    Product       *DataProductAsset 'json :"Product" '
    Lineage       [] MetadataAsset  'json :"Lineage" '  }
type ConsumptionRequestAsset struct {
    ID               string           'json :"ID" '
    Name             string           'json :"Name" '
    RequestTime      time . Time      'json :"RequestTime" '
    Owner            string           'json :"Owner" '
    RequestedAsset   *MetadataAsset   'json :"RequestedAsset" '
    Rights           [] Right         'json :"Rights" '  }
type ConsumptionResponseAsset struct {
    ID            string                     'json :"ID" '
    Name          string                     'json :"Name" '
    Owner         string                     'json :"Owner" '
    ResponseTime  time . Time                'json :"ResponseTime" '
    Request       *ConsumptionRequestAsset   'json :"Request" '
```

```
ProductInfo    *MetadataAsset                    'json:"ProductInfo"'
GrantedRights  [] Right                          'json:"GrantedRights"' }
type ConsumptionResponseUpdateAsset struct {
    ID            string                         'json:"ID"'
    AutoApprove   bool                           'json:"AutoApprove"'
    Name          string                         'json:"Name"'
    Owner         string                         'json:"Owner"'
    ProductInfo   *MetadataAsset                 'json:"ProductInfo"'
    Responses     [] ConsumptionResponseAsset    'json:"Request"'
    UpdateTime    time.Time                      'json:"UpdateTime"' }
```

References

1. Androulaki, E., et al.: Hyperledger fabric: a distributed operating system for permissioned blockchains. In: Proceedings of the Thirteenth EuroSys Conference, pp. 1–15 (2018)
2. Cherradi, M., EL Haddadi, A., Routaib, H.: Data lake management based on DLDS approach. In: Ben Ahmed, M., Teodorescu, H.-N.L., Mazri, T., Subashini, P., Boudhir, A.A. (eds.) Networking, Intelligent Systems and Security. SIST, vol. 237, pp. 679–690. Springer, Singapore (2022). https://doi.org/10.1007/978-981-16-3637-0_48
3. Child, A.W., Hinds, J., Sheneman, L., Buerki, S.: Centralized project-specific metadata platforms: toolkit provides new perspectives on open data management within multi-institution and multidisciplinary research projects. BMC. Res. Notes 15(1), 106 (2022)
4. Dehghani, Z.: Data Mesh: Delivering Data-Driven Value at Scale. O'Reilly (2022)
5. Demichev, A., Kryukov, A., Prikhodko, N.: The approach to managing provenance metadata and data access rights in distributed storage using the hyperledger blockchain platform. In: Ivannikov Ispras Open Conference. IEEE (2018)
6. Dolhopolov, A., Castelltort, A., Laurent, A.: Exploring the benefits of blockchain-powered metadata catalogs in data mesh architecture. In: Proceedings of the 15th International Conference on Management of Digital EcoSystems. Springer (2023). https://conferences.sigappfr.org/medes2023/
7. Driessen, S., Monsieur, G., van den Heuvel, W.J.: Data product metadata management: an industrial perspective. In: Troya, J., et al. (eds.) ICSOC 2022. LNCS, vol. 13821, pp. 237–248. Springer, Cham (2023). https://doi.org/10.1007/978-3-031-26507-5_19
8. Halevy, A.Y., et al.: Managing Google's data lake: an overview of the goods system. IEEE Data Eng. Bull. 39(3), 5–14 (2016)
9. Hillmann, D.I., Marker, R., Brady, C.: Metadata standards and applications. Ser. Libr. 54(1–2), 7–21 (2008)
10. Hooshmand, Y., Resch, J., Wischnewski, P., Patil, P.: From a monolithic PLM landscape to a federated domain and data mesh. Proc. Des. Soc. 2, 713–722 (2022)
11. Inmon, W., Strauss, D., Neushloss, G.: DW 2.0: The Architecture for the Next Generation of Data Warehousing. Elsevier (2010)
12. Koscina, M., Manset, D., Negri-Ribalta, C., Perez, O.: Enabling trust in healthcare data exchange with a federated blockchain-based architecture. In: International Conference on Web Intelligence-Companion Volume (2019)
13. Laurent, A., Laurent, D., Madera, C.: Data Lakes. Wiley, Hoboken (2020)
14. Liu, L., Li, X., Au, M.H., Fan, Z., Meng, X.: Metadata privacy preservation for blockchain-based healthcare systems. In: Bhattacharya, A., et al. (eds.) DASFAA

2022. LNCS, vol. 13245, pp. 404–412. Springer, Cham (2022). https://doi.org/10.1007/978-3-031-00123-9_33

15. Machado, I.A., Costa, C., Santos, M.Y.: Data mesh: concepts and principles of a paradigm shift in data architectures. Procedia Comput. Sci. **196**, 263–271 (2022)
16. Majchrzak, J., Balnojan, S., Siwiak, M., Sieraczkiewicz, M.: Data Mesh in Action. Manning Publishing (2022)
17. Meadows, D.H.: Leverage points: places to intervene in a system (1999)
18. Newman, S.: Building Microservices. O'Reilly Media, Inc. (2015)
19. Sawadogo, P., Darmont, J.: On data lake architectures and metadata management. J. Intell. Inf. Syst. **56**(1), 97–120 (2021)
20. Sawadogo, P.N., Darmont, J., Noûs, C.: Joint management and analysis of textual documents and tabular data within the AUDAL data lake. In: Bellatreche, L., Dumas, M., Karras, P., Matulevičius, R. (eds.) ADBIS 2021. LNCS, vol. 12843, pp. 88–101. Springer, Cham (2021). https://doi.org/10.1007/978-3-030-82472-3_8
21. Zhao, Y.: Metadata management for data lake governance. Ph.D. thesis, Univ. Toulouse 1 (2021)

Workshop on Beyond the Promises of Web3.0: Foundations and Challenges of Trust Decentralization (WEB3-TRUST)

Workshop on Beyond the Promises of Web3.0: Foundations and Challenges of Trust Decentralization (WEB3-TRUST)

Web 3.0 has arisen as the next step into the configuration of a more secure and trustworthy internet. This new stage in the deployment of web technologies is based on the implementation of new architectures for trust decentralization. Distributed ledger technologies and, in specific, blockchain are used as the main components to manage trust without any central authority. Nonetheless, the deployment of such technologies in real and practical scenarios is of problematic nature, and on many occasions, this leads to the re-centralization of decision taking. This being the case, governance and the supposed equality provided by blockchain and DLT are hindered, which eventually determine cybersecurity risks of major impact than those in web 2.0.

This workshop is devoted to discussing these shortcomings and the associated cyber-risks, taking into account the current state of maturity of blockchain, smart contracts and the new governance schemes in the blockchain era.

Workshop Organization

Organizing Committee

David Arroyo   CSIC, Spain
Jesús Díaz Vico   IOHK (EE.UU.)

Program Committee

Luca Nizzardo   Protocol Labs, Spain
Antonio Nappa   Universidad Carlos III de Madrid, Spain
Andrés Marín López   Universidad Carlos III de Madrid, Spain
Andrea Vesco   Head of cybersecurity at LINKS Foundation, Italy
Pedro López   CSIC, Spain
Mayank Dhiman   Dropbox, USA

Decentralized Oracle Networks (DONs) Provision for DAML Smart Contracts

Iqra Mustafa[1]([⊠]) [ID], Bart Cant[2] [ID], Alan McGibney[1] [ID], and Susan Rea[1] [ID]

[1] Munster Technological University, Nimbus Research Centre, Cork, Ireland
iqra.mustafa@mycit.ie
[2] Rethink Ledgers, Charlotte, NC, USA
https://www.nimbus.cit.ie/, https://rtledgers.com/

Abstract. Decentralized oracle networks (DONs) provide a mechanism for interoperability within blockchain networks, additionally, they can address potential security vulnerabilities associated with smart contracts reliant on external data feeds. Decentralized oracle networks (DONs), allow smart contracts to access data from multiple sources without the need for a centralized authority, which can reduce the risk of manipulation or data tampering. However, not all platforms support DONs, including Digital Asset Modelling Language (DAML). DAML is a privacy-focused smart contract language that is designed to support multi-party applications, but it does not have native support for external data feeds. Furthermore, its applicability is limited by the lack of an integrated trust mechanism for external oracle feeds. Hence, this study proposes a DAML-DONs system which is comprised of five components: DON Broker Contract, Service Registry, Ranking Engine, Web-service, and an Interactive User Interface with Chatbot Assistance. By combining these five components, the proposed DAML-DON system can provide a robust and trustworthy solution for integrating DAML Ledger with external oracle networks. Additionally, it increases the interoperability of DAML smart contracts and contributes to enhancing the effectiveness and quality of using the services provided by Oracles.

Keywords: Digital Asset Modeling Language (DAML) · Decentralized Oracle Networks · Smart Contracts · Blockchain · API · Reputation Management System

1 Background and Related Work

Decentralized oracle networks (DONs), also known as blockchain middleware, are crucial elements of a blockchain ecosystem as they extend the functionality of blockchains by connecting smart contracts to real-world data, events, off-chain computations, etc., in a tamper-resistant and reliable manner [2]. Without

This research work is supported by Science Foundation Ireland Centre for Research Training focused on Future Networks and the Internet of Things (AdvanceCRT), under Grant number 18/CRT/6222.

oracles, smart contracts could only access the transactions recorded on their own networks, limiting their scope and potential functionality. However, by providing a universal gateway to off-chain resources while maintaining the blockchain's core security features, DONs enhance smart contract functionality by enabling them to securely access external data or send data to off-chain systems. DONs operate as a data source and as a layer that employs different Oracle decision models to query, examine, verify, and validate resources from external services for use on the blockchain [1, 2].

DONs compile multiple external data streams using a variety of methods, such as Application Programming Interfaces (APIs), Software Development Kits (SDKs), tamper-proof data, IoT sensor data, automation functions, big data, or others and provide these as a signed message on the blockchain. The external data can be of any type, for example, weather conditions, successful payments, and price fluctuations. To request data from the outside world, a smart contract needs to be triggered, and network resources are *spent*. This results in network transaction charges, such as "gas," in the context of Ethereum [3–5]. Some oracles can send data back to external sources in addition to relaying it to smart contracts. The provision of smart contract data feeds is also impacted by three crucial issues: trustee concerns, communication issues, and data reliability [6]. For blockchain-based distributed service providers, trust and communication issues are simple to resolve as blockchain provides traceability and anonymity in our communications, whereas the quality of the data source has a direct impact on the data's reliability.

The parameter values collected from external feeds have a significant impact on the viability of smart contracts. When a smart contract relies on a new trusted intermediary, the oracle in this case, it could be violating the security and reduced-trust models of blockchain applications. This may result in malicious smart contracts, which provide criminals with the opportunity to maximize their gains through engaging in illegal activity and seriously undermine the security of cyberspace. Criminals have the incentive to corrupt the parameters of these contracts to increase their legitimacy [7]. As such, ensuring the reliability of data sources has become a priority for oracle providers in recent years, leading to the development of numerous types of reliable decentralized oracles, including Provable, Witnet, Cryptocompare, the Band protocol, DIA, API3, and Chainlink [8]. These oracle providers have support for multiple platforms, e.g., Ethereum, Bitcoin, etc. However, Digital Assets Modelling Language (DAML), which is a platform-independent smart contract language, cannot leverage these external oracles because calling external APIs in response to contract execution is not supported. This means that developers who want to use external data in their DAML smart contracts would need to use a different mechanism, such as a centralized oracle, to access that data. This limits the applicability of DAML-based contracts[1]. Additionally, DAML has no consensus mechanism, which is indeed the utmost factor required to utilize the external oracle feed services. So, this issue raises three major questions:

---

[1] https://blog.digitalasset.com/developers/calling-any-api-through-daml.

1. *What is the profile of DAML Clients that require access to oracle data feeds?*
   Chainlink's existing data feeds, i.e., BTC-ETH can be a critical component for non-enterprise and DAML small clients for real-time asset trading, parametric insurance, and online gaming. Large DAML clients such as Goldman Sachs will use price data from costly sources such as Bloomberg, but small start-ups do not have the budget and may be more interested in other solutions, for example, a Chainlink price feed as it is more cost-effective.

2. *Is it possible to execute an external service inside DAML Smart Contracts?*
   At present, no such mechanism exists to achieve this from a DAML perspective. The DAML ledger can be considered a "passive" ledger where external clients are permitted to add entries in accordance with certain rules. The rules are expressed in the DAML language (DAML logic encoded in choices). The rules must be followed to enter information or write something on the ledger, but even more crucially, the rules must be proven to have been followed. This means that other DAML users must be able to view the ledger's state prior to that new entry/transaction and are able to replay the computation of all the rules involved. If those rules, aka the DAML code, allowed us to reach out to an external API, there would be no way for other users to double-check that the specific rules have been followed, as they would not know what response was received from the API.

3. *Under what conditions can DAML clients utilize external feeds?*
   To use oracle data, a DAML user needs to decide to trust the oracle provider. In other words, a DAML user must declare which oracle to trust when consuming reference data feeds. This trust decision is important because the accuracy and reliability of the oracle data can impact the integrity of the entire DAML contract. One way to establish trust is through a reputation system. In the context of oracles, a reputation system can help to quantify the trustworthiness of oracle providers based on their past performance. Moreover, according to [9–11], reputation is defined as a party's prior behaviours about intents and norms. Thus, mutual trust between users and service providers is crucial for both parties. The users need to trust oracle providers, and providers also need to trust users to uphold their contractual obligations. Therefore, a mechanism is required to quantify the trustworthiness between both the service provider and the data consumer.

Hence, to reconcile with these issues a secure service stream (i.e. API) for the ledger's Active Contract Set needs to be built along with a reputation management system to ensure trust between both parties i.e., DAML users and oracle providers. Thus, this paper presents a DAML-DON system to connect the DAML Ledger with external oracle feeds in a secure and reliable manner. The proposed contributions of the research presented in this paper are outlined as follows:

1. Extending DAML to support DONs via a secure web service. This service connects DONs and a DAML broker contract, i.e., for real-time trading tailored to their requirements.

2. The proposed system is equipped with a ranking engine that provides capabilities such as a service registry (store logs), oracle service ranking, eval-

uating user/client reputation, and rewarding users, which are implemented automatically during the evaluation stage of the proposed DAML-DONs system component. It enables trust between DAML clients and Oracle service providers. This can save clients time and cost in finding the best fit for their application prior to any client encounter with any oracle provider.
3. An interactive user interface, including chatbot assistance aids users in intelligently submitting data request forms by reducing errors. Additionally, displaying the chosen data provider's reputation analytics. Due to the cloud-based nature of the solution, DAML clients are freed from the burden of information storage and computation.

The remaining paper is organised as follows: Sect. 2 summarises the need for decentralized oracles for DAML. Section 3 describes the components of the proposed DAML-DONs System. While Sect. 4 deals with the experimental evaluation, case study along with results, and analysis of the proposed system. Finally, Sect. 5 concludes the DAML-DONs system.

2   Decentralized Oracles for DAML

Decentralized apps (dApp) require current, up-to-date data to be consumed in order to function effectively. A modern illustration of this is DeFi apps, which require timely price feeds to trade assets in real-time. A DeFi protocol or dApp may also occasionally require the historical price of a certain asset at a specific point in time. A financial product that demands price comparison over time, such as Bitcoin insurance, is an example of this use case. This insurance service requires historical data to determine market volatility and to dynamically adjust premiums[2]. To support scenarios such as this, there are various categories of blockchain oracles available, such as centralized, distributed, contract-specific, software, hardware, human, inbound/outbound, and consensus-based oracles. The functionality of each oracle is entirely reliant on the purpose for which it is intended for [12,13].

However, for private-permissioned BCs, DAML goes beyond a smart contract modelling language; it is a complete platform for building full-stack applications, the smart contracts can work with many distributed ledger technologies (DLTs) and databases such as VMWare, Hyperledger Sawtooth, Amazon Aurora, Amazon QLDB etc. In this context, the key difference between DAML and any other platform is its sub-transaction privacy, which ensures that each participant involved in a transaction only sees the parts of the transaction that they are authorised to see. On the surface, the DAML language allows users to specify which parties view which parts of a transaction, and the Canton protocol allows the smart contract to plug into the consensus layer of different distributed ledgers and databases while preserving privacy guarantees. This is particularly

---

[2] https://blog.chain.link/historical-cryptocurrency-price-data/.

relevant for financial institutions and others who require transaction confidentiality[3]. The language, in particular, adds a layer of security by including privacy and authentication as language features.

Apart from its benefits, the integration of oracles with the DAML ledger poses challenges because, by design, DAML restricts direct interaction with external APIs. This is crucial, as the execution of DAML needs to be fully deterministic in order for different parties to independently verify the result. If we enabled interaction with external APIs, we could get different results each time it is invoked, making it hard to verify execution. Hence, there is a need to develop a mechanism that offers an integration of external oracles with DAML without compromising its security and deterministic nature.

3   Components of DAML-DONs System Architecture

The system architecture shown in Fig. 1 consists of a Broker Contract, Registry Contract, Ranking Engine, DAML Hub (a cloud platform for deploying DAML contracts), DONs, a Community Cloud Governance (CCG) hosting the proposed DAML-DON Web service, and an Interactive User Interface (UI) with chatbot assistance. The descriptions of each of these components are given below:

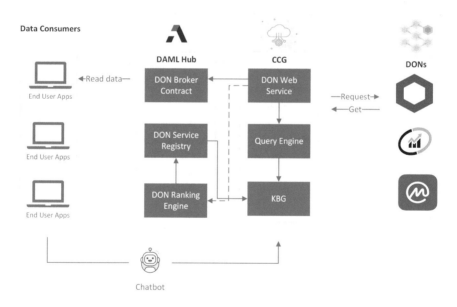

Fig. 1. DAML-DONs System Architecture

---

[3] https://docs.daml.com/.

3.1   Broker Contract

The DAML broker contract provides an information gateway for other contracts running on the same ledger. It enables parties to negotiate and execute financial transactions efficiently. It includes the terms of the transaction, such as the type of asset being traded, the price, and any other relevant details. The broker contract updates its state in response to transactions sent to it, in this case, data feeds retrieved from the oracle. This updated state can be accessed by various contracts (i.e., contracts for services or goods) so they can collaborate and work together to achieve a common goal such as the creation or execution of an agreement. It creates a more interconnected and integrated decentralized system which acts as a secure and trusted intermediary between the ledger and external feeds. This can help to improve the efficiency and effectiveness of the system as a whole.

3.2   Service Registry Contract

The service registry contract that resides on-chain maintains a record of a registry mapping for each oracle name, address, description, and ranking as *(name, address, description, rating)*. This is an interconnected metadata structure in which nodes represent the oracle name, address, description and rating where edges indicate the relationship between these nodes. Before invoking the Broker Contract, the Service Registry Contract enables efficient, secure, and transparent service discovery and usage within a decentralized network based on the oracle reputation. This helps to ensure that only reliable and trustworthy oracle services are used by the DAML smart contracts. The Ranking Engine Sect. 3.3 component of the system is responsible for updating the registry each time a user interacts with a service. This is done to reflect the most up-to-date reputation of the oracle services in the network. The Knowledge Base Graph (KBG) is stored off-chain and is periodically updated based on the revised results from the Ranking Engine. The system intends to create a robust architecture for service discovery, ensuring the use of reliable oracle services within the decentralized network by integrating the on-chain service registry contract and the off-chain Knowledge Base Graph.

3.3   Ranking Engine

To use external oracle feeds, a DAML user needs a mechanism to evaluate the trustworthiness of the service provider. In other words, any contract consuming that reference data needs to declare which oracles to trust. Thus, this trust between DAML users and oracles is established only when a whole group of parties endorse an oracle service provider. For this purpose, a ranking engine component has been introduced. Its functions include evaluating user reputation, rating oracle services, and rewarding users. The proposed ranking engine facilitates oracle selection to establish a reliable relationship between DAML users and oracle service providers. Its objective is to improve the robustness of service providers by leveraging user feedback to identify high-quality service providers and eliminate those that are unreliable or of low quality. Hence, users can avoid spending gas fees on substandard oracles prior to their interaction with any service.

3.3.1 Reputation Evaluation Mechanism.
The reputation evaluation mechanism includes oracle's reputation, user reputation, and reward incentivization to the user.

3.3.1.1 Oracle Service Provider Reputation (R_{O_k}).
The DAML user accesses the oracle service to get the required data feed to perform the task. After encountering the service, the user is prompted to rate it against the following factors such as *platform support, price, timeliness of data feeds, compliance with industry standards & regulations, and information accuracy*. Based on their experience and reputation, the oracle provider reputation is calculated by the designed Algorithm 1. As an input we have taken a number of users U_i, previous timestamp epoch p_{ts}, current timestamp epoch c_{ts}, the oracle O_k, user previous reputation score PR_{u_i} and the trust score T_s. The trust score T_s is a real number in a range $[1, 5]$ that indicates the DAML user satisfaction towards oracle service provider O_k. A trust rating of 5 indicates complete trust between the oracle service provider O_k and the user U_i and a value of 1 shows that the oracle provider is not reliable. The new feedback, however, carries more weight in the trust score, which is indicated by the symbol α.

Algorithm 1. Oracle Service Provider Reputation

1: **procedure** ORACLE REPUTATION
2:     **Input:** $U_i, T_s, p_{ts}, c_{ts}, PR_{u_i}, \alpha, O_k$
3:     **Output:** R_{O_k}
4:     $T_s \in R[1, 5]$
5:     $\alpha \leftarrow p_{ts}/c_{ts}$ $where$ $0 < \alpha < 1$
6:     $R_{O_k}^u = [tanh(\alpha \times PR_{u_i} \times T_s) + 1] * 2$
7:     $\bar{R}_{O_k} = \sum_{r=0}^{1} R_{O_k}^u / \sum_{r=1}^{r'} U_i$
8: **end procedure**

$PR_{u_i} \in R[1, 5]$ and its value is initially set to 1 for all users whereas $\bar{R}_{O_k}$ represents the average of each oracle reputation ranging $R[0, 1]$.

3.3.1.2 User Reputation RP_{u_i}.
The user's reputation score computation in relation to the oracle service requires three parameters namely type of service f_r, transaction cost β, and the user's past reputation score PR_{u_i} in order to prevent reputation attacks such as collusion attack (see Sect. 4.2.3).

1. **Type of Service f_r:** The user rating is restricted based on the type of service. For instance, if an application is consuming a service type that updates

frequently i.e., currency or weather feed then for that type of service the user can give a rating once a day. However, if a fluctuation factor associated with data feed is not high i.e., trading medical or surgical equipment then they are bound to rate a service once a week. If a user/data consumer U_k uses the same service S_k from the oracle provider O_k daily, it is possible that the user U_k and the oracle O_k collude to submit a good rating. The proposed system aims to reduce the probability of false or biased ratings by restricting the frequency of ratings f_r.

2. **Transaction cost β:** The transaction cost is associated with the service gas charges spent on each transaction i.e., the transaction costs associated with the gas charges required to execute each transaction when using Ethereum-supported platforms such as Decentralized Oracle Networks (DONs) to retrieve data feeds. If a user U_k utilized the same service S_k from an oracle O_k by spending a gas worth of $\geq 10,000 USD$. Then the weighting factor $\beta \in R[0,1]$ of transaction fee computed as follows:

$$\beta = tanh(p_s/10000) \tag{1}$$

Normalize the value of β to scale in the range $[0,1]$:

$$\beta = [tanh(p_s/10000) + 1]/2 \tag{2}$$

where $p_s \in Z$ represents the gas fee spent on each transaction. The β will increase if the value of gas spent p_s is higher.

3. **User previous reputation score PR_{u_i}:** The rating of an oracle service ought to be connected to the service providers' or users' prior reputation scores, as legitimate users typically provide more fair ratings than the malicious users. Therefore, the previous reputation score of the user is also considered as the weighting factor of the rating, denoted as $PR_{u_i} \in R[1,5]$. Where the value of PR_{u_i} is initially set to 1 for all users. The user reputation score can be computed as follows:

$$RP_{u_i} = PR_{u_i} \times \beta \times T_s \tag{3}$$

Normalize the value of RP_{u_i} to scale in the range $R[1,5]$:

$$RP_{u_i} = [tanh(PR_{u_i} \times \beta \times T_s) + 1] * 2 \tag{4}$$

A data consumer can decide whether or not to trust an oracle provider service based on the user-provided feedback. Notably, in the user reputation computation, the time interval factor has been ignored because an external service client can be a large enterprise or a small enterprise (SME). The possibility is that large enterprises can be more frequent users of the particular service, which might not be the case with SMEs. Therefore, each user's ratings for the oracle service, whether they emanate from a large or small enterprise, will take into account equally every time.

4. **Reward System:** The reward computation, Eq. 5, is dependent on the weighting factor of the gas spent p_s, previous user reputation PR_{u_i}, and initial reward r when utilizing a service i.e., S_k from an oracle provider O_k.

$$Reward = PR_{u_i} \times p_s \times r \tag{5}$$

where the value of initial reward r is in USD and for all users, it is set to 2% as a baseline for evaluation and fine-tuning. This percentage serves as a starting point to assess the system's effectiveness and can be adjusted based on observed outcomes and system goals.

The user can provide feedback and ratings after using the service and based on the fee spent on utilizing the oracle service and their prior reputation score, the user will get a reward. This incentivizes users to actively engage with the service, contribute valuable feedback, and maintain a positive reputation, as it directly impacts the reward they receive.

3.4   DAML-DON Web Service

The proposed system architecture is designed to support DAML interaction with external oracles without compromising the deterministic nature of DAML. Therefore, the DAML-DONs system will ensure trust between DAML users and external oracles via our web service. It is a critical component of the system which act as a bridge between the DAML ledger and DONs. It serves the following purposes:

– Integration with regular oracle APIs for current and historical data feeds i.e., Chainlink[4].
– DAML contract call to DAML-DONs API to initiate two-way communication between DAML and external oracle data feed on contract creation.

The sequence flow of the web service component interaction in the DAML-DONs system is illustrated in Fig. 2. To initiate the data request process, the DAML broker contract makes a request to the proposed web service using the provided JWT (JSON Web Tokens) token of the broker contract party (i.e., controller) to authenticate the request. Once the web service receives the request, it directs it to the appropriate oracle(s) to retrieve the requested data feed. Consequently, in return, the web service receives the data from the oracle, which it transmits to the broker contract for further processing or storage on the DAML ledger. By using a JWT token to authenticate the request, ensures that only authorized parties are able to access the requested data. Hence, the web-service flow has the potential to provide a secure and decentralized method for accessing external oracles while maintaining the DAML properties such as privacy and authentication.

---

[4] https://chain.link/.

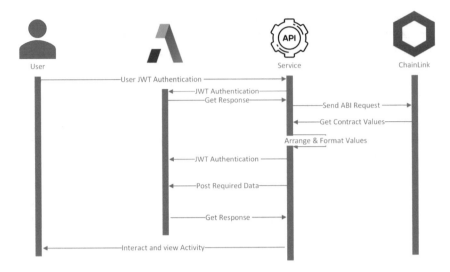

Fig. 2. DAML-DONs Service Sequence Flow

3.5 Interactive User Interface

The user interface (UI) component with chatbot assistance has been designed to enable users' interaction with the DAML-DONs system components via the DAML-DON service. Additionally, users have access to oracle rankings and feedback to aid in selecting an oracle service. They can vote for a mutually agreed-upon oracle and digitally sign their agreement to utilize the chosen service, ensuring transparency, consensus, and a secure commitment. To do so, the UI component provides users with a data request form in which they can specify their data requirements such as an authentication token, an oracle address (agreed upon oracle service), and an HTTP payload to exchange the data feeds between service providers and brokers via web service (detailed later in Sect. 4.2). If the request has been accepted by the service provider and broker contract, it also offers a dashboard where data consumers can check the status of those requests and view fetched results. If the user requires assistance in completing the data request form, a chatbot service provides the necessary guidance to help them with customized queries in the request forms. The chatbot service relies on the information from a knowledge base to generate the conversation responses.

4 Experimental Evaluation and Results

This section will present an experimental setup first, followed by a case study used to evaluate the proposed study's methodology and findings.

4.1 Experimental Environment

The experimental settings for evaluating the proposed DAML-DON system are listed in Table 1: 1) DAML Smart Contracts; 2) DAML HUB for deployment and testing of DAML Contracts (i.e., broker, registry, ranking engine) with the proposed DONs Web-service. It is a cloud platform that enables DAML users to build simpler and more scalable serverless backends for their applications. Also, to design, create, and test new features, it allows interaction with live ledgers; 3) .NET 6 used to develop the DONs web service; 4) Infura to connect the DON service to the Ethereum platforms for testing and 5) Chatbot developed using Power Virtual Agent (PVA). To implement the experiment Windows 10 operating system with Visual Studio Code, Visual Studio 2022 and DAML v 2.2.0 has been used.

Table 1. Experimental Environment for testing DAML-DONs System

Name	Tool and Version
Programming Language	DAML v2.2.0; .NET 6.0
Network Environment and Testing	DAML HUB; Infura
Editor	Visual Studio Code; Visual Studio 2022
Chatbot Assistance	Microsoft Power Virtual Agent
Operating System	Windows 10

4.2 Case Study and Results

In order to evaluate the proposed DAML-DONs system workflow, a case study is implemented Sect. 4.2.2. For this study, the following oracle platforms are listed in the registry for the DAML Ledger support, namely, Chainlink[5], Cryptocompare[6], and CoinMarketCap[7] to fetch the historical and up-to-date data feed for DAML clients. The proposed system can be extended to other platforms too.

4.2.1 Use Case Scenarios. In order to analyse and evaluate the proposed workflow's functionality the following scenarios have been used:

– **Price validation:** In this model, $User_A$ is suggesting a trade/transfer to $User_B$ for a Transfer of $Asset_X$. $User_A$ is proposing a $Price_Y$ for $Asset_X$. In this case, $User_B$ could leverage an Oracle Request for getting an independent Price for $Asset_X$ before it agrees to the trade/transfer.

---

[5] https://chain.link/.
[6] https://www.cryptocompare.com/.
[7] https://coinmarketcap.com/.

- **Trade based on agreed oracles:** In this model, $User_A$ is suggesting a trade/transfer to $User_B$ for $Asset_X$ and $User_A$ and $User_B$ have previously agreed upon to use an Oracle Services. When the Trade is proposed for $Asset_X$, $User_A$ is requesting the Price and the Oracle provides a $Price_Y$. This is the price that is used as part of the trade/transfer.
- **Conditional trade based on price X:** In this model $User_A$ and $User_B$ have agreed upon a $Price_Y$ for $Asset_X$. An oracle service will send an event to the DAML contract if the Oracle Price is matching to $Price_Y$, then the trade/transfer is automatically executed.

The DAML Contract and Oracle have a bilateral agreement represented by the first model, whereas the single centralized oracle is depicted by the second model. The third model, however, reflects the DAML contract's on-request data feed. As a result, three asset trading Broker DAML smart contracts have been designed to cover each of the above-mentioned scenarios and are deployed on the DAML Hub.

4.2.2   Case Study. There are many reliable and authenticated DONs platforms available in the crypto marketplace that offer services such as time & interval data, capital markets data, benchmark reference data, geolocation, flight statistics data, etc.[8]. On decentralized oracle platforms, there are sellers and buyers. Sellers are those who sell their services to the DON, whereas buyers are those who utilize them. As such DONs' services are grouped in the knowledge base (KB), as there are multiple service providers offering the same service that has been made available for our system users with ratings and reviews based on price, platform support, and customer satisfaction. If a DAML client leverages the DAML-DON system for utilizing the external data feed. They need a relevant broker contract. For instance: for the experimental purpose we have included the second scenario namely *"Trade price based on agreed upon Oracle"* related Broker Contract whose step-by-step working and evaluation has been illustrated. The contract is intended to automatically initiate asset trades based on the most recent ETH/USD price feed between parties from the three above-mentioned platforms i.e., Chainlink, Cryptocompare, and CoinMarketCap; individually.

Three different templates make up the broker contract namely,

- Wallet
- OracleRequest (params, oracle link fee)
- OracleResponse (Store oracle data feed)

- *Step A:* In Fig. 3, the broker contract wallet has three parties i.e., owner, issuer, and oracleparty, where an owner can only see their own wallet; the issuer can see everybody's, and an oracleparty is a listener to any call out for requesting the price for ETH and update the TokenPrice in USD by applying the ETH Price. Thus, if the DAML Clients are interested in utilizing the price

---

[8] https://cointelegraph.com/.

```
template Wallet
  with
    owner: Party
    issuer: Party
    oracleparty : Party
    tokens: Int
    tokenpriceUSD : Optional Decimal
    tokenpriceETH : Decimal
    tokenexchangerate : Optional Decimal
  where
    signatory issuer
    observer owner, oracleparty
    key issuer: Party
    maintainer key
```

Fig. 3. Wallet Template

feeds from an external oracle for asset trading via DAML-DONs system they can use the UI component to trigger/invoke the DON web service, see Fig. 1. To do so, the data requester (authenticated party of the Broker Contract, i.e., the controller of the choice - oracleparty) enters the JWT token also referred to as a ledger access token, that complies with the $OAuth2.0$ standard for authentication in the data request form. Now the interaction between the broker contract and web service has started as well as the chatbot assistance activated.

– *Step B:* Following successful authentication, the user *issuer* is directed to a request form with data request fields such as the data source URL to be queried or the oracle address. At this step, to fill out the request form the user that is *issuer* in this scenario is assisted by the interactive chatbot.
– *Step C:* When a user requests data, the chatbot receives the request-data intent through API calls, and this retrieves a response from the knowledge base that facilitates retrieval and sharing of the previously collected knowledge of the Oracle services by service registry as well as their reputation ratings computed by the ranking engine. As a result, the chatbot aids the requester by providing them with relevant information utilising the knowledge base and guiding them, for instance, when they are having trouble filling out the request form or in oracles selection.
– *Step D:* After receiving the results in *Step C*, the user will request the digital signatures of other involved parties to confirm their agreement to utilize the selected oracle service. This step ensures that the user actively chooses and commits to using a specific oracle within the network, and it involves obtaining the digital signatures of relevant parties to ensure consensus and commitment. In this scenario, for the ETH/USD token price update, the

Chainlink oracle service has been selected based on its high reputation compared to other oracles. Once the *issuer* and the involved parties agree on the selected best-fit oracle, the remaining parameters of the request form, such as the payload, need to be filled in. These parameters include the *oracle USDE-THexhange* (a DAML contract variable that requires an assigned value), the choice-name to be exercised (*UpdateTokenPrice*), the template-id, and the key of the OracleRequest (refer to Fig. 4a and 4b). Thus, our Query Engine will send a query on behalf of the OracleRequest contract to the specific DON through our REST API web service to retrieve the results. These results will be appended to the DAML ledger, allowing other data consumers to utilize the data for asset trading.

Notably, we have used the contract key instead of using the contract-id because the contract-id changes each time on exercising the choice. Whereas a contract key is similar to a primary key for a database, it specifies a method of uniquely identifying contracts using the parameters of the template. This primarily serves to offer uniqueness assurances, which are useful in many scenarios. One is that they can serve to de-duplicate data coming from external sources.

– *Step E:* The Wallet is created/updated and the trade is placed after getting the data feed from the oracle, see Fig. 4c. However, we won't terminate the session until we received the client's feedback about the quality of the oracle service. Then based on their reputation score the monetary incentive has granted and the service registry has updated. The chatbot processor takes customer feedback and directed it to the ranking engine Sect. 3.3 which in turn evaluates the results and updates the knowledge base in accordance with the scoring metrics.

Figure 4 shows that the interaction between external oracle and broker contract has been successfully established via DON's service. Since the proposed service is a REST API so, we cannot invoke it from within the DAML contract due to no HTTP PUSH or PULL support in DAML ledger. However, it can be achieved by gRPC API. The difference between the proposed service and gRPC is the service invocation procedure. The only way to trigger an external application from DAML is to create a ledger event without creating a contract by the choice observer mechanism. In this case, a party on the ledger exercises a non-consuming choice on a contract, which does nothing. A client application, running on behalf of a choice observer can see the event. This pattern can only be used on the gRPC API.

4.2.3   Reliability Analysis. In order to demonstrate the proposed DAML-DONs System's reliability, this subsection first addresses how resistant it is to modification and common attacks, after which it evaluates the effectiveness of the reputation evaluation system and the reward incentivization method.

– Firstly, the proposed system can protect the oracle service information and reputation scores from modification by storing the oracle provider data on

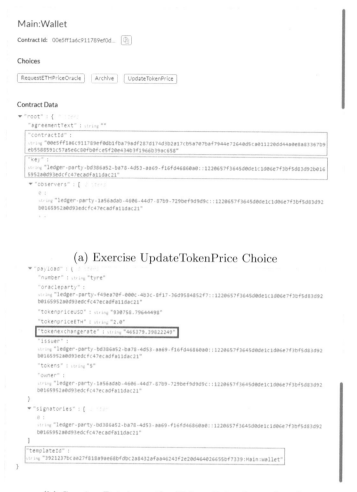

(a) Exercise UpdateTokenPrice Choice

(b) Service Retrieves the Token Price from Oracle

(c) Result

Fig. 4. DAML-DONs Interaction Outcome

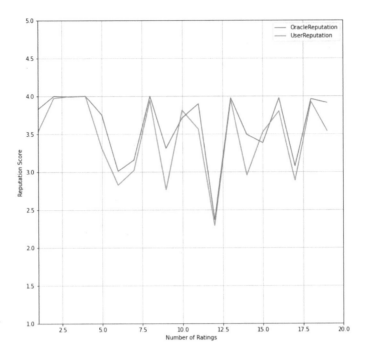

Fig. 5. Reputation scores of users and oracles with the increase of the number of ratings given by others

the DAML ledger and using the Knowledge base graph (KBG) to access the results, see Sect. 3.2. The data is tamper-evident since the registry contract is kept on the DAML ledger. This can help to guarantee the integrity of the data stored in the registry. Moreover, by using the KBG the users can discover trusted oracle providers without repeatedly calling the service registry. This may reduce the load on the registry and increase system efficiency.

– The ranking engine component of the system effectively resists the unfair rating attack and collusion attack. The unfair rating attack involves negative and fake reviews with low ratings to decrease the reputation scores of others (i.e., oracle service provider competitors). This attack is damaging and can lead to a decrease in oracle provider revenue. In the proposed system, malicious attackers will find it difficult and expensive to decrease the reputation ratings of others because of the following factors i.e., 1) Restricted rating frequency based on service type and 2) Transaction Cost. They have to endure an extended wait to rate the service (see Sect. 3.3.1.2) and spend a lot of money using other oracle services to offer unfair ratings. The consequences of malicious user ratings on other users' reputation scores are quite minimal because the prior reputation scores of users are also taken into account when calculating reputation. However, in a collusion attack, multiple users collude to rate each other favourably in order to establish a high reputation. For the following reasons, it is difficult to have a major impact on these users' reputation scores under the proposed approach. They need to conduct costly transactions in order to give each other positive ratings and to re-rate the service they need to wait for long hours (i.e., at least 24 hrs, depending on the service type).

However, the performance of the proposed ranking engine (reputation evaluation mechanism) is illustrated in Fig. 5: In Fig. 5, the number of ratings for the user and oracle is shown according to the reputation score. This image clearly illustrates that the oracle reputation score varies according to the rating provided by the user. The influence of rating is calculated based on the users' reputation and the value consumed (i.e., gas spent) by them. The results show that the proposed system is robust enough to respond abruptly based on the user rating. The proposed system will absorb the impact of malicious users and incentivise the trusted user with the system-calculated reward (Eq. 5).

5   Conclusion

The proposed DAML-DON system is built on a five-component architecture and is intended to offer a complete solution for integrating DAML Ledger with external oracle networks. These elements consist of the DON Broker Contract, Service Registry, Ranking Engine, Web-service, and an Interactive User Interface (UI) with Chatbot Assistance. The broker contract acts as an information gateway for other contracts running on the same ledger. It ensures that the provided data feed by oracle is usable within DAML client applications, thereby increasing the overall reliability and security of the system. The service registry component provides a registry of available external oracle services along with a ranking that can be used by the DAML users before encountering any service. This helps to simplify the process of finding and selecting trusted oracle services for specific use cases. However, the ranking engine is in charge of evaluating user reputation and implementing reward incentivization for users in addition to ranking the external oracle services based on their performance history and other relevant metrics. It quantifies mutual trust between the user and an oracle service provider. Whereas the web service offers a standard interface for utilizing broker contracts to access external oracle services. The interactions between the DAML smart contracts and the external oracle services are managed by the interactive (UI) with Chatbot Assistance. In conclusion, the proposed DAML-DONs system offers a reliable, secure and trusted mechanism for integrating the DAML Ledger to external oracle networks and enhancing the interoperability of DAML smart contracts.

References

1. Mammadzada, K., Iqbal, M., Milani, F., García-Bañuelos, L., Matulevičius, R.: Blockchain oracles: a framework for blockchain-based applications. In: Asatiani, A., et al. (eds.) BPM 2020. LNBIP, vol. 393, pp. 19–34. Springer, Cham (2020). https://doi.org/10.1007/978-3-030-58779-6_2
2. Al-Breiki, H., et al.: Trustworthy blockchain oracles: review, comparison, and open research challenges. IEEE Access **8**, 85675–85685 (2020)
3. Cai, Y., et al.: Truthful decentralized blockchain oracles. Int. J. Netw. Manag. **32**(2), e2179 (2022)
4. Pasdar, A., Lee, Y.C., Dong, Z.: Connect API with blockchain: a survey on blockchain oracle implementation. ACM Comput. Surv. **55**(10), 1–39 (2022)
5. Kaleem, M., Shi, W.: Demystifying pythia: a survey of chainlink oracles usage on ethereum. In: Bernhard, M., et al. (eds.) FC 2021. LNCS, vol. 12676, pp. 115–123. Springer, Heidelberg (2021). https://doi.org/10.1007/978-3-662-63958-0_10
6. He, J., et al.: SDFS: a scalable data feed service for smart contracts. In: 2019 IEEE 10th International Conference on Software Engineering and Service Science (ICSESS). IEEE (2019)
7. Ma, L., et al.: Reliable decentralized oracle with mechanisms for verification and disputation. In: 2019 Seventh International Symposium on Computing and Networking Workshops (CANDARW). IEEE (2019)
8. Shi, P., et al.: Blockchain-based trusted data sharing among trusted stakeholders in IoT. Softw. Pract. Exp. **51**(10), 2051–2064 (2021)
9. Jøsang, A.: Trust and reputation systems. In: Aldini, A., Gorrieri, R. (eds.) FOSAD 2006-2007. LNCS, vol. 4677, pp. 209–245. Springer, Heidelberg (2007). https://doi.org/10.1007/978-3-540-74810-6_8
10. Zhou, Z., et al.: Blockchain-based decentralized reputation system in E-commerce environment. Future Gener. Comput. Syst. **124**, 155–167 (2021)
11. Xiong, A., et al.: A truthful and reliable incentive mechanism for federated learning based on reputation mechanism and reverse auction. Electronics **12**(3), 517 (2023)
12. Beniiche, A.: A study of blockchain oracles. arXiv preprint arXiv:2004.07140 (2020)
13. Pasdar, A., Lee, Y.C., Dong, Z.: Connect API with blockchain: a survey on blockchain oracle implementation. ACM Comput. Surv. **55**(10), 1–39 (2023)

Analyzing the Effectiveness of Native Token Airdrop Campaigns in NFT Marketplaces

Paul Kuhle[1]([✉]) [iD], David Arroyo[2] [iD], and Pablo de Andrés[3] [iD]

[1] Finance and Marketing Department, Universidad Autónoma de Madrid,
C/Francisco Tomás y Valiente, 5, 28049 Cantoblanco, Madrid, Spain
paul.kuhle@estudiante.uam.es

[2] Consejo Superior de Investigaciones Científicas, Instituto de Tecnologías Físicas y
de la Información, Serrano 144, 28006 Madrid, Spain
david.arroyo@csic.es

[3] Finance and Marketing Department, Universidad Autónoma de Madrid and ECGI,
C/Francisco Tomás y Valiente, 5, 28049 Cantoblanco, Madrid, Spain
p.andres@uam.es

Abstract. NFT marketplaces are increasingly using free token airdrops as a customer acquisition strategy. This paper presents a descriptive analysis of the airdrop campaigns of LooksRare and X2Y2 as a case study. Through this analysis, we identify key research questions about studying airdrops as a marketing strategy in this context. In the absence of a comprehensive financial model to analyze airdrop campaign effectiveness, we propose a set of key metrics. We find that while both campaigns managed to attract new customers, the LooksRare campaign was overall more successful by most metrics. However, customer conversion and retention rates are low on both platforms, suggesting that the airdrop campaigns may not be sustainable in the long term.

Keywords: Airdrops · NFT · DeFi · Ethereum · Blockchain

1 Introduction

Non-Fungible Tokens (NFTs) have become one of the most talked-about topics in the world of blockchain and cryptocurrency in recent years. NFTs are unique digital assets that represent ownership of a particular item, such as artwork, music, videos, and other types of digital content. Unlike traditional cryptocurrencies, NFTs cannot be exchanged for an equivalent value, as each NFT is unique and has its own intrinsic value.

NFTs are typically traded on NFT marketplaces, which feature increasingly complex tokenomics, incentive schemes and customer acquisition strategies. Many marketplaces have launched their own native ERC-20 token to facilitate these mechanisms. As part of their initial customer acquisition campaign, some of these marketplaces distributed their native tokens to eligible users for free

J. M. Machado et al. (Eds.): BLOCKCHAIN 2023, LNNS 778, pp. 383–393, 2023.
https://doi.org/10.1007/978-3-031-45155-3_37

through *airdrops*. In this paper, we present a descriptive analysis of the airdrop campaigns of LooksRare and X2Y2, two recent campaigns of marketplaces that gained significant marketshare.

In the following section, we describe the dynamics of token airdrops. We then present case studies of LooksRare and X2Y2 and propose key performance indicators to compare the two campaigns. We conclude with a discussion of the key research questions that arise from our analysis.

2   Dynamics of Token Airdrops

The term *airdrop* refers to the free distribution of a certain number of tokens or NFTs to a group of users. The delivery mechanism of the airdrop depends on operational and financial considerations [6]. Although the purpose of an airdrop can vary, some common reasons for airdropping tokens are marketing, decentralization efforts and liquidity provision [1].

While any tradeable token can be airdropped, we focus on the airdrops of ERC-20 tokens created by NFT marketplaces. NFT marketplaces may launch an ERC-20 token for a variety of reasons, including fundraising, governance, incentive/reward schemes, founding team incentives and treasury management. However, not all NFT marketplaces have done so. In fact, Opensea, the largest NFT marketplace during most of 2022, does not have a native token. Perhaps this can be attributed to the fact that the marketplace was one of the first open marketplaces for NFTs when it was founded in 2017 and did not need a native token as a differentiator or marketing tool.

In this paper, we focus on the airdrop campaigns LooksRare and X2Y2, which both launched in early 2022 and managed to gain significant marketshare in the NFT ecosystem. These marketplaces were chosen due to their comparability and concurrent launch. While other NFT marketplaces have also implemented airdrop campaigns, they were excluded from our analysis for various reasons, such as ongoing status (e.g., Blur), lack of momentum, or non-comparability with our selected cases. We prioritize recent campaigns that align with LooksRare and X2Y2 in terms of launch timeframe to ensure an accurate and up-to-date analysis.

3   Related Work

Our paper is directly related to the study of NFTs. Existing research papers often focus on specific NFT collections, their characteristics and userbase [7,11]. Some of these papers focus on early NFT collections such as CryptoPunks, which are not representative of the whole market and do not use the newer ERC-721 and ERC-1155 standards. As it is shown in [13], the collection-based research can be expanded by characterizing the entire Opensea marketplace. Empirical research with a financial focus has also been conducted, as it is conducted in [2,5] by studying NFT pricing and related questions. Other work on NFTs tends to focus on fraudulent activity such as wash-trading [12], pump-and-dump schemes [8] or other security issues [4].

The literature on token airdrops, especially in the context of governance tokens and decentralized communities, is also closely related to our work. Several scholars analyze the benefits and motivation behind launching an airdrop campaign and how they fit into the governance of decentralized organizations [1,9]. Others study airdrops as a mechanism to promote financial inclusion in the DeFi ecosystem [3]. Lastly, some papers focus on technical aspects and cost estimations for airdrop campaigns [6].

4   Dataset

In this paper, we work with on-chain data extracted from our own self-hosted full-archive Erigon node, using TrueBlocks to query the Ethereum blockchain. The latest block we have extracted data from is block 15 965 524. Our dataset includes the airdrop transactions of LooksRare and X2Y2 as well as all of their ERC-20 transfer events that were registered on-chain. In total, we have extracted 1 822 605 transactions and their logs.

5   Case Studies

In this section, we present case studies of the airdrop campaigns of LooksRare and X2Y2. We describe the tokenomics of each native token and provide descriptive statistics of the airdrop campaigns.

LooksRare

LooksRare is an NFT marketplace that launched in 2022 with the aim of creating a decentralized platform for digital art and collectibles. The platform uses its native token, $LOOKS, to incentivize users to stake, trade and participate in the ecosystem.

Tokenomics. The $LOOKS token is an ERC-20 token with a fixed total supply of 1 billion. The token allocation is as follows: 12% for airdrops, 3.3% for strategic sale, 1.7% for liquidity management, 44.1% for volume rewards, 18.9% for staking rewards, 10% for the team, and 10% for the treasury. The airdrops, strategic sale and liquidity management tokens are available on launch, while the rest are released linearly over 720 days. The token has two functions, as it can be used both as a utility token to be used within the LooksRare ecosystem, e.g. for staking rewards, and a tradeable token. The trade functionality of the team, treasury and strategic sale tokens is unlocked gradually over time to prevent dumping.

The $LOOKS token is designed to encourage participation in the ecosystem by rewarding users for staking and trading on the platform. Users can stake their tokens in a smart contract and receive a variable amount of $LOOKS tokens per block as staking rewards. Stakers also receive a percentage of all trading fees paid by users on the platform. In addition to staking rewards, users are also rewarded with $LOOKS tokens for listing and trading NFTs on the platform. The token has no governance functions at the moment, although the website states that it is planned to introduce them in the future.

Airdrop Campaign. The LooksRare airdrop campaign was divided into nine tiers, with each tier receiving a different amount of tokens. Users were eligible for the airdrop based on their trading volume in ETH on the Opensea marketplace[1]. A minimum volume of 3 ETH was required to be eligible for the lowest tier (Tier 9, 125 tokens). The required volume for the highest tier was 1000 ETH (Tier 1, 10 000 tokens). Users were also required to either list or buy an NFT on the LooksRare marketplace.

During the airdrop campaign, 124 665 unique wallets claimed $LOOKS tokens. The total amount of $LOOKS tokens distributed is 98 366 440. This means that 81.97% of the $LOOKS tokens allotted to the airdrop campaign (120 000 000 or 12% of the total supply) were distributed. Figure 1a shows the distribution of the airdropped $LOOKS tokens by tier and the maximum amount of tokens available in each tier. The remaining 21 633 518 $LOOKS tokens were transfered to the LooksRare treasury.

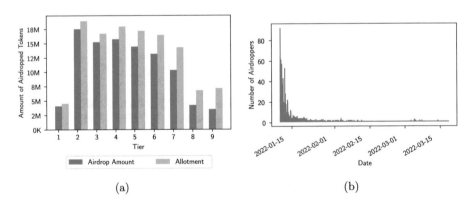

Fig. 1. LooksRare airdrop analysis: (a) airdrop distribution and allotment by tier

The campaign was more successful in the tiers that required the most trading volume. While the mid-tiers still had a relatively high distribution rate, a much lower rate was achieved in tiers 7, 8 and 9. The low distribution rate cannot be attributed to an oversupply of available tokens in those tiers because the eligibility criteria were static and only considered activity before the campaign had started.

---

[1] Trades between blocks 12 642 194 and 13 812 868 (only Ethereum) were considered.

As shown in Fig. 1b, the majority of the $LOOKS tokens were distributed in the first 10 days of the campaign. The campaign closed on March 18, 2022.

X2Y2

X2Y2 is a decentralized platform that aims to create a fair and transparent marketplace for NFTs. Its native token, $X2Y2, is an ERC-20 token that serves as the utility and reward token of the platform.

Tokenomics. The $X2Y2 token has a fixed total supply of 1 billion tokens, which are allocated as follows: 12% for the airdrop, 1.5% for the Initial Liquidity Offering (ILO) for the $X2Y2/$ETH pair on Uniswap, 1.5% for liquidity management, 20% for staking rewards, 45% for NFT staking rewards, 10% for the team, and 10% for the treasury.

To prevent "wash trading and whale domination", X2Y2 does not have any trading rewards or private sale. Instead, the marketplace offered an ILO. The $X2Y2 token does not have any governance functions at the moment.

Airdrop Campaign. The airdrop campaign was open to all users who had traded on Opensea's smart contract from its deployment until block 13 916 166. Users were also required to list NFTs or NFT collections on X2Y2's marketplace.

The userbase was divided into two categories: *Whales* were defined as wallets with more than 30 ETH in eligible trading volume on Opensea. Those users could claim up to 1000 $X2Y2 tokens. A total of 41 592 000 tokens were allotted to this group.

All other users were classified as *people* and were allotted a total of 78 408 000 tokens. The amount of tokens that each wallet categorized as *people* could claim was based on the contribution of each wallet to the total eligible trading volume on Opensea, excluding *whales*. Additionally, the amount of tokens that every wallet could claim was limited by the number of NFTs or NFT collections that they listed on X2Y2's marketplace.

As shown in Table 1, the campaign was notably more successful in reaching *whales* rather than *people*, as demonstrated by both the total number of tokens distributed and the percentage of available tokens claimed. Overall, only 23.7% of the tokens dedicated to the airdrop campaign were claimed.

Table 1. X2Y2 airdrop campaign statistics

	tokens			
	# airdroppers	# available	# claimed	% claimed
Whales	17 324	41 592 000	17 324 000	41.1%
People	25 443	78 408 000	10 707 293	13.7%
Total	42 767	120 000 000	28 031 293	23.3%

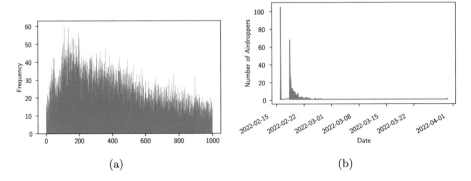

Fig. 2. X2Y2 airdrop analysis: (a) X2Y2 airdrop distribution by tokens claimed, excluding *whales*, (b) Timeline of the X2Y2 airdrop campaign

As shown in Fig. 2a, the X2Y2 airdrop campaign did not define tiers for the *people* category. The resulting distribution of the amount of claimed tokens is slightly skewed towards the lower end of the spectrum, with a median of 380 and a mean of 421.

Similarly to LooksRare, the majority of all claimed tokens were distributed at the beginning of the airdrop campaign, as shown in Fig. 2b. Due to technical difficulties with the eligibility verification process and overall insatisfactory user experience, the airdrop campaign was paused briefly by the team on February 16th, 2022. Any unclaimed \$X2Y2 tokens were permanently locked in the smart contract when the campaign ended on April 1st, 2022. The last airdrop was claimed in block 14 487 427 on March 30, 2022.

6 Results

Despite the popularity of airdrops, there is currently no comprehensive financial model for analyzing the success of airdrop campaigns. To gain insight into the results of the campaigns, we evaluate the initial distribution of tokens, token sell-off behavior, conversion rates, and engagement with rewards systems as key performance indicators (Table 2).

Initial Distribution. While LooksRare was able to distribute 82% of the tokens reserved for the airdrop, X2Y2 only distributed 23.3% of the available tokens and consequently had fewer airdroppers. Both campaigns were notably more successful in attracting *power users*, i.e. users with a high trading volume on Opensea.

Sell-Off Behavior. As of November 2022, the majority of all airdropped tokens from both campaigns have been sold off. We define a wallet as having sold all of its airdropped tokens if its balance is 0 \$LOOKS or \$X2Y2 tokens in its own wallet

Table 2. Summary of airdrop campaign results

		LooksRare	X2Y2
Token distribution	Total token supply	120 000 000	120 000 000
	Tokens claimed	98 366 440	28 031 293
	Tokens claimed (%)	82.0	23.3
	Airdroppers	124 665	42 767
Token sell-off	Sold airdrop tokens	84 166 645	25 962 097
	Sold airdrop tokens (%)	85.6	92.6
	Wallets that sold all airdrop tokens (%)	77.5	92.1
Customer conversion	Conversion (%)	24.2	37.1
	Total NFT trade volume in USD	1550.1	714.7
	Fraction of trade volume attributed to airdroppers (%)	42.1	20.5
Engagement	Total staking volume in USD	13 418 605 254	589 164 935
	Fraction of staking volume attributed to airdroppers (%)	17.7	29.7
	Total native token trade volume in USD	18 371 393 926	1 425 257 214
	Fraction of native token trade volume attributed to airdroppers (%)	19.5	16.1
	Airdroppers that interacted with rewards system	49 589	7040
	Airdroppers that interacted with rewards system (%)	40.2	17.4

as well as the respective fee sharing or staking contract. As shown in Fig. 3, the airdroppers' trading volume of the ERC-20 tokens has also declined significantly. Given that airdroppers generally sold their tokens quickly after receiving them from the airdrop contract, a large portion of the early trading activity can be attributed to the dumping of those tokens.

Conversion. The eligibility criteria of both campaigns only required users to list a token on the exchange. Only 24% of all airdroppers on LooksRare and 37% of all airdroppers on X2Y2 have actually completed a trade on the respective marketplace, which is a low conversion rate.

Nevertheless, airdroppers are responsible for a significant portion of the total NFT trading volume on both exchanges. Even though the number of active airdroppers has been declining on both platforms, they make up for more than

40% of the total NFT trading volume on LooksRare and 20% of the total NFT trading volume on X2Y2.

Engagement with Rewards Systems. LooksRare was notibly more successful in engaging airdroppers with its rewards system. More than 40% of all LooksRare airdroppers have interacted with the fee sharing scheme, whereas only 17.4% of all X2Y2 airdroppers have interacted with the X2Y2 rewards system. The total ERC-20 trading volume and staking volume also shows better engagement with the LooksRare ecosystem compared to X2Y2. However, X2Y2's airdroppers are responsible for a larger share of the staking volume than LooksRare's airdroppers.

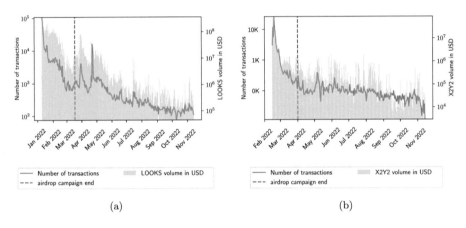

(a)   (b)

Fig. 3. (a) timeline of the LooksRare airdrop campaign, (c) daily trading volume of `$LOOKS` only for airdroppers (logarithmic scale), (b) daily trading volume of `$X2Y2` only for airdroppers (logarithmic scale)

7 Discussion and Future Research

We find that the X2Y2 campaign was less successful than the LooksRare campaign, with fewer tokens claimed and less trading volume generated compared to non-airdroppers. This could be due to the campaign's complex eligibility criteria, technical difficulties during the airdrop campaign, or lower interest in the marketplace and its native token.

We also observe that airdroppers tend to sell their tokens quickly after claiming them, indicating low engagement with the marketplace's incentive systems. This is supported by low conversion and reward system participant numbers, as well as the decline in the native token's trading volume by airdroppers. However, further research is necessary to understand the dynamics of the selling process and airdroppers' motivations for participating in the campaign.

Despite these limitations, both airdrop campaigns successfully attracted users and generated NFT trading volume. However, further research is needed to evaluate the sustainability of that trading activity. It is possible that the trading volume was driven by misaligned staking and fee sharing incentives, as observed in previous studies. [10] Future research should also include the gamified airdrop campaign of the Blur marketplace, which incentivizes loyalty throughout the campaign and has been able to gain significant market share since its launch.

While the proposed categories of performance indicators in this paper are useful for analyzing the success of airdrop campaigns, individual metrics may not work for all campaigns. For example, a campaign that requires users to complete an NFT trade on the platform would have a conversion rate of 100%, but the stricter eligibility criteria would likely affect token distribution metrics. Thus, it is important to tailor performance indicators to the specific objectives and requirements of each airdrop campaign.

Based on our findings, we propose the following future research lines:

Analyzing the Impact of Airdrops on Customer Acquisition in NFT Marketplaces

One important research line is to investigate why airdrop campaigns primarily appeal to power users and whether such users are merely financial speculators or loyal customers who engage with the platforms incentive system in a sustainable manner. Moreover, we have yet to understand the differences between airdroppers and regular customers.

Research on ERC-20 tokens has shown that airdrops can have a positive effect on overall trading volume and token price [3]. However, it is unclear if the same effect applies to the trading volume of NFT marketplaces and not just the native token.

Examining the Tokenomics and Incentive Schemes of NFT Marketplaces

Another line of research could focus on the tokenomics and incentive schemes of NFT marketplaces. It has yet to be analyzed if the tokenomics models of LooksRare and X2Y2 are sustainable in the long run and whether the features that supposedly protect the marketplace from fraudulent activity work as intended.

Future research could investigate whether the airdrop campaign contributed to the wash trading activity on LooksRare and X2Y2.

Examining the Financial Viability of Airdrop Campaigns

A central research question is whether an airdrop campaign is a financially viable customer acquisition strategy for an NFT marketplace. One way to approach this question is to evaluate the lifetime value of the airdropped users against the costs of the campaign.

However, estimating the costs of the campaign is challenging. In addition to the direct costs of implementing and marketing the airdrop campaign, which are generally not made public, the value of the ERC-20 token given away for free must be considered. The token's price is subject to change over time and is also influenced by the airdrop campaign itself. It may be necessary to aim for conservative estimates or ceilings.

8 Conclusions

Both airdrop campaigns succeeded in attracting new users to the two NFT marketplaces, although the LooksRare campaign was more successful by most metrics like token distribution and participation in the rewards system. On LooksRare, airdroppers are responsible for over 42% of the NFT trading volume, compared to X2Y2's 20%.

However, we also find that both campaigns had very low customer retention and conversion rates. The majority of airdroppers sold their tokens within a few days of claiming them and did not participate in the rewards system. The trading volume of airdroppers peaked at the beginning of the airdrop campaign and has since declined significantly.

We have also identified several research questions that warrant further investigation. In particular, we propose to analyze the impact of airdrop campaigns on customer acquisition in NFT marketplaces, the tokenomics and incentive schemes of NFT marketplaces, and the financial viability of airdrop campaigns.

Acknowledgements. We acknowledge the financial support provided by the Doctorados Industriales project financed by the Comunidad de Madrid (Spain) (IND2020/SOC-17536), and the NEOTEC program by the Spanish CDTI-E.P.E. (SNEO-20201032). This work has also been supported by the SPIRS Project with Grant Agreement No. 952622 under the EU H2020 research and innovation programme and by Madrid regional CYNAMON project (P2018/TCS-4566), co-financed by European Structural Funds ESF and FEDER. We also wish to extend a special thanks to Thomas Jay Rush (TrueBlocks LLC) for his valuable contributions.

References

1. Allen, D.W., Berg, C., Lane, A.M.: Why airdrop cryptocurrency tokens? J. Bus. Res. **163**, 113945 (2023). https://doi.org/10.1016/j.jbusres.2023.113945
2. Ante, L.: The non-fungible token (NFT) market and its relationship with bitcoin and ethereum. SSRN Electron. J. (2021). https://doi.org/10.2139/ssrn.3861106
3. Cong, L.W., Tang, K., Wang, Y., Zhao, X.: Inclusion and democratization through Web3 and DeFi? Initial evidence from the Ethereum ecosystem. Working Paper 30949, National Bureau of Economic Research (2023). https://doi.org/10.3386/w30949. http://www.nber.org/papers/w30949
4. Das, D., Bose, P., Ruaro, N., Kruegel, C., Vigna, G.: Understanding security issues in the NFT ecosystem. In: ACM Conference on Computer and Communications Security (CCS) (2021). https://doi.org/10.48550/ARXIV.2111.08893. https://arxiv.org/abs/2111.08893

5. Dowling, M.: Is non-fungible token pricing driven by cryptocurrencies? Finance Res. Lett. **44**, 102097 (2022). https://doi.org/10.1016/j.frl.2021.102097
6. Fröwis, M., Böhme, R.: The operational cost of Ethereum airdrops. In: Pérez-Solà, C., Navarro-Arribas, G., Biryukov, A., Garcia-Alfaro, J. (eds.) DPM/CBT -2019. LNCS, vol. 11737, pp. 255–270. Springer, Cham (2019). https://doi.org/10.1007/978-3-030-31500-9_17
7. Jiang, X.J., Liu, X.F.: CryptoKitties transaction network analysis: the rise and fall of the first blockchain game mania. Front. Phys. **9**, 631665 (2021). https://doi.org/10.3389/fphy.2021.631665
8. Kshetri, N.: Scams, frauds, and crimes in the nonfungible token market. Computer **55**(4), 60–64 (2022). https://doi.org/10.1109/MC.2022.3144763
9. Makridis, C.A., Fröwis, M., Sridhar, K., Böhme, R.: The rise of decentralized cryptocurrency exchanges: evaluating the role of airdrops and governance tokens. J. Corp. Finance **79**, 102358 (2023). https://doi.org/10.1016/j.jcorpfin.2023.102358
10. Morgia, M.L., Mei, A., Mongardini, A.M., Nemmi, E.N.: NFT wash trading in the Ethereum blockchain (2022)
11. Nadini, M., Alessandretti, L., Giacinto, F.D., Martino, M., Aiello, L.M., Baronchelli, A.: Mapping the NFT revolution: market trends, trade networks, and visual features. Sci. Rep. **11**(1) (2021). https://doi.org/10.1038/s41598-021-00053-8
12. Serneels, S.: Detecting wash trading for nonfungible tokens. Finance Res. Lett. **52**, 103374 (2023). https://doi.org/10.1016/j.frl.2022.103374
13. White, B., Mahanti, A., Passi, K.: Characterizing the OpenSea NFT marketplace. In: Companion Proceedings of the Web Conference 2022, WWW 2022, pp. 488–496. Association for Computing Machinery, New York (2022). https://doi.org/10.1145/3487553.3524629

A Novel DID Method Leveraging the IOTA Tangle and Its Integration into OpenSSL

Alessio Claudio and Andrea Vesco[✉]

LINKS Foundation - Cybersecurity Research Group, 10138 Turin, Italy
{alessio.claudio,andrea.vesco}@linksfoundation.com

Abstract. This paper presents, for the first time, a novel DID Method called Over-The-Tangle and discusses its design and working principles that leverage the IOTA Tangle as the Root-of-Trust for identity data. The results of a long lasting experimental test campaign in real-world settings suggests the adoption of a private gateway node synchronized with the IOTA Tangle on the mainnet for efficient DID control. Moreover, the paper promotes the integration of the Decentralized IDentifier technology into OpenSSL through the use of Providers. A novel DID Operation and Provider is presented as a solution for building DID Method agility in OpenSSL.

Keywords: Self-Sovereign Identity (SSI) · OTT · IOTA · OpenSSL

1 Introduction

The Self-Sovereign Identity (SSI) [9] is a decentralized digital identity paradigm that gives both human beings and things full control over the data they use to build and to prove their identity. The overall SSI stack, depicted in Fig. 1, enables a new model for trusted digital interactions.

The Layer 1 is implemented by means of any Distributed Ledger Technology (DLT) acting as the Root-of-Trust (RoT) for identity data. In fact, DLTs are distributed immutable means of storage by design [5]. A Decentralized IDentifier (DID) [13] is a globally unique identity designed to verify a subject. The DIDs are Uniform Resource Identifiers (URIs) in the form *did:method-name:method-specific-id* where *method-name* is the name of the DID Method used to interact with the DLT and *method-specific-id* is the pointer to the DID Document stored on the DLT. Thus, DIDs associate a DID subject with a DID Document [13] to enable trustable interactions with that subject. The DID Method [13,15] is a software implementation to interact with a specific ledger technology. In accordance with W3C recommendation [13], a DID Method must provide the primitives (i) to create the DID, that is, generate a key pair (sk_{id}, pk_{id}) for authentication purposes, the corresponding DID Document containing the public key of the pair and store the DID Document into the ledger at the *method-specific-id* pointed by the DID, (ii) to resolve a DID, that is, retrieve the DID

© The Author(s), under exclusive license to Springer Nature Switzerland AG 2023
J. M. Machado et al. (Eds.): BLOCKCHAIN 2023, LNNS 778, pp. 394–404, 2023.
https://doi.org/10.1007/978-3-031-45155-3_38

Document from the *method-specific-id* on the ledger pointed to by the DID and verify the validity of the DID, (*iii*) to update the DID, that is, generate a new DID and corresponding DID Document while revoking the previous one and (*iv*) to revoke a DID, that is, provide an immutable evidence on the ledger that a DID has been revoked by the owner of the DID. The DID Method implementation is ledger-specific and it makes the upper layers independent from the specific ledger technology.

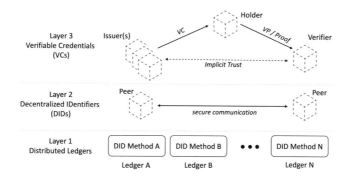

Fig. 1. The Self-Sovereign Identity stack.

The Layer 2 makes use of DIDs and DID Documents to establish a secure channel between two peers. In principle, both peers prove the ownership of the private key sk_{id} bound to the public key pk_{id} in the DID Document stored on the ledger. While the Layer 2 leverages DID technology (*i.e.* the security foundation of the SSI stack) to begin the authentication procedure, the Layer 3 finalizes it and deals with authorization to services and to resources by means of Verifiable Credentials [14]. VCs are secure and machine verifiable digital credentials. A VC is an unforgeable digital document that contains further characteristics of the digital identity of a peer than a simple key pair, a DID and the related DID Document. The addition of digital signatures makes VCs tamper-evident and trustworthy.

The combination of the key pair (sk_{id}, pk_{id}), a DID, the related DID Document and at least one VC forms the digital identity in the SSI world. This composition of the digital identity reflects the decentralized nature of SSI. There is no authority that provides all the components of the identity to a peer and no authority is able to revoke completely the identity of a peer. Moreover, a peer can enrich its identity with multiple VCs issued by different issuers.

The Layer 3 works in accordance with the Triangle-of-Trust depicted in Fig. 1. Three different roles coexist:

- **Holder** is the peer that possesses one or more VCs and that generates a Verifiable Presentation (VP) to request a service or a resource from a Verifier;
- **Issuer** is the peer that asserts claims about a subject, creates a VC from these claims, and issues the VC to the Holders. Multiple Issuers can coexist.

– **Verifier** is the peer that receives a VP from the Holder and verifies the two signatures made by the Issuer on the VC and by the Holder on the VP before granting him/her access to a service or a resource based on the claims.

A VC contains the metadata to describe properties of the credential (*e.g.* context, ID, type, Issuer of the VC, issuance and expiration dates) and most importantly, the claims about the identity of the subject in the `credentialSubject` field. The digital signature is made by the Issuer to make the VC an unforgeable and verifiable digital document. The Holder requests access to services and/or resources from the Verifier by presenting a VP. A VP is built as an envelop of the VC issued by an Issuer where signature is made by the Holder. Issuers are also responsible for VCs revocation for cryptographic integrity and for status change purposes [14]. On top of these three layers, it is possible to build any ecosystem of trustable interactions among human beings and things.

This paper focuses on Layer 1 of the SSI stack and presents for the first time a novel DID Method called Over-The-Tangle (OTT). OTT works directly on top of the IOTA Tangle [8], a DLT offering feeless transactions to attach any type of data to the structure of the distributed ledger. The paper provides the following novel contributions: (*i*) design and implementation of the OTT DID Method and (*ii*) statistically relevant results of the overall performance of the solution in real world settings. Then, supported by the results achieved, the paper discusses some future perspectives and proposes to work on the design of a Transport Layer Security (TLS) handshake making use of DIDs and DID Documents while maintaining the interoperability with public key certificates. As first step toward this objective, the paper provides two further contributions, (*iii*) design and implementation of the logic in OpenSSL [6] for agile integration of DID Methods through a novel DID Provider. A Provider, in OpenSSL term, is a unit of code that provides one or more implementations for various operations for diverse algorithms that one might want to perform [7]; in the context of this work, the Provider collects a series of DID Methods and make them available to LibSSL, a sub-module of OpenSSL implementing the TLS handshake; this novel DID Operation universally defines the templates (*i.e.* inputs and return values) for the DID Method functions *Create*, *Resolve*, *Update*, and *Revoke*, and (*iv*) an easy to follow procedure to add other DID Methods in an agile fashion and open the way for further adoption of the solution.

2   A DID Method for the IOTA Tangle

OTT is a novel DID Method that leverages the IOTA Tangle as the RoT for DID Documents. The main working principles of the IOTA Tangle are detailed in [8]. OTT is designed to work on top of the Chrysalis version of the IOTA protocol running on the IOTA mainnet. OTT makes use of indexation features of the IOTA Tangle. A peer can issue a transaction to attach a data to the Tangle while associating the data with a string 32 byte long. The Tangle indexes the strings and allows any peer to search for a data based on this string value; for this reason the string is called *index*. Recalling the DID form and DID Document purpose detailed in Sect. 1, in OTT the DID has the form *did:ott:index* and a DID Document containing the public key pk_{id} has the following structure:

```
"@context": ["https://www.w3.org/ns/did/v1"],
"id":"did:ott:index",
"created": "---Date-Hour---",
"authenticationMethod": {
    "id": "did:ott:index#keys-0",
    "type": "---Verification-Method-Type---",
    "controller": "did:ott:index",
    "publicKeyPem": "---Public-Key---" }
```

2.1 Working Principles

OTT uses two different messages to control a DID on the IOTA Tangle, see Fig. 2. The create message (left) and the revoke message (right) to create and revoke a DID respectively. An OTT message wraps the DID Document and it represents the unit of data attached to the Tangle through a single transaction. OTT also associates an *index* to the messages for making them searchable.

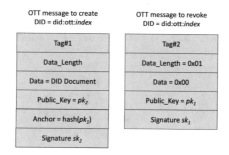

Fig. 2. Structure of the OTT create message (left) and revoke message (right) that OTT uses in the *Create, Resolve, Update* and *Revoke* functions.

- **Tag#1** and **Tag#2** [32 byte]: to identify OTT messages;
- **Data_Length** [2 byte]: the length of the Data field in byte;
- **Data** [up to 31600 byte]: the DID Document or 0x00 in the case of an OTT revoke message;
- **Public_Key** [32 byte]: ephemeral public key instrumental to establish the cryptographic binding between the create and the revoke messages associated with the same *index*;
- **Anchor** [32 byte] (only in create messages): the value instrumental to verify the cryptographic binding between the create and the revoke messages associated with the same *index*;

– **Signature** [64 byte]: digital signature of the message only for integrity purpose; made with Edwards-curve Digital Signature Algorithm (EdDSA) [4] over the edwards25519 curve.

Figure 3 depicts the process for generating two ephemeral key pairs (sk_1, pk_1) and (sk_2, pk_2) instrumental to generate the *index* that OTT associates to create and revoke messages before attaching them to the Tangle. The process starts from two random seeds ($seed_1$ and $seed_2$) to generate the two ephemeral elliptic key pairs (sk_1, pk_1) and (sk_2, pk_2) over the edwards25519 curve. Then, the public key (pk_2) is concatenated with the digest, computed with the hash [11], of the first public key (pk_1). The resulting value is finally hashed to generate the *index*. The two ephemeral key pairs are bound to the value of *index* and anyone who proves to own the first key pair also proves to own the second key pair. Moreover, also knowing the *index* and the Anchor value, the derivation of pk_1 is impracticable because the hash function is assumed to be not reversible. This construction is designed to ensure that (i) the DID has been generated by the peer in accordance with OTT principles and (ii) only that peer is able to revoke its own DID. In a permissionless DLT anyone can build and attach an OTT message to the Tangle and associate it with a previously selected DID, but only the owner, in OTT terminology, of the DID can build a valid OTT revoke message revealing the right pk_1 value. For the sake of clarity, the purposes of key pair (sk_{id}, pk_{id}) and the key pairs (sk_1, pk_1), (sk_2, pk_2) are different. The first key pair is a piece of the digital identity of the peer, whereas the second couple of key pairs are ephemeral keys at the core of OTT working principles.

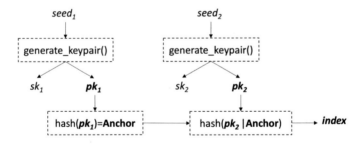

Fig. 3. Generation of the *index* and the ephemeral key pairs.

2.2   Implementation of the OTT Method Functions

OTT has been implemented leveraging the C language IOTA client library [1] and Libsodium cryptographic library [12]. The IOTA client library provides the primitives to attach and retrieve messages to/from the Tangle. The primitive *send_indexation_msg()* and the *indexation payload* to attach a DID Document to the Tangle associating it with the specific DID URI, *i.e. did:ott:index*. The

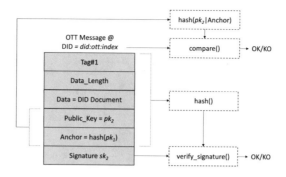

Fig. 4. Validation of an OTT create message.

primitive *find_message_by_index*() to retrieve the DID Document from a given DID. OTT provides the implementation of the four functions to create, update and revoke DIDs and to resolve DIDs.

– **Create** function is responsible for generating a new DID; this means generating an *index* as in Fig. 3 and the related DID Document to be attached to the IOTA Tangle. This function encapsulate the DID Document into an OTT create message as in Fig. 2 (left), before sending it to the IOTA Tangle.
– **Resolve** function is responsible for resolving a DID, that is, extract the corresponding DID Document from the IOTA Tangle while verifying the validity of the OTT message carrying the DID Document.
  In case of an OTT create message, the function checks its integrity by verifying the integrity signature field against the hash of the rest of the message as depicted in Fig. 4 and, then, it checks that the DID has been generated in accordance with the OTT method, *i.e.* $hash(pk_2|Anchor) \equiv index$.
  In case of an OTT revoke message, the function checks its integrity by verifying the signature field against the hash of the rest of the message as depicted in Fig. 5 and, then, it checks whether the peer who revoked the DID is the same who created it by verifying that the relations among the public key pk_1 in the revoke message, pk_2 in the create message, and the DID holds: $Anchor \equiv hash(pk_1) \wedge hash(pk_2|hash(pk_1)) \equiv index$.
  The Resolve function returns the DID Document in case of a valid DID, an empty document in case of a revoked DID and a proper error code in other possible cases.
– **Revoke** function is responsible for revoking a DID by means of an OTT revoke message as in Fig. 2 (right).
– **Update** function is responsible for updating the DID as a combination of a Revoke and a Create. The function revokes the current DID and then generates the new one from new random seeds (*e.g.* $seed'_1$ and $seed'_2$).

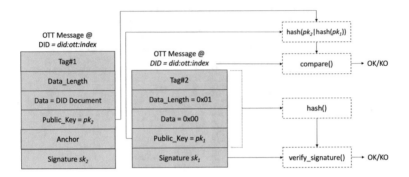

Fig. 5. Validation of an OTT revoke message.

3   Experimental Setup and Results

The tests has been designed to assess the performances of OTT functions over a long period of time. A single target node is requested to interact with the Tangle on the IOTA mainnet to create, update and revoke its own DID Document and to resolve the DID of a peering node. The setup is pretty simple, the target node interacts with the Tangle through a gateway node [2] integral part of the distributed ledger. The same tests have been performed by exploiting an available public node [3] and a private node deployed and operated in our lab. Both nodes are synchronized with the IOTA mainnet and they maintain the same version of the Tangle by design. However, the private gateway node is configured to serve only the requests from the target node to evaluate the influence of competing requests on the overall performances of OTT. The overall test campaign lasted one month during which the four OTT functions have been tested 1000 times each. The target node runs on an Ubuntu 20 LTS machine with kernel 5.15, with Intel(R) Core(TM) i7-10510U CPU @ 1.80 GHz 2.30 GHz, 24.0 GB RAM.

The graphs in Fig. 6 show the empirical cumulative distribution function (CDF) of the execution time of the four OTT functions. Each value y represents the CDF at x of the empirical distribution calculated from the tests, *i.e.* $F_T(t) = P(T \leq t)$ where t is the overall execution time of the specific OTT function. Each graph depicts the empirical CDF obtained with the public and the private gateway node for comparison. The empirical CDFs of the *Create* function are comparable. The 0.95 quantile, *i.e.* the estimated value such that 95% of the execution times is less than (or equal to) that value, is equal to 58.44 s with the public node and 68.02 s with the private node. Both are of the order of tens of seconds, as expected. The *Create* function lasts 19.47 s and 23.16 s in average with the public and the private node respectively. The same considerations stand for the *Revoke* and *Update* functions. A *Revoke* last 6.03 s and 8.19 s with the public and private gateway node, respectively, whereas an *Update* needs 24.25 s and 32.82 s by using the public and private gateway node respectively. The sum of the average time for the execution of a *Create* and a *Revoke* is about the same of an *Update*, as expected by design of these functions. These results suggest that

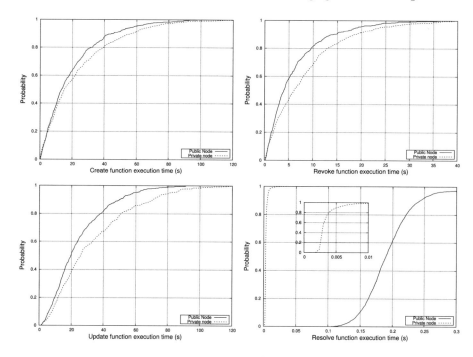

Fig. 6. Empirical CDF of the execution time of the four OTT functions; figure in figure in the Resolve graph shows a zoom of the empirical CDF with the private node setup.

competing requests at a public gateway node do not influence the performances of the OTT functions that require a write on the IOTA Tangle. It is worth noting that the creation, the revoke and the update of a DID are performed only once or periodically at a low frequency, hence the choice on the adoption of a public or a private gateway node has a limited impact on the overall performance. The empirical CDFs of the *Resolve* function, shows, as expected, that the performance with the private gateway node outperforms the one with the public gateway node. Resolving a DID by leveraging an unloaded private gateway node is about one hundred times faster than with a public node. Resolving the same DID lasts 3.49 ms and 216 ms by querying the private and the public gateway node, respectively. This important result opens up new perspectives.

4   Future Perspectives

Most of the discussions about the DID-based authentication in the SSI world consider an implementation at the application layer of the TCP/IP stack. The primary reason for this choice is that most of the community comes from the WWW. The Authors of this paper believe in the implementation of the DID-based authentication at the transport layer of the TCP/IP stack through and directly within the Transport Layer Security (TLS) v1.3 [10]. Implementing this option requires to modify the current TLS handshake to work with DIDs and DID Documents. This paper proposes to integrate the DID technology into OpenSSL, one of the most adopted cryptographic library, and made it transparently available in LibSSL for the adoption by the TLS protocol. The very fast execution of a DID resolution with the private setup, shown in Fig. 6, is an important result and it supports the proposed option. In fact, a DID-based TLS handshake skips the time for public key certificate chain verification (*i.e.* the most computational demanding operation) but it pays a DID resolution, hence a fast DID resolution is desirable. As a first step toward the adoption of DID technology into the TLS handshake, it is necessary to provide OpenSSL the implementation of the DID Method. The paper here presents for the first time, the design and implementation of a DID Provider specifically conceived to ease the integration of DID methods into OpenSSL, *i.e.* makes it pluggable and avoids code refactoring.

The integration of DID technology into OpenSSL begins with the definition of a novel Operation called DID. As shown in the code listing below, the definition of the DID Operation requires three main steps. First, assigning an integer value to the Operation through the identifier prefixed by the *OSSL_ OP_* string. Second, declaring the required template functions of the Operation through *OSSL_ FUNC_ * _ * * strings that univocally defines the function's identifiers. Finally, using the macro *OSSL_ CORE_ MAKE_ FUNC*() to define the OpenSSL name of the functions (*i.e.* *did_ create, did_ update, did_ resolve* and *did_ revoke*), the input and the return values. Developers calls those functions from within OpenSSL through a set of APIs.

```
#define OSSL_OP_DID              24
#define OSSL_FUNC_DID_CREATE     1
#define OSSL_FUNC_DID_RESOLVE    2
#define OSSL_FUNC_DID_UPDATE     3
#define OSSL_FUNC_DID_REVOKE     4
OSSL_CORE_MAKE_FUNC(void *, did_create,
    (void *sig, size_t siglen, int type))
OSSL_CORE_MAKE_FUNC(int, did_resolve,
    (char * index, DID_DOCUMENT* did_doc))
OSSL_CORE_MAKE_FUNC(int, did_update,
    (char * index, void *sig, size_t siglen, int type))
OSSL_CORE_MAKE_FUNC(int, did_revoke, (char * index))
```

This DID Operation definition, allows the developers of DID Methods to make them available through OpenSSL via a Provider in an *agile* fashion. The term agile means that developers are provided with a stable logic to simply plug their own specific implementations and make them available in OpenSSL without major code refactoring. Thus, an OpenSSL application can load the Provider offering the DID Operation and ask it to supply the implementation of the chosen DID Method functions. In case of a DID Provider offering multiple different DID Method implementations, *e.g.* DOM for Ethereum, and OTT for IOTA Tangle, as shown in the listing below:

```
operation_did[] = {
 {"DOM","provider=didprovider", didprovider_dom_functions},
 {"OTT","provider=didprovider", didprovider_ott_functions}}
```

an application can request OTT method by calling `DID_fetch(didctx,"OTT")` where `didctx` is the OpenSSL context, to receive the pointers to the implementations of the *did_create, did_update, did_resolve* and *did_revoke* functions of the OTT method at run time.

Finally, OTT method has been plugged into OpenSSL via the DID Provider and the new tests to evaluate the overhead introduced by OpenSSL when calling the OTT functions showed no significant statistical variation with respect to the ones presented in Sect. 3; hence they are not reported here for conciseness.

5   Conclusions and Future Works

The paper has presented, for the first time, a novel DID Method called OTT and has discussed its design and working principles that leverage the IOTA Tangle as the RoT for identity data. The results of a long lasting experimental test campaign in real-world settings has suggested the adoption of a private gateway node synchronized with the IOTA Tangle on the mainnet for efficient DID control, in particular for reducing of about 100x the time for a DID resolution. This result has suggested the Authors to deal with the integration of DID technologies into OpenSSL and to start addressing DID-based TLS handshake in SSI framework while maintaining the interoperability with public key certificates. The results of the implementation of a DID-based TLS v1.3 handshake and its security analysis will be provided in a future paper.

Acknowledgements. This work has been developed within the SEDIMARK project https://sedimark.eu/. SEDIMARK is funded by the European Union under the Horizon Europe framework programme [grant no. 101070074]

References

1. IOTA Foundation: Client Library in C. https://github.com/iotaledger/iota.c
2. IOTA Foundation: HORNET. https://github.com/iotaledger/hornet
3. IOTA Foundation: Public Node. https://chrysalis-nodes.iota.org/api/v1/info
4. Josefsson, S., Liusvaara, I.: Edwards-Curve Digital Signature Algorithm (EdDSA). RFC 8032 (2017). https://www.rfc-editor.org/rfc/rfc8032.html
5. Kannengießer, N., et al.: Trade-offs between distributed ledger technology characteristics. ACM Comput. Surv. **53**(2), 1–37 (2020)
6. OpenSSL: Cryptography and SSL/TLS toolkit. https://www.openssl.org
7. OpenSSL: Operation implementation providers. https://www.openssl.org/docs/man3.0/man7/provider.html
8. Popov, S.: The Tangle (2018). https://assets.ctfassets.net/r1dr6vzfxhev/2t4uxvsIqk-0EUau6g2sw0g/45eae33637ca92f85dd9f4a3a218e1ec/iota1_4_3.pdf

9. Preukschat, A., Reed, D.: Self-Sovereign Identity - Decentralized Digital Identity and Verifiable Credentials. Manning, Shelter Island (2021). https://www.manning.com/books/self-sovereign-identity

10. Rescorla, E.: The Transport Layer Security (TLS) Protocol Version 1.3. Internet Request for Comments (2018). https://www.rfc-editor.org/info/rfc8446

11. Saarinen, M., Aumasson, J.: The BLAKE2 Cryptographic Hash and Message Authentication Code (MAC). RFC 7693 (2015). https://www.rfc-editor.org/rfc/rfc7693.html

12. Sodium: Libsodium. https://doc.libsodium.org/

13. W3C: Decentralized Identifiers (DIDs) v1.0. Core architecture, data model, and representations. W3C Recommendation (2022). https://www.w3.org/TR/did-core/

14. W3C: Verifiable Credentials Data Model v1.1. W3C recommendation (2022). https://www.w3.org/TR/vc-data-model/

15. W3C: DID Specification Registries. The interoperability registry for decentralized identifiers (2023). https://www.w3.org/TR/did-spec-registries/

Beyond the Chain: Workshop on DAG-Based Distributed Ledger Technologies (DAG-DLT)

Beyond the Chain: Workshop on DAG-based Distributed Ledger Technologies (DAG-DLT)

The promise of increased scalability and improved performance is pushing more projects to adopt a Direct Acyclic Graph (DAG) structure for their distributed ledgers instead of a traditional chain of blocks. The Workshop on DAG-based Distributed Ledger Technologies (DLTs) aims to bring together researchers and practitioners from academia and industry to discuss the latest developments and future directions in this very active field.

The workshop focuses on the DAG-based DLTs based on the Nakamoto consensus as well as on the newer development of the Byzantine Fault Tolerance (BFT) consensus. In addition, the event explores the various modes of writing access, such as permissionless or permissioned, leaderless, or leader-based, and how they impact the fairness, security, and performance of the resulting DLTs. Special attention is given to developing appropriate and meaningful performance measures and test scenarios to compare and evaluate the performance of the most prominent DAG-based DLTs. Furthermore, due to its prominence, it includes contributions that pertain to the topic of Maximum Extractable Value (MEV), especially concerning robustness and wealth distribution in DAG-based DLTs.

Overall, the workshop provides a platform for participants to exchange ideas, share insights, and engage in lively discussions on the latest advances and challenges in DAG-based DLTs, focusing on the tradeoffs between fairness, security, and performance.

Workshop Organization

Organizing Committee

Sebastian Müller	Université de Marseille
Pietro Ferraro	Imperial College London
Luigi Vigneri	IOTA Foundation

Program Committee

Isabel Amigo	IMT Atlantique
Darcy Camargo (Senior Research Scientist)	IOTA Foundation
Teodoro Montanaro	Università del Salento
Luigi Patrono	Università del Salento
Serguei Popov	University of Porto
Maria Potop-Butucaru	Sorbonne
Mayank Raikwar	Oslo
Alexandre Reiffers-Masson	IMT Atlantique
Olivia Saa (Senior Research Scientist)	IOTA Foundation
Robert Shorten	Imperial College London
Yan Tao	UZH Zürich
Nicolò Vallarano	UZH Zürich
Roman Vitenberg	University of Oslo

An Attack Resilient Policy on the Tip Pool for a DAG-Based Distributed Ledger

Lianna Zhao[1]([✉]), Andrew Cullen[2], Sebastian Mueller[2,3], Olivia Saa[2], and Robert Shorten[1]

[1] Dyson School of Design Engineering, Imperial College London, London, UK
l.zhao20@imperial.ac.uk
[2] IOTA Foundation, 10405 Berlin, Germany
[3] Aix Marseille Universite, CNRS, Centrale Marseille, I2M - UMR 7373, 13453 Marseille, France

Abstract. This paper discusses congestion control and inconsistency problems in DAG-based distributed ledgers and proposes an additional filter to mitigate these issues. Unlike traditional blockchains, DAG-based DLTs use a directed acyclic graph structure to organize transactions, allowing higher scalability and efficiency. However, this also introduces challenges in controlling the rate at which blocks are added to the network and preventing the influence of spam attacks. To address these challenges, we propose a filter to limit the tip pool size and to avoid referencing old blocks. Furthermore, we present experimental results to demonstrate the effectiveness of this filter in reducing the negative impacts of various attacks. Our approach offers a lightweight and efficient solution for managing the flow of blocks in DAG-based DLTs, which can enhance the consistency and reliability of these systems.

Keywords: DAG-based DLT · Tip selection · Attacks analysis · Past-cone confirmation time (PCT) condition

1 Introduction

Distributed ledger technologies (DLTs) have gained significant attention for their potential to revolutionize how we store, manage and transfer digital assets. DLTs, such as blockchains, have traditionally relied on a linear chain of blocks. However, this approach has limited scalability and efficiency, particularly when faced with high volumes of blocks. Directed Acyclic Graph (DAG)-based DLTs offer an alternative solution to this problem by using a DAG structure to organize blocks. This more flexible structure allows for higher scalability and lower latency, as each block can reference multiple previous blocks rather than relying on a linear chain. One of the benefits of DAG-based DLTs is the less restrictive writing access enabled by their DAG structure. Ideally, this can make the mempool redundant

This work is funded by the IOTA Foundation.

and allow for a more efficient design of the block dissemination. However, this also introduces challenges in controlling the rate at which blocks are added to the ledger and preventing spam attacks. Writing access can be controlled by Proof of Work (PoW) [1], Proof-of-Stake (PoS)-based lottery, or permissioned setup. We will focus our attention in this work on a specific DAG-based DLT architecture, based on IOTA [2,3]. This architecture does not restrict writing access to special validator nodes but enables all nodes to participate through congestion control on a different layer. This congestion control mechanism must be resilient against malicious actors who wish to compromise the throughput of honest actors by spamming the network with their blocks. To address this challenge, we propose a filter, in addition to the scheduler and drop-head policy proposed in [4], which prevents referencing of old blocks. This mitigation strategy corresponds to the two following components that seem vulnerable to liveness and consistency attacks.

First, the proposed protocol uses a congestion control algorithm on the underlying P2P layer. Each honest node adjusts its issuing rate using an AIMD rate setter; for more details, see [4,5]. Every node keeps a separate queue for each issuer inside the scheduler that regulates the gossiping of the blocks. The queues of the malicious or faulty nodes that exceed their quota of blocks, will blow up (in the absence of a mitigation strategy) and eventually lead to inconsistencies in the ledger. Second, in the proposed protocol, the *tip selection algorithm* (TSA), which determines how new blocks are attached to the existing DAG, plays a crucial role. Each time a node issues a new block, it *approves* previous existing blocks from the DAG which have not yet received any approvals. These unapproved blocks are called tips. The set of tips eligible for selection is called the tip pool, and the size of this set is a critical metric of the DAG-based DLTs, e.g. [5–7]. The situation where the tip pool size becomes excessively large due to faulty or malicious behaviour is considered the most problematic, as it eventually leads to liveness issues and inconsistencies.

Our main focus of this paper is mitigating spamming attack scenarios and inconsistency problems in tip selection. Spamming attackers send a large number of blocks into the network and inflate the buffer size of honest nodes. The consequence of this attack is causing a large delay of blocks issued by honest nodes, resulting in a large tip pool size and affecting the liveness of the network. We will also consider spammers that send different streams of blocks to each of their neighbors, causing inconsistency issues. The TSA is a key component of the protocol for mitigating the negative effects of these attacks by ensuring honest nodes attach their blocks to the correct part of the DAG. The goal of this paper is to design a secure TSA for DAG-based ledgers to provide resilience against tip pool inflation and inconsistency problems caused by spammers. To reduce the network burden, we adopt the buffer management component proposed in [4], also known as drop-head policy, as the first filter to deal with spamming blocks. Furthermore, our solution for the tip pool regulation (named the *past-cone confirmation time (PCT)* condition) checks the validity of blocks after being scheduled, effectively regulating the size of the tip pool and ensuring the consistency of the ledger. Specifically, the PCT condition is introduced to manage the inconsistency problem caused by the previous tip selection and limits the size of the tip pool, especially in the presence of attacks.

2 Related Work

Although the design of congestion control algorithms has been a widely studied topic in the domain of computer networking when it comes to distributed ledgers, this topic is relatively new, and developing access control algorithms in this specific context poses unique and unprecedented challenges. This is primarily due to the competitive and hostile nature of the environments in which ledgers are designed to operate. A novel design paradigm for DAG-based DLTs is introduced in [4] to enable the integration of reputation-based access control for the first time. The access control algorithm designed in [4] enables the system to be resilient against malicious agents attempting to acquire a greater portion of network resources than their fair allocation. An enhanced access control algorithm specifically for DAG-based IOTA Tangle is proposed in [5] to improve the network's security and resilience. Specifically, to address issues related to spamming and multi-rate malicious attackers, a blacklisting algorithm that employs a reputation-weighted threshold is introduced.

Tip selection has been a central part of the IOTA protocol since its inception, and numerous prior works have studied TSAs and their implications. The authors in [3] show under reasonable assumptions that a growing tip pool size grows results in larger confirmation times. The authors in [8] presented an analysis of processing two classes of blocks and the resulting delay time and further proposed a more general model for predicting tip pool size. In [9], to tackle the tip inflation attack, the author provided a fluid model to analyse the attack and designed a feedback control to regulate the number of tips that should be selected. According to the papers [8], the average number of tips in each node's tip pool is approximately two times the product of the average time delay of blocks and the block arrival rate. Here, the time delay is measured from when blocks are issued to when blocks arrive at the tip pool, while the delay is averaged over all blocks issued by all nodes. A first mitigation strategy for tip pool attacks was recently studied in [10], which limits the size of the local tip pools and drops the oldest tips if the maximal tip pool size is achieved. These previous methods are based on the dynamic increase of the number of references for each block. While these work efficiently to some extent, it increases the block sizes and, hence, the communication overhead and does not allow control to which parts of the Tangle blocks are attached. In our work, we propose a different approach that also allows controlling to which parts of the DAG new blocks are attached. This feature is crucial for the construction of finality gadgets and to prevent "parasite-chain" attacks as discussed in [6].

3   System Model

We begin by describing the relevant components of the data flow of a node, m (More information can be found in [4] [5]). According to the network's congestion condition, as indicated by buffer lengths, each node adjusts its issuing rate using an AIMD rate setter (the interested reader can refer to [4] [5]). In node m's inbox buffer, the inbox is split into N queues to identify blocks issued by different nodes. For example, for blocks issued by node i, its blocks are assigned to Q_{m_i} in node m's buffer. The Deficit Round Robin (DRR)- scheduler is used for scheduling the next message to be forwarded and added to the tip pool (the interested reader can refer to [4] Given this background, we introduce four possible states of a node in the network [4] [5]. We assume the issuing rate of nodes is defined as λ_m and $\tilde{\lambda}_m = \frac{\nu \cdot rep_m}{\sum_{i \in \mathcal{N}} rep_i}$ is the guaranteed allowed rate, where rep_m is a numeric reputation value which is associated to a node m and ν is the scheduling rate of node m [4,5].

(1) A node m is said to be *inactive* if the issuing rate $\lambda_m = 0$.
(2) A node m is said to be *content* if it issues blocks that can be approximated as a Poisson process with a fixed rate parameter $\lambda_m \leq \tilde{\lambda}_m$.
(3) A node is said to be *best-effort* if it issues at rate $\lambda_m > \tilde{\lambda}_m$ under the rate control policy imposed by the access control algorithm.
(4) A node is said to be *malicious* if it deviates from the designed protocol, including the rate setting algorithm and the forwarding algorithm [4,5].
   - For attackers deviating from the rate-setting algorithm, they issue blocks at a rate far above the allowed protocol. We name this type of attacker spamming attackers.
   - For attackers deviating from the forwarding algorithm, they send a different stream of blocks to different neighbours while each stream obeys the rate control policy. We name this type of attacker multi-rate attackers.

Normally, scheduled blocks are then added to the tip pool directly. But in the presence of attacks, such as spamming attacks, a large number of blocks are issued by a malicious node, and this burst of traffic inflates the node inbox. The consequence of this congestion is that the delay of each block increases, many old tips are added to the tip pool, and then the tip pool size keeps increasing. Finally, regarding the ledger, both the width of the DAG structure and the confirmation time (Refer to Definition 5) for blocks increase, affecting the confirmation efficiency of DAG-based DLTs. Further, as mentioned in Sect. 1, the ledger might become inconsistent when some nodes pick up blocks that other nodes dropped. Hence, we propose the PCT condition to protect against adversarial attacks and ensure the ledger's consistency. To facilitate exposition, we present some further notations and definitions that will be used in the remainder of the paper.

Definition 1 (Cumulative weight). *The cumulative weight of block A is one plus the number of scheduled blocks that directly or indirectly approve block A.*

Definition 2 (Disseminated block). *A block is defined as being disseminated when all honest nodes in the network receive it.*

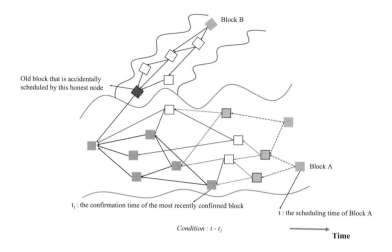

Fig. 1. PCT condition.

Definition 3 (Dissemination rate [4] [5]**).** *We denote the rate of dissemination of all blocks issued by node i as DR_i. The dissemination rate of all blocks is denoted DR where $DR \triangleq \sum_{i=1}^{N} DR_i$.*

Definition 4 (Scaled dissemination rate [4] [5]**).** *The reputation-scaled dissemination rate is defined as the dissemination rate of this node divided by the node's reputation value.*

Definition 5 (Confirmed block). *If the cumulative weight of a block in node m, is larger than a preset cumulative weight threshold, we define this block as confirmed in node m. But only if the block is confirmed by all honest nodes in the network, we say this block is* fully *confirmed. For example, after a block A is scheduled, its weight is added to the cumulative weight named CW_A of itself and to all blocks in this block A's past-cone. If CW_A is larger than a preset cumulative weight threshold CW_T defined in Table 1 in every network node, then we mark this block as fully confirmed.*

Definition 6 (Confirmation rate). *The rate of blocks issued by node i becoming* fully *confirmed is denoted by CR_i and $CR \triangleq \sum_{i=1}^{N} CR_i$.*

Definition 7 (Scaled confirmation rate). *The reputation-scaled confirmation rate of node i is defined as the confirmation rate of node i divided by the node's reputation value, and it is calculated by CR_i/rep_i[1].*

---

[1] The significance of studying the reputation-scaled metric is to ensure the fairness of the system. The allocation is said to be max-min fair of the confirmation rate if the confirmation rate increase of node i deceases another node m's confirmation rate with equal or smaller reputation-scaled confirmation rate [4].

As previously mentioned, there are mainly two components used in this paper to deal with the spamming issue and the inconsistency problem. The first one is borrowed from [4], in which a buffer management algorithm is used as a filter for spamming attacks. Namely, for node m, when the total buffer size is larger than the maximum buffer size W_{Max}, the eldest block issued by the node with the largest reputation-scaled queue length will be dropped. This process is executed repeatedly until the total buffer size is smaller than W_{Max}. The second one, which is the focus of this paper, is the past-cone confirmation time (PCT) condition.

3.1 PCT Condition

Definition 8 (Past-cone confirmation time (PCT)). *The PCT of a block is defined as the difference between the scheduling time of the block and the confirmation time of the most recently confirmed block in its past-cone.*

To regulate tip pool inflation and inconsistency, we introduce the PCT condition as described above. As illustrated in Fig. 1, the orange blocks represent confirmed blocks, the blue blocks are current tips, and the white blocks denote unconfirmed ones. The grey blocks symbolize the newly scheduled blocks, with one designated as block A and another as block B. The PCT mechanism operates as follows: if the most recent confirmed block in the past-cone of the referenced block is significantly dated, the block is not included in the tip pool. Adhering to the PCT condition, a new incoming block is incorporated into the tip pool only if the time difference between block A's scheduling time, denoted as t, and the confirmation time of the most recently confirmed block's time, t_1, within block A's past-cone is less than the predetermined PCT condition threshold, PCT_{th}.

Conversely, it may be feasible that a particular block (identified as the red block within the subbranch) could be scheduled by only a small subset of honest nodes, added to their tip pool, and subsequently selected as a tip. Such a block cannot accumulate sufficient weight quickly, and tips that approve the red block will eventually fail the PCT condition. Calculation of a block's PCT requires a search for the most recent confirmed block within the incoming block's past-cone, which is conducted using a breadth-first search (BFS) algorithm. It is important to note that the search within this branch will be terminated once the issuing time of the block under investigation surpasses the predetermined maximum depth of BFS, denoted as T_{Max_i}.

4   Simulations

The simulation[2] results are produced using a Python simulator. The number of nodes in the network is set to be $N = 20$. Nodes are connected in a random 4-regular graph topology. The communication delay between each pair of nodes i, j is uniformly distributed with values between 50ms and $150ms$. The

---

[2] The code can be found in https://github.com/Mona566/TSC-condition.git.

Table 1. Parameters and selected values for the simulation.

Parameter	Definition	Value (units)
N	The total number of nodes	20
CW_T	Cumulative weight threshold	25
T_{Max_i}	Maximum depth of BFS	80 (s)
PCT_{th}	PCT threshold	25 (s)
ν	The scheduling rate of nodes	20 (blocks per second)

node reputations are assumed to follow a Zipf distribution (as measured from account balances in the IOTA network). The simulation results are averaged over 10 Monte Carlo simulations, each of which is 800 seconds. The other parameters of the simulations are outlined in Table 1. The values of CW_T, T_{Max_i}, and Max_{Inbox} are chosen in such a way that various phenomena can be observed. Their choices in a real implementation would depend on various tradeoffs such as liveness vs. safety and efficiency vs. robustness. It is important to note that the specific performance figures indicated in our results are highly dependent on the parameters chosen, so they do not reflect actual DLT network performance. For example, by simply increasing the scheduling rate, one can immediately increase the confirmation rates and reduce the confirmation latency. We focus instead on steady-state behaviours and comparative results here.

We consider the following adversarial scenarios.

A_1: Attackers deviating from the rate setting algorithm, which are spamming attackers.

A_2: Attackers deviating from the forwarding algorithm, which are multi-rate attackers.

A_3: As a contrast experiment, we consider the same attack scenario as A_2, but without implementing the PCT condition.

4.1 A_1: Single Spamming Attacker

Here we consider the attack A_1, in which a spamming attacker is in the network.

The reputation-scaled dissemination rate is depicted in Fig. 2(a). The colour of each line denotes the mode of the node to which it corresponds, as indicated in the legend, the thickness of each line is set to be proportional to its reputation. The scaled dissemination rate of best-effort and content nodes converge to a relevant constant value respectively. Hence, fair access to the network is ensured for each node. The reason that the malicious node also has a constant value is it can still issue blocks and attempt to forward its blocks to other nodes, although most of its blocks get dropped.

The reputation-scaled confirmation rate is depicted in Fig. 2(b), which shows that the confirmation rate of nodes is also proportional to the node's reputation for each mode of issuer. There are more confirmed blocks issued by high-reputation nodes because high-reputation nodes issue (or are allowed to issue)

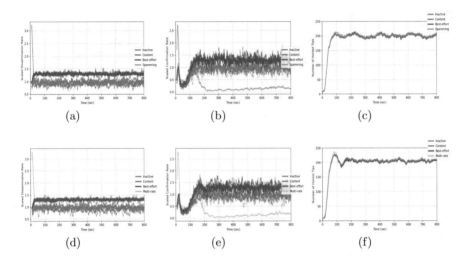

Fig. 2. A_1: Single spamming attacker (a): Scaled dissemination rate across all nodes. (b): scaled confirmation rate across all nodes. (c): The number of tips from honest nodes; A_2: Single multi-rate attacker (d): Scaled dissemination rate across all nodes. (e): scaled confirmation rate across all nodes. (f): The number of tips from honest nodes.

more blocks. The scaled confirmation rate values of best-effort and content nodes converge to constant values, while the confirmation rate for the malicious node is around 0, as most of its blocks are either dropped or get severely delayed and hence do not get selected as tips due to the PCT condition. As a result, these blocks do not gain cumulative weight to become confirmed. Figure 2(c) depicts the number of tips for honest nodes. As can be observed, the equilibrium point for all nodes' tip pool size is around 200, even in the presence of attackers. This ensures that honest nodes get selected quickly by tip selection and do not experience significant delays in becoming confirmed.

4.2 A_2: Single Multi-rate Attacker

Here we consider the attack A_2, in which a multi-rate attacker is in the network. The corresponding results for this attack scenario are shown in Fig. 2(d), which shows the scaled dissemination rates, Fig. 2(e) which shows the scaled confirmation rates, and Fig. 2(f), which shows the number of tips from honest nodes. It is clear by comparing these results to those in the previous subsection that our PCT condition is just as effective at dealing with a multi-rate attacker as it is for a spammer. The plots for A_2 are almost identical to those of A_1, so we will not repeat our analysis from the previous subsection.

4.3 A_3: Benchmark Experiment (Multi-rate Attacker, No PCT)

We evaluate and benchmark the effectiveness of our approach by repeating attack scenario A_2, but without the PCT condition. As illustrated in Fig. 3, when the

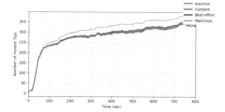

Fig. 3. A_3: benchmark experiment: the number of tips from honest nodes.

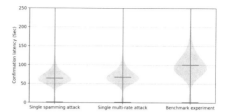

Fig. 4. The confirmation latency of all blocks in the network under above different scenarios.

PCT condition is not implemented, the number of tips increases continuously rather than converging to a steady state. With an extended simulation time, we can expect the tip pool size to continue to grow significantly. As a result, we can expect confirmation times to increase due to old tips being selected, further slowing the growth of cumulative weight for honest nodes. Indeed, this expected behaviour is observed in Fig. 4, which shows the distribution of confirmation latency for scenarios A_1, A_2 and the benchmark A_3. In A_1, the single spammer with PCT condition used, the average confirmation latency is approximately 64 s, and the maximum latency is around 110 s. In scenario A_2, the single multi-rate attacker with PCT condition used, the confirmation latency is slightly longer, with an average value of around 70 s and a maximum value of approximately 120 s. In the benchmark experiment, however, it is evident that in the absence of the PCT condition, the average and maximum confirmation latency increase substantially to around 100 s and 190 s, respectively.

5 Conclusions and Future Work

In this paper, we presented a past-cone confirmation time (PCT) condition for DAG-based distributed ledgers. Under the congestion control algorithm architecture, the proposed algorithm ensures the security, robustness and consistency of the network even in the presence of attacks. Furthermore, simulations are provided to illustrate the efficacy of the proposal. As noted in Sect. 4, our results demonstrate steady-state behaviour and provide a comparative analysis, but the performance figures, such as confirmation rates and latencies, are highly dependent on parameter choices and are not indicative of real network performance. As such, future work should focus on verifying these results in real networks and obtaining performance figures. For example, IOTA's GoShimmer prototype[3] could be a suitable testbed for our proposal. Another focus of this future implementation work should be on quantifying the computational burden of the PCT check and improving the efficiency of the solution, if necessary. Moreover, it is also interesting to consider how to combine the algorithm proposed here with works in [8–10].

---

[3] https://github.com/iotaledger/goshimmer.

References

1. Sompolinsky, Y., Sutton, M.: The DAG KNIGHT protocol: a parameterless generalization of nakamoto consensus. Cryptology ePrint Archive, Paper 2022/1494 (2022). https://eprint.iacr.org/2022/1494
2. Popov, S., et al.: The Coordicide, pp. 1–30 (2020)
3. Müller, S., Penzkofer, A., Polyanskii, N., Theis, J., Sanders, W., Moog, H.: Tangle 2.0 leaderless Nakamoto consensus on the heaviest DAG. IEEE Access **10**, 105807–105842 (2022)
4. Cullen, A., Ferraro, P., Sanders, W., Vigneri, L., Shorten, R.: Access control for distributed ledgers in the internet of things: a networking approach. IEEE Internet Things J. **9**(3), 2277–2292 (2021)
5. Zhao, L., Vigneri, L., Cullen, A., Sanders, W., Ferraro, P., Shorten, R.: Secure access control for DAG-based distributed ledgers. IEEE Internet Things J. **9**(13), 10792–10806 (2021)
6. Penzkofer, A., Kusmierz, B., Capossele, A., Sanders, W., Saa, O.: Parasite chain detection in the IOTA protocol. *arXiv preprint* arXiv:2004.13409 (2020)
7. Müller, S., Amigo, I., Reiffers-Masson, A., Ruano-Rincón, S.: Stability of local tip pool sizes. *arXiv preprint* arXiv:2302.01625 (2023)
8. Penzkofer, A., Saa, O., Dziubałtowska, D.: Impact of delay classes on the data structure in IOTA. In: Garcia-Alfaro, J., Muñoz-Tapia, J.L., Navarro-Arribas, G., Soriano, M. (eds.) DPM/CBT -2021. LNCS, vol. 13140, pp. 289–300. Springer, Cham (2022). https://doi.org/10.1007/978-3-030-93944-1_19
9. Ferraro, P., Penzkofer, A., King, C., Shorten, R.: Feedback control for distributed ledgers: an attack mitigation policy for DAG-based DLTs. *arXiv preprint* arXiv:2204.11691 (2022)
10. Camargo, D., Penzkofer, A., Müller, S., Sanders, W.: Mitigation of liveness attacks in DAG-based ledgers. *arXiv preprint* arXiv:2305.01207 (2023)

Welcome to the Tangle: An Empirical Analysis of IOTA Ledger in the Chrysalis Stage

Peilin He[1], Tao Yan[2(✉)], Chuanshan Huang[1], Nicolò Vallarano[2,3],
and Claudio J. Tessone[2,3]

[1] Department of Informatics, University of Zurich, Zürich, Switzerland
[2] Blockchain and Distributed Ledger Technologies Group,
Department of Informatics, University of Zurich, Zürich, Switzerland
yan@ifi.uzh.ch
[3] UZH Blockchain Center, University of Zurich, Zürich, Switzerland

Abstract. The Tangle, the distributed ledger of IOTA, is built on a directed acyclic graph (DAG) where the nodes represent messages, and the edges correspond to reference relations. This feature of the Tangle makes it suitable for analysis using network science methods. While previous research on the Tangle has mainly focused on stability, security, scalability, and the forthcoming IOTA Coordicide update, there is limited empirical research on the evolution of the Tangle, particularly during the Chrysalis stage (IOTA 1.5). To address this gap in the literature, our study aims to develop workflows for collecting IOTA Tangle data and transaction data in the Chrysalis stage, conduct a comprehensive analysis to examine the features of messages and reconstruct the Tangle graph to study its characteristics from a network science viewpoint. Our study is the first to illuminate activities and information transferred on the Tangle, as well as its graph characteristics during the Chrysalis stage, which can enhance our comprehension of its functionality and contribute to the improvement of the Tangle's performance in the upcoming upgrades.

Keywords: IOTA · Tangle · Networks · DAG

1 Introduction

IOTA is a distributed ledger technology that facilitates secure, feeless, and scalable transactions, primarily intended for machine-to-machine micropayments in

The authors would like to express their gratitude to Bing-Yang Lin, Luigi Vigneri, and Andreas Penzkofer from IOTA Foundation who provided valuable assistance in the data collection process. Additionally, we thank Benjamin Kraner for his careful reviews and insightful comments. Finally, Tao Yan acknowledges support from the China Scholarship Council (No. 202006980012), Nicolò Vallarano acknowledges support from IOTA Foundation.

J. M. Machado et al. (Eds.): BLOCKCHAIN 2023, LNNS 778, pp. 419–431, 2023.
https://doi.org/10.1007/978-3-031-45155-3_40

the Internet of Things (IoT) domain. This technology is built on a directed acyclic graph (DAG) called the Tangle [21], the reason why it is directed and acyclic is that only new messages which are vertices in the Tangle graph can reference previous messages. In contrast to traditional blockchain-based systems, IOTA does not have blocks, chains, miners, or transaction fees. Existing research on the Tangle can be mainly classified into four categories. The first category of research focuses on examining the stability and security of the Tangle. Several studies [8–10,21] have investigated different types of attacks in IOTA network, including parasite chain attacks, splitting attacks, double-spending attacks, and lazy tips attacks. The second category investigates tip selection algorithms, as seen in studies like [13,24,25], various tip selection methods of the Tangle are described in those research, such as random walk algorithm, weighted random walk algorithm, and dynamic balance algorithm. The third category examines the scalability of the Tangle, including studies such as [7,11]. Finally, the fourth category focuses on the future version of IOTA, known as Coordicide [23], which aims to achieve full decentralization. The research on Coordicide encompasses new consensus mechanisms [19,22] and autopeering [20]. However, there is limited research on the evolution of the Tangle using empirical data. One study [14] analyzes the topological features of the Tangle using real-world data and finds that the features of the real Tangle are topologically different from the simulated Tangle, but this research was conducted on the older IOTA 1.0 version. In order to present an up-to-date characteristic of the Tangle graph, we conduct a comprehensive analysis of the current IOTA Tangle using statistical and network analysis methods.

Currently, IOTA is in the Chrysalis stage (IOTA 1.5), as outlined in the IOTA roadmap [12]. The Chrysalis is an intermediate stage before the implementation of Coordicide stage (IOTA 2.0) [17]. The purpose of Chrysalis is to improve the scalability and reliability of the IOTA Mainnet. The Chrysalis stage consists of two phases: the first phase, which commenced in August 2020, improved the tip selection algorithm and adopted milestone selection and White Flag mechanisms. The second phase of Chrysalis began on April 28th, 2021, and it introduced significant changes such as the switch to UTXO transaction model from the account model and the adoption of reusable addresses [16]. During this development phase, there is still a client called Coordinator [18] which sends signed messages called milestones. In this work, we focus on analyzing the evolution of the Tangle, examining the characteristics of different message types, and studying the network features of the Tangle graph during the second phase of Chrysalis stage.

Our research makes significant contributions in two ways. First, we develop a workflow to collect and parse the Tangle data and transaction data. Second, our study focuses on the characteristics of the Tangle at the Chrysalis stage, and to the best of our knowledge, we are the first to undertake such an investigation, and our work provides novel insights into information transferred and the graph features of the Tangle. Overall, this work can enhance the comprehension of functionality and performance of Chrysalis and contribute to the improvement of the Tangle's performance in the upcoming upgrades.

The article is structured as follows: Sect. 2 outlines the workflow for data collection. Section 3 elaborates on the statistical characteristics of various message types. In Sect. 4, we describe the process of constructing the Tangle graph and discuss its graph features. Lastly, we discuss the implications of our findings and future research directions in the conclusion section.

2 Data Description

In this section, we will provide a detailed explanation of our data collection method. The data we use come directly from IOTA's log files hosted on a database[1], which are regularly updated by the IOTA foundation. Our data span exactly one year, starting from April 28th, 2021 when the Chrysalis version (IOTA 1.5) was officially launched and concluding on April 20th, 2022. The dataset comprises approximately 600 million JSON files, where each JSON represents a message. Totally, there are $677,604,217$ messages and $2,714,382,251$ references during this period of the Tangle.

2.1 Data Collection Workflow

The big picture of our data collection workflow is illustrated in Fig. 1. Log files, the sole source of Chrysalis data, are hosted on IOTA's database. We download them and undertake different measures to process them based on different purposes. To obtain data for the Tangle graph construction, we import the log files to a Cassandra database leveraging IOTA's Chronicle node, a solution to manipulate and warehouse the Tangle data. We opt for the database method to facilitate rapid query and frequent reuse. To obtain usable data, we unload tables from the database using Apache Spark and an open-source library called Spark-Cassandra-Connector[2]. Given the massive volume of data and the inability to perform joins in Cassandra, Spark has proven to be a practical solution to unload data. The data collected from the Chronicle node are messages with an "included" ledger state, meaning that these messages are attached to the Tangle and approved. On the other hand, there is some data in the log files that can not be imported to the Cassandra by Chronicle. In order to obtain those data, namely conflicting and indexation messages, we directly parse the log files.

2.2 Extract Conflicting and Indexation Messages

Log files are in nature a collection of JSON files. To locate the messages we want, we need to rely on some fields in JSON file. We distinguish conflicting messages via ledger states. To parse the message, we locate the "ledgerInclusionState" field which contains two possible values: conflicting or included. We use this field to filter out conflicting messages. To extract indexation messages from the

---

[1] https://chrysalis-dbfiles.iota.org/?prefix=chronicle/.
[2] http://datastax.github.io/spark-cassandra-connector/.

Fig. 1. Our workflow to collect data

entire message pool, we utilize the field "Payload type". We parse the log files with an indexation payload and retrieve the raw content of indexation messages. The parsed content is in nature a string of decimal numbers. We first need to turn them into hexadecimal numbers, then further convert them into Unicode characters so that they become human-readable.

3   Message Characteristics

The vertices of the Tangle are messages. A message [5] is an object broadcast in the Tangle which comprises information such as ledger state, payload and message identifier. There are three types of messages which are transaction payload, indexation payload, and milestone payload [6]. The type of message is determined by the payload field. When the value of payload type is 0, the message is a transaction. A transaction sends IOTA tokens among addresses, as in the second phase of Chrysalis IOTA switches to UTXO transaction model, each transaction satisfies the UTXO transaction data structure. When the value of payload field is 1, the message is a milestone. A milestone [3] is only issued by the Coordinator, its objective is to guarantee finality of messages, a message is confirmed only if it is approved directly or indirectly by a milestone. When the value of payload type is 2, the message is an indexation. An indexation [4] payload is a payload type that can be used to attach arbitrary data and a key index to a message.

Figure 2 shows the number of messages of all types and the total number of messages from April 28th, 2021, to April 20th, 2022. The number of indexations is the highest among all message types, causing its curve to slightly overlap with the total number curve. The number of milestones is roughly consistent with the Coordinator's description, which says that it sends one milestone every 10 s [18]. The milestone curve approaches a horizontal line with a value of 8,640, occasionally dropping below, such as 7,513 on Oct 24th, 2021, or 6,632 on May 30th, 2021. The number of transactions shows a general upward trend, it decreased from 8k at the beginning of the second phase of Chrysalis to around 200 per day in June 2021, then it began to rise, reaching around 30,000 on April 15th, 2022.

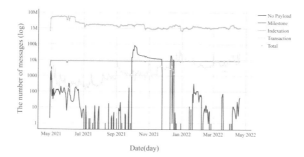

Fig. 2. The number of messages per day for each payload type over a year

3.1 Transactions

Among transactions, there are included transactions and conflicting transactions. Transactions are conflicting if they spend the same amount of money from the same input (a unique Transaction Id) of a UTXO transaction, but the input does not include enough money to satisfy both at the same time. IOTA 1.5 eliminates the conflict spamming attack [1] through the White Flag mechanism [27]. When the confirmation cone of the milestone leads to double spending, only the first transaction will be applied, and the other transactions will be ignored, not only within one milestone, but across all milestones afterward. The order of transactions is determined by the milestone merkle validation [26]. Ignored transactions are called conflicting transactions, here is an example of conflicting transaction[3]. In the metadata of a conflicting transaction message, the "ledgerInclusionState" is "conflicting" rather than "included".

Figure 3 proves that the percentage of conflicting transactions is generally small compared to the number of included transactions. However, there are periods when conflicting transactions exhibit a significantly larger number, from December 9th, 2021 to January 21st, 2022, the ratio of conflicting transactions to included transactions exceeds 0.2. Notably, the number of conflicting transactions surpasses the number of included transactions on December 30th and December 31st, 2022. Except for the period when IOTA 1.5 was just released, when the number of transactions surged, the proportion and number of conflicts also surged. Although these conflicting transactions are ignored by the Tangle, the messages containing them are kept on the Tangle. Generally, as the ratio of conflicting transactions to included transactions increases, more double-spend attempts are included as part of the Tangle graph.

When analyzing conflicting transactions, it is found that many conflicting transactions were issued by the same addresses. As shown in Fig. 4, it can be

---

[3] https://explorer.iota.org/mainnet/message/
01eb7b1a070d89f8116e2c84473e6caabda6a52980e4888abad560eba4e613b3.

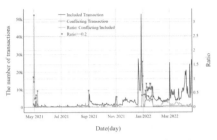

(a)   Transaction   numbers   (conflicting transactions vs included transactions)

Fig. 3. The number of conflicting and included transactions

found that the top 10 addresses have issued a large number of conflicting transactions, while most addresses only issue one. The top four addresses each issues more than ten thousand conflicting transactions.

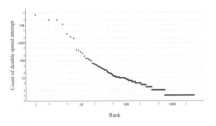

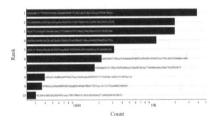

(a) The number of conflicting transactions sent by the same input address

(b) The top 10 addresses that send the most conflicting transactions

Fig. 4. The number of conflicting transactions sent by addresses

3.2   Milestones

One function of the Milestone is to improve the throughput of the network. Here, we analyze the number of messages directly confirmed by the Milestone and the total number of messages confirmed (either directly or indirectly). According to Guo et al. [15], the proposed range for the number of parental messages is between 2 and 8. However, through analysis of empirical data, we have observed that the actual range is between 1 and 8. Some messages just reference 1 parent, here is an example of a message that only has a parent[4]. In Fig. 5a, the index of milestone ranges from 1 to 3,084,565, and the count of their parents (messages directly referenced by milestones) varies from 1 to 8, with a concentration in the range of 6 to 8. In the later stage, most milestones directly approve 6 messages.

---

[4] https://explorer.iota.org/mainnet/message/e6112392f7d8f98414fecfb0948838e
64386860e902e974c4b0b1c6561bd928f.

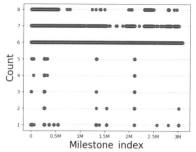

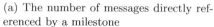

(a) The number of messages directly referenced by a milestone

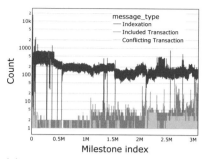

(b) The total number of different messages referenced by a milestone

Fig. 5. Confirmation rate analysis

Milestones can directly or indirectly confirm previous messages. Messages confirmed by the preceding milestones will indirectly confirm by the subsequent milestones. Messages that have not been confirmed by preceding milestones, will be directly confirmed by future milestones. In this study, we define the total number as the count of messages directly or indirectly confirmed by a milestone. Currently, the median number of total messages confirmed by milestones is 126, and the median number of total transactions confirmed by milestones is 1, this shows that the Tangle has the potential to confirm more transactions.

3.3   Indexations

Indexations can contain arbitrary data, and thus, we analyze what contents are included in the indexation messages. In the data field "data", there is an "index" for all indexation messages. After parsing data using the methods described in Sect. 2.2, it can be found that most of the indexes overlap. As shown in Fig. 6, indexes are mainly divided into three categories. "SPECIAL" represents indexes that start with a fixed phrase like "ATE-" or "R360". The total number of such addresses has been relatively stable since June 2021, at around 10,000 per day. There are a lot of indexation messages with "Spammer" type index, reaching millions of messages every day, among which "HORNET Spammer" and "Testnet Spammer" account for the most. We presume that signed spam messages are probably due to the use of spammers in order to test IOTA throughput capacity. Except for "SPECIAL" and "Spammer", there are many standalone indexes, which we classify as "Others".

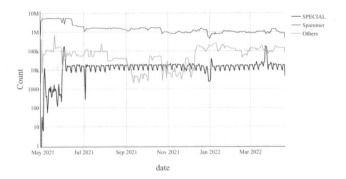

Fig. 6. The number of indexation messages of different index types per day

4   Tangle Graph Characteristics

In the Tangle graph, vertices represent messages and edges represent references. If message B is directly referenced by message A, then we say message B is a parent of message A and arc (A, B) is considered to be directed from A to B.

4.1   Tangle Network Construction

The Tangle graph data was extracted by Spark from the Cassandra database. The data volume of the Tangle graph in one year is astonishing, with 677, 604, 217 nodes and 2, 714, 382, 251 directed edges, and the data size reaches 80 GB. Considering limited memory and running speed, we choose to divide the data into 12 parts (0–11 months). Month 0 represents dates from April 28th, 2021 to May 28th, 2021 and other months can be done in the same manner. The only exception is that month 11 represents the time period from March 28th, 2022 to April 20th, 2022. A simple sample of the Tangle graph we constructed is shown as Fig. 7.

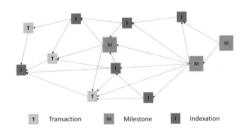

Fig. 7. The Tangle graph

4.2   Network Features

The Tangle graph each month is a fully connected graph that has only one component. The component size was larger in the first two months, and the number of messages in the following months decreased and fluctuated between 20 million and 50 million. The average out-degree and average in-degree of the Tangle graph fluctuate around 4, and the fluctuation range does not exceed 0.02. The Tangle graph is sparse: its density ranges from $2.50e-08$ to $3.03e-05$, as we could expect from the Tangle growing mechanism. Although this value shows an upward trend, it only reached $3.03e-05$ in the 12th month.

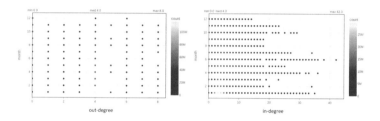

Fig. 8. In-degree and Out-degree of the Tangle graph over a year

Figure 8 shows the distribution of the Tangle graph out-degree and in-degree for each month respectively. In-degree represents how many times the vertex was referenced by other vertices. Analyzing the in-degree of messages for a whole year, we found that the mode of in-degree is 3, the median is 4, and the first quartile and third quartile are 3 and 5 respectively. There is also a message with an in-degree of 1 in Fig. 8, which is mainly because the messages that reference it are in a later month, so its children can not be found in the Tangle graph of this month. This result is different from the result of IOTA 1.0 [14]. In IOTA 1.0, the mean and median in-degree are around 1 or 2, although there are some exceptional cases with higher degrees in the range of $[10, 10^3]$. According to the data structure of the Tangle graph, each message can only have 1 to 8 out-degrees. It can be seen from Fig. 8 that the mean, median, and mode values of the out-degree are all 4. This result is inconsistent with the number of parents (out-degree) of milestones whose mode and median are both 6 according to Fig. 5a.

Figure 9a and Fig. 9b show the degree assortativity coefficient of monthly and daily Tangle graphs. Assortativity [28], which ranges between $[-1, 1]$, refers to the preference for a graph's vertices to attach to other vertices with similar in-degree or out-degree. If the assortativity index is close to 1, the graph is said to be assortative, meaning that nodes of similar degree tend to be connected; otherwise, if the assortativity index is close to -1, it suggests that the graph is disassortative, meaning that vertices of very different degrees tend to be connected.

The subscript of $Assortativity_{out\_in}$ represents the degree type for the source node (directed graphs only) is out-degree and the degree type for the target node is in-degree. It can be seen from Fig. 9a that except for $Assortativity_{out\_out}$, other assortativity coefficients are close to 0, which means that vertices with large out-degrees tend to link to vertices with large out-degrees. $Assortativity_{out\_in}$ is significantly less than 0 in month 0, indicating that vertices with large out-degrees tend to be connected to vertices with small in-degrees. Figure 9b shows assortativity daily in month 11. The daily results are similar to the monthly results. This phenomenon is related to the tip selection algorithm deployed on IOTA 1.5 [2]. Because of the White Flag confirmation mechanism, IOTA 1.5 can use a simpler and better performance tip-selection algorithm, thereby increasing transaction throughput. To maximize the confirmation rate, a message does not attach to a cone of transactions which is too far in the past. Thanks to this mechanism, messages that have not been referenced are more likely to be confirmed by new messages or high out-degree messages.

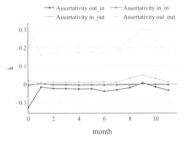 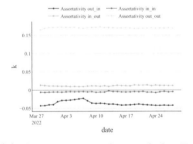

(a) Assortativity indexes of the Tangle graph per month

(b) Assortativity indexes of the Tangle graph for month 11 per day

Fig. 9. Assortativity analysis

5   Conclusion

This paper introduces two solid approaches for collecting various types of data from the IOTA Chrysalis stage, which have not been examined in prior research. We leverage the Chronicle node to construct a database of confirmed Tangle data, while a proprietary parser is employed to capture unconfirmed data and extract conflicting transaction and indexation data. A comprehensive analysis of the messages and graph characteristics is subsequently conducted based on this data.

This one-year Tangle graph has over 677 million vertices, and around 2.7 billion directed edges. There are three types of messages on the Tangle graph, namely milestone, transaction and indexation. Indexation accounts for the largest proportion and many indexation messages are issued by the index "Spammer". The number of milestones remains relatively stable at 8640 per day. While the total number of messages and reference relations has been decreasing each month due to a reduction in indexation messages, there is an upward trend in the number of transaction messages. Empirical data shows that there are lots of double-spending attacks on Tangle, and the number of conflicting transactions also increases over time, even more than the number of included transactions sometimes. In addition, we find that the majority of conflicting transactions actually are issued by a small number of addresses. In the present work we identified these addresses and we believe this may be useful for future research investigating the matter.

By reconstructing the monthly Tangle we are able to study its in- and out-degree distributions. An examination of the out-degree distribution reveals that milestones primarily reference six messages as parents, while all messages, on average, reference four messages. Based on IOTA 1.5, the mean and median in-degree of the Tangle graph are both approximately four, which differs from the values observed in IOTA 1.0 [14]. According to the assortativity analysis, the vertexes with large out-degree tend to reference vertexes with small in-degree, which helps develop the Tangle with few orphans. Unexpectedly, we also observe the presence of one out-degree messages which is not in agreement with the protocol reference rules.

Our work can be extended in several directions. First, apply our methods to the Tangle of IOTA 1.0 and 2.0, to determine whether the Tangle topology changes when different protocols are applied. Second, in the forthcoming Coordicide update (IOTA 2.0), the Coordinator role and milestones will be removed, presenting an opportunity to investigate the impact of its absence on the Tangle graph features. Lastly, in this paper we collected a dataset of addresses issuing conflicting transactions: this dataset may prove useful for future research on double-spending attacks on the IOTA network.

Author contributions. Conceptualisation, C.J.T., N.V., T.Y.; methodology, T.Y. and N.V.; data collection, C.-S.H.; and T.Y.; formal analysis, P.-L.H., C.-S.H. and T.Y.; writing-original draft preparation, P.-L.H.,T.Y. and C.-S.H.; writing-review and editing, C.J.T., T.Y., N.V., C.-S.H. and P.-L.H.. All authors have read and agreed to the published version of the manuscript.

References

1. Conflict spamming attack-research/security/attacks - govern.iota (2019). https://govern.iota.org/t/conflict-spamming-attack/232. Accessed 6 Apr 2023
2. Improved tip selection (URTS): weighted-uniform-random-tip-selection (2020). https://github.com/luca-moser/protocol-rfcs/blob/rfc-urts-tip-sel/text/0008-weighted-uniform-random-tip-selection/0008-weighted-uniform-random-tip-selection.md. Accessed 4 Apr 2023
3. IOTA glossary - IOTA wiki (2021). https://wiki.iota.org/learn/glossary. Accessed 27 Mar 2023
4. Send a data message - IOTA wiki (2021). https://wiki.iota.org/iota.rs/examples/data_message/. Accessed 27 Mar 2023
5. Data transfer: what is an IOTA message - IOTA wiki (2022). https://wiki.iota.org/learn/about-iota/data-transfer/#what-is-an-iota-message. Accessed 27 Mar 2023
6. Messages - IOTA wiki (2022). https://wiki.iota.org/learn/about-iota/messages/. Accessed 28 Mar 2023
7. Alsboui, T., Qin, Y., Hill, R., Al-Aqrabi, H.: Towards a scalable IOTA tangle-based distributed intelligence approach for the Internet of Things. In: Arai, K., Kapoor, S., Bhatia, R. (eds.) SAI 2020. AISC, vol. 1229, pp. 487–501. Springer, Cham (2020). https://doi.org/10.1007/978-3-030-52246-9_35
8. Bramas, Q.: The stability and the security of the tangle (2018)
9. Brighente, A., Conti, M., Kumar, G., Ghanbari, R., Saha, R.: Knocking on tangle's doors: security analysis of IOTA ports. In: 2021 IEEE International Conference on Blockchain (Blockchain), pp. 433–439. IEEE (2021)
10. Conti, M., Kumar, G., Nerurkar, P., Saha, R., Vigneri, L.: A survey on security challenges and solutions in the IOTA. J. Netw. Comput. Appl. **203**, 103383 (2022)
11. Cullen, A., Ferraro, P., Sanders, W., Vigneri, L., Shorten, R.: Access control for distributed ledgers in the Internet of Things: a networking approach. IEEE Internet Things J. **9**(3), 2277–2292 (2021)
12. IOTA Foundation: IOTA roadmap (2022). https://roadmap.iota.org/. Accessed 2 Apr 2023
13. Gardner, R., Reinecke, P., Wolter, K.: Performance of tip selection schemes in DAG blockchains. In: Pardalos, P., Kotsireas, I., Guo, Y., Knottenbelt, W. (eds.) Mathematical Research for Blockchain Economy. SPBE, pp. 101–116. Springer, Cham (2020). https://doi.org/10.1007/978-3-030-37110-4_8
14. Guo, F., Xiao, X., Hecker, A., Dustdar, S.: Characterizing IOTA tangle with empirical data. In: GLOBECOM 2020–2020 IEEE Global Communications Conference, pp. 1–6. IEEE (2020)
15. Guo, F., Xiao, X., Hecker, A., Dustdar, S.: A theoretical model characterizing tangle evolution in IOTA blockchain network. IEEE Internet Things J. **10**(2), 1259–1273 (2022)
16. IOTA Foundation: path to chrysalis (2021). https://wiki.iota.org/introduction/explanations/update/path_to_chrysalis. Accessed 1 Apr 2023
17. IOTA Foundation: what is chrysalis? (2021). https://wiki.iota.org/introduction/explanations/update/what_is_chrysalis. Accessed 2 Apr 2023
18. IOTA Foundation: the coordinator (2022). https://wiki.iota.org/learn/about-iota/coordinator/. Accessed 29 Mar 2023
19. Lin, B.Y., Dziubałtowska, D., Macek, P., Penzkofer, A., Müller, S.: Robustness of the tangle 2.0 consensus. In: Hyytiä, E., Kavitha, V. (eds.) VALUETOOLS 2022. LNICST, vol. 482, pp. 259–276. Springer, Cham (2023). https://doi.org/10.1007/978-3-031-31234-2_16

20. Müller, S., et al.: Salt-based autopeering for DLT-networks. In: 2021 3rd Conference on Blockchain Research & Applications for Innovative Networks and Services (BRAINS), pp. 165–169. IEEE (2021)
21. Popov, S.: The tangle. White paper **1**(3), 30 (2018)
22. Popov, S., Buchanan, W.J.: FPC-BI: fast probabilistic consensus within byzantine infrastructures. J. Parallel Distrib. Comput. **147**, 77–86 (2021)
23. Popov, S., et al.: The coordicide, pp. 1–30 (2020)
24. Purohit, S., Champaneria, T.: Evaluating tip selection algorithms for IOTA blockchain. In: Singh, P.K., Wierzchoń, S.T., Chhabra, J.K., Tanwar, S. (eds.) FTNCT 2021. LNEE, vol. 936, pp. 253–264. Springer, Cham (2022). https://doi.org/10.1007/978-981-19-5037-7_17
25. Wang, J., Yang, J., Wang, B.: Dynamic balance tip selection algorithm for IOTA. In: 2021 IEEE 5th Information Technology, Networking, Electronic and Automation Control Conference (ITNEC), vol. 5, pp. 360–365. IEEE (2021)
26. Welz, W.: Milestone merkle validation (2020). https://iotaledger.github.io/tips/tips/TIP-0004/tip-0004.html. Accessed 1 Apr 2023
27. Welz, W.: White flag ordering (2020). https://iotaledger.github.io/tips/tips/TIP-0002/tip-0002.html. Accessed 2 Apr 2023
28. Newman, M.E.J.: Assortative Mixing in Networks. Physical Review Letters textbf89(20) (2002). https://doi.org/10.1103/physrevlett.89.208701

Performance Analysis of the IOTA Chrysalis on Heterogeneous Devices

Muhammad Waleed$^{(\boxtimes)}$, Knud Erik Skouby , and Sokol Kosta

Aalborg University, Copenhagen, Denmark
{muhammadw,skouby,sok}@es.aau.dk

Abstract. Existing Distributed Ledger Technologies (DLTs) based models like blockchain pose scalability and performance challenges for IoT systems due to resource-demanding Proof of Work (PoW), slow transaction confirmation rates, and high costs. Against a need to adopt a viable approach, especially for low-power IoT devices, IOTA emerges as a promising technology, leveraging the Direct Acyclic Graph (DAG) based approach called Tangle for IoT-focused applications. In this paper, we design a system enabling secure data exchange between IoT devices on IOTA Chrysalis, the latest version. We perform extensive experiments on two machines, a Workstation PC and Raspberry Pi, to demonstrate the performance gap between powerful and low-power devices. Our findings show that even low-power devices, such as Raspberry Pi, perform well with small payload sizes on the Chrysalis network but face challenges with larger payloads. We observe that variation in transmission time increases as payload size grows, indicating the impact of PoW complexity, but it still is feasible for Raspberry Pi. We further validated our experimental setup to ensure the validity and accuracy of our approach through discussions with the IOTA Foundation's technical team.

Keywords: IoT · Distributed Ledger Technology · IOTA Chrysalis · Heterogeneous Devices · Scalability and Data Security

1 Introduction

Internet of Things (IoT) is growing in our everyday life by connecting people or things with advanced devices to communicate and achieve a common goal in different applications. However, such a rampant explosion of these devices raises concerns for the security and privacy of IoT devices' data. The IoT systems cannot afford a malicious device to participate in the communication leading

This work was supported by the IoTalentum research program funded by the European Union Horizon 2020 research and innovation program within the framework of Marie Skłodowska-Curie Actions (MCSA) ITN-ETN with grant number 953442. We would also like to thank Thoralf Müller and other members of the IOTA Foundation for their valuable discussion to enhance experiments and good reviews, which led to further improvements.

J. M. Machado et al. (Eds.): BLOCKCHAIN 2023, LNNS 778, pp. 432–441, 2023.
https://doi.org/10.1007/978-3-031-45155-3_41

to the necessity of a secure exchange of information. Further, most devices are resource constrained, where they cannot run complex cryptographic algorithms or store large amounts of data due to low power capabilities. Therefore, these device data must be stored securely, so it cannot be accessible to any malicious device to prevent data abuse.

Initial models for securing IoT devices' data often rely on centralized authorities, introducing vulnerabilities and security risks due to single points of failure [24]. Distributed Ledger Technologies (DLTs), such as Blockchain, have been adopted to eliminate the need for intermediaries in IoT applications [3]. However, scalability issues arise from limited block capacity, and high computation power requirements for Proof of Work (PoW) may not always be applicable, as demonstrated by emerging Proof of Stake (PoS) protocols like Ethereum [23]. Furthermore, this limitation makes traditional blockchain architectures less suitable for deployment on resource-constrained IoT devices. Therefore, developing a lightweight system where IoT devices can communicate and securely store data is demanded. The system should not be confined to powerful devices but also acclimate low-power devices to store their data securely.

The IOTA protocol proposed by the IOTA foundation[1] is a DLT that functions as a Direct Acyclic Graph (DAG) instead of a single chain as in the blockchain [18]. IOTA employs the DAG structure to improve scalability and develop a keystone technology term as a Tangle, a permissionless scalable design focusing on IoT applications [14]. The technology mitigates the mining race and transaction cost by allowing the new transaction to reference two existing tips (i.e., newly attached transactions).

1.1 Contributions

Our contribution involves designing a system based on IOTA Chrysalis to facilitate secure information exchange for resource-constrained IoT devices. The design system utilizes the new IOTA client library *iota.rs* replacing the old one *iota.js* to facilitate communication with the Chrysalis Tangle. We conducted extensive experiments on a real testbed to assess the performance gap between low-power and powerful devices. The experiments stage in two phases: payload creation and payload broadcasting, with 1000 tests performed for each payload size on two machines separately, a workstation PC and Raspberry Pi (RPi). Our analysis includes comprehensive performance data, ranging from small to large payloads (approximately 32 KB), the maximum limit allowed in IOTA Chrysalis. These experiments represent the first real testbed evaluation using the maximum payload size for 1000 tests on the Chrysalis network.

The paper is structured as follows: Sect. 2 provides a background study on DLTs. Section 3 describes the proposed system and its implementation details. Section 4 presents the experimental results, and Sect. 5 concludes the paper with a discussion and proposed future directions.

---

[1] https://www.iota.org/.

Table 1. Analysis of different DLTs compared to proposed system for IoT.

Reference	DLT (Structure, Consensus)	Analysis	Results	Main Weaknesses
Marzouqi et al. (2022) [16]	Ethereum (Blockchain, PoW)	Scalability	Shows efficiency of transactions from different devices, PC perform better mining to add new blocks fast compared to RPi	High resource demanding, performed only 10 tests, tradeoff between performance and memory consumption
Xuan et at. (2020) [7]	Private Ethereum (Blockchain, PoW)	Scalability	Shows latencies in different workloads, achieved average time of 63.92 ms (without mining time) for one transaction	High resource demanding, tested only with 100 transactions, mining is performed on powerful PC
Zia et al. (2022) [21]	Private Ethereum (Blockchain, PoW)	Scalability	Introduced proof-of-authority to reduce transaction time average transaction time achieved is 487.6 ms	High resource demanding, one transaction degraded claim of high throughput, not feasible for small devices
Kumar et al. (2021) [13]	Ethereum and Hyperledger Fabric (Blockchain, No info)	Scalability	Internet of Forensic (IoF) framework for secure evidence chain, High transaction throughput, latency achieved is 9.6 sec	High communication cost, low scalability, consensus affect the performance, tested with only 100 transactions, no info on the consensus mechanism provided
Jiaping and Hao (2019) [22]	Monoxide (Blockchain, PoW)	Scalability/ Security	Experiments on testbed for Ethereum and Bitcoin Network, the authors claim 1,000× throughput and 2,000× on each network	High resource demanding, susceptible to attacks due distribution of mining power across zones
Fengyang et al. (2020) [11]	IOTA early version-1.0 (DAG, Tangle reference)	Scalability/ Security	Key findings are that most transactions take around 10 min to attach Tangle, also authors claim that the confirmation rate of 1–5% transaction is exceptionally long	Low transaction confirmation rate (1–10 min), not feasible for delay-sensitive devices, susceptible to attacks such as parasite chain attack
Akhtar, M.M et al. (2021) [2]	IOTA early version-1.0 (DAG, Tangle reference)	Scalability	Enhanced MAM protocol for better communication of IoT data, transactions (small in this case), confirmation time reduced to constant time (5.3 sec)	Experiments performed with small transactions, not on real testbed, MAM library creates overhead in restricted mode
Caixiang et al. (2019) [9]	Private IOTA, IRI 1.5.3 (DAG, Tangle reference)	Scalability	Shows good efficiency in processing transactions, average TPS can reach 1 sec for one transaction, the average time for confirmation of transaction is 0.83 sec	Not tested with maximum data, lack of synchronization time (no info on shifting from private to main Tangle), transaction speed reduces with increase in transactions
Sabah et al. (2020) [19]	IOTA early version-1.0 (DAG, Tangle reference)	Scalability	Shows good throughput for supply chain entities, average time for sending data is 3 sec for a local node, RPi is used to analyze the feasibility of energy consumption on low-power devices	Experiments performed with smaller payloads up to 1000 characters on RPi, large message attach time (average 23.1 sec)
Our Designed System	IOTA 1.5 (Chrysalis) (DAG, Tangle reference)	Scalability	Shows good efficiency in attaching data to Tangle, performed experiments up to maximum data, average time for largest payload (30,000 char) on PC is 55.7 sec and for RPi it is 1457 sec, implemented on a real testbed, accommodate low-power devices	Need to extend it for real sensors, to perform experiments for a large time (24 hrs for each payload) to analyze behaviour of Chrysalis in different timings, support up to 32 KB of data, Note: These limitations are considered as part of the future work

2   Related Work

DLTs are introduced in recent IoT environments to enable secure exchange of information. One motivation behind bringing attention to DLTs is the problem of a single point of failure with the centralized authority; however, these models also pose various issues, such as scalability and performance are the critical ones.

The initial explosion of Blockchain as a DLT has minimal use in IoT like bitcoin and Litecoin are not viable for IoT due to limited block size, low transaction rate and high transaction costs [17]. Following that, in [12], the authors proposed a model based on Ethereum for IoT devices; however, the concept is based on a limited number of devices, which needs to be clarified in relation to more transactions. Further, it is not feasible for low-power IoT device [7].

However, many researchers have recently worked towards improving performance and throughput in traditional Blockchain. An interesting work is presented in [22], called Monoxide for IoT systems. They made different zones, where an independent PoW is assigned to each zone to enable running PoW in parallel. However, the interaction of different zones with deploying cross-zone algorithms leads to security issues between zones as it weakens the mining power for sub-blockchains in the single system [15]. In one of the other works [20], the authors proposed an optimal node algorithm to analyze Blockchain's throughput and transaction success rate for IoT systems. However, the model needs to show feasibility for low-power devices.

Previous research has examined DAG-based IOTA to assess transaction confirmation time and system performance. Notably, a study on an early version of IOTA observed delays in attaching transactions to the Tangle [11]. Another simulation-based investigation highlighted the superior performance of IOTA compared to traditional blockchains, particularly in terms of transaction confirmation rate and computational requirements [5]. However, these experiments focused on smaller data sizes and made certain unrealistic assumptions [11]. In a separate study, an offline IOTA network was deployed to analyze performance with varying data amounts. However, the maximum data used and synchronization with the online Tangle were not explicitly addressed [9]. Table 1 provides an analysis of different DLTs related to the research. Recent experiments have focused on energy consumption in the future network of IOTA [10], known as Coordicide. However, these experiments are limited to energy analysis and do not cover the variations in data sizes. It is important to note that Coordicide is still in the development stage.

To the best of our knowledge, the performance gap between low-power and powerful devices on the recent version of IOTA, the Chrysalis, is not investigated in the literature, which is the main contribution addressed in this work.

2.1 IOTA Chrysalis

The IOTA foundation recently released the updated version of IOTA, named Chrysalis[2], with potential improvements to optimize the protocol and enhance the usability of the IOTA legacy for IoT applications. Chrysalis is improved in numerous aspects: *i)* the tip selection algorithm has been substituted with a new algorithm called Weighted Uniform Random Tip Selection (W-URTS), significantly reducing the time in nominating new tips [8]; *ii)* The Winternitz One Time Signature (W-OTS) scheme is replaced with the Edwards-curve (Ed25519) signature scheme to reduce signatures size and time for validation [4]; *iii)* the unspent transaction output (UTXO) model; and *iv)* the integration of atomic transactions with the protocol [8]. This work aims to analyze the impact of these new significant additions in terms of performance overhead on different types of devices.

3 System Architecture

The proposed system is designed on the lightweight DLT IOTA Chrysalis to secure the IoT devices' data. The system aims to evaluate its potential for creating and broadcasting data from powerful and low-power IoT devices. We have set up a powerful workstation PC and a low-power device like RPi. We use the Chrysalis public Hornet node, which is fully functional with the Chrysalis network. The proposed system architecture is shown in Fig. 1, while Table 2 provides more details about the components' specs.

IOTA has been utilized in several applications [1,6]; however, they were based on the IOTA legacy version, where the old libraries have been used, inducing complications in employing them with different programming languages. Further, the IOTA legacy version also raises issues while including new messages, such as the Random Walk tip selection algorithm's inefficiency in selecting new tips (i.e., messages), later improved with Chrysalis's new tip selection algorithm called Restricted Uniform Random Tip Selection (R-URTS) algorithm [14].

---

[2] https://chrysalis.iota.org/.

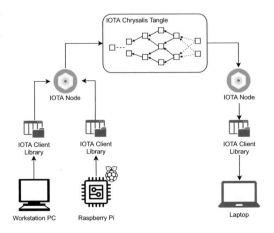

Fig. 1. Designed system architecture.

Table 2. System description.

System and Environment	Description
Workstation	Intel® Xeon(R) W-2133 CPU @ 3.60 GHz×12, Ubuntu 22.04.1 LTS, 32 GB RAM, 512 GB
Language for development	JavaScript (Node.js)
IoT device	RPi 4 Model B (ARMv7), Raspbian GNU/Linux 11

3.1 Implementation Details and Procedures

The communication of IoT devices with the Chrysalis network is facilitated by utilizing the new IOTA client library, specifically the `iota.rs`[3] library. This library allows for direct integration with Rust or enables binding with other programming languages. In our system, we have integrated the IOTA client library with Node.js. This integration allows us to convert requests into REST API format and send them to the node for processing. It provides a convenient way to interact with the IOTA network. For this research, we utilize the Chrysalis public Hornet node[4], a robust and capable node that supports full node functionalities. It is designed to be compatible with low-power devices such as the RPi. The selected Hornet node interacts with the Chrysalis network, including the operational network (mainnet) and the network for testing purposes (devnet).

In our design system, we leverage the client libraries and the Hornet node to interact specifically with the Chrysalis devnet. This enables us to conduct thorough testing and research activities in a controlled environment. Additionally, we include a laptop, as shown on the right side in Fig. 1 of the setup. Its role is to verify the reception of data transmitted by the IoT devices. Although our primary focus is on data creation and broadcasting, we include the laptop

---

[3] https://github.com/iotaledger/iota.rs.

[4] https://wiki.iota.org/hornet/welcome.

to ensure that data is correctly attached to the Tangle and can be received on the other side. By using the address and message indexation, we verify the successful transmission and retrieval of data. It serves as a validation step in our experimental setup. Further, in our testbed, we have set the devices to perform PoW locally. This allows us to assess the feasibility of executing PoW on low-power devices within the IOTA network. Additionally, the validation process verifies and confirms transactions on the IOTA Tangle. It checks transaction structure, data consistency, and completion of the required PoW. The IOTA nodes, including the Hornet node we used, perform this validation to maintain network integrity and security.

4   Experimental Results

In our study, we conducted two phases of experiments to assess the impact of IOTA Chrysalis on both powerful and low-power devices. In the first phase, we created data payloads with varying sizes, ranging from 10 to 30,000 characters, on both the workstation PC and RPi. This allowed us to examine the performance of each device in generating payloads of different magnitudes. In the second phase, we transmitted these payloads and attached them to the Chrysalis Tangle. By doing so, we analyzed the effectiveness and suitability of the IOTA Chrysalis technology for small, low-power devices. Through these experiments, we aimed to gain insights into how the IOTA Chrysalis implementation meets different devices' requirements and capabilities, providing a comprehensive understanding of its support and applicability.

4.1   Measuring Overhead of Data Creation

The first phase of experiments is set up to create various data/payloads. Different payloads are created, ranging from small (10 characters) to large (30,000 characters). Two devices, a workstation PC and a RPi (detailed description in Table 2), are used to conduct the experiments. The payloads are created on both machines using the IOTA Node.js client library.

For each of the six payload creations, we perform 1000 tests on each machine. The results of the experiments for the workstation PC and the RPi are shown in Fig. 2. Furthermore, statistical details of the results are shown in Table 3.

From the experiments, we notice that, when considering small payloads (up to 1000 characters), the difference in creation time between the workstation and the RPi is relatively small. Both devices are able to create payloads in less than 1 ms, with the workstation being slightly faster due to its higher processing capabilities. However, a significant performance gap becomes apparent as the payload size increases beyond 1000 characters. The workstation consistently outperforms the RPi, completing the payload creation process 3–5 times faster. Additionally, the time variations for small payloads are generally lower compared to large payloads. The workstation exhibits greater stability in all measurements, particularly with larger payloads. This is evident from the smoother curves and smaller standard deviation values, indicating a more consistent performance.

Table 3. Experimental results on Workstation (left) and RPi (right) for creation different payload sizes (1000 times).

Payload (nr. chars)	Min. Time (ms)	Max. Time (ms)	Avg. Time (ms)	St. dev. (ms)	Payload (nr. chars)	Min. Time (ms)	Max. Time (ms)	Avg. Time (ms)	St. dev. (ms)
10	0.034	0.184	0.047	0.011	10	0.028	0.389	0.076	0.031
100	0.030	0.901	0.066	0.058	100	0.044	0.579	0.129	0.045
1000	0.084	0.984	0.226	0.139	1000	0.264	2.268	0.644	0.246
10000	0.594	6.127	1.679	0.347	10000	2.317	29.925	6.008	1.473
20000	1.407	7.135	3.713	0.446	20000	4.688	32.727	10.724	3.435
30000	2.229	11.251	5.020	0.644	30000	7.159	48.021	16.509	4.987

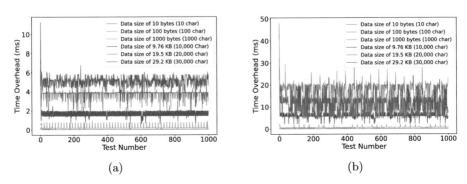

(a)   (b)

Fig. 2. Data/payload creation (1000 tests performed for each payload) on (a) Workstation and (b) RPi.

4.2   Overhead of Data Transmission on the IOTA Chrysalis Tangle

In our study's second phase, we conduct experiments to measure the time overhead of broadcasting payloads to the Chrysalis Tangle on both machines. We capture accurate time variance by repeating the process and attaching each payload to the Tangle 1000 times. The time taken for each message depends on the complexity of the required PoW. In the legacy IOTA implementation, users manually set the PoW difficulty when issuing messages. However, in Chrysalis, the PoW complexity is automated and adjusted based on message size. This dynamic PoW mechanism ensures efficient and scalable PoW computations tailored to each message. Looking forward, IOTA Coordicide introduces Adaptive PoW as a further advancement. Adaptive PoW dynamically adjusts the PoW process based on the rate of message issuance. It serves as a safeguard against transaction bursts, preventing spam attacks and congestion within the network.

Figure 3 shows the results of these experiments on the workstation and RPi. The results of the 1000 measurements are presented as boxplots showing the minimum, the first quartile, the median, the third quartile, and the maximum. The statistical facts of the outcomes are shown in Table 4.

Based on our experiments, several key findings emerged; firstly, the broadcasting time for data on the IOTA Tangle is significantly faster for powerful devices compared to resource-constrained IoT devices. This can attribute to the faster execution of PoW required for each payload attachment by the more capa-

Table 4. Experimental results on Workstation (left) and RPi (right) for broadcasting different payload sizes (1000 times).

Payload (nr. chars)	Min. Time (s)	Max. Time (s)	Avg. Time (s)	St. dev. (s)
10	0.117	1.985	0.421	0.248
100	0.136	5.648	0.825	0.633
1000	0.145	12.280	2.164	1.974
10000	0.211	120.131	17.250	16.586
20000	0.256	130.552	18.624	17.409
30000	0.312	386.629	55.786	55.534

Payload (nr. chars)	Min. Time (s)	Max. Time (s)	Avg. Time (s)	St. dev. (s)
10	0.147	21.304	3.269	3.207
100	0.211	90.639	12.315	13.092
1000	0.235	257.607	17.667	25.591
10000	1.126	2302.348	295.868	324.819
20000	2.460	3446.051	402.049	391.866
30000	6.574	6080.719	1457.376	1399.201

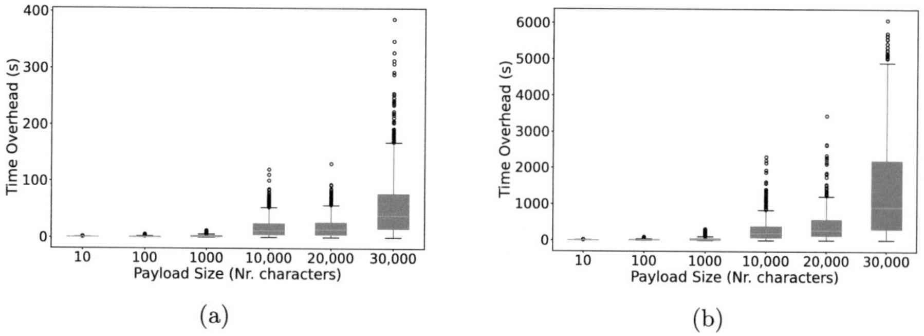

Fig. 3. Broadcasting payloads (1000 tests performed for each payload) from (a) Workstation and (b) RPi on Chrysalis Tangle.

ble devices. When considering small payloads (up to 1000 characters), the RPi can perform the PoW and transmit the data without significant hindrance. However, the RPi remains slower than the workstation by an average factor of ten. However, for larger payloads (more than 1000 characters), the performance gap becomes more pronounced, with the RPi being 17–26 times slower than the workstation on average. In terms of stability, smaller payloads exhibit minimal time variation and higher system stability. However, as the payload size increases, the variance in transmission time also increases for both the workstation and the RPi.

These findings provide valuable insights into the performance differences between powerful and low-power devices when broadcasting data on the IOTA Tangle.

5   Conclusion and Future Work

Secure storage of IoT devices' data is imperative in numerous IoT applications. In this paper, we have designed a system based on the nascent DLT IOTA Chrysalis that enables communication and securely stores heterogeneous devices' data. The system is not limited to powerful devices but can accommodate low-power devices such as RPi. We conducted detailed experiments by creating and broadcasting different payloads from two machines to compare and analyze the support of IOTA Chrysalis. Results show practical compatibility of IOTA Chrysalis for low-power devices that can effectively create and broadcast different payloads, especially small ones typical for IoT devices, on the Chrysalis network without hindrance.

In the future, we will extend the designed system to accommodate more low-power IoT devices, such as sensors (i.e., CC2650 sensortag), to collect a large amount of real data and see its impact on the system.

References

1. Abdullah, S., Arshad, J., Khan, M.M., Alazab, M., Salah, K.: PRISED tangle: a privacy-aware framework for smart healthcare data sharing using IOTA tangle. Complex Intell. Syst. **9**, 3023–3041 (2023). https://doi.org/10.1007/s40747-021-00610-8

2. Akhtar, M.M., Rizvi, D.R., Ahad, M.A., Kanhere, S.S., Amjad, M., Coviello, G.: Efficient data communication using distributed ledger technology and IOTA-enabled Internet of Things for a future machine-to-machine economy. Sensors **21**(13) (2021). https://doi.org/10.3390/s21134354. https://www.mdpi.com/1424-8220/21/13/4354

3. Bera, B., Saha, S., Das, A.K., Vasilakos, A.V.: Designing blockchain-based access control protocol in IoT-enabled smart-grid system. IEEE Internet Things J. **8**(7), 5744–5761 (2021). https://doi.org/10.1109/JIOT.2020.3030308

4. Bernstein, D.J., Duif, N., Lange, T., Schwabe, P., Yang, B.Y.: High-speed high-security signatures. J. Cryptogr. Eng. **2**(2), 77–89 (2012). https://doi.org/10.1007/s13389-012-0027-1

5. Bottone, M., Raimondi, F., Primiero, G.: Multi-agent based simulations of block-free distributed ledgers. In: 2018 32nd International Conference on Advanced Information Networking and Applications Workshops (WAINA), pp. 585–590 (2018). https://doi.org/10.1109/WAINA.2018.00149

6. Brogan, J., Baskaran, I., Ramachandran, N.: Authenticating health activity data using distributed ledger technologies. Comput. Struct. Biotechnol. J. **16**, 257–266 (2018). https://doi.org/10.1016/j.csbj.2018.06.004. https://www.sciencedirect.com/science/article/pii/S2001037018300345

7. Chen, X., Nguyen, K., Sekiya, H.: Characterizing latency performance in private blockchain network. In: Loke, S.W., Liu, Z., Nguyen, K., Tang, G., Ling, Z. (eds.) MONAMI 2020. LNICST, vol. 338, pp. 238–255. Springer, Cham (2020). https://doi.org/10.1007/978-3-030-64002-6_16

8. Conti, M., Kumar, G., Nerurkar, P., Saha, R., Vigneri, L.: A survey on security challenges and solutions in the IOTA. J. Netw. Comput. Appl. **203**, 103383 (2022). https://doi.org/10.1016/j.jnca.2022.103383. https://www.sciencedirect.com/science/article/pii/S1084804522000467

9. Fan, C., Khazaei, H., Chen, Y., Musilek, P.: Towards a scalable DAG-based distributed ledger for smart communities. In: 2019 IEEE 5th World Forum on Internet of Things (WF-IoT), pp. 177–182 (2019). https://doi.org/10.1109/WF-IoT.2019.8767342

10. IOTA Foundation: Energy consumption of IOTA 2.0: continued energy efficiency with new protocol components (2022). https://blog.iota.org/energy-consumption-of-iota-2-0/. Accessed 20 Dec 2022

11. Guo, F., Xiao, X., Hecker, A., Dustdar, S.: Characterizing IOTA tangle with empirical data. In: GLOBECOM 2020–2020 IEEE Global Communications Conference, pp. 1–6 (2020). https://doi.org/10.1109/GLOBECOM42002.2020.9322220

12. Huh, S., Cho, S., Kim, S.: Managing IoT devices using blockchain platform. In: 2017 19th International Conference on Advanced Communication Technology (ICACT), PyeongChang, Korea (South), pp. 464–467 (2017). https://doi.org/10.23919/ICACT.2017.7890132

13. Kumar, G., Saha, R., Lal, C., Conti, M.: Internet-of-forensic (IoF): a blockchain based digital forensics framework for IoT applications. Future Gener. Comput. Syst. **120**, 13–25 (2021). https://doi.org/10.1016/j.future.2021.02.016. https://www.sciencedirect.com/science/article/pii/S0167739X21000686

14. Kusmierz, B., Sanders, W., Penzkofer, A., Capossele, A., Gal, A.: Properties of the tangle for uniform random and random walk tip selection. In: 2019 IEEE International Conference on Blockchain (Blockchain), pp. 228–236. IEEE (2019)

15. Li, Y., et al.: Direct acyclic graph-based ledger for Internet of Things: performance and security analysis. IEEE/ACM Trans. Network. **28**(4), 1643–1656 (2020). https://doi.org/10.1109/TNET.2020.2991994

16. Marzouqi, S.A., Baddeley, M., Lopez, M.A.: Benchmarking performance of Ethereum blockchain on resource constrained devices. In: 2022 Workshop on Benchmarking Cyber-Physical Systems and Internet of Things (CPS-IoTBench), pp. 12–16 (2022). https://doi.org/10.1109/CPS-IoTBench56135.2022.00009

17. Okegbile, S.D., Cai, J., Alfa, A.S.: Performance analysis of blockchain-enabled data-sharing scheme in cloud-edge computing-based IoT networks. IEEE Internet Things J. **9**(21), 21520–21536 (2022). https://doi.org/10.1109/JIOT.2022.3181556

18. Popov, S.: The tangle. White paper **1**(3) (2018)

19. Suhail, S., Hussain, R., Khan, A., Hong, C.S.: Orchestrating product provenance story: when iota ecosystem meets electronics supply chain space. Comput. Ind. **123**, 103334 (2020). https://doi.org/10.1016/j.compind.2020.103334. https://www.sciencedirect.com/science/article/pii/S0166361520305686

20. Sun, Y., Zhang, L., Feng, G., Yang, B., Cao, B., Imran, M.A.: Blockchain-enabled wireless Internet of Things: performance analysis and optimal communication node deployment. IEEE Internet Things J. **6**(3), 5791–5802 (2019). https://doi.org/10.1109/JIOT.2019.2905743

21. Ullah, Z., Raza, B., Shah, H., Khan, S., Waheed, A.: Towards blockchain-based secure storage and trusted data sharing scheme for IoT environment. IEEE Access **10**, 36978–36994 (2022). https://doi.org/10.1109/ACCESS.2022.3164081

22. Wang, J., Wang, H.: Monoxide: scale out blockchains with asynchronous consensus zones. In: NSDI 2019, pp. 95–112 (2019)

23. Wang, P., Xu, N., Zhang, H., Sun, W., Benslimane, A.: Dynamic access control and trust management for blockchain-empowered IoT. IEEE Internet Things J. **9**(15), 12997–13009 (2022). https://doi.org/10.1109/JIOT.2021.3125091

24. Wang, S., Li, H., Chen, J., Wang, J., Deng, Y.: Dag blockchain-based lightweight authentication and authorization scheme for IoT devices. J. Inf. Secur. Appl. **66**, 103134 (2022). https://doi.org/10.1016/j.jisa.2022.103134. https://www.sciencedirect.com/science/article/pii/S2214212622000242

Enhancing Social Compliance with an IOTA Tangle-Enabled Smart Mask System

Lianna Zhao[✉], Pietro Ferraro, and Robert Shorten

Dyson School of Design Engineering, Imperial College London, London, UK
l.zhao20@imperial.ac.uk

Abstract. In this paper, we present an overview of a wearable smart-mask prototype that is fully described in [1]. The purpose of the mask is to encourage people to comply with social distancing norms, through the use of incentives. The smart mask is designed to monitor Carbon Dioxide and Total Volatile Organic Compounds concentrations. The IOTA Tangle ensures that the data is secure and immutable and here IOTA Tangle acts as a communication backbone for the incentive mechanism. A hardware-in-the-loop simulation based on indoor positioning, paired with Monte-Carlo simulations, is developed to demonstrate the efficacy of the designed prototype.

Keywords: Distributed Ledger Technology · Security and Privacy · Secure Communications · Network Architecture

1 Background

During the pandemic, especially the Covid-19 and its mutations, it is reasonable to expect that masks and social distancing norms play a significant role in certain sensitive situations. The need to wear face-masks was very important in many aspects of daily life, such as in passenger planes, buses, and trains. In such situations, enforcement of mask wearing is the responsibility of *observers*, such as flight attendants, rather than the mask wearer. In this paper, we explore ways to encourage people to wear masks properly, especially in confined and crowded spaces (for example: airplanes), without the need for observers. Importantly, we wish to design mechanisms where compliance with mask wearing (or more general social contracts) remains with the mask wearer, rather than with *observers*.

To achieve the aforementioned goals, we build on our previous work described in [2], in which the authors discuss a general framework, based on control theory, to regulate compliance to social contracts in the sharing economy domain. We develop personalised dynamic pricing strategy to assist in ensuring compliance

This work is funded by the IOTA Foundation.

with social contracts. Here, by a social contract, we mean guidelines that must be followed to ensure the proper wearing a mask correctly. While other works have discussed the issue of compliance [3,4], our work addresses aspects of the compliance problem not considered elsewhere. For example, the systems advocated in these papers are often centralised and lack anonymity and other privacy guarantees. Accordingly, the main contributions of this paper are the following. First, we propose a digital bonding system, based on the IOTA tangle technology and using ideas from feedback theory, to encourage adherence of a population to a social contract; in this case wearing a smart mask [5,6]. Second, we also present a smart mask design. The smart mask is designed to detect people's mask wearing status. A prototype based on a Raspberry Pi 3B hardware platform is described, complete with integration of a digital bonding system. The architecture of the system is depicted in Fig. 1. Here users deposit a bond based on the contract, which is fully or partially returned based in a personalised level of compliance. See [2] for full details.

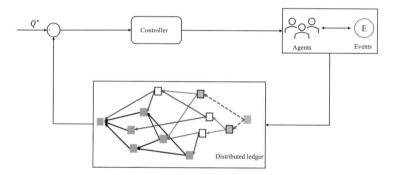

Fig. 1. The three components of the proposed compliance architecture are: the Distributed Ledger that acts as the communication backbone of the infrastructure, the compliance policy and the feedback controller [2].

2 Architecture and Compliance Model

As we have mentioned our work builds on [2], in which digital tokens are used to encourage people to comply with a social contract. The basic idea is that agents deposit tokens as a bond when they put on a mask, in areas where it is expected for them to wear one (e.g., an airplane). These tokens are then returned in full if the person does not remove the mask during the journey. A pricing algorithm is used to determine the number of tokens that are deposited (based on the level of previous compliance). The architecture of this system in [2] is depicted in Fig 1.

The algorithm in [2] is organized around three functional components: a DLT (more specifically the IOTA Tangle) is used as a communication layer (i.e., to record the deposit and the withdrawal of the tokens); the physical layer represents the agents' interaction in engaging with the social contract; the controller is used to adjust the number of tokens deposited and achieve expected compliance level. By adopting smart contracts, the whole process including the deposit and return of the tokens can be automatically operated through the use of Smart Contracts. All operations are recorded on the IOTA Tangle which is an immutable ledger and data are shared anonymously among agents (since each agent's identity is represented by an encrypted address). The rationale for the choice for the IOTA Tangle as the communication backbone of the whole system is twofold: firstly, due to the private nature of the data that are being collected, it is important that the information is kept anonymous and individual's privacy is safe-guarded. Secondly, although Blockchain might in theory provide the same kind of service that the IOTA Tangle provides, the inherent sequential structure in Blockchain to add new transactions often results in a limited transaction rate and the heavy cryptography based Proof-of-Work (PoW) consensus mechanism leads to expensive computation power. The IOTA Tangle on the other hand, due to its DAG structure and lightweight PoW can accommodate the throughput required for an Internet of Things (IoT) application such as the one proposed in this paper.

2.1   Compliance Algorithm

There are four policies that could be adopted: fixed penalty policy, adaptive penalty policies, adaptive penalty policies with return and event driven policies. For more details about these policies, interested readers please refer to [2].

A feedback mechanism for designing a proper value of the bond can be constructed as follows:

$$C(k+1) = C(k) + \alpha(Q^* - n^{-1} \sum_{i=1}^{n} M_i(k-m) \qquad (1)$$

$$c_i(k+1) = c_i(k) + \beta \left(Q^* - \overline{M_i(k-m)} \right) \qquad (2)$$

where $C(k)$ and $c_i(k)$ represent, respectively, a global and an individual feedback signal whose purpose is to regulate the behaviour of each agent to achieve the desired level of compliance. The sum of $C(k)$ and $c_i(k)$ represents the value of the token bond staked by agent i at time-step k. $\alpha > 0$ and $\beta > 0$ are two constants, $\forall k \in \mathbb{N}$ and $\forall i \in \{1, \ldots, n\}$, $Q^* \in [0, 1]$ is the desired level of compliance, $\overline{M_i(k)}$ is a windowed time average of the compliance of agent i, which is defined as

$$\overline{M_i(k)} = (1 - \gamma) \sum_{j=1}^{k} \gamma^{k-j} M_i(j) \qquad (3)$$

where $(1 - \gamma)^{-1}$ is the length of the window for the average, with $\gamma < 1$.

Intuitively, this means that the value of the bond staked by agent i depends **not only** on the overall compliance of all agents, **but also** and on how agent

i behaved in the past. The use of both a global and an individual control signal brings several advantages, such as *fairness* (a polluted-pays approach), the ability for *distributed trading of compliance levels*, and resiliency from *pricing attacks*. Interested readers can refer to [2] for further discussion of these topics.

Remark: Notice that no information is recorded on the DLT other than an average level of compliance, and all information is private and annonomysed. Specifically, even though our mask uses Co2 and TVOC information to detect whether a mask is worn correctly, none of this, or any user specific biological information, is recorded on the DLT.

3   Mask Design

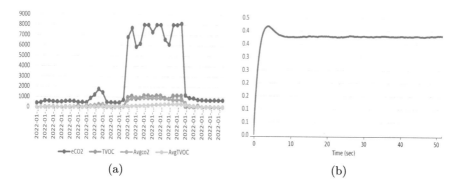

Fig. 2. (a) The level of $eCo2$, $TVOC$, $AvgCo2$ and $AvgTVOC$. We use blue line for $eCo2$, orange line for $TVOC$, grey line for $AvgCo2$ and yellow line for $AvgTVOC$. (b) The variation of the individual cost.

The system operates as follows.

– When an agent wishes to be present in a space where mask wearing is mandatory, the agent must use a disposable mask with a detachable integrated chip[1] connected to the agents phone or computer. This chip is connected to the customer's digital wallet.
– Each mask has a unique identifier, a detachable sensor, and it is connected to a Raspberry Pi which acts as the main computing unit (in a more realistic setting, the role of the Raspberry Pi would be carried out by other smaller computing units).
– Immediately after an agent enters a confined or crowded space, the agent deposits a certain amount of tokens through a smart contract (this might be based on the past level of compliance of the agent and on the current average level of compliance as described above).

---

[1] The detachable integrated chip should include a main board, gas sensor for monitoring, and battery for power.

- For the prototype mask, the IOTA *firefly wallet*[2] is used as the agents account and the tool with which an agent interacts with the IOTA Tangle.
- A Raspberry Pi 3 is used as the main processing element to collect sensor data and to send and publish data to IOTA Tangle. For context, the Raspberry Pi 3 consumes about 260 mA of current at 5.0 V (which is about 1.3–1.4 W) when no USB devices are connected to it and it's in idle state and weight is about 45 g. One of the main tasks of the Pi is to implement the controller given in equation (1). Note that the computational, memory, and energy cost of implementing this equation is low due to its recursive nature. The controller could also be delegated to the smart-phone.
- A gas sensor, CJMCU-811 is used to monitor the density of Carbon-dioxide (CO2) and Total Volatile Organic Compounds (TVOC) exhaled by an agent. The CJMCU-811 sensor is an ultra-low power digital gas sensor and its weight is 60 g.

The power requirements arsing from the system software is determined as follows. Communication is based on Message Queuing Telemetry Transport (MQTT) [7] and Masked Authenticated Messaging (MAM) [8]. MQTT is a standard IoT protocol, and MAM is the IOTA communication protocol (See remark below). Both are lightweight, efficient, and low-power consumption protocols. The main amount of power consumed by the system is due to the data transferred through the network. Finally we note that all electronic modules including sensors and the main board are reusable in our designed system. For convenience, these modules are designed to be easy to remove and reattach.

Remark: The mask presented in the paper is a proof-of-concept device and we use a Raspberry Pi as part of the prototype for convenience. In an actual implementation, one would use a small chip integrated into the mask for detecting data (to be mounted on the mask) and the majority of the computing, related to computing and analysing data, would be delegated to other devices, such as a smart phone. To give some examples, recently some smart masks have been developed [9]. All these prototypes use lightweight wearable electronics that do not affect the wearability of the masks. For example, a microcontroller named AtTiny402 with a 500 mAh lithium polymer (LiPo) battery is used in [10] to detect people's body temperature and breath rate. A commercial version of the smart mask presented in this paper might use an integrated PCB board with AtTiny402 or ESP32 as controllers and a 500 mAh LiPO battery.

4   Experimental Set-Up

To illustrate the effectiveness of the designed algorithm, we set up the hardware-in-the-loop simulation as follows: a user wears the mask prototype and its position and mask wearing status are respectively recorded to the IOTA Tangle and sent to the agent-based simulator in real time, as if the user was one of the agents

---

[2] https://wiki.iota.org/wallets/firefly/general.

of the simulations. This simulation is used to give real agents (people) the feeling of using a smart mask in a crowed area and to show that the interaction of the hardware and the control strategy works seamlessly.

The position and the status of the mask are both sent to the IOTA Tangle and the python simulation in which a number of virtual agents are simulated. Furthermore, the mask status from the simulated agents is appended on the IOTA Tangle, from which is read by the FPCA to generate the cost signals. To detect the smart mask position we make use of indoor positioning technology based on Ultra-Wide Band (UWB) [11]. Experiments were performed in a 5 m by 3 m room at Imperial College London. There are four anchors named DW4105, DW9B10, DW4A2F (as initiator), DW4C15, and one Tag named DW181C. Fig 2(a) depicts the data collected by the gas sensor, to be specific, the values for Co2 and for TVOC detected from the mask. It is possible to appreciate, even by visual inspection, that the reading from the sensors increase drastically when somebody is wearing a mask, making it straightforward to detect, whether or not a user is complying. Finally, Fig 2(b) depicts the variation of individual cost, paid by the specific user according to the compliance.

5 Conclusion

In this paper, a smart mask prototype, described in [1], is discussed. The use of DAG based DLT – IOTA Tangle, severs as both a communication layer for the control algorithm as well as ledger ensuring the security and immutability of data. The designed mechanism is validated through extensive simulations including a python-based one and a hardware-in-the-loop one.

Acknowledgement. The authors acknowledge support from DecaWave.

References

1. Zhao, L., Ferraro, P., Shorten, R.: A DLT enabled smart mask system to enable social compliance. *arXiv preprint* arXiv:2205.13256 (2022)
2. Ferraro, P., Zhao, L., King, C., Shorten, R.: Personalised feedback control, social contracts, and compliance strategies for ensembles. IEEE Internet Things J. (2023)
3. Ashby*, S., Chuah, S.-H., Hoffmann, R.: Industry self-regulation: a game-theoretic typology of strategic voluntary compliance. Int. J. Econ. Bus. **11**(1), 91–106 (2004)
4. Agarwal, P., Hunt, K., Srinivasan, S., Zhuang, J.: Fire code inspection and compliance: a game-theoretic model between fire inspection agencies and building owners. Decis. Anal. **17**(3), 208–226 (2020)
5. Ferraro, P., King, C., Shorten, R.: Distributed ledger technology for smart cities, the sharing economy, and social compliance. IEEE Access **6**, 62728–62746 (2018)
6. Ferraro, P., King, C., Shorten, R.: On the stability of unverified transactions in a DAG-based distributed ledger. IEEE Trans. Autom. Control **65**(9), 3772–3783 (2019)
7. Al-Masri, E., et al.: Investigating messaging protocols for the internet of things (IoT). IEEE Access **8**, 94880–94911 (2020)

8. Brogan, J., Baskaran, I., Ramachandran, N.: Authenticating health activity data using distributed ledger technologies. Comput. Struct. Biotechnol. J. **16**, 257–266 (2018)
9. Pazienza, A., Monte, D.: Introducing the monitoring equipment mask environment. Sensors **22**(17), 6365 (2022)
10. Lazaro, M., Lazaro, A., Villarino, R., Girbau, D.: Smart face mask with an integrated heat flux sensor for fast and remote people's healthcare monitoring. Sensors **21**(22), 7472 (2021)
11. Zhu, X., Yi, J., Cheng, J., He, L.: Adapted error map based mobile robot UWB indoor positioning. IEEE Trans. Instrum. Meas. **69**(9), 6336–6350 (2020)

Topology and the Tangle: How the Underlying Network Topology Influences the Confirmation of Blocks in IOTA

Benjamin Kraner[1](✉)(iD), Nicolò Vallarano[1,2](iD), and Claudio J. Tessone[1,2](iD)

[1] Blockchain & Distributed Ledger Technologies Group, Department of Informatics,
University of Zurich, Zurich, Switzerland
{benjamin.kraner,nicolo.vallarano,claudioj.tessone}@uzh.ch
[2] UZH Blockchain Center, University of Zurich, Zurich, Switzerland

Abstract. We present an agent-based model of the IOTA system to provide an estimate of block finality on the Tangle. Our simulation framework is based on the Gillespie agent-based model. Our approach provides insights into the behavior of the IOTA system and can be used to explore different scenarios and test potential improvements. The results of our simulation show that the finality and confirmation time of blocks on the Tangle are affected by the underlying topology of the peer-to-peer network. Depending on the topology, the blocks of higher Mana nodes may have an advantage in gaining approval weight.

Keywords: IOTA · Tangle · Directed Acyclic Graph · Consensus · Latency · Distributed Ledger

1 Introduction

Distributed Ledger technologies (DLT) revolutionized how we store and exchange value online without relying on central authorities or third-party trust providers. This started with the seminal Bitcoin Paper from the internet persona Satoshi Nakamoto [9]. The core of every DLT is the consensus protocol, which is the set of rules peers follow when participating in the validation and storage of the DLT data. The consensus protocol provides the participants with the means to agree on the DLT content in a secure way, removing the need for trust issuers (central banks, financial third parties, etc.).

A central aspect to keep in mind when analyzing a consensus protocol or a DLT, in general, is the idea of finality. Finality, somewhat of an elusive concept, refers to that point in time when we know, beyond any reasonable doubt, that a transaction is permanent. Finality does not coincide with the inclusion of a

The authors would like to express their gratitude to Olivia Saa, Luigi Vigneri and the IOTA foundation.

J. M. Machado et al. (Eds.): BLOCKCHAIN 2023, LNNS 778, pp. 449–458, 2023.
https://doi.org/10.1007/978-3-031-45155-3_43

transaction into a block (or more in general on the ledger); it doesn't even have a fixed definition valid for all DLTs. Finality timing, eventuality, nature, and definition strictly depend on the rules of the underlying DLT we consider. For example, when we talk about Bitcoin [9] or pre-Merge Ethereum [3], or more generally in the context of Proof-of-Work DLTs, we talk about probabilistic finality: the probability of reversion of a transaction reversely correlates with the time the transaction has been included in the ledger (the block in this case), implying that the longer the transaction was written on the blockchain, the deeper the corresponding block is in the past, the less likely it is that the transaction may be excluded from the blockchain [9,13]. On Proof-of-Stake Ethereum [2], we talk about multi-slot finality, ensured by the introduction of Casper FFG [1], a somewhat slower finality gadget that relies on the eventuality (still a probabilistic approach) of reaching a certain threshold of votes (called attestations), in a specific amount of time (defined by slots), supporting a block to consider that block final. In the context of the Coordicide [12], a need for a re-definition of the concept of finality emerges. While yet not officially defined, the consensus protocol of IOTA 2.0 is drifting towards the implementation of a PoS-like system [8,10], which definitely prompts us to consider a finality measure taking into account the approval of peers/users supporting the specific block. In opposition to Ethereum PoS thought, no time partition is hard-coded into the system, rejecting the imposition of a consensus deadline. Agent-based models are a consolidated tool to explore and understand the behavior of complex socioeconomic systems [4,15]. In recent years, multiple applications were presented of Agent-based models for DLTs; agent-based models proved particularly useful because of the intrinsic dual nature of peer-to-peer consensus protocols: on one hand, there is the cooperative process of block diffusion among the agents who share information in order for the common Ledger record to emerge, and on the other hand, there exists the competition game between users/validators/miners to achieve individual rewards (competing incentives) or even to achieve consensus degradation or even failure (from malicious actors) [13,18]. In line with recent applications, such as miners' competition modeling [19], agent strategies under particular conditions, such as rewards absence [6,16], or even malicious attacks [7,17], we propose an agent-based model to simulate the growth of the Tangle [11], IOTA DAG Ledger. The simulation framework is based on the Gillespie chemical reaction model [5], which provides a lightweight, efficient algorithm based on an underlying Poissonian assumption on the agents' latencies.

The paper is structured as follows: in Sect. 2, there is a high-level description of the IOTA system, with particular attention to how new blocks are attached to the Tangle. In Sect. 3, we describe in detail how the agent-based model works, with respect to agent definition, parameter setting, and event description. Finally, in Sect. 4, we explain the main experiments we ran and their theoretical implications.

2   The IOTA System

IOTA is a decentralized distributed ledger technology designed to facilitate secure and feeless transactions and data exchange between devices in the Internet of Things (IoT) ecosystem. Unlike traditional blockchain systems that rely on a linear chain of blocks, IOTA uses a directed acyclic graph (DAG) called the Tangle. In the Tangle, every block is linked to at least two previous blocks, and the verification of new blocks depends on the confirmation of previous blocks. Blocks can be initiated by any node and do not require transaction fees, making it an ideal solution for microtransactions in the IoT ecosystem.

The Tangle is IOTA's equivalent of Bitcoin's blockchain, the collection of all blocks issued in the IOTA system on which IOTA users reached consensus. A DAG is a type of graph data structure in which the edges between nodes have a specific direction, and no cycles (loops) are allowed. In the context of distributed ledgers, DAGs have been used as an alternative to traditional blockchain structures to achieve scalability and faster transaction processing times. The vertices represent blocks, bundles of transactions, and/or data, and the edges represent block approvals.

In the Tangle, each block must approve (at least) two previous blocks as a form of validation. This process results in a network of DAG blocks, where each block is indirectly connected to all others via a series of approvals. Key components of the Tangle include:

- **Block**: In the Tangle, a block represents the transfer of value or data. Each block has several fields, including the transaction's hash, the two blocks it approves, a timestamp, a value, and a signature.
- **Tip**: A tip is a block that has not yet been approved by any other blocks. New blocks must select and approve at least two tips to join the Tangle.
- **Block Creation**: When a user wants to create a block, they must first assemble the block data, including the sender's address, the recipient's address, and the value in a block. The user then signs the block using their private key.
- **Tip Selection**: After creating the block, the user must select between two and eight unapproved blocks (tips) to reference and validate. The Tangle uses a tip selection algorithm to help users identify tips to approve. The algorithm considers factors like block's age, cumulative weight, and other heuristics to ensure a fair and efficient selection process.
- **Mana**: Mana can be considered a very basic reputation system in itself. Every user in the IOTA network may be pledged a certain quantity of mana, and this endowment serves both as a Sybil protection mechanism and as a congestion mechanism. We do not pretend to provide a full cover of Mana's roles in IOTA in the present work, which would take way more space than what we can devote to it. For the reader, it is sufficient to know that the Mana endowment regulates the user's access to writing rights, and its assignment process is safe enough to work as a guarantee against identity attacks.

3   The Model

We will explain our model along three dimensions. The first dimension concerns the *objects* of the simulated system and how they are connected to each

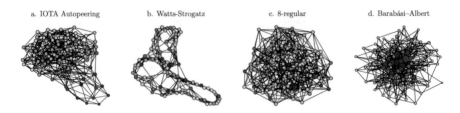

a. IOTA Autopeering   b. Watts-Strogatz   c. 8-regular   d. Barabási–Albert

Fig. 1. The four different topologies studied, the colours of nodes indicated the Mana endowment while the size represents the degree.

other. Along the second dimension, we want to explain what *events* are possible within the system. Events will alter the state of objects in the system. The final dimension regards how the events are ordered, in our case, using the Gillespie algorithm [5], and how the occurrence of one or another influences the system as a whole.

The model of IOTA we propose here consists of two main objects: nodes and blocks. Both of them are part of a bigger structure. Nodes are embedded in a peer-to-peer network, and blocks bond together to form the bigger data structure that IOTA knows as the Tangle, a DAG.

- **Nodes**: A node represents an entity actively participating in the creation of the IOTA system, i.e., some form of computer that is storing data, issuing blocks, verifying blocks, etc. It is connected to other nodes (via the internet infrastructure) - commonly called peers - to whom it communicates data. In addition to its peers, a node possesses certain idiosyncratic characteristics: it possesses *Mana*, a numerical value representing either a reputation or a stake in the system. It possesses a *database*, a set of blocks that ultimately form the personal Tangle of a node. Finally, it possesses a gossiping module - the *Scheduler* - responsible for the order in which blocks are exchanged with the peers of the node.
- **Mana**: In IOTA, Mana is intended to serve two *main* purposes: protection against Sybil attacks (rate control) and the voting-based consensus mechanism (On-Tangle Voting (OTV)). We represent the Mana in node i as a constant number, μ_i. In our current setting, without malicious actors in the network, we are mostly interested in the rate control aspect of Mana. Mana should be more prevalent in nodes that contribute positively to the network. Thus, we assume that Mana and the usage of the system (issuing blocks, appending information to the Tangle) are correlated.
- **Database**: The database of a node refers to the local copy of the Tangle a node has. It consists of a set of blocks, which by their inherent structure, form the DAG-based ledger innate to IOTA. We will denote the database of node i as T_i.
- **Scheduler**: Nodes keep track of blocks that are arriving at them. These blocks will be attached to the node's database and will then be propagated to its peers. Scheduler refers here to a stack of blocks q_i^j that have to be

relayed from node i to node j (where j is a peer of i). Upon arrival, the blocks get added to the queue. Should the node communicate with one of its neighbors, q_i^j will be consulted, and a block will be sent. The blocks leave the queue in a first-in, first-out (FIFO) manner.

- **Blocks**: blocks are the atomic element of the Tangle. In our model, they represent the atomic unit created by a node of the network and reference a number of blocks $k \geq 2$. We will assume that blocks connect to (if available) eight tips, blocks of the local Tangle that have no incoming edges (i.e., references).
- **Peer-to-Peer Network**: The nodes of the model are connected to each other to represent the underlying peer-to-peer network. They are embedded in a topology. Each node possesses neighbors and is aware of them. The nodes will exchange blocks with these neighbors.

We consider the set of links connecting the node-set N: a directed link from i to j exists when $(i, j) \in E$. The graph of the peer-to-peer network is completely described by the couple $\mathcal{G} = (N, E)$.

We assume a static topology, i.e., no nodes leave or enter the system, and their links to each other persist. Node churn and node renewal happen on timescales much larger than those of finality and block creation [19].

The peer-to-peer topology is one of the main focuses of our study. We want to analyze how the different structure of the network influences the finality time of the system in relation to the Mana distribution across the nodes. We will consider IOTA's proposed *Autopeering*[12] solution, a Watts-Strogatz small-world graph, a k-regular graph, and a Barabási-Albert graph. Figure 1 shows visualizations of the different topologies used in the simulations.

3.1   Simulation

In order to model the evolution of the system, we can distinguish two key ideas: *events* and their *ordering*. First, we will describe the events that change the state of the system. Second, we will describe the Gillespie algorithm, responsible for the ordering of the event and evolution of time within the model.

We define two (stochastic) events: *block creation* and *block gossip*. When a block is created, a random (proportional to its mana) node i will create a block. To do so, it selects up to eight tips (depending on their availability) that the new block references. Then, it attaches the newly created block to its personal Tangle (T_i) and queues the block to be gossiped to its neighbor (as it cannot yet know about it).

When a block is gossiped, a queue j of the scheduler of node i is selected, and the block is attached to the Tangle of j. The block is deleted from the queue q_i^j but added to the queue of j's scheduler (except the one where it is originating from, to avoid circulating blocks back and forth).

To each kind of event, we associate a parameter τ_{creation} and τ_{gossip}, respectively. We assume that each node is producing blocks according to an underlying Poisson process. The inter-event time between block creation is thus distributed according to an exponential distribution with parameter $1/\tau_{\text{creation}}$.

Taking advantage of the superposition property of the Poisson distribution, we may control the parameter on a global level. Thus, we assume a global rate of block creation, which consists of single processes scaled by the Mana of a node. Put simply, nodes produce blocks proportional to their Mana endowment. In all of our simulations, we will fix the block creation average total rate to 100 blocks per second (bps).

The exchange of blocks follows the same logic. The time between two gossip events is going to be distributed according to an exponential distribution with parameter $1/\tau_{\text{gossip}}$. This parameter will adapt depending on the total number of blocks in the queues of nodes: $Q = \sum_i \sum_j q_i^j$. We will vary the parameter associated with the latency in the system to explore how the confirmation time changes under different latency scenarios (i.e., higher, lower, average).

Because both waiting time processes (block creation and block gossip) follow exponential distributions, we take advantage of the superposition property of exponential distributions, which implies that the random variable defined as the minimum of a collection of exponential distributions distributes as an exponential distribution as well, where the parameter is equal to the sum of the single exponential distributions composing the minimum [14].

4   Results

We want to assess the performance of the system in relation to finality and how the topology and mana distribution change its behavior. Finality is a spiky subject; here we do not want to approach the ontological nature of finality so we will limit ourselves to interpreting the finality of a block as a measure of the number of IOTA users who actually observed the block and trusted it enough to reference it(directly or indirectly) with their own blocks, a measure weighted by user's mana, therefore the common interpretation of Mana as reputation.

For this reason, we recall the notion of Approval Weight. Given a block i on the Tangle with an in-degree larger than zero, we can define the approval weight AW_i as:

$$AW_i = \frac{\sum_{f \in fc(i)} Mana(f)}{\sum_b Mana(b)} \tag{1}$$

where $fc(i)$ is the future cone of block i, meaning the set of all blocks referencing block i either directly or indirectly, or more formally the set of all blocks in the tangle from which you can reach block i following a directed path of links. The denominator in Eq. 1 is a normalization factor, the sum of all the mana of the node(IOTA peer) actually contributing to the Tangle. It is important to observe that by $Mana(i)$ we denote the mana associated with the node that released block i.

Upon reaching a predefined percentage of total Mana that is *approving* a block, the block is confirmed (as a complete restructuring of the future cone becomes less likely). The *confirmation threshold* can be set arbitrarily, representing a trade-off between security, finality, and timeliness. We will consider

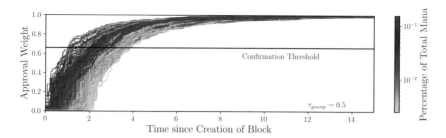

Fig. 2. Trajectories of the approval weight of single blocks (approximately 10'000) from one simulation with the IOTA Autopeering topology.

a confirmation threshold of 66 percent for the evaluation of our simulations. We will first focus on the trajectories of approval weight (Mana) of individual blocks in a single system and then aggregate the results of multiple simulations to compare them across different topologies.

The baseline parameters of the simulations will be the following: We will consider $N = 100$ agents, with an average of eight neighbors. The creation rate of blocks is set to $\tau_{\text{creation}} = 0.01$ (i.e., 100 bps), and the latency is set to $\tau_{\text{gossip}} = 0.5$ (i.e., 500 ms). Most importantly, the Mana distribution will be distributed according to Zipf's law, with its controlling parameter set to 0.9, displaying a moderate concentration of Mana within some nodes. We will stop the simulations after 100 s have been simulated.

4.1   Approval Weight Trajectories

Figure 2 shows the approval weight trajectories of approximately 10,000 blocks obtained from a simulation with an underlying topology obtained by an implementation of the IOTA Autopeering implementation[1]. The color of the trajectories refers to the percentage of total Mana the creator of a block possesses. Immediately visible is the contrast of the trajectories from low Mana nodes and high Mana nodes. While the trajectories of high Mana nodes' blocks (red) expose a steep (concave) trajectory in the beginning, blocks issued by low Mana nodes show a flatter (convex) trajectory during the initial phase. Naturally, blocks issued by high Mana nodes have a higher initial weight (as the issuer pledges its Mana to the block), but the speed at which Mana is pledged to the blocks is also higher compared to low Mana nodes. The topology plays a crucial role in this dynamics, as high mana nodes are connected to each other (Mana homophily), the blocks reach them faster and thus the accrual of Mana (approval weight) is faster in the first few seconds after a block is released. Initial results suggest that this effect becomes more pronounced for higher latencies. We plan on exploring this parameter space further in the future.

---

[1] Original implementation: https://github.com/iotaledger/autopeering-sim. We used a Python implementation of the original code.

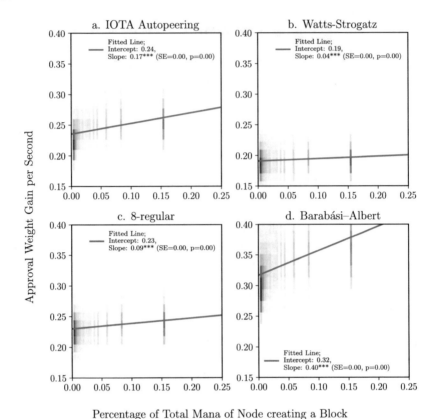

Fig. 3. Regressions results for different topologies, blue shaded areas indicate the prevalence of observation. Data aggregated over 10 simulations for each topology (ca. 100000 observations) per panel.

This relationship poses a question that lies at the core of IOTA's system design: are faster confirmation times for higher mana nodes a given, or are they a result of the underlying peer-to-peer topology?

4.2   Approval Weight Gain and Different Topologies

In order to quantify the differences across high and low Mana nodes across different topologies, we want to quantify the trajectories of single blocks in a simple way. We define the *Approval Weight Gain per Second* (AWG) as the slope of the approval weight a block gets until it is considered confirmed:

Approval Weight Gain per Second = Approval Weight ÷ Confirmation Time

To better understand the relation between Mana and Approval Weight Gain, we use a simple Ordinary Least Squares regression (OLS), where we estimate:

$$AWG_i = \alpha + \beta * \text{Issuer Mana}_i + \epsilon_i$$

The intercept (α) will serve as a baseline to compare the AWG for low mana nodes, and the slope (β) will inform us how much faster blocks of higher mana nodes gain approval weight.

Figure 3 shows the results for four different topologies. The average degree is kept constant at 8 for all of them. While Mana has no influence on the location of the nodes in the Watts-Strogatz and 8-regular topology, it does by design in the IOTA Autopeering topology and by choice in the Barabási-Albert, as we correlated high Mana with high degree (a reasonable assumption as highly invested nodes may choose to connect more often to peers).

Comparing the intercepts, we notice the high value with the Barabási-Albert topology, as supposedly high mana nodes act as hubs and enable the whole system to confirm blocks faster. We find that the most interesting result is represented by the stark difference in the slopes of OLS results. The systems in which the location within the topology is dictated by the Mana endowment of the nodes reveal a steeper slope, and thus blocks of high Mana nodes gain approval weight faster than their lower Mana colleagues. This may be explained by the homophily of high Mana nodes under those settings. As they are in the direct neighborhood of each other, blocks sent from them reach other high Mana nodes faster, while blocks sent by low Mana nodes have to traverse a larger part of the network before the *influential* high Mana nodes approve them.

5   Conclusion

Our results show how the topology of the underlying peer-to-peer network may influence the speed at which blocks issued by high, respectively low, Mana nodes are approved. We reveal differences in the approval weight trajectories and quantify differences across different topologies. Most importantly, we find that the structure of the peer-to-peer network may benefit certain groups of nodes (i.e., high Mana nodes) more than others. This is a design dimension DAG-based protocols may want to take into account more strongly. In the case of IOTA, depending on the exact implementation details and mechanics of Mana, this may lead to the centralization of certain services where fast confirmation is required. If IOTA wants to hold on and eventually implement the Autopeering solution, it will be essential to understand how the Mana works to prevent the favoritism of high Mana nodes in the system.

The simplicity of our model allows for various extensions across multiple dimensions. In future work, we would like to increase the parameter space and explore different configurations (e.g., expanding the size, different topologies, different latencies, etc., connectivity). A further very interesting extension would be the possible inclusion of Mana dynamics to study its evolution in the system.

References

1. Buterin, V., Griffith, V.: Casper the friendly finality gadget. arXiv preprint arXiv:1710.09437 (2017)
2. Buterin, V., et al.: Combining ghost and casper. arXiv preprint arXiv:2003.03052 (2020)
3. Buterin, V., et al.: A next-generation smart contract and decentralized application platform. White paper **3**(37), 2–1 (2014)
4. De Marchi, S., Page, S.E.: Agent-based models. Annu. Rev. Polit. Sci. **17**, 1–20 (2014)
5. Gillespie, D.T.: A general method for numerically simulating the stochastic time evolution of coupled chemical reactions. J. Comput. Phys. **22**(4), 403–434 (1976)
6. Kraner, B., Li, S.N., Teixeira, A.S., Tessone, C.J.: Agent-based modelling of bitcoin consensus without block rewards. In: 2022 IEEE International Conference on Blockchain (Blockchain), pp. 29–36. IEEE (2022)
7. Kraner, B., Vallarano, N., Tessone, C.J., Schwarz-Schilling, C.: Agent-based modelling of ethereum consensus. In: IEEE International Conference on Blockchain and Cryptocurrency, ICBC 2023, Dubai, EAU, 1–6 May 2023. IEEE 2023 (2023)
8. Müller, S., Penzkofer, A., Kuśmierz, B., Camargo, D., Buchanan, W.J.: Fast probabilistic consensus with weighted votes. In: Arai, K., Kapoor, S., Bhatia, R. (eds.) FTC 2020. AISC, vol. 1289, pp. 360–378. Springer, Cham (2021). https://doi.org/10.1007/978-3-030-63089-8_24
9. Nakamoto, S.: Bitcoin: A Peer-to-Peer Electronic Cash System (2008). https://bitcoin.org/bitcoin.pdf. Accessed 17 July 2019
10. Popov: FPC on a set
11. Popov, S.: The tangle. White paper **1**(3), 30 (2018)
12. Popov, S., et al.: The coordicide, pp. 1–30, January 2020
13. Rosenfeld, M.: Analysis of hashrate-based double spending. arXiv preprint arXiv:1402.2009 (2014)
14. Ross, S.M.: Introduction to Probability Models. Academic press, Cambridge (2014)
15. Samanidou, E., Zschischang, E., Stauffer, D., Lux, T.: Agent-based models of financial markets. Rep. Prog. Phys. **70**(3), 409 (2007)
16. Schwarz-Schilling, C., Li, S.N., Tessone, C.J.: Agent-based modelling of strategic behavior in pow protocols. In: 2021 Third International Conference on Blockchain Computing and Applications (BCCA), pp. 111–118. IEEE (2021)
17. Schwarz-Schilling, C., Li, S.N., Tessone, C.J.: Stochastic modelling of selfish mining in proof-of-work protocols. J. Cybersecur. Priv. **2**(2), 292–310 (2022)
18. Schwarz-Schilling, C., Neu, J., Monnot, B., Asgaonkar, A., Tas, E.N., Tse, D.: Three attacks on proof-of-stake Ethereum. In: Eyal, I., Garay, J. (eds.) Financial Cryptography and Data Security. FC 2022. LNCS, vol. 13411, pp. 560–576. Springer, Cham (2022). https://doi.org/10.1007/978-3-031-18283-9_28
19. Tessone, C.J., Tasca, P., Iannelli, F.: Stochastic modelling of blockchain consensus. arXiv preprint arXiv:2106.06465 (2021)

FPCS: Solving n-Spends
on a UTXO-Based DLT

Rafael Cizeski Nitchai[1,2][✉] (ID), Serguei Popov[1,2] (ID), and Sebastian Müller[3,4] (ID)

[1] Universidade do Porto, 4099-002 Porto, Portugal
up202008679@up.pt, serguei.popov@fc.up.pt
[2] Centro de Matemática da Universidade do Porto, 4169-007 Porto, Portugal
[3] Aix Marseille Université, CNRS, Centrale Marseille, I2M - UMR 7373,
13453 Marseille, France
sebastian.muller@univ-amu.fr
[4] IOTA Foundation, 10405 Berlin, Germany

Abstract. The fast probabilistic consensus (FPC) is a leaderless voting consensus protocol that allows a set of nodes to agree on a value of a single bit. FPC is robust and efficient in Byzantine infrastructures and presents a low communicational complexity. In this paper, we introduce a modification of the Fast Probabilistic Consensus protocol (FPC) capable of achieving consensus on a maximal independent set of a graph —hence named Fast Probabilistic Consensus on a Set (FPCS)— that still preserves the robustness, effectiveness, and low communicational complexity of FPC.

This paper shows that FPCS effectively resolves the problem (with high probability) of achieving consensus on a maximal independent set of a graph of conflicts (i.e. a maximal set of nonconflicting transactions) in the particular case of n-spend conflicts, even when a significant (up to 1/3) proportion of nodes is malicious. These nodes intend to delay the consensus or even completely break it (meaning that nodes would arrive at different conclusions about the maximal independent set). Furthermore, the paper provides explicit estimates on the probability that the protocol finalizes in the consensus state in a given time.

Our study refers to a specific implementation of cryptocurrencies, but the results hold for more general majority models.

Keywords: Distributed systems · Consensus protocols · Byzantine infrastructures

1 Introduction

In this paper, we study a modification of the Fast Probabilistic Consensus (FPC) protocol, proposed by Popov and Buchanan in [1] and explored in detail in other works (e.g. [2–4]). FPC is a protocol of low communicational complexity, which allows a set of nodes to come to a consensus on a value of a bit. The novelty of FPC compared to previous variants of leaderless voting-based consensus protocols is its robustness in faulty or Byzantine infrastructure as FPC reaches

© The Author(s), under exclusive license to Springer Nature Switzerland AG 2023
J. M. Machado et al. (Eds.): BLOCKCHAIN 2023, LNNS 778, pp. 459–471, 2023.
https://doi.org/10.1007/978-3-031-45155-3_44

consensus even if a substantial (but minor) part of the nodes is malicious and tries to break the protocol actively.

An important property of FPC is that it is not guaranteed (even in the loose sense, i.e., with high probability) to finalize on the initial majority opinion; this is only so if this majority is *very significant* (for example, something like 90%). For typical cryptocurrency-related applications, this means that, in the case when it is independently applied to several conflicting transactions, it may occur that none of those transactions will be considered valid by the protocol. In most cases, such an outcome is not of a big concern — conflicts usually arise out of malicious activities (such as attempting to double-spend funds), so, in principle, it is not a problem if both those transactions are invalidated (and then forgotten) by the system. However, one can still imagine situations when conflicting transactions arise in a "non-malicious" way. As an example, consider a situation when a transaction is to be signed by a 2-out-of-3 threshold signature: that is, Alice, Bob, and Charlie are to create a transaction, and, to sign it properly, at least two of them are needed. Alice signs her part and passes it to Bob, who receives it, but then goes offline for some reason. Alice, seeing no response from Bob, passes it to Charlie who signs it and publishes it to the network. At the same time, Bob comes back online and publishes his version of the transaction; those transactions will be formally different (because signed by different validators), and, depending on the system's rules, may be considered conflicting.

The purpose of this paper is to modify the FPC protocol in such a way that, with high probability, it would select a nonempty[1] subset of non-contradicting transactions from a larger transaction set where some elements may contradict each other. We call it Fast Probabilistic Consensus on a Set, FPCS for short. Similarly to FPC, the FPCS protocol makes use of a sequence of random numbers, which are either provided by a trusted source or generated by the nodes themselves using some decentralized random number generating protocol. Another interesting feature of FPCS is that it is still likely to work well in situations when nodes have slightly different "visions" of the conflicting transactions and possibly new conflicts arrive during the runtime of the protocol.

2   Description of the Protocol

2.1   Notations

Consider a set of N nodes denoted by $\mathcal{N} = \{1, \ldots, N\}$ and a set of conflicting transactions $\mathbb{T}_t$ for $t \in \mathbb{N}$, that we call the *conflict set*. As the time dependency suggests, we assume that the conflict set can change over time. The exact dynamics of how new transactions emerge will not be explored and we consider it can happen arbitrarily. In particular, we assume that nodes themselves can issue new conflicting transactions and this can potentially become a vector of attack. For our purposes, we assume that a transaction is composed of a unique transaction identifier ("Id", for short), a set of inputs – often referred to as UTXOs (unspent

---

[1] and, preferably, *maximal* — we explain the meaning of that later.

transaction outputs) – and a set of outputs (see Fig. 1). The sums of the input and output values must always be equal.

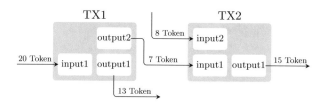

Fig. 1. Diagram representing the general structure of two transactions in the IOTA network.

We say two transactions $x, y \in \mathbb{T}_t$ are in conflict if they consume the same UTXO (or, in other words, if at least one of their inputs is the same) and denote this by $x \leftrightarrow y$. A transaction x is in conflict with a set B if it is in conflict with every element of B and it is represented by $x \leftrightarrow B$. It is natural to represent the set $\mathbb{T}_t$ and its conflicts as a graph $G_t = (\mathbb{T}_t, E_t)$, where given $x, y \in \mathbb{T}_t$ an edge $(x, y) \in E_t$ denotes that $x \leftrightarrow y$.

In this paper, we restrict our attention to the special case of a *complete* conflict graph with n transactions. This means that for every $i, j \in \mathbb{T}_t$ there exists an edge (i, j), or, in other words, that every transaction shares at least one input. This problem is popularly known as a *n-spend* problem in the context of cryptocurrencies.

Depending on the network throughput, the set $\mathbb{T}_t$ can be very large and it is convenient to find a global (i.e., known to all nodes) way to totally order it. A popular way to do this is through the aid of a *cryptographic hash function* (CHF for short). Following the definition introduced by Merkle [5], a CHF is a function F such that:

1. F can be applied to any argument of any size and produces a fixed-size output.
2. Given F and x, it is easy to compute $F(x)$.
3. Given F, it is computationally infeasible to find any pair x, x' such that $x \neq x'$ and $F(x) = F(x')$. We call this the *no-collision condition*.

For the purpose of the paper, we will also assume that the CHF is a pseudo-random function. Pseudo-random means that the outcome of the function is deterministic but it "appears" (passes a wide class of statistical tests) to be uniformly distributed in the interval $[0, 1]$. In other words, our hash function satisfies the property that any random perturbation in the input results in a uniformly distributed independent new output. We refer to this as the *diffusion property*.

Let us note that the hash function allows us to define an order on some arbitrary data x, y: one can say that $x < y$ if $\text{hash}(x) < \text{hash}(y)$.

Considering discrete time $t = 0, 1, 2 \ldots$ (we refer to it as the round t), we define by $A_t^{(n)}$ the set of transactions known by the node n at time t and call it the *node's vision.* The (arbitrary) process of new transactions' arrival at time t can be seen as the inclusion of such transaction into $A_t^{(n)}$ for at least one $n \in \mathcal{N}$. This way, the conflict set can be written as $\mathbb{T}_t = \cup_{n \in \mathcal{N}} A_t^{(n)}$. Furthermore, we assume that the node never forgets transactions, i.e., we assume that $A_0^{(n)} \subset A_1^{(n)} \subset \ldots$ for any $n \in \mathcal{N}$, which implies that $\mathbb{T}_0 \subset \mathbb{T}_1 \subset \ldots$

We define *node n's opinion* at round t as the collection $O_t^{(n)} = \left\{ \theta_t^{(n,x)}; \ x \in A_t^{(n)} \right\}$, where $\theta_t^{(n,x)}$ assumes the value:

$$\theta_t^{(n,x)} = \begin{cases} 1, & \text{if node } n \text{ likes transaction } x \text{ at time } t, \\ 0, & \text{otherwise.} \end{cases}$$

We also assume there exists a sequence of random numbers $X_t \sim U[\beta, 1 - \beta]$ which is either provided by a trusted source or generated by the nodes themselves using some decentralized random number generating protocol. This approach is referred to as a *global coin* in many works on Byzantine consensus, for example, in [6–9]. We assume that all messages between the nodes and the random numbers are delivered on time in every round.

Finally, our goal is to propose a protocol that achieves consensus among the nodes $\mathcal{N}$ about one preferred maximal independent (in the complete case, a singleton) set contained in $\mathbb{T}_t$.

2.2   Protocol

For $t = 0$, the initial visions, $A_0^{(n)}$, and opinions, $O_0^{(n)}$, can be arbitrary, as long as the set of liked transactions forms a maximal independent set of the conflict graph induced by $A_0^{(n)}$. Then, the following protocol should be executed iteratively once for every round $t \geq 1$ until the stop criterion is met:

(i) At the beginning of every round t, each node n stores any new transaction it became aware of, call this set $N_t^{(n)}$.

(ii) Each node then proceeds to randomly query (with uniform distribution) k nodes about their liked transactions.

(iii) The node then also stores any shared-by-others transactions it was not aware of, call this set $M_t^{(n)}$.

(iv) The node's vision at round t is then given by $A_t^{(n)} = A_{t-1}^{(n)} \cup N_t^{(n)} \cup M_t^{(n)}$

(v) It then stores the collection $\left\{ \eta_t^{(n,x)}; \ x \in A_t^{(n)} \right\}$, where $\eta_t^{(n,x)}$ corresponds to the number of 1-opinions the node n received from the queries in round t with respect to the transaction x.

(vi) $X_t \sim U[\beta, 1 - \beta]$ is made public.

(vii) It defines an auxiliary collection of opinions $\left\{ \theta'(x); \ x \in A_t^{(n)} \right\}$, that will not be shared and will last only until the end of the round (hence we omit the dependence on n and t), using the following rule:

$$\theta'(x) = \begin{cases} 1, & \text{if } \eta_t^{(n,x)}/k > X_t \\ 0, & \text{otherwise.} \end{cases}$$

(viii) Let $B := \left\{ x \in A_t^{(n)};\ \theta'(x) = 1 \right\}$. The node must find a way to assign 1 only to the opinions of a maximal independent subset of $A_t^{(n)}$. To do so, it iteratively removes from B the transaction $x \in B$ with the largest hash(Id_x, X_t) (this means the hash of Id_x concatenated with the random number X_t) until it obtains an independent set. Note, using the "largest hash" is not crucial, as any deterministic rule leading to unpredictable results is sufficient. Explicitly, it performs the following algorithm:

Algorithm 1. elim(U, X_t)

1: $W = U$
2: **while** W is not an independent set **do**
3:    Compute
$$y = \operatorname*{argmax}_{x \in W : \exists z \in W : z \leftrightarrow x} \mathrm{hash}(\mathrm{Id}_x, X_t),$$
4:    $W = W \setminus \{y\}$
5: **end while**
6: **return** W

Consider $B' := \mathrm{elim}(B, X_t)$. While this set is independent by construction, it may not be maximal. Then, starting with B', the node includes iteratively the non-conflicting transaction with the smallest hash(x, X_t) until a maximal independent set is obtained. In other words, denote by $N(U, V)$ the neighborhood of a set U with respect to a set V, i.e.,

$$N(U, V) := \{z \in V : z \leftrightarrow y \text{ for least one element } y \in U\},$$

then, the node executes the following:

Algorithm 2. compl(U, V, X_t)

1: $W = U$
2: **while** W is not a maximal independent set **do**
3:    Compute
$$y = \operatorname*{argmin}_{x \in V \setminus N(W, V)} \mathrm{hash}(x, X_t)$$
4:    $W = W \cup \{y\}$
5: **end while**
6: **return** W

Let $B'' := \mathrm{compl}(B', A_t^{(n)}, X_t)$. Finally, the node assigns value 1 to the opinion $\theta_{t+1}^{(n,x)}$ of every transaction $x \in B''$ and zero to the others.

For each transaction independently, if the node's opinion does not change for ℓ rounds, then it is considered final and will not be further modified in the subsequent rounds (Fig. 2).

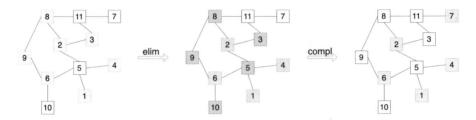

Fig. 2. Representation of step (viii) of the protocol. The graph represents the node's vision at time t. The numbers represent the order according to the hash function given some X_t. In the first subfigure, the transactions in yellow represent the set B; in the second subfigure, transactions in green represent B' and transactions in red represent $N(B', A_t^{(n)})$; in the last subfigure, transactions in blue represent B'', the chosen maximal independent set of the round.

We assume that a proportion q of the nodes may choose not to follow the proposed protocol and, instead, have an arbitrary behaviour. We refer to these nodes as *malicious* and the remaining as *honest nodes*. We consider that malicious nodes can work together in order to achieve a certain goal or even be controlled by the same entity. Without loss of generality, suppose that $\mathcal{N}$ is ordered in such a way that the first $(1-q)N$ nodes[2] are honest; then, for a round t and a transaction $v \in \mathbb{T}_t$, the proportion of likes among the honest nodes is given by

$$p_t^{(v)} := \frac{1}{(1-q)N} \sum_{j=1}^{(1-q)N} \theta_t^{(j,v)}. \tag{1}$$

Finally, we define the *Interval of Control* of the malicious nodes over a transaction v at round t as

$$\mathcal{I}_{q,t}^{(v)} := [(1-q)p_t^{(v)}, (1-q)p_t^{(v)} + q].$$

The lower/upper boundary of this interval is precisely the overall proportion of likes (i.e., considering both honest and malicious' opinions or, in other words, taking the sum in (1) from 1 to N) that the transaction has when all malicious nodes dislike/like it. We want our protocol to be resistant to the largest possible proportion of malicious nodes, but, unfortunately, q cannot be made arbitrarily large. In fact, we will need to impose restrictions to q based on the parameter β of our system in order to guarantee consensus.

3   Results

As we shall see, the topology of complete graphs allows us to assume very weak conditions regarding their size. In particular, we consider that $|\mathbb{T}_t| = f(t)$ for any

---

[2] For the sake of better presentation we will always assume that qN and $(1-q)N$ are integers.

nondecreasing function f taking the naturals into themselves. We also consider that every node is aware of all the conflicting transactions at every round, i.e., $A_t^{(n)} = \mathbb{T}_t$ for every $t \in \mathbb{N}$ and all $n \in \mathcal{N}$.

For the remaining part of this work assume that $q < \beta$ (this is the first restriction to q, and is motivated by Fig. 3) and abbreviate

$$\mu = \frac{\beta - q}{2(1 - q)}. \tag{2}$$

For complete graphs, we say that the system is in a *pre-consensus* state at round t if there is a transaction $v \in \mathbb{T}_t$ such that $p_t^{(v)} > 1 - \mu$. In this state, the intervals $\mathcal{I}_{q,t}^{(u)}$ are separated from the support of the random variable X_t for every transaction $u \in \mathbb{T}_t$ (see Fig. 3). Moreover, let $\mathcal{R}$ be the number of rounds until all nodes come to conclusion on their final opinion and let $\mathcal{C}$ be the event that all honest nodes have achieved consensus.

Our main result states that with high probability (depending on k and N), for any distribution of initial opinions, a pre-consensus state is achieved and maintained for enough consecutive rounds so that event $\mathcal{C}$ happens in the round $\mathcal{R} \leq r_0 + \ell u$ for some $r_0, u \in \mathbb{N}$.

Theorem 3.1. *For any $u, r_0 \in \mathbb{N}$ and sufficiently large k and N, given an arbitrary distribution of the initial opinions of the honest nodes it holds that*

$$\mathbb{P}[\mathcal{C} \cap \{\mathcal{R} \leq r_0 + \ell u\}] \geq 1 - W(N, k, r_0, \ell, u) - (\Phi(k, N))^{r_0},$$

where

$$W(N, k, r_0, \ell, u) := (1 - q)N \left[(1 - (1 - e^{-\frac{k}{2}(\beta - q)^2})^\ell)^u + \frac{e^{-\frac{k}{2}(\beta - q)^2}}{1 - e^{-\frac{k}{2}(\beta - q)^2}}^{(\ell-1)} \right]$$
$$+ (r_0 + \ell u) \exp\left\{ -2(1 - q)N\varphi_1^2 \right\},$$

$$\varphi_1 := \mu - \exp\left\{ -\frac{k}{2}(\beta - q)^2 \right\}, \tag{3}$$

$$\Phi(k, N) := \min\left\{ \frac{h}{2(1 - 2\beta)} \left(\exp\left\{ -2N(1 - q)\varphi_2^2 \right\} \right), \right.$$
$$\left. \frac{k\beta h}{2(1 - 2\beta)(N + k\beta)} \left(1 - \exp\left\{ -2N(1 - q)\varphi_3^2 \right\} \right) \right\},$$

$$\varphi_2 := \mu + \left(1 - \exp\left\{ -\frac{kh^2}{2} \right\} \right)^{m_2} - 1, \quad \varphi_3 := \mu - \exp\left\{ -\frac{kh^2}{2} \right\}, \tag{4}$$

$$h := (1 - q - 2\beta)/2, \quad m_2 := 2N/[(2 - 2\beta - h)k]. \tag{5}$$

Fig. 3. The intervals of control for a transaction u which has a proportion of likes among honest nodes $p_t^{(u)} > 1 - \mu$ (in blue), and for a transaction w such that $p_t^{(v)} < \mu$ (in red). Notice that these intervals are separated from the support $[\beta, 1 - \beta]$ of X_t. (Color figure online)

In order to prove the theorem, we need first to prove some additional lemmas. The first one estimates the probability of a pre-consensus state being maintained after one iteration of our algorithm.

Lemma 3.2. *Suppose that, at round t, the system is in a pre-consensus state and let $v_0 \in \mathbb{T}_t$ be the transaction such that $p_t^{(v_0)} > 1 - \mu$. Then, for a sufficiently large k the probability of a proportion of at least $1 - \mu$ honest nodes liking transaction v_0 at round $t + 1$ is bounded by*

$$\mathbb{P}[p_{t+1}^{(v_0)} > 1 - \mu] \geq 1 - \exp\{-2N(1-q)\varphi_1^2\}.$$

Proof. Note that if $p_t^{(v_0)} > 1 - \mu$ the lower boundary of $\mathcal{I}_{q,t}^{(v_0)}$ is greater than $1 - (\beta + q)/2$. For a node n we consider the event $G_1 := \{\eta_t^{(n,v_0)} k^{-1} > 1 - \beta\}$ that a proportion larger than $1 - \beta$ of the queried nodes like v_0. Hence,

$$\mathbb{P}\left[G_1 \mid p_t^{(v_0)} > 1 - \mu\right] \geq \mathbb{P}[k^{-1} S_k \geq 1 - \beta],$$

where $S_k \sim \mathcal{B}(k, 1 - (\beta + q)/2)$. This latter probability can be estimated using the Hoeffding inequality

$$\mathbb{P}\left[k^{-1} S_k \geq 1 - \beta\right] \geq 1 - \exp\left\{-\frac{k}{2}(\beta - q)^2\right\}.$$

Under the event G_1, every other transaction is liked by a proportion of the queried nodes smaller than β and, consequently, every other auxiliary opinion is equal to zero. Then, the probability that an honest node likes transaction v_0 at round $t + 1$ is estimated by

$$\mathbb{P}\left[\theta_{t+1}^{(n,v_0)} = 1 \mid p_t^{(v_0)} > 1 - \mu\right] \geq \mathbb{P}[\theta'(v_0) = 1, \theta'(v) = 0 \quad \forall v \neq v_0 \mid p_t^{(v_0)} > 1 - \mu]$$

$$\geq \mathbb{P}[G_1 \mid p_t^{(v_0)} > 1 - \mu] \geq 1 - \exp\left\{-\frac{k}{2}(\beta - q)^2\right\}.$$

$$(6)$$

Now, the probability that a proportion of at least $1 - \mu$ of the honest nodes likes transaction v_0 at round $t + 1$ can be estimated by the probability of a random variable $S_{N(1-q)} \sim \mathcal{B}(N(1-q), 1 - \exp\{-(k/2)(\beta - q)^2\})$ being larger than $N(1-q)(1-\mu)$. Considering that k can be taken sufficiently large so that the probability of success of this binomial is larger than $1-\mu$, using again the Hoeffding inequality we find that

$$\mathbb{P}\left[p_{t+1}^{(v_0)} > 1 - \mu\right] \geq 1 - \exp\left\{-2N(1-q)\varphi_1^2\right\}.$$

$\square$

We have shown that the pre-consensus state is maintained with high probability if N and k are sufficiently large. Our next objective is to prove that such a pre-consensus state can, in fact, be achieved relatively fast, given any distribution of honest nodes' likes. For this purpose, we distinguish between the two following cases; Case 1 represents the situation where the proportion of likes assigned to the favorite transaction is smaller or equal to $1/2$, whereas Case 2 represents the complementary case.

First notice that, since intervals of control have length q, an interval that is centered around $1/2$ will be completely contained in the support of the random variable X_t as long as $q < 1-2\beta$. Considering this condition together with $q < \beta$ introduced before and maximizing for q^3 gives us the restriction $q < \beta < 1/3$ which shall be used for the remaining of this work.

Now given a transaction u such that $p_t^{(u)} = 1/2$, its interval of control $\mathcal{I}_{q,t}^{(v)} = [(1-q)/2, (1+q)/2]$ is not only contained in the support of X_t but is separated from the boundary of this domain by a distance $h_c = (1-q-2\beta)/2$ (see Fig. 4). Consequently, in Case 1/Case 2, the distance between the interval of control of the favorite transaction and the lower/upper boundary of $[\beta, 1-\beta]$ is always larger than h.

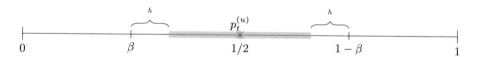

Fig. 4. The interval of control (in blue) of a transaction $u \in \mathbb{T}_t$ with $p_t^{(u)} = 1/2$ is always separated from the boundary of $[\beta, 1-\beta]$ by a distance $h = (1-q-2\beta)/2$. (Color figure online)

Denote by v_1 the transaction with the highest proportion[4] of honest likes at round t; in other words, $p_t^{(v_1)} \geq p_t^{(w)}$ for any $w \in \mathbb{T}_t$. We start with Case 1. The main idea, in this case, is that with a probability $h/(2-4\beta)$, the random number X_t belongs to the interval $(1-\beta-h/2, 1-\beta)$ that is separated from $\mathcal{I}_{q,t}^{(v_1)}$ by a distance larger than $h/2$, which means that an honest node will likely assign 0 to the auxiliary opinion of v_1 (and every other transaction, since they all have a smaller proportion of likes) and then, following our protocol, choose the transaction with the smallest hash as the favorite (see Fig. 5).

---

[3] Remember we want our protocol to be resistant to the largest q possible.

[4] We assume that this maximum is unique, otherwise we could choose the one with largest hash.

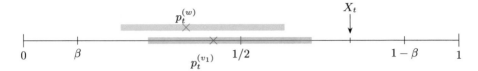

Fig. 5. In Case 1, every time X_t falls on the right of $\mathcal{I}_{q,t}^{(v_1)}$ (in blue), an honest node is likely to assign 0 to the auxiliary opinion of every transaction. (Color figure online)

Lemma 3.3. *Let $p_t^{(v_1)} \leq 1/2$ be the largest proportion of honest likes among all transactions at round t and $m_2 = 2N/[(2 - 2\beta - h)k]$. Then, for a sufficiently large k, there exists a transaction $u \in \mathbb{T}_t$ (not necessarily v_1) such that the probability that a proportion of at least $1 - \mu$ of honest nodes likes u at round $t + 1$ satisfies*

$$\mathbb{P}\left[p_{t+1}^{(u)} > 1 - \mu\right] \geq \frac{h}{2(1 - 2\beta)}\left(1 - \exp\left\{-2N(1 - q)\varphi_2^2\right\}\right).$$

Proof. Let G_2 be the event that the random threshold X_t lies in the interval $(1 - \beta - h/2, 1 - \beta)$ and $u = \text{argmin}_{v \in \mathbb{T}_t} \text{hash}(v, X_t)$. Since the upper boundary of $\mathcal{I}_{q,t}^{(v_1)}$ is smaller than $1 - \beta - h/2$, we consider $S_k \sim \mathcal{B}(k, 1 - \beta - h)$. Now, we use Hoeffding's inequality to bound the probability that an honest node assigns 0 to the auxiliary opinion of v_1:

$$\mathbb{P}\left[\theta'(v_1) = 0 \mid p_t^{(v_1)} \leq 1/2, G_2\right] \geq \mathbb{P}\left[k^{-1}S_k < X_t\right] \geq 1 - \exp\left\{-\frac{kh^2}{2}\right\}.$$

While this estimate holds for all transactions, in some cases it is excessively conservative. Notice, for example, that in general, the auxiliary opinion for a transaction v can take value 1 only if at least βk nodes like v. Under G_2 this number is increased even further to $(1 - \beta - h/2)k$. This implies that under G_2, only $N/[(1 - \beta - h/2)k] = m_2$ transactions could possibly be liked, while all others will be disliked with a probability 1. Consequently, the probability of an honest node assigning $\theta'(v) = 0$ for every transaction and then, following the protocol, liking transaction u at round $t + 1$ satisfies:

$$\mathbb{P}\left[\theta_{t+1}^{(n,u)} = 1 \mid p_t^{(v_1)} \leq 1/2, G_2\right] \geq \left(1 - \exp\left\{-\frac{kh^2}{2}\right\}\right)^{m_2}.$$

Considering $S_{N(1-q)} \sim \mathcal{B}(N(1 - q), [1 - \exp\{-kh^2/2\}]^{m_2})$ then, analogously to the previous lemma, it follows that the probability of a proportion of at least $1 - \mu$ of the honest nodes liking transaction u at round $t + 1$ satisfies

$$\mathbb{P}\left[p_{t+1}^{(u)} > 1 - \mu \mid p_t^{(v_1)} \leq 1/2\right] \geq \mathbb{P}[G_2]\mathbb{P}\left[p_{t+1}^{(u)} > 1 - \mu \mid p_t^{(v_1)} \leq 1/2, G_2\right]$$

$$\geq \frac{h}{2(1 - 2\beta)}\left(1 - \exp\left\{-2N(1 - q)\varphi_1^2\right\}\right).$$

$\square$

We now consider Case 2, i.e., if $p_t^{(v_1)} > 1/2$. In this situation, a random threshold smaller than $p_t^{(v_1)}(1-q)$ (the lower boundary of $\mathcal{I}_{q,t}^{(v_1)}$, the blue interval in Fig. 6) will not only favor v_1, but can also favor another transaction w, which overall proportion can be increased to $p_t^{(w)}(1-q)+q$ (the upper boundary of $\mathcal{I}_{q,t}^{(w)}$, the red interval in Fig. 6). This is not necessarily a problem, but in order to guarantee that a large proportion of honest nodes like the same transaction we rely on the diffusion property of the hash function.

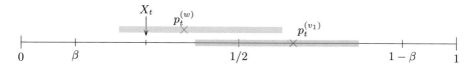

Fig. 6. In Case 2, every time X_t falls on the left of $\mathcal{I}_{q,t}^{(v_1)}$ (in blue), an honest node might also like a transaction w if the upper boundary of $\mathcal{I}_{q,t}^{(w)}$ (in red) is also on the right of X_t. (Color figure online)

Lemma 3.4. *Let $p_t^{(v_1)} > 1/2$ be the largest proportion of honest likes among all transactions at round t. Then, for a sufficiently large k the probability that at least a proportion $1 - \mu$ of honest nodes likes the transaction v_1 at round $t+1$ satisfies*

$$\mathbb{P}[p_{t+1}^{(v_1)} > 1 - \mu] \geq \frac{k\beta h}{2(1-2\beta)(N+k\beta)}\left(1 - \exp\left\{-2N(1-q)\varphi_3^2\right\}\right).$$

Proof. Let G_3 be the event that the random threshold X_t belongs to the interval $(\beta, \beta + h/2)$. Now, the lower boundary of $\mathcal{I}_{q,t}^{(v_1)}$ is greater than $\beta + h/2$. Let $S_k \sim \mathcal{B}(k, \beta+h)$, then, the probability that an honest node likes transaction v_1 conditioned on G_3 is estimated again with Hoeffding's inequality by

$$\mathbb{P}[\theta'(v_1) = 1 \mid p_t^{(v_1)} > 1/2, G_3] \geq \mathbb{P}[k^{-1}S_k \geq X_t] \geq 1 - \exp\left\{-\frac{kh^2}{2}\right\}.$$

As stated in the previous lemma, the number of auxiliary opinions that are 1 is bounded. In this case, at most $m_3 := N/(k\beta)$ of those transactions can exist. Instead of studying every possible combination of likes and dislikes of v_1 and its potential competitors, we consider the event G_4 that v_1, at round t, has the smallest hash[5] among them. Then, by the design of the protocol, if an honest node assigns 1 to the auxiliary opinion of v_1, it will be chosen as the node's favorite independently of its auxiliary opinion on the competitors. By the diffusion property of the hash function, the event G_4 occurs roughly with probability $1/m_3$. This means that, for an honest node n, it holds that

$$\mathbb{P}\left[\theta_{t+1}^{(n,v_1)} = 1 \mid p_t^{(v_1)} > 1/2, G_3, G_4\right] \geq \mathbb{P}[\theta'(v_1) = 1 \mid p_t^{(v_1)}, G_3]. \qquad (7)$$

---

[5] Rigorously, that $v_1 = \operatorname{argmin}_{u \in \mathbb{T}} \operatorname{hash}(\mathrm{Id}_u, X_t)$.

It follows that the probability of a proportion $1 - \mu$ of honest nodes liking transaction v_1 at round $t + 1$ is estimated by

$$\mathbb{P}[p_{t+1}^{(v_1)} > 1 - \mu \mid p_t^{(v_1)} > 1/2] \geq \mathbb{P}[G_3, G_4]\mathbb{P}[p_{t+1}^{(v_1)} > 1 - \mu \mid p_t^{(v_1)} > 1/2, G_3, G_4]$$
$$\geq \frac{k\beta h}{2(1 - 2\beta)(N + k\beta)} \left(1 - \exp\{-2N(1 - q)\varphi_3^2\}\right).$$

$\square$

Proof of Theorem 3.1 First, we define the random variable

$$\Psi := \min\{r \geq 1 : p_t^{(v_1)} > 1 - \mu\},$$

that is the first round in which the system is in the pre-consensus state. It follows directly from Lemmas 3.3 and 3.4 that for every $s \geq 1$

$$\mathbb{P}[\Psi > s] \leq \Phi(k, N)^{s-1} \tag{8}$$

Next, we define the variable Z that is the first round after Ψ for which the system leaves the pre-consensus; or explicitly

$$Z := \min\{r > \Psi : p_t^{(v_1)} < 1 - \mu\}.$$

Finally, the random variable

$$\hat{\tau}_n^{(i)} := \min\{r \geq r_0 + \ell : \theta_t^{(n,v_1)} = \cdots = \theta_{t-\ell+1}^{(n,v_1)} = i\}$$

for $i \in \{0, 1\}$, is the first round after $r_0 + \ell$ where v_1 was disliked/liked by an honest node n for ℓ consecutive rounds, respectively.

Now consider the events

$$D_1 = \{\Psi \leq r_0\},$$
$$D_2 = \{Z \geq r_0 + \ell u\},$$
$$D_3 = \left\{\hat{\tau}_n^{(1)} \leq r_0 + \ell u;\ \hat{\tau}_n^{(1)} < \hat{\tau}_n^{(0)},\ \forall n = 1, \ldots, (1 - q)N\right\}.$$

Notice that
$$\mathcal{C} \cap \{\mathcal{R} \leq r_0 + \ell u\} \supset D_1 \cap D_2 \cap D_3. \tag{9}$$

The probabilities $\mathbb{P}[D_1]$, $\mathbb{P}[D_2]$ and $\mathbb{P}[D_3]$ can be estimated using (8), Lemma 3.2 and Lemma 5.4 from [1] respectively. Finally, the result follows from (9) and the union bound.

$\square$

4   Conclusion and Future Work

This paper shows that FPCS resolves n-spend conflicts effectively even when a substantial (but not major) proportion of nodes is malicious. Given any arbitrary initial condition of likes by honest nodes, we presented bounds for the probability of consensus at time t.

An essential, but not surprising, characteristic of our protocol is that it relies on the existence of a sequence of random numbers X_t, similar to FPC and other papers in classic consensus.

In future work, we pretend to relax the condition that all visions are the same, allowing them to be slightly different while still holding the same robustness and effectiveness, and to relax the synchronicity assumptions on the underlying communication. The core essence of our voting protocol, as viewed from a heuristic standpoint, naturally exhibits resilience to perception differences and message loss, essentially due to its dependence on randomly sampling opinions.

Finally, a more ambitious goal is to prove that similar results hold in the case of an arbitrary conflict set when malicious nodes can interfere not only with the voting dynamics but also with the topology of the graph itself.

References

1. Popov, S., Buchanan, W.J.: FPC-BI: fast probabilistic consensus within byzantine infrastructures. J. Parallel Distrib. Comput. **147**, 77–86 (2021). ISSN 0743–7315, https://doi.org/10.1016/j.jpdc.2020.09.002
2. Müller, S., Penzkofer, A., Kuśmierz, B., Camargo, D., Buchanan, W.J.: Fast probabilistic consensus with weighted votes. In: Arai, K., Kapoor, S., Bhatia, R. (eds.) FTC 2020. AISC, vol. 1289, pp. 360–378. Springer, Cham (2021). https://doi.org/10.1007/978-3-030-63089-8_24
3. Capossele, A., Müller, S., Penzkofer, A.: Robustness and efficiency of voting consensus protocols within byzantine infrastructures. Blockchain Res. Appl. **2**(1), 00007 (2021). ISSN 2096–7209, https://doi.org/10.1016/j.bcra.2021.100007
4. Müller, S., Penzkofer, A., Camargo, D., Saa, O.: On fairness in voting consensus protocols. In: Arai, K. (ed.) Intelligent Computing. LNNS, vol. 284, pp. 927–939. Springer, Cham (2021). https://doi.org/10.1007/978-3-030-80126-7_65
5. Merkle, R.: One Way Hash Functions and DES, pp. 428–446 (1989). https://doi.org/10.1007/0-387-34805-0_40
6. Cachin, C., Kursawe, K., Shoup, V.: Random oracles in constantinople: practical asynchronous byzantine agreement using cryptography. J. Cryptol. **18**(3), 219–246 (2005). https://doi.org/10.1007/s00145-005-0318-0
7. Canetti, R., Rabin, T.: Fast asynchronous Byzantine agreement with optimal resilience. In Proceedings of the 25th Annual ACM Symposium on Theory of Computing, pp. 42–51 (1993). https://doi.org/10.1145/167088.167105
8. Aguilera, M.K., Toueg, S.: The correctness proof of Ben-Or's randomized consensus algorithm. Distrib. Comput. **25**(5), 371–381 (2012). https://doi.org/10.1007/s00446-012-0162-z
9. Friedman, R., Mostefaoui, A., Raynal, M.: Simple and efficient oracle-based consensus protocols for asynchronous Byzantine systems. IEEE Trans. Dependable Secure Comput. **2**(1), 46–56 (2005). https://doi.org/10.1109/RELDIS.2004.1353024

Bridging Protocols in DAG-Based DLTs: Facilitating Interaction Between IOTA and Other DLT Networks

Carlos Álvarez López[1], Yeray Mezquita[1,2]([✉]), and Diego Valdeolmillos[2]

[1] BISITE Research Group, University of Salamanca, Salamanca, Spain
carlos_alvarez@usal.es
[2] AIR Institute, Valladolid, Spain
{ymezquita,dval}@air-institute.com

Abstract. While Distributed Ledger Technologies offer (DLT) increased security, transparency, and resilience, the decentralized ecosystem is currently fragmented, lacking interoperability between different DLT platforms. Bridging protocols act as a bridge between different DLT networks, enabling the transfer of assets and data across disparate systems. This paper explores the development of bridging protocols in DAG-based DLTs, with a focus on facilitating interaction between IOTA and other networks. This paper offers insights into the characteristics and requirements of effective bridging protocols, which are crucial for realizing the full potential of DLTs and unlocking new use cases for decentralized systems.

Keywords: Interoperability · Bridging protocols · DLT platforms

1 Introduction

Distributed Ledger Technologies (DLTs) have revolutionized the way we handle transactions and store data in a decentralized, tamper-proof manner. From cryptocurrencies to supply chain management, the potential applications of DLTs are vast and varied. However, the current landscape of DLTs is fragmented, with multiple DLT platforms operating independently without any means of communication between them. This lack of interoperability poses a significant challenge to the wider adoption of these technologies.

The lack of interoperability between different DLT platforms has become a major hindrance to the growth and development of the industry. With each platform operating in isolation, there is no way for them to share data, transactions, or assets. This makes it challenging to build complex systems that require multiple platforms to work together. It also limits the ability of users to move assets and data across different platforms, which makes it difficult to achieve their widespread adoption.

One potential solution to this problem is the use of bridge protocols. These protocols act as intermediaries between different DLTs platforms, facilitating

J. M. Machado et al. (Eds.): BLOCKCHAIN 2023, LNNS 778, pp. 472–481, 2023.
https://doi.org/10.1007/978-3-031-45155-3_45

communication and the exchange of data, transactions, and assets. In the case of IOTA, which is a directed acyclic graph (DAG)-based DLT, bridging protocols could enable interoperability with other DLT networks.

This paper aims to benefit researchers and professionals in the field of DLTs by exploring the concept of bridging protocols in DAG-based DLTs and their potential role in facilitating interaction between IOTA and other DLT networks. The concept of bridge protocols and their use in making DLT platforms interoperable with IOTA is introduced in Sect. 2. The findings and conclusions of this research, highlighted in Sect. 4, will help guide future efforts to develop effective bridging protocols, promoting interoperability and unlocking the full potential of DLT-based platforms.

The introduction of bridge protocols is crucial in establishing interoperability between DLT platforms and IOTA, as highlighted in Sect. 2. These bridging protocols play a significant role in facilitating seamless interaction and data transfer between different DLT networks. The comprehensive discussion in Sect. 3 explores the considerations and challenges involved in designing bridging protocols, focusing on the distinct topology and consensus algorithm of IOTA's DAG structure. By addressing these challenges, the research findings and conclusions, emphasized in Sect. 4, aim to serve as a guiding resource for future endeavors in developing effective bridging protocols. By promoting interoperability, these protocols have the potential to unlock the full capabilities of DLT-based platforms and propel the growth and innovation within the field.

2 Study of Protocol Characteristics to Ensure Interoperability

The development of cross-chain bridges has become increasingly important in the world of distributed ledger technology (DLT) as it enables seamless communication between different networks, allowing them to work together and share information without friction. IOTA is a distributed ledger that aims to facilitate machine-to-machine communication and value transfer, and it is evolving into an intelligent contract and cross-chain platform to facilitate greater interoperability. In this section, we will detail the characteristics found in the literature that any model should have to act as a bridge between different blockchain systems.

2.1 Security and Realibility

First and foremost, secure and trustworthy communication channels between the networks involved must be established. Implementing cryptographic protocols, such as Secure Socket Layer (SSL) or Transport Layer Security (TLS), can ensure that transmitted data is encrypted and protected from unauthorized access or tampering [5,21]. These protocols provide a strong foundation for secure communication, as they have undergone rigorous testing and validation in various industries and applications. Also, the use of secure multi-party computation (MPC) [27], a cryptographic technique that allows multiple parties to compute a function over their inputs while keeping them private [24], in the context of cross-chain communication, MPC can be used to generate and verify cryptographic proofs for transactions that occur between blockchains, ensuring data integrity and confidentiality. Additionally, MPC nodes can be employed to secure the transfer of information across chains, verifying the correctness of the information and preventing malicious actions. As an example, Multichain, (also known as Anyswap) a cross-chain router protocol for Web3 Interoperability, can enable seamless asset transfer and smart contract execution between multiple chains, further enhancing communication efficiency [27].

Another important aspect of security in any bridge model is the authentication and verification of transactions [22]. Utilizing digital signatures and public key cryptography can confirm the legitimacy and ownership of entities involved in transferring assets or data between IOTA and other DLT networks [7] using non-custodial solutions, which eliminates the need for trusted third parties in managing assets and transactions. Non-custodial solutions ensure that users maintain control over their assets throughout the process, reducing the risk of theft or fraud. This approach aligns with the decentralized nature of blockchain networks, promoting trustless transactions without intermediaries [27].

Anycall [1] is another example of a feature that is being tested on the Shimmer network [27]. This feature allows users to interact with any DLT network through a single call, providing a unified interface for communication across multiple networks. Anycall makes it easier for users to access different DLT networks without needing to understand the specific communication protocols for each network. The anyCall function, a cross-chain messaging router protocol, can be used to facilitate seamless communication between different blockchains. The anyCall function allows contracts on one chain to be called directly from another chain, sharing information like commands, state, data, or messages. This generalized cross-chain communication is secured by the MPC network, which verifies the transfer of information across chains [27].

As IOTA continues to evolve into a smart contract and cross-chain platform, ensuring security and reliability becomes increasingly important. To achieve a high level of security and reliability, the bridge model proposed for IOTA and other DLT networks must address various aspects, such as secure communication, authentication, consensus mechanisms, fault tolerance, data transfer mechanisms, attack vector mitigation, privacy preservation, scalability, risk management, and adaptability to technological changes. To further enhance security, the proposed bridge model should incorporate a consensus mechanism to validate and confirm transactions. This is particularly important when dealing with networks that have different native consensus algorithms. One potential solution is to implement a cross-chain consensus mechanism, which allows for transactions to be validated by nodes from both networks. This approach provides increased security by requiring multiple confirmations before a transaction is considered valid, reducing the risk of double-spending or other fraudulent activities [11].

In addition to security, the bridge model must also ensure a high level of reliability. This can be achieved by implementing fault-tolerant protocols that can maintain the integrity and consistency of the system even in the presence of failures or malicious behavior [12]. For example, the bridge model can incorporate Byzantine Fault Tolerance (BFT) algorithms, which are designed to function correctly even if some nodes in the network are unresponsive or behaving maliciously [4]. This ensures that the system remains operational and can continue to process transactions despite potential disruptions.

Another aspect of reliability is the need for efficient and robust data transfer mechanisms. The bridge model should employ strategies to handle data transfer between IOTA and other DLT networks, while minimizing the potential for data loss or corruption [10]. One possible solution is the use of atomic swaps, which are smart contracts that enable the exchange of assets between different networks without the need for a trusted intermediary [18]. Atomic swaps can be designed to ensure that the transfer of assets is only executed if all conditions of the swap are met, ensuring that no party can cheat or manipulate the transaction [29].

Addressing potential attack vectors and ensuring the privacy of users are other essential aspects of the bridge model's security and reliability. To mitigate possible attack vectors, such as Sybil attacks or eclipse attacks, the model can implement strategies like reputation systems or require nodes to demonstrate proof of resources [6,23]. These measures would make it more challenging for adversaries to compromise the system or control a significant portion of the network.

Furthermore, any bridge model should address privacy concerns, as the exchange of assets or data between IOTA and other DLT networks might expose sensitive information about users and transactions. One solution is the implementation of privacy-enhancing technologies such as zero-knowledge proofs (ZKPs) or confidential transactions [8]. These technologies allow for the verification of transactions without revealing specific details, such as the sender, receiver, or amount being transferred, effectively preserving user privacy [8].

Scalability is another critical aspect that needs to be addressed by the bridge model. As the number of cross-chain transactions increases, the model must be able to handle the growth in volume without compromising performance or security. Techniques such as sharding or off-chain solutions like state channels can be employed to improve the scalability of the bridge model [19,28]. These approaches allow for transactions to be processed more efficiently, reducing the load on the main chain and improving overall throughput [19].

Lastly, the bridge model should be adaptable to technological changes and the evolving landscape of distributed ledger technologies. This could involve incorporating support for emerging technologies like quantum-resistant cryptography or implementing modular architectures that allow for easy updates and improvements. By being adaptable, the bridge model can continue to provide secure and reliable cross-chain interoperability as the DLT ecosystem evolves.

In conclusion, any bridge model for IOTA and other DLT networks must address numerous aspects to ensure security and reliability. This includes secure communication, authentication, consensus mechanisms, fault tolerance, data transfer mechanisms, attack vector mitigation, privacy preservation, scalability, risk management, and adaptability to technological changes. By addressing these aspects, the bridge model can provide a robust and secure solution for cross-chain interoperability, enabling IOTA to become an intelligent contract and cross-chain platform.

2.2   Communication Efficiency

In this section, we focus on the efficiency of communication in proposed bridge models between IOTA and other DLTs, such as Bitcoin and Ethereum [3]. Efficient communication is essential to ensure that data transfer and cross-chain transactions are executed in a timely manner and that network resources are optimally utilized. To achieve communication efficiency, the proposed model relies on several key techniques and design choices, including optimized data structures, efficient consensus algorithms, parallel processing, network-aware communication routing, parallelized Directed Acyclic Graph (DAG) ledgers, the anyCall function, sharding techniques, and the adoption of common interoperability standards and protocols. In the following subsections, we elaborate on these aspects and discuss their contributions to the overall communication efficiency of any bridge protocol [20].

Optimized Data Structures. The choice of data structures has a significant impact on communication efficiency in bridge protocols. As the primary data storage for the protocol, efficient data structures enable rapid access to and retrieval of relevant information. The proposed bridge model adopts the use of Merkle trees and Merkle proofs for efficient data storage and verification [13]. Merkle trees facilitate compact representation of data and allow for efficient validation of data integrity across different DLTs, reducing the communication overhead in the protocol [29].

Efficient Consensus Algorithms. Efficient consensus algorithms are crucial to maintaining communication efficiency in the bridge protocol. Since the bridge protocol involves interactions between multiple DLTs, the consensus algorithm must be able to handle cross-chain transactions and reach consensus on the state of the connected networks. Depending on the specific requirements of the cross-chain communication framework, various consensus algorithms, such as Proof of Stake (PoS), Delegated Proof of Stake (DPoS), or Byzantine Fault Tolerance (BFT), can be explored to strike the right balance between efficiency, security, and decentralization [15,26].

Parallel Processing and Dependency Tracking. Parallel processing is a technique that enhances communication efficiency by processing multiple tasks simultaneously. The proposed bridge model employs parallel processing to facilitate concurrent validation and execution of cross-chain transactions, allowing for increased throughput and reduced latency. Dependency tracking of smart contract transactions is a crucial aspect of parallel processing in the protocol, as it ensures that transactions are executed in the correct order, maintaining consistency across the connected DLTs [14].

Network-Aware Communication Routing. Network-aware communication routing is a technique that optimizes communication paths based on the network topology and congestion status. This approach improves communication efficiency by selecting the most optimal routes for data transfer between the connected DLTs [17]. The proposed bridge model incorporates network-aware communication routing to minimize latency and maximize throughput when transferring data between IOTA and other DLTs.

The AnyCall Function. The anyCall function is a technique that can be employed to improve communication efficiency in the bridge protocol [1]. The anyCall function allows for the execution of arbitrary function calls between smart contracts on different DLTs [2]. By incorporating the anyCall function into the bridge protocol, it becomes possible to facilitate more efficient and direct communication between IOTA and other DLTs, further enhancing the overall communication efficiency of the protocol.

Sharding Techniques. Sharding techniques can be adopted to improve communication efficiency in the bridge protocol. In a traditional blockchain, all nodes in the network must maintain a complete copy of the chain, which can become a bottleneck as the network grows in size and complexity. Sharding involves partitioning a blockchain network into smaller, more manageable pieces or "shards" that can process transactions and data independently [28]. Each shard maintains its own subset of the blockchain's transaction history, so the entire blockchain can be reconstructed by combining the individual shards. This allows for greater scalability while maintaining the security and decentralization benefits of the blockchain. The application of sharding techniques for the enhancement of IOTA bridging protocols could have a major impact on scalability. One possible approach would be to split the cross-chain network into n parts as well as to split the validation nodes into n groups. In this way, each group of nodes would store its corresponding part of the network, while still being able to validate transactions in parallel. In this way a theoretical speed-up of n would be achieved in the cross-chain bridge.

Adoption of Common Interoperability Standards and Protocols. Adopting common interoperability standards and protocols is crucial for the efficient communication between IOTA and other DLTs [16]. Such standards and protocols can facilitate seamless integration and communication between different DLTs, ensuring that data transfer and cross-chain transactions are executed efficiently and with minimal overhead [9]. Any bridge model should actively engage with ongoing standardization efforts, such as the work being done by the Enterprise Ethereum Alliance (EEA) and the Interledger Protocol (ILP), to ensure compatibility and efficient communication with other DLTs [25].

3 Discussion

In this section, we explore the considerations and challenges involved in designing a bridging protocol between IOTA and other Distributed Ledger Technology (DLT) platforms, with a particular focus on more traditional Proof of Stake (PoS) and Proof of Work (PoW) based blockchains. While the scope of DLT platforms is broad, the emphasis on PoS and PoW blockchains is justified by their widespread adoption and established consensus mechanisms. The unique topology of IOTA's Directed Acyclic Graph (DAG) structure, known as the Tangle, and its consensus algorithm introduce complexities that require careful attention when developing an effective bridging protocol. By addressing these challenges, our goal is to enable seamless interaction and data transfer between IOTA and PoS/PoW blockchains, fostering enhanced interoperability and unlocking the full potential of DLT-based platforms.

The topology of IOTA's Tangle differs from the block-based structure of PoS/PoW blockchains, posing a challenge when establishing interoperability. Traditional blocks and miners are absent in the Tangle, which affects transaction validation and confirmation processes. To enable seamless interaction and data transfer between the two systems, the bridging protocol must account for these structural differences.

Moreover, the consensus algorithms used by IOTA and PoS/PoW blockchains exhibit significant variations. IOTA employs a unique consensus mechanism known as the Coordinator (or Coordicide in its upcoming version), while PoS/PoW blockchains rely on their respective consensus algorithms. Integrating these disparate consensus mechanisms presents challenges in ensuring the security and reliability of cross-chain transactions. Therefore, the bridging protocol should explore the implementation of a cross-chain consensus mechanism that enables transaction validation by nodes from both IOTA and PoS/PoW blockchains, enhancing security and reducing the risk of fraudulent activities like double-spending.

To address these challenges, the bridging protocol should consider comprehensive security and reliability measures, incorporating authentication and verification mechanisms, integrating cross-chain consensus mechanisms, addressing scalability concerns, preserving privacy, and ensuring adaptability to technological changes. By considering these aspects and overcoming the complications introduced by the distinct characteristics of IOTA, the bridging protocol can facilitate seamless interaction and data transfer between IOTA and more traditional PoS/PoW-based blockchains, paving the way for enhanced interoperability and frictionless exchange of assets and information across different DLT networks.

The effective design and implementation of a bridging protocol promote increased security, reliability, and privacy preservation. Establishing secure communication channels, utilizing authentication and verification mechanisms, integrating cross-chain consensus mechanisms, and adopting privacy-enhancing technologies that ensure trustless and efficient cross-chain transactions. Moreover, scalability and efficiency considerations address the growing demands of cross-chain communication, allowing the bridge model to handle higher transaction volumes while maintaining performance and integrity. By staying attuned to emerging technologies and adopting modular architectures, any bridge model should remain adaptable and future-proof, accommodating evolving DLT landscapes and advancements.

In conclusion, any bridging protocol, designed by considering these aspects and addressing the complications introduced by the distinct characteristics of IOTA, empowers seamless interaction and data transfer between IOTA and other DLT networks. It fosters secure, reliable, and efficient cross-chain transactions while promoting scalability, privacy preservation, and adaptability. Through its implementation, the full potential of DLT-based platforms, including IOTA, can be unlocked, opening up new opportunities for collaboration and innovation in the decentralized ecosystem.

4   Conclusion and Future Lines of Work

DLT platforms offer the industry increased security, transparency, and resilience, making it an attractive alternative to traditional centralized systems. However, the decentralized ecosystem based on DLT is currently fragmented, with various platforms operating independently and lacking interoperability. This fragmentation limits the potential benefits of DLT, as users face challenges in seamlessly interacting with different platforms and applications.

Bridging protocols play a crucial role in achieving interoperability between DLT platforms. These protocols act as bridges, facilitating the transfer of assets and data across different DLT networks. By establishing common standards and protocols, bridging protocols enable seamless communication and collaboration between diverse DLT platforms, enhancing the user experience and unlocking the full potential of DLT technologies. Without effective bridging protocols, the DLT ecosystem would remain fragmented, limiting the advantages of decentralization and impeding the development of innovative use cases. Therefore, the development of robust bridging protocols is essential for achieving interoperability and realizing the full potential of DLT-based platforms.

This study explores the characteristics and requirements that enable interoperability between DLT platforms, with a specific focus on IOTA and other DLT platforms. The investigation emphasizes the importance of efficient communication, taking into account factors such as optimized data structures, efficient consensus algorithms, parallel processing, network-aware communication routing, parallelized DAG ledgers, the anyCall function, sharding techniques, and the adoption of common interoperability standards and protocols. By leveraging these identified characteristics, the study aims to enhance the design and development of cross-chain transactions while optimizing network resources. The efficient communication facilitated by these characteristics contributes to the overall success of an interoperability framework, fostering seamless interaction between IOTA and other DLT platforms.

Looking ahead, there are several promising avenues for future research in the field of bridging protocols and interoperability between DLT platforms. One potential area of focus is the development of cross-chain bridge and router integration mechanisms that enable seamless transfer of data and assets across diverse DLT networks. Such mechanisms would facilitate interoperability between different DLT platforms, breaking down barriers and enabling fluid collaboration. Additionally, there is a need to delve into the security implications of bridging protocols and to devise mechanisms that ensure the secure transfer of data and assets across different DLT platforms. Exploring innovative security measures will foster trust and confidence in cross-chain transactions, bolstering the adoption of interoperable DLT solutions.

In conclusion, the development of effective bridging protocols is pivotal for realizing the full potential of DLT and unlocking new use cases for decentralized systems. By fostering interoperability, DLT platforms can seamlessly communicate and collaborate, creating a unified ecosystem that maximizes the advantages of decentralization. Future research efforts should continue to refine and enhance bridging protocols, pushing the boundaries of interoperability and enabling transformative applications across the DLT landscape.

Acknowledgements. This research has been supported by the project "COordinated intelligent Services for Adaptive Smart areaS (COSASS), Reference: PID2021-123673OB-C33, financed by MCIN /AEI /10.13039/501100011033/FEDER, UE.

References

1. Anycall: What is anycall? (2022). https://anycall.gitbook.io/anycall
2. Buterin, V.: Chain interoperability. R3 Res. Pap. **9**, 1–25 (2016)
3. Buterin, V., et al.: A next-generation smart contract and decentralized application platform. White Pap. **3**(37), 2–1 (2014)
4. Castro, M., Liskov, B., et al.: Practical byzantine fault tolerance. In: OsDI, vol. 99, pp. 173–186 (1999)
5. Dierks, T., Rescorla, E.: The transport layer security (tls) protocol, version 1.1. internet request for comments 4346 (rfc 4346) (2006)
6. Douceur, J.R.: The Sybil attack. In: Druschel, P., Kaashoek, F., Rowstron, A. (eds.) IPTPS 2002. LNCS, vol. 2429, pp. 251–260. Springer, Heidelberg (2002). https://doi.org/10.1007/3-540-45748-8_24
7. ElGamal, T.: A public key cryptosystem and a signature scheme based on discrete logarithms. IEEE Trans. Inf. Theory **31**(4), 469–472 (1985)
8. Goldwasser, S., Micali, S., Rackoff, C.: The knowledge complexity of interactive proof-systems. In: Providing Sound Foundations for Cryptography: On the Work of Shafi Goldwasser and Silvio Micali, pp. 203–225 (2019)
9. Hardjono, T., Lipton, A., Pentland, A.: Towards a design philosophy for interoperable blockchain systems. arXiv preprint arXiv:1805.05934 (2018)
10. Herlihy, M.: Atomic cross-chain swaps. In: Proceedings of the 2018 ACM Symposium on Principles of Distributed Computing, pp. 245–254 (2018)

11. Herlihy, M., Liskov, B., Shrira, L.: Cross-chain deals and adversarial commerce. arXiv preprint arXiv:1905.09743 (2019)
12. Lamport, L., Shostak, R., Pease, M.: The byzantine generals problem. In: Concurrency: The Works of Leslie Lamport, pp. 203–226 (2019)
13. Laurie, B.: Certificate transparency. Commun. ACM **57**(10), 40–46 (2014)
14. Luu, L., Narayanan, V., Zheng, C., Baweja, K., Gilbert, S., Saxena, P.: A secure sharding protocol for open blockchains. In: Proceedings of the 2016 ACM SIGSAC Conference on Computer and Communications Security, pp. 17–30 (2016)
15. Mezquita, Y., González-Briones, A., Casado-Vara, R., Chamoso, P., Prieto, J., Corchado, J.M.: Blockchain-based architecture: a MAS proposal for efficient agrifood supply chains. In: Novais, P., Lloret, J., Chamoso, P., Carneiro, D., Navarro, E., Omatu, S. (eds.) ISAmI 2019. AISC, vol. 1006, pp. 89–96. Springer, Cham (2020). https://doi.org/10.1007/978-3-030-24097-4_11
16. Mezquita, Y., Podgorelec, B., Gil-González, A.B., Corchado, J.M.: Blockchain-based supply chain systems, interoperability model in a pharmaceutical case study. Sensors **23**(4), 1962 (2023)
17. Network-Fast, R.: Cheap, scalable token transfers for ethereum (2018). Accessed 7 July 2020
18. Nolan, T.: Alt chains and atomic transfers. In: Bitcoin Forum (2013)
19. Poon, J., Dryja, T.: The bitcoin lightning network: Scalable off-chain instant payments (2016)
20. Popov, S.: The tangle. White Pap. **1**(3), 30 (2018)
21. Rescorla, E.: The transport layer security (TLS) protocol version 1.3. Technical report (2018)
22. Rivest, R.L., Shamir, A., Adleman, L.: A method for obtaining digital signatures and public-key cryptosystems. Commun. ACM **21**(2), 120–126 (1978)
23. Singh, A., Ngan, T.W., Druschel, P., Wallach, D.S., et al.: Eclipse attacks on overlay networks: threats and defenses (2006)
24. TheLuWizz: How iota is becoming an intelligent contract and cross-chain platform (2022). https://theluwizz.medium.com/how-iota-is-becoming-an-intelligent-contract-cross-chain-platform-fb418c2b5ec4
25. Thomas, S., Schwartz, E.: A protocol for interledger payments (2015). https://interledger.org/interledger.pdf
26. Valdeolmillos, D., Mezquita, Y., González-Briones, A., Prieto, J., Corchado, J.M.: Blockchain technology: a review of the current challenges of cryptocurrency. In: Prieto, J., Das, A.K., Ferretti, S., Pinto, A., Corchado, J.M. (eds.) BLOCKCHAIN 2019. AISC, vol. 1010, pp. 153–160. Springer, Cham (2020). https://doi.org/10.1007/978-3-030-23813-1_19
27. Wiki, S.: The complete reference for shimmer (2022). https://wiki.iota.org/shimmer/
28. Zamani, M., Movahedi, M., Raykova, M.: Rapidchain: scaling blockchain via full sharding. In: Proceedings of the 2018 ACM SIGSAC Conference on Computer and Communications Security, pp. 931–948 (2018)
29. Zohar, A.: Bitcoin: under the hood. Commun. ACM **58**(9), 104–113 (2015)

Workshop on Building the Potential of Blockchain in Farming (FARMS4CLIMATE)

Workshop on Building the Potential of Blockchain in Farming (FARMS4CLIMATE)

Rural economies and smallholder communities are struggling to sustain production, as low incomes, inefficient practices and limited access to markets and information are widespread. To overcome these challenges, holistic farming models built around a sustainable integration of human activities within the supporting ecosystem are required, especially in the fragmented Mediterranean agriculture. Unfortunately, it is difficult to bring to life alternative, theoretically beneficial organizational models, as the drivers of agricultural transformation are multidimensional, interrelated, location-specific, and change over time.

This workshop includes papers addressing the challenge of developing digital enablers to make carbon farming and other digital practices operational for smallholders, specially by means of the use of distributed ledger technologies. The workshop organisation is partially supported by the FARMS4CLIMATE project, funded by the Partnership for Research and Innovation in the Mediterranean Area (PRIMA) and the European Union.

Workshop Organization

Organizing Committee

Juan M. Corchado Rodríguez

University of Salamanca, Spain, AIR Institute, Spain

Javier Prieto Tejedor

University of Salamanca, Spain and AIR Institute, Spain

Program Committee

Yeray Mezquita Martín   AIR Institute, Spain

Massimo Vecchio   Fondazione Bruno Kessler, Italy

Miguel Pincheira   Fondazione Bruno Kessler, Italy

Carla Gonzales Gemio   AIR Institute, Spain

Development of an Open Source IoT-Blockchain Platform for Traceability of Fresh Products from Farm to Fork

Amaia Pikatza⊕, Nekane Sainz(✉)⊕, Iván González⊕, and Mikel Emaldi⊕

Faculty of Engineering, University of Deusto, Bilbao, Spain
{a.pikatza,nekane.sainz,gonza-lez.i,m.emaldi}@deusto.es

Abstract. Food traceability is now being given more and more importance, especially due to its benefit in food safety and sustainability. This article will carry out the research, design, development, and validation of a web application based on blockchain technology for food traceability from farm to fork. Thanks to the security, reliability and immutability offered by blockchain, it will be possible to know the status of the product anywhere in the food chain by all the agents involved, among which are farmers, distributors, producers, and end customers. Therefore, greater reliability will be provided in the layout of fresh products and greater security and transparency throughout the process, in addition to protecting all parties involved against fraud and final consumers against the health risks that can cause the consumption of food in poor condition. To do this, a study and design of a blockchain has been carried out, the data model has been designed and the *chaincode* implementing the functionality has been programmed. On the other hand, a frontend to interact with the backend system through a REST API has been designed and implemented. In addition, a study on the Internet of Things (IoT) devices and their integration with the blockchain is proposed. Last, an evaluation over the functionality of the *chaincode*, the performance of the REST API and the validity of the solution has been carried out.

Keywords: Blockchain · IoT · Food traceability · Food Supply Chain (FSC)

1 Introduction

Since the arrival of new technologies such as IoT, artificial intelligence or Distributed ledger technologies (DLTs) as blockchain, more and more companies have been forced to adapt and evolve towards what is now known as an intelligent company, which is summarized in having a high capacity to adapt to changes thanks to the use of technology in a way that responds quickly to the demands of a challenging environment [1]. Companies related to the FSC are not outside of the current challenges the World is facing, e.g., climate and demographic crisis. According to the UN [2], the global population could grow to around 8.5 billion in 2030, 9.7 billion in 2050 and 10.4 billion in 2100, which will create a significant increase in the global demand for food. On the other hand, the FAO estimates that up to 14% of food produced globally is lost between the post-harvest and retail stages, while the 17% of total food production may be wasted at the retail, food-service and consumer stages [3]. Direct causes of Food Loss and Waste (FLW) include inadequate inputs in production operations; poor scheduling and timing of harvesting operations; inappropriate production, harvesting and handling practices; and poor storage conditions and temperature management around perishable products. Apart from the zero waste objective, according to the barometer of the climate of confidence of the agri-food sector [4], consumers are increasingly interested in knowing the origin of the products they are purchasing.

In this paper, we tackle the problematic of the traceability at the FSC. Specifically, we present a use case covering the supply chain of fourth range products (ready-to- eat fresh-cut vegetables). The work combines IoT and blockchain to try to reduce food waste and losses in the agri-food chain and, to ensure a trustworthy traceability of different ingredients and products involved on the supply chain. In this article, we present the design and implementation of a system for the tracing and monitoring of fresh products from harvest to table. The objective of this job is therefore the development of a blockchain platform for the realization of a system capable of verifying, among other things, the origin and safe conditions of storage and transfer of food. In Sect. 2 a brief background of blockchain solutions for traceability is presented. In Sect. 3 the development of the platform is explained, Sect. 4 summarizes the requirements achieved and the unit tests performed and in Sect. 5 some conclusions and future lines are explained.

2   Background

Blockchain is a distributed, transparent, and immutable ledger that helps the process of recording transactions and tracking assets. There are several types of blockchain depending on the accessibility to it, i.e., public, permissioned or based on software as a service (BaaS) [4–9]. The development of the blockchain network is not going to be a problem for the integration of DLT services in companies, so they should focus on the definition of a data model adjusted to the requirements and the development and integration of a webapp like the one presented in this article.

2.1   Blockchain for Food Traceability

The fame of the blockchain is mainly due to its application in cryptocurrencies; how-ever, the possibilities offered by this technology go much further. Another sector in which blockchain can be used is the food industry, thanks to blockchain the supply chain can be completely traced.

Walmart along with its technology partner IBM created IBM Food Trust, involving leading companies in the food industry such as Nestlé and Unilever [10, 11]. There are other successful use cases such as those implemented by Carrefour and Starbucks. The success of these small initiatives and the potential that is clearly being seen is resulting in more complicated use cases involving the integration of IoT devices with limited computing capabilities and therefore limitations [12–16].

3 Development

3.1 Requirements Definition

The project aims to trace fresh food from farm to fork, which depicts the four distinct stages of the process: ingredient cultivation, factory production, retail distribution, and transportation. To better understand the requirements and data model, an initial meeting has been held with all the agents involved, a first draft of the model has been developed and it has been discussed with each of the agents in another meeting, introducing the appropriate modifications. In Table 1 requirements are shown.

Table 1. Functional and Non functional requirements

Code	Description	Priority
FR_01	All type of users, including final users, will be able to consult the information about the product and its raw materials	High
FR_02	All type of users except final users will be able to log in to the system	High
FR_03	Producer, manufacturers and couriers may update the storage parameters of the raw materials	Medium
FR_04	Manufacturers, couriers and retailers will be able to update the storage parameters of the products	Medium
FR_05	Producers, manufacturers and couriers may update the location of raw materials	Medium
FR_06	Manufacturer, couriers and retailers will be able to update the location of the products	Medium
FR_07	Producer and manufacturers may invalidate a raw material	Medium
FR_08	The manufacturer may invalidate a product	Medium
FR_09	The producer may begin the shipment of a raw material	High
FR_10	The manufacturer may begin the shipment of a product	High
FR_11	The producer will be able to register batches of new raw materials	High
FR_12	The manufacturer will be able to register batches of new products	High
FR_13	Couriers may change the shipping stage of raw materials and products	High
FR_14	Couriers may complete the shipment of raw material and products	High
FR_15	The manufacturer will be able to confirm the receipt of a raw material	High

(continued)

Table 1. (*continued*)

Code	Description	Priority
FR_ 16	The retailer will be able to confirm the receipt of a product	High
NFR_ 01	Simple and intuitive interface to use	High
NFR_ 02	The application must work in a fluid way	High
NFR_ 03	Responsive application, the visualization adapts to different devices such as mobiles, tablets, or computers	High

3.2 Design and Development of the Data Model for Food Traceability

There are 4 well separated parts on that process, those are ingredients obtained by cultivation, products made on a factory, retail distributors as supermarkets and transportation, according to that, the blockchain is composed of the following organizations and components: TLS, Deusto, Manufacturer, Agr1, Courier1, Courier2 and Retailer. Except for the TLS (Transport Layer Security) that encrypts data sent over the Internet to ensure that eavesdroppers and hackers are unable to see what it is transmitted [17], the rest of the organizations are formed by three components, the orderer, the CA and the peer. Ordering service, orderer, provides a shared communication channel to clients and peers, offering a broadcast service for messages containing transactions [18]. The Certificate Authority (CA) provides a number of certificate services to users of a blockchain. More specifically, these services relate to user enrollment, transactions invoked on the blockchain, and TLS-secured connections between users or components of the blockchain, and, peer nodes host ledgers and smart contracts [19]. The design and development has been carried out according to the bibliography [20, 21].

The data model obtained is shown in Fig. 1. It indicates that the holder can belong to any of the organizations mentioned above, while the producer belongs to Agr1, and the manufacturer belongs to the company. Parameters refer to any of the parameters, such as humidity or temperature, that need to be stored for sustainability and food safety. The name and description are selected by the organization, while the lot ID is used by the blockchain, not the company. Active and status refer to the state of the asset, with Active being set to true or false and Status being either idle, lost or destroyed, validated or in transit. The list of ingredients in the product is the composition of the ingredients, while longitude and latitude refer to the coordinates of the location that can be static or dynamic, depending on the type of location, whether it is a vehicle or a warehouse.

In the context of a particular design shown in Fig. 1, several modifications were made during development to achieve higher performance. The use of two separate keys: one for fixed parameters such as name and description, and another for changing parameters like location and holder. This was achieved using composite keys. The first key consists of the asset type (ingredient, product, or location) and the asset ID, while the second key adds the word 'STATUS' to the first and is used to store the variable parameters of the asset. In the case of location assets, all parameters are stored under a single key. To store assets, the putState function is used and, to retrieve assets, the getState function is used. To use this function, you need to pass the key of the asset you want to retrieve

Fig. 1. Data model schema

as a parameter. The chaincode is composed of both internal and external functions. The chaincode and all functions explanation can be found in https://github.com/amaiiapiiki/fabric-chaincode-traceability.

3.3 Web App Development

ReactJS has been selected for the front-end as the preferred option. React is a component-based framework that operates on a single page, creating SPAs. However, unlike Multi-Page Applications (MPAs), not all content is displayed at once, and different views can be obtained for the same page. To build the website, Create React App was used to set up the front-end construction flow. The initial project structure was created and all dependencies were installed by running the command "npx create-react-app web app". After setting up the initial structure and dependencies, the web app was developed primarily in the "src" folder which contains six subfolders and five files. The most important file is "index.js," which executes as soon as the web app is started and uses the router to define all the routes. The router is located in the "Router" folder of the project, and the "react-router-dom" library was used for redirects and redirection routes. The "/" route redirects to the App view, but more complex routes can be created using variable parameters, which are denoted by a word preceded by a colon (e.g., "/:parameter"). The views, which are equivalent to pages, consist of 12 components describe in Table 2. These views also use a set of 13 reusable components, some of which receive functions as parameters to allow for different behaviours depending on the view that uses them.

The react-components-form library has been utilized to develop all the forms mentioned in the table, while the "/Bootstrap" has been added while importing the library so that the form's appearance adapts to the rest of the page. The "react-qr-code" and "react-qr-reader" libraries have been used for reading and generating QR codes. The Scan view serves as a prelude to other views not mentioned in the table, such as InvalidateItem, ValidateShipment, StartShipment, ShipmentStep, and FinishShipment views.

Table 2. React components

Component	Description	Views
InfoDisplay	It is responsible for displaying the common information among all the items, i.e., the name, QR code, description, and lot	Iteminfo
InfoIngredient	Visualize the producer of the ingredient	Iteminfo
InfoProduct	Visualize the manufacturer and the ingredients of which the product is composed	Iteminfo
TraceGraph	Gets the item's historical temperature and humidity data and displays it in two different graphs. In addition, the graphics are interactive so you can also know the place and the hands of who was the item at a specific moment in time	Iteminfo
Login Form	It consists of a form composed of two text fields, one for the email and one for the password; and a button to send the information	Login
Produce Ingredient Form	It consists of a form composed of 4 text fields for the id, name, batch, and location; a text area for the description; a list of objects for the parameters; and a button to submit the form	ProduceIngredient
Manufacture Product Form	It consists of a form composed of 4 text fields for the id, name, batch, and location; a text area for the description; a list of objects for the parameters; a list of ingredients captured by reading their QR codes; and a button to submit the form	ManufactureProduct
Create Location Form	It consists of a form composed of 4 text fields for the id, name, latitude, and longitude; a selection field for the location type; a list of objects for the parameters; and a button to submit the form	CreateLocation
Start Shipment Form	It consists of a form composed of 5 text fields for the id, the type of item, the id of the Courier that will begin the shipment, the id of the vehicle with which the product will be moved, and the id of the destination; and a button to submit the form	StartShipment

(*continued*)

Table 2. (*continued*)

Component	Description	Views
Shipment Step Form	It consists of a form composed of 4 text fields for the id, the type of item, the id of the Courier that will continue with the shipment, and the id of the vehicle with which the product will be moved; and a button to submit the form	ShipmentStep
Finish Shipment Form	It consists of a form composed of 4 text fields for the id and type of the item, the id of the receiving organization and the current location; and a button for submitting the form	FinishShipment
QRReader	It displays a window connected to the camera, reads QR codes and performs with that information the action that the component receives as a parameter	App, Scan
NavBar	It creates a navigation menu at the top that varies depending on whether there are any users who are logged in and whether that is the case depends on the user as well	All the views

Scan reads a QR code and redirects to one of the previous views, including the information obtained by reading. This view serves as an example to explore the programming mode of the web page with React.

In addition to views and components, the "src" folder also contains the "assets," "lib," and "styles" folders. "Assets" contains all the images used on the page, "styles" contains all the style sheets, and "lib" contains global functions and parameters that can be used in different views, such as the API URL for interaction with the Blockchain or the "get-Menu()" function. The website's appearance has been created using the Moon template, which has a Creative Commons 3.0 license and was designed and developed by GetTemplates.co and FreeHTML5.co. After completing the page's development, two commands are needed for the Linux environment to start it: "export SET NODE_OPTIONS= --openssl-legacy-provider npm start". Any errors will be displayed if there are any. If everything is correct, the website will be accessible for local use at http://localhost:3000 and at http://192.168.1.74:3000 for the network. In Fig. 2, view of TraceGraph component are shown as example of the webapp apparieance.

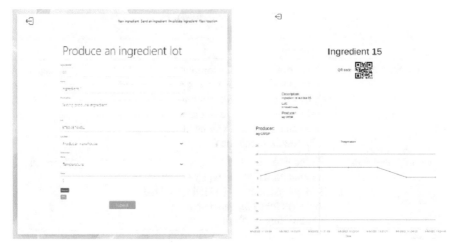

Fig. 2. Views for Produce Ingredient Lot and TraceGraph

3.4 Development of the Connection with the Blockchain

To connect with the blockchain, an API was used. The 'config.yaml' file in the '/volumes/config' folder was modified to allow the previously described website to make requests. The two necessary URLs mentioned in the previous section were added to the file, and the docker containers hosting the API were rebuilt and relaunched.

After completing these steps, the webpage has permission to use the API, and the axios library can be used to make requests. The URL for interacting with the API was specified and exported in the 'lib' folder, and can be imported anywhere necessary.

While most API calls throughout the project are not made during the login process, the post and get functions are used in this example. The login process consists of three steps: first, a post request is made to the login URL with the data collected by the login form. Second, if the login is successful, another request is made to retrieve the logged user's information, which is saved in session storage. Third, the user is redirected to their homepage using the organization to which they belong. The most commonly used URL of the API is for executing chaincode transactions. To submit any other transaction, only the body needs to be changed to match the required transaction and parameters, and the use of axios and the request will remain the same.

3.5 Development and Interconnection of the Hardware Data Acquisition Module

In this section, we will discuss how to use IoT technology to capture and store temperature and humidity data in the blockchain. To accomplish this, we will use three components: a Raspberry Pi, a grove base hat for the RasPi, and the DHT11 sensor. The base must be connected to the Raspberry Pi, and the sensor must be connected to port 12 of the base. We need to install several libraries and dependencies to obtain the data. These include grove.py, which is a Python library for the GrovePi and Grove Sensors from Seeed Studio, and Seeed_Python_DHT, which is a Python library for the DHT series of temperature and humidity sensors from Seeed Studio.

Once we have confirmed that the sensor is working correctly, we can develop the script to capture and send the data to the blockchain. The script uses the same API as the web app to send the collected data. The URLs used to carry out the transactions are the same as those used in the web app, with the body of the call being the only thing that changes each time. We can obtain the login credentials by using the getCredentials function, which creates a form. Through the script, the temperature and humidity data are obtained in real-time every 10 s, and the parameters of the location with ID 'L14,' which is a warehouse belonging to the company in charge of manufacturing the fresh products, are updated.

4 Evaluation

4.1 Chaincode Validation

Chaincode validation was ensured through unit tests. 126 unit tests were developed to check the largest number of declarations, branches, functions, and lines for almost total code coverage. It was noted that it took only one second to perform all the tests, which was one of the great advantages of unit testing. To validate the correct operation when all the conditions were met, the creation of a new location was taken as an example. It was necessary to ensure that the calls to the putState function were made with the correct parameters and that the function returned the correct object. In the case of the CreateLocation function, a string was returned with the object containing all the location information.

To execute the tests correctly, the context, including the stub and the client identity, had to be mocked. Sinon.js was used to create test spies, stubs, and mocks for JavaScript. For some tests, additional faked calls were necessary. The unit tests aimed to ensure the correct functioning of the code in parts. Thus, it did not make sense to allow calls to other functions of the chaincode that would be checked separately. By using false calls to other functions, it was easier to cause errors and check their correct functioning. An example of this was the test that checked the correct functioning of the ProduceIngredientLot function, where the user had to be a producer. By using the false client identity, it was possible to verify whether the function failed as it should when it was not a producer. Similarly, each of the prerequisites to carry out the transaction could be checked. To validate the correct operation when all the conditions were met, it was necessary to ensure that the calls to the putState function were made with the correct parameters and that the function returned the correct object.

4.2 Requirements Fulfilment Checking

In Table 1, requirement fulfilment is checked (Table 3).

Table 3. Requirements Fulfilment

Code	Fulfilled	Status	Code	Fulfilled	Code	Fulfilled
FR_ 01	Yes		FR_ 08	Yes	FR_ 15	Yes
FR_ 02	Yes		FR_ 09	Yes	FR_ 16	Yes
FR_ 03	Yes		FR_ 10	Yes	NFR_01	Yes
FR_ 04	Partially	The transactions needed to	FR_ 11	Yes	NFR_02	Yes
FR_ 05	Partially	perform those actions are	FR_ 12	Yes	NFR_03	Yes
FR_ 06	Partially	developed at the chaincode,	FR_ 13	Yes		
FR_ 07	Yes		FR_ 14	Yes		

5   Conclusions and Future Lines

The article studies the integration of a track system based on blockchain with IoT devices based on a raspberry that allows capture, through temperature and humidity measurement sensors among others, and automatic storage of data in the blockchain network. In this way, it could be ensured, for example, that the cold chain of food is not broken at any point or in the case where it was so, measures could be taken in a much more efficient way. In addition, thanks to the immutability offered by blockchain, it would be impossible to falsify data, offering consumers real food security. Then, a suitable framework will be selected for the development of a web page that serves for the different actors during the supply chain to interact with the blockchain. Finally, through some validation tools, the verification of the correct functioning of the system will be carried out. In conclusion, the use of a raspberry with blockchain in the supply chain of fresh products has been validated, providing greater reliability in its layout and greater security and transparency throughout the process, and it also protects all parties involved against fraud and end consumers against the health risks that can be caused by the consumption of spoiled food.

This article has been funded by "Cátedra de Telefónica de la Universidad de Deusto" and the European project Horizon 2020 FoodRUS (GA 101000617).

References

1. Md Mehedi, H.O.: Performance analytical comparison of blockchain-as-a-service (BaaS) platforms. In: Miraz, M., Excell, P., Ware, A., Soomro, S., Ali, M. (eds.) Emerging Technologies in Computing. iCETiC 2019. LNICS, Social Informatics and Telecommunications Engineering, vol. 285, pp. 3–18. Springer, Cham (2019). https://doi.org/10.1007/978-3-030-23943-5_1
2. United Nations Department of Economic and Social Affairs, Population Division (2022). World Population Prospects 2022: Summary of Results. UN DESA/POP/2022/TR/NO. 3
3. FAO. 2022. Voluntary Code of Conduct for Food Loss and Waste Reduction. Rome. FAO. 2017. Food Traceability Guidance. Santiago. ISBN 978-92-5-109876-9, https://doi.org/10.4060/cb9433en

4. Government. https://www.mapa.gob.es/es/alimentacion/temas/consumo-tendencias/barome
   tro-del-clima-de-confianza-del-sector-agroalimentario/. Accessed 19 Oct 2021
5. IBM: https://www.ibm.com/topics/hyperledger. Accessed 03 Nov 2021
6. Hyperledger Fabric. Chaincode. https://hyperledger-fabric.readthedocs.io/en/v1.0.0-beta/cha
   incode.html. Accessed 19 Oct 2021
7. Ethereum. Welcome to Ethereum. https://ethereum.org/en/. Accessed 28 Oct 2021
8. Frankenfield, J.: Blockchain-as-a-Service (BaaS). https://www.investopedia.com/terms/b/blo
   ckchainasaservice-/aas.asp. Accessed 03 Nov 2021
9. Amazon. https://aws.amazon.com/es/blockchain/?nc1=h_ls. Accessed 03 Nov 2021
10. Hyperledger. Case Study. https://www.hyperledger.org/learn/publications/walmart-case-
    study. Accessed 12 Nov 2021
11. Supply chain movement. https://www.supplychainmovement.com/nestle-trials-blockchain-
    to-improve-food-ingredient-supply-chain-transparency/. Accessed 12 Nov 2021
12. Sadawi, A.A., Hassan, M.S., Ndiaye, M.: A survey on the integration of blockchain with
    IoT to enhance performance and eliminate challenges. IEEE Access **9**, 54478–54497 (2021).
    https://doi.org/10.1109/ACCESS.2021.3070555
13. Hussain, M., et al.: Blockchain-based IoT devices in supply chain management: a systematic
    literature review. Sustainability. **13**(24), 13646 (2021). https://doi.org/10.3390/su132413646
14. Feng, H., Wang, X., Duan, Y., Zhang, J., Zhang, X.: Applying blockchain technology to
    improve agri-food traceability: a review of development methods, benefits and challenges.
    J. Clean. Prod. **260**, 121031 (2020). ISSN 0959-6526, https://doi.org/10.1016/j.jclepro.2020.
    121031
15. Tsang, Y.P., Choy, K.L., Wu, C.H., Ho, G.T.S., Lam, H.Y.: Blockchain-driven IoT for food
    traceability with an integrated consensus mechanism. IEEE Access **7**, 129000–129017 (2019).
    https://doi.org/10.1109/ACCESS.2019.2940227
16. Bhat, S.A., Huang, N.-F., Sofi, I.B., Sultan, M.: Agriculture-food supply chain management
    based on blockchain and IoT: a narrative on enterprise blockchain interoperability. Agriculture
    12(1), 40 (2022). https://doi.org/10.3390/agriculture12010040
17. Hyperledger Fabric. The Ordering Service. https://hyperledger-fabric.readthedocs.io/en/rel
    ease/arch-deep-dive.html. Accessed 23 June 2022
18. IBM Corp. CA Setup. https://openblockchain.readthedocs.io/en/latest/Setup/ca-setup/.
    Accessed 23 June 2022
19. Role of Peers in Hyperledger Fabric. Accessed 23 June 2022
20. Honar Pajooh, H., Rashid, M., Alam, F., Demidenko, S.: Hyperledger fabric blockchain for
    securing the edge internet of things. Sensors **21**(2), 359 (2021). https://doi.org/10.3390/s21
    020359
21. Grecuccio, J., Giusto, E., Fiori, F., Rebaudengo, M.: Combining blockchain and IoT: food-
    chain traceability and beyond. Energies **13**(15), 3820 (2020). https://doi.org/10.3390/en1315
    3820

Blockchain-Based Platforms for Carbon Offsetting: A Survey of Existing Approaches and Their Potential to Promote Carbon Farming for Smallholders

Yeray Mezquita[1,2]([✉]) [ID], Carlos Álvarez[1] [ID], Diego Valdeolmillos[2] [ID], and Javier Prieto[1] [ID]

[1] BISITE Research Group, University of Salamanca, Salamanca, Spain
{yeraymm,carlos_alvarez,javierp}@usal.es
[2] AIR Institute, Valladolid, Spain
dval@air-institute.com

Abstract. This paper examines the use of blockchain technology in improving the carbon offset market and promoting carbon farming for smallholders. The paper presents a survey of six blockchain-based platforms for carbon offsetting, namely Blockchain for Climate Foundation, Open Climate Collabathon, Aircarbon Exchange, Cryptocarbon NFT, Toucan, and KlimaDAO. The platforms were evaluated based on their impact on the carbon offset market, and their potential to make carbon farming operational for smallholders was discussed. Overall, this paper demonstrates that blockchain technology has the potential to enhance the carbon offset market and empower smallholders in carbon farming.

Keywords: Carbon offset market · Blockchain · Survey

1 Introduction

The carbon offset market has emerged as a popular mechanism to reduce carbon dioxide (CO_2) emissions and mitigate climate change. Carbon offsetting involves funding projects that capture or reduce greenhouse gas (GHG) emissions to compensate for one's own carbon footprint. These projects can include renewable energy installations, reforestation, and energy-efficient technology adoption. However, the carbon offset market faces several challenges, including the difficulty of verifying carbon emissions reductions, concerns about additionality and permanence of carbon credits, and the complexity of trading carbon offsets.

One of the significant challenges in the current centralized carbon offset market is that it does not effectively address the needs of smallholders who are interested in participating in carbon capture and sequestration projects. These smallholders often lack access to the resources necessary to develop and implement

J. M. Machado et al. (Eds.): BLOCKCHAIN 2023, LNNS 778, pp. 498–506, 2023.
https://doi.org/10.1007/978-3-031-45155-3_47

carbon offset projects, such as technical expertise, financing, and market access. As a result, they are often unable to benefit from carbon capturing projects, while larger corporations and organizations are the main beneficiaries of the carbon offset market. Therefore, it is essential to explore alternative mechanisms that can make carbon capture and sequestration projects accessible to smallholders and enable them to participate in the carbon offset market.

Blockchain technology offers a potential solution to the challenges faced by smallholders in the current carbon offset market. The decentralized and transparent nature of blockchain allows for the creation of a more accessible and inclusive marketplace that enables smallholders to participate in carbon capture and sequestration projects. Blockchain can help to improve transparency and accountability in the carbon offset market by enabling the tracking of carbon credits from their creation to their retirement. This can help to address concerns about the additionality and permanence of carbon credits and make it easier to verify carbon emissions reductions. Additionally, blockchain can facilitate the development of peer-to-peer trading platforms that connect smallholders with buyers directly, bypassing the need for intermediaries and reducing transaction costs. These platforms can provide smallholders with access to financing, technical expertise, and market access, enabling them to participate in the carbon offset market and benefit from carbon capture and sequestration projects. Overall, blockchain technology has the potential to transform the carbon offset market and make it more accessible and inclusive for smallholders, promoting the development of carbon sequestration projects and contributing to the mitigation of climate change.

The objective of this paper is to study the various blockchain-based platforms that have been developed to improve the carbon offset market and promote smallholder participation in carbon capture and sequestration projects. We will evaluate the different approaches that have been taken to address the challenges faced by smallholders in accessing the carbon offset market and assess the impact of these platforms on the carbon offset market as a whole. The study aims to provide a comprehensive analysis of the benefits and limitations of these blockchain-based platforms and identify opportunities for further development and improvement. By presenting the results of our study, we hope to contribute to the ongoing discussion on how to make the carbon offset market more accessible and inclusive for smallholders and promote the development of carbon sequestration projects. Ultimately, we believe that this study will be of interest to academics, policymakers, and practitioners working in the fields of blockchain, climate change, and sustainable agriculture, and will provide valuable insights into the potential benefits of blockchain technology in addressing the challenges faced by smallholders in the carbon offset market.

Overall, this paper provides a comprehensive analysis of the potential benefits of blockchain technology in promoting smallholder participation in carbon offsetting projects and contributes to the ongoing discussion on how to make the carbon offset market more accessible and inclusive. The rest of the paper is organized as follows. In the literature review section, Sect. 2, we provide an overview of the carbon offset market and its challenges, as well as an introduction to blockchain technology and its potential in addressing these challenges. We then present a review of the existing blockchain-based platforms for carbon offsetting. In the results and discussion section, Sect. 3 we present and evaluate the different blockchain-based platforms that we studied, and discuss their potential to promote smallholder participation in carbon capture and sequestration projects. Finally, in Sect. 4, we summarize our findings, discuss their implications, and identify opportunities for further research.

2   Literature Review

In this section we provide an overview of the carbon offset market and its challenges, as well as an introduction to blockchain technology and its potential to address these challenges. We will begin by discussing the carbon offset market, development, and current state. We will then examine the challenges that the carbon offset market faces, such as additionality, verification, and accounting. Next, we will introduce blockchain technology and explain how it can help to address the challenges faced by the carbon offset market. We will discuss the key features of blockchain, such as decentralization, immutability, and transparency, and explain how they can be applied to the carbon offset market. Finally, we will review existing blockchain-based platforms for carbon offsetting and evaluate their potential to promote smallholder participation in carbon capture and sequestration projects. Through this literature review, we aim to provide a comprehensive understanding of the carbon offset market and the potential benefits of blockchain technology in addressing its challenges.

2.1   Overview of Carbon Offset Market and Its Challenges

Carbon offsetting is a market-based mechanism that allows organizations to compensate for their GHG emissions by purchasing carbon credits from carbon capture and sequestration projects. The carbon offset market has grown rapidly in recent years, with a global value of $2 billion in 2022, expecting to grow to $100 billion in 2030 [10]. However, the carbon offset market faces several challenges that limit its effectiveness and impact, including additionality, verification, accounting, accessibility, and transparency.

One of the main challenges of the carbon offset market is additionality, which refers to the need for carbon credits to represent emissions reductions that would not have occurred without the offset project. This is important to ensure that carbon offsets actually result in a net reduction of GHG emissions. Verification and accounting are also significant challenges for the carbon offset market. Carbon credits must be accurately quantified and tracked, and the accounting systems used must be transparent and credible to ensure trust and confidence in the market.

Accessibility is another challenge of the carbon offset market, particularly for smallholders who lack the resources and expertise to participate in carbon capture and sequestration projects. This can lead to a concentration of carbon offset projects in certain regions and exclude smallholders who may have significant carbon sequestration potential. Finally, there is a lack of transparency and trust in the carbon offset market, with concerns over double counting and fraud.

Addressing these challenges is critical to ensuring the effectiveness and credibility of the carbon offset market. In the next section, we will explore how blockchain technology can help to address these challenges and promote the development of carbon capture and sequestration projects with transparency and accountability.

2.2   Blockchain Technology and Its Potential in Addressing Carbon Offset Market Challenges

Blockchain technology offers several potential benefits for addressing the challenges facing the carbon offset market. One of the key advantages of blockchain technology is its ability to provide transparency and accountability [13] in the carbon offset market. By using a decentralized ledger, all participants in the market can view and verify the transaction history of carbon credits [8]. This can help to prevent fraud and double counting, and increase trust and confidence in the market.

Another advantage of blockchain technology is its potential to facilitate the tracking and verification of carbon credits [6]. Smart contracts can be used to automatically execute transactions based on predefined conditions, such as the verification of emissions reductions [9]. This can help to simplify the verification process and reduce costs.

Blockchain technology can also improve the accessibility of the carbon offset market, particularly for smallholders. By reducing transaction costs and providing a transparent and trusted platform for carbon offset projects, blockchain can help to incentivize smallholders to participate in carbon capture and sequestration projects [7]. This can lead to a more distributed and equitable carbon offset market, with benefits for both smallholders and the environment.

In addition to these benefits, blockchain technology can also help to address other challenges facing the carbon offset market, such as additionality and accounting. By providing a transparent and auditable ledger of carbon credits, blockchain can help to ensure that carbon credits represent genuine emissions reductions and that accounting systems are credible and accurate.

Overall, blockchain technology offers significant potential for addressing the challenges facing the carbon offset market and promoting the development of carbon capture and sequestration projects. In the next section, we will review some of the existing blockchain-based platforms for carbon offsetting and assess their effectiveness in addressing these challenges.

2.3   Existing Blockchain-Based Platforms for Carbon Offsetting

Several blockchain-based platforms have emerged in recent years to address the challenges facing the carbon offset market. In this section, we will review some of the most prominent platforms and assess their effectiveness in promoting the development of carbon capture and sequestration projects.

The Blockchain for Climate Foundation [2] is a non-profit organization that aims to use blockchain technology to accelerate climate action. The foundation has developed a platform that allows individuals and organizations to purchase and sell carbon credits directly, without the need for intermediaries. The platform uses smart contracts to automatically verify and track carbon credits, providing transparency and accountability in the carbon offset market. The foundation has also developed a token-based incentive system to encourage smallholders to participate in carbon capture and sequestration projects. While the platform is still in its early stages, it has the potential to significantly improve the accessibility and transparency of the carbon offset market.

The Open Climate Collabathon [3] is a global collaboration that brings together individuals and organizations to develop solutions for climate change using blockchain technology. The initiative is designed to generate ideas, prototypes, and solutions for addressing climate change and supporting the transition to a more sustainable world. The Collabathon's website provides a platform for participants to collaborate and submit their ideas and projects, which are then evaluated and developed further during various hackathons and events. The ultimate goal of the Open Climate Collabathon is to create an open-source, decentralized infrastructure that enables the world to collaborate and tackle the challenges of climate change together

The Aircarbon Exchange (ACX) [4] is a Singapore-based platform that is committed to facilitating carbon trading to combat climate change. The platform is built on blockchain technology, which allows for transparency and traceability of carbon credits. ACX operates a marketplace that enables carbon buyers and sellers to trade carbon credits and manage their carbon portfolios. ACX also offers a unique feature called "Carbon Negative", which allows individuals and organizations to offset their carbon footprint by purchasing carbon credits that have a negative carbon footprint. These credits are generated by carbon-negative projects, which remove more carbon from the atmosphere than they emit. By offering this feature, ACX aims to create a carbon-neutral future by incentivizing and facilitating the adoption of carbon-negative practices.

Crypto Carbons [1] is a project that aims to provide carbon offsets through the sale of unique and collectible pixel art NFTs (non-fungible tokens) on the Ethereum blockchain. The project uses the concept of "proof-of-offset" to verify that each NFT represents a specific amount of carbon offset. Each NFT is backed by a carbon offset certificate that has been retired to ensure that the corresponding carbon credits cannot be double-counted or resold. The NFTs can be traded on marketplaces like OpenSea, providing a new avenue for individuals and companies to participate in the carbon offset market.

The Crypto Carbons project takes a unique approach to carbon offsetting, leveraging the growing popularity of NFTs and the blockchain's ability to provide transparency and immutability. The project also aims to raise awareness of carbon offsetting and encourage individuals to take action to reduce their carbon footprint. By creating a fun and collectible way to offset carbon, Cryptocarbon NFTs could help to make carbon offsetting more accessible to a wider audience, helping the carbon offset market to gain popularity and increase its market capitalization.

Toucan [12] is a blockchain-based platform that aims to promote sustainable development by incentivizing carbon capture and sequestration projects, particularly those led by smallholders. The platform uses a token-based reward system to encourage smallholders to participate in carbon offset projects, which historically have been dominated by larger players in the market. By providing an accessible and transparent ledger of carbon credits, Toucan allows smallholders to benefit from the carbon market and participate in sustainable development efforts. The platform also uses smart contracts to ensure the additionality and verification of carbon credits, further increasing the integrity of the market. Toucan's use of Verra registries [14] for carbon credit tokenization is another example of its commitment to transparency and accountability. Overall, Toucan has the potential to significantly improve the accessibility and transparency of the carbon offset market, while also promoting sustainable development and incentivizing the participation of smallholders.

KLIMADAO [5] is a decentralized autonomous organization (DAO) that aims to create a price floor for carbon by using a token called KLIMA. The KLIMA token is designed to act as a "decentralized carbon bank" that can be used to purchase carbon offsets on the Klima platform. KLIMADAO is backed by a reserve of USDC stablecoins and a portion of the KLIMA tokens themselves. The reserve is used to purchase and hold carbon credits, which are then used to back the value of the KLIMA token [11]. This creates a feedback loop where demand for carbon credits on the Klima platform drives up the price of KLIMA, which in turn increases the value of the reserve and allows KLIMADAO to purchase more carbon credits.

The goal of KLIMADAO is to create a self-sustaining ecosystem where the price of carbon is determined by market forces rather than government regulation. By using blockchain technology and a decentralized governance structure, KLIMADAO aims to promote transparency and accountability in the carbon offset market while empowering individuals and smallholders to participate.

3   Results and Discussion

In this paper, we have conducted a survey of six existing blockchain-based platforms for carbon offsetting. These platforms include the Blockchain for Climate Foundation, Open Climate Collabathon, AirCarbon Exchange, Cryptocarbon NFT, Toucan, and KLIMADAO. Each platform takes a unique approach to addressing the challenges of the carbon offset market.

Our evaluation of these platforms reveals that they have the potential to significantly impact the carbon offset market by increasing transparency, reducing transaction costs, and enabling greater participation by individuals and smallholders. For example, Toucan's use of NFTs to represent carbon offsets creates a more tangible and accessible asset that can be easily traded on the blockchain. KLIMADAO's creation of a decentralized bank backed by carbon credits helps to stabilize the price of carbon and make it more attractive to investors.

However, the impact of these platforms on the carbon offset market is still limited. Most of the platforms are still in their early stages of development and adoption, and their effectiveness in addressing the challenges of the carbon offset market has yet to be fully tested.

Despite the limitations of the studied platforms, we believe that they have the potential to make carbon farming operational for smallholders. By reducing transaction costs and increasing transparency, blockchain-based platforms can make it easier for smallholders to participate in carbon offset projects and receive fair compensation for their efforts. Additionally, the use of NFTs and other blockchain-based assets can create more tangible and accessible representations of carbon offsets, which can make it easier for smallholders to market and sell their carbon credits.

Furthermore, platforms like KLIMADAO can help stabilize the price of carbon and create a more attractive investment opportunity for smallholders. This can help to incentivize greater participation in carbon offset projects and promote the growth of the carbon offset market overall.

In conclusion, while the impact of blockchain-based platforms on the carbon offset market is still limited, we believe that they have significant potential to promote carbon farming for smallholders and create a more transparent, accessible, and sustainable carbon offset market. Further research and development in this area is needed to fully realize this potential.

4   Conclusion and Future Directions

This survey paper provided an overview of various blockchain-based platforms that aim to address challenges in the carbon offset market. Our analysis of the platforms shows that they have the potential to significantly impact the carbon offset market by providing a more transparent, secure, and efficient way of tracking and trading carbon credits.

We evaluated the platforms based on their impact on the carbon offset market, and found that each platform has its own unique features that contribute to addressing specific challenges in the market. The Blockchain for Climate Foundation and Open Climate Collabathon provide a collaborative platform for developers and stakeholders to create innovative solutions for climate action. Aircarbon Exchange and Toucan provide a marketplace for carbon offset trading, while Cryptocarbon NFT and Klima provide a way for individuals and organizations to invest in carbon offset projects.

We discussed how these platforms can make carbon farming operational for smallholders by providing them with access to carbon markets that were previously inaccessible. These platforms can help smallholders to monetize their carbon sequestration projects, thereby promoting sustainable land management practices and contributing to climate mitigation efforts.

This survey paper contributes to the literature on blockchain technology and the carbon offset market by providing an overview of existing blockchain-based platforms and their potential impact on the market. Our analysis of the platforms highlights the importance of transparency, security, and efficiency in the carbon offset market.

The implications of our findings suggest that blockchain-based platforms can play a significant role in promoting sustainable land management practices and contributing to climate mitigation efforts. By providing a more accessible and efficient way of trading carbon credits, these platforms can encourage more individuals and organizations to invest in carbon sequestration projects.

One limitation of this survey paper is the limited number of platforms studied. There are numerous blockchain-based platforms currently being developed, and future research could expand on the analysis provided in this paper. Additionally, the impact of these platforms on the carbon offset market is yet to be fully realized, and more research is needed to evaluate the long-term effectiveness of these platforms.

Future research directions could include evaluating the environmental and social impacts of carbon offset projects, as well as the effectiveness of blockchain technology in promoting sustainable land management practices. Furthermore, research could focus on the role of these platforms in promoting equitable access to carbon markets and ensuring that smallholders benefit from carbon sequestration projects.

Acknowledgements. This work has been partially supported by the project FARMS4CLIMATE, co-funded by the Partnership for Research and Innovation in the Mediterranean Area (PRIMA) and the European Union.

References

1. Carbons, C.: Crypto carbons (2023). https://opensea.io/collection/crypto-carbons. Accessed 8 May 2023
2. for Climate Foundation, B.: The BITMO platform (2023). https://www.blockchainforclimate.org/the-bitmo-platform. Accessed 8 May 2023
3. Collabathon, O.C.: An open climate accounting system (2023). https://www.collabathon.openclimate.earth/open-climate-project. Accessed 8 May 2023
4. Exchange, A.: AirCarbon exchange (2023). https://www.aircarbon.co/ Accessed 8 May 2023
5. KLIMADAO: Introducing klimadao (2021). https://docs.klimadao.finance/. Accessed 8 May 2023
6. Mezquita, Y., Casado-Vara, R., González Briones, A., Prieto, J., Corchado, J.M.: Blockchain-based architecture for the control of logistics activities: pharmaceutical utilities case study. Logic J. IGPL **29**(6), 974–985 (2021)
7. Mezquita, Y., González-Briones, A., Casado-Vara, R., Chamoso, P., Prieto, J., Corchado, J.M.: Blockchain-based architecture: a mas proposal for efficient agri-food supply chains. In: Novais, P., Lloret, J., Chamoso, P., Carneiro, D., Navarro, E., Omatu, S. (eds.) ISAmI 2019. AISC, vol. 1006, pp. 89–96. Springer, Cham (2020). https://doi.org/10.1007/978-3-030-24097-4_11

8. Mezquita, Y., Podgorelec, B., Gil-González, A.B., Corchado, J.M.: Blockchain-based supply chain systems, interoperability model in a pharmaceutical case study. Sensors **23**(4), 1962 (2023)

9. Mezquita, Y., Valdeolmillos, D., González-Briones, A., Prieto, J., Corchado, J.M.: Legal aspects and emerging risks in the use of smart contracts based on blockchain. In: Uden, L., Ting, I.-H., Corchado, J.M. (eds.) KMO 2019. CCIS, vol. 1027, pp. 525–535. Springer, Cham (2019). https://doi.org/10.1007/978-3-030-21451-7_45

10. Stanley, M.: Where the carbon offset market is poised to surge (2021). https://www.morganstanley.com/ideas/carbon-offset-market-growth. Accessed 8 May 2023

11. Strauf, F.: Tokenomics 101: Klima dao (2021). https://banklesspublishing.com/tokenomics-101-klima-dao/. Accessed 8 May 2023

12. Toucan: Web3 infrastructure for regenerative finance - refi (2023). https://docs.toucan.earth/toucan/introduction/readme. Accessed 8 May 2023

13. Valdeolmillos, D., Mezquita, Y., González-Briones, A., Prieto, J., Corchado, J.M.: Blockchain technology: a review of the current challenges of cryptocurrency. In: Prieto, J., Das, A.K., Ferretti, S., Pinto, A., Corchado, J.M. (eds.) BLOCKCHAIN 2019. AISC, vol. 1010, pp. 153–160. Springer, Cham (2020). https://doi.org/10.1007/978-3-030-23813-1_19

14. Verra: Who we are? (2021). https://verra.org/about/overview. Accessed 8 May 2023

Doctoral Consortium

Blockchain Oracles for Asset Ownership Verification and Trust Establishment in the Metaverse

Hadi Nowandish$^{(\boxtimes)}$ ⓘ, Alex Norta ⓘ, and Peeter Normak ⓘ

Tallinn University, Narva mnt 25, 10120 Tallinn, Estonia
{hadinow,peeter.normak}@tlu.ee, alex.norta.phd@ieee.org

Abstract. Metaverse encompasses the ability to transform the way we interact with technology and with each other, opening up new avenues for socializing, trading, communicating, entertaining, and learning. As the metaverse ecosystems continue to expand and assets within them become increasingly valuable, the issue of trust in transferring ownership of these assets becomes a significant concern. The objective of this research is to design and develop a blockchain oracle system for metaverses that address the challenges of ownership verification, identity theft, and trust establishment in the metaverse while transferring ownership of assets in order to enhance the security and trustworthiness of virtual transactions. This blockchain oracle has the ability to securely and impenetrably confirm ownership in the metaverse, increase trust in the ownership transfer of assets and ensure that ownership transfers are carried out correctly without the risk of fraud or other malicious behaviors. The study employs a design science research method to design, develop and evaluate a novel blockchain oracle for trust establishment and asset ownership verification in the metaverses.

Keywords: Blockchain · Smart Contract · Oracle · Metaverse · Trust · Ownership

1 Problem Statement and Motivation

Metaverse, originally a product of science fiction [1,2], is now a tangible concept with the help of advanced and innovative technologies. The term "metaverse" is formed by combining the words "meta" and "universe", and describes a virtual world that exists in three dimensions and allows users to engage in a variety of activities related to politics, economics, society, and culture through their avatars [3]. The metaverse, enabled by advanced technologies such as 3D graphics, networking, artificial intelligence, blockchain, virtual- and augmented reality, IoT, cloud computing, and machine learning [2], provides an immersive and interconnected virtual environment that allows users to engage in a wide range of activities and interactions, creating new opportunities for socialization, entertainment, education, commerce [4], and so on.

J. M. Machado et al. (Eds.): BLOCKCHAIN 2023, LNNS 778, pp. 509–514, 2023.
https://doi.org/10.1007/978-3-031-45155-3_48

Digital asset exchange/transformation is one of the key issues and activities of the metaverse ecosystem, allowing users to buy, sell, and trade virtual assets within virtual worlds, between- and across the metaverses. An example of digital asset exchange and verification is that users of the metaverse platform designing, creating, and selling unique digital assets within a metaverse in the form of non-fungible tokens (NFTs) through a smart contract that verifies ownership [5]. Another recent study [6] discusses the use of self-sovereign identity (SSI) technology to promote interoperability between virtual worlds and ensure trust in the metaverse using a blockchain-based SSI management system. On the other hand, the economy created in a metaverse may be under threat from a variety of attacks, including those that target economic justice, digital asset ownership, and service trust [7]. Moreover, As the metaverse develops, there are an increasing risk of security and privacy threats, such as identity theft, broken authentication, fake data and data breaches, difficulties in tracking ownership/provenance, violations of intellectual property rights, unauthorized access, phishing, eavesdropping, personal information leakage, and cyber attacks[1, 8–11]. However, these studies lack trust establishment, identity authentication, and ownership verification when transferring assets between metaverses.

Blockchain is necessary in the metaverse to ensure accountability and secure the digital content of all users, enabling user integrity, privacy, and reputation [12]. Blockchain is a tamper-evident and -resistant digital ledger for processing transactions without the need for a trusted third-party [13] with advantageous features such as integrity, high transparency, efficiency, decentralization, distributed structure, immutability, highly secured structure by design, programmability, incontestability, traceability, real-time accessibility, built-in-trust, and resilience against censorship and tampering [14]. A smart contract is the key use case of blockchain technology [15] being a script, or computer program that is self-verifying, self-executing, tamper-resistant, and executes on a blockchain platform to enable trustworthy transactions without the involvement of a third-party [16]. EigenLayer is a collection of Ethereum-based smart contracts that enables consensus layer Ether (ETH) stakers to opt in for validating new software constructed on top of the Ethereum ecosystem [17]. Furthermore, EigenLayer has the potential to develop an oracle for Ethereum that can incorporate price feeds, with the prerequisite of requiring majority trust on ETH that has been restaked with EigenLayer and the condition that it is an opt-in layer.

Blockchain oracles are required to transfer trusted data from off-chain to on-chain. An oracle is a third-party agent, yielding a very promising interoperability technique to connect on-chain and off-chain technologies [18], converting non-deterministic data into a format that a blockchain can interpret, supplies smart contracts with external information, and validates and verifies real-world events [19, 20]. Furthermore, an oracle needs to provide guarantees that the data obtained is tamper-proof. In addition, oracles are designed with specific characteristics and functionalities for particular domains and use cases [19].

The state of the art shows that the ownership transfer of assets between metaverses faces challenges in ownership verification, identity authentication, and trust establishment. In addition, although blockchain is one of the technologies in the metaverse for digital asset management, the blockchain oracle and metaverse have not been integrated for trust establishment and asset ownership verification. To address these challenges, there is a need to design and develop a blockchain oracle that can facilitate secure and reliable ownership transfer in the metaverse ecosystems. Thus, this research study rigorously designs and develop high-utility blockchain oracles for establishing trust, ensuring secure ownership transfer of assets, and identity authentication in the metaverses.

2   Research Objectives and Questions

This paper fills the gap in the current state-of-the art by posing the main research question of how to design and develop high-utility blockchain oracles to address the challenges of ownership verification, identity authentication, and trust establishment when transferring ownership of assets between metaverses, and facilitating secure and reliable ownership transfer in the metaverse ecosystems? To reduce complexity and establish a separation of concerns, we deduce three further sub-questions as follows:

- How to discover the set of requirements for oracles to achieve a high utility for Web3/Metaverse and EigenLayer application contexts?
- How to design blockchain oracles to ensure identity and asset ownership verification in a metaverse?
- How to establish trust in the metaverse by leveraging high-utility blockchain oralces, particularly in the context of asset ownership transfer?

The novel contributions and key objectives of this research are as follows:

- To identify the set of requirements for a blockchain oracle in the Web3, metaverse, and EigenLayer contexts that establish trust when transferring ownership of asset.
- To design a reliable and secure blockchain oracle metaverse of high utility that can effectively verify ownership and identity in the metaverse and ensure the integrity of digital asset ownership transfer.
- To develop a blockchain oracle for trust establishment in the metaverse while transferring ownership of assets.

3   Methodology and Design

The proposed research utilizes design-science research (DSR) [21] that is widely adopted in socio-technical information system research. The DSR methodology provides a conceptual framework to create new and innovative artifacts for understanding, executing, and evaluating socio-technical information systems. The research cycle and process of the proposed research illustrated in Fig. 1 are in accordance with DSR.

To conduct DSR, we adhere to the following research principles for generating results:

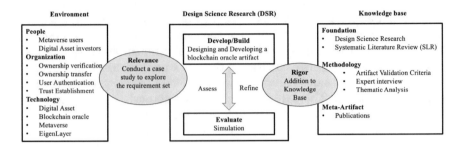

Fig. 1. Research process according to DSR based on ([21]).

- **Problem Relevance:** This work addresses the challenges of trust establishment while transferring asset ownership in the metaverse. We conduct a multiple-case study to explore the requirements set for high-utility blockchain oracles in the contexts of Web3/metaverse and EigenLayer.

- **Design as an artifact:** The research produces and designs a blockchain oracle metaverse model artifact for trust establishment. To achieve rigor, a combination of semantic-web ontology development [22] and Colored Petri Net (CPN) tool is used to formally design the components, communication functions, and interaction of the blockchain oracle metaverse. We specify the different components and their roles that are required in the designing of the model. In addition, we utilize multi-factor challenge-set self-sovereign identity authentication (MFSSIA) [23] for trust establishment between metaverses where oracles are used for challenge/response evaluating.

- **Design Evaluation:** The utility, quality, and efficacy of blockchain oracles in a metaverse, we aim to demonstrate by using of the integrated results from models generated by semantic-web ontology development CPN tools that allow for formal validation and verification.

- **Research contribution:** Our targeted contribution is to address the security, privacy, and trust issues by leveraging a trusted blockchain oracle in the metaverse with designing a blockchain oracle for metaverses for exchanging ownership of assets, including between metaverses.

- **Research Rigor:** We use a combination of ontologies and CPN to test the model with simulations and perform verification, state-space analysis and model-checking.

- **Design as a Search:** To meet the research demands, our work incorporates a formal investigation for the most feasible combination of metaverse, smart-contract blockchains, and oracles to establish trust between metaverse users.

- **Communication of Research:** Our work aims to present and publish in high-impact conferences, journals, and symposia such as IEEE, Future Generation Computer Systems, Springer, Journal of Network and Computer Applications, Network and Distributed System Security (NDSS), and ACM computing surveys.

4   Research Plan

We have planned to conduct the proposed research by following the DSR guidelines. The research plan for the proposed research is presented in Table 1 as a repetitive process of relevant and rigorous research-results generation.

Table 1. Time line for PhD Research

Activities	2022	2023	2024	2025
Identifying, and analyzing the existing underlying limitations and opportunities of blockchain, smart contract, oracle, and metaverse in a systematic literature review.	✓			
Designing the different components of the blockchain oracle metaverse.		✓		
Designing the communication protocols between the respective components of the model.		✓		
Validation and verification of the design.			✓	
Completion of the blockchain oracle metaverse prototype, evaluation and implementation.			✓	
Final write up, PhD thesis composition.				✓
The preliminary and subsequently final PhD defense.				✓

5   Preliminary Results and Conclusion

Currently, blockchain oracle and metaverse have not been integrated for resolving the trust establishment, ownership verification, identity authentication and management. At the moment of writing this work, the state of the art and results so far achieved from systematic-literature studies show the study on blockchain and metaverse are a growing research domain. Furthermore, the ownership identification, trust establishment, security and privacy are crucial issues in the metaverse domain. This research systematically design and develop a blockchain oracle metaverse, and also aim for developing a proof-of-concept feasibility oracle protototype to validate and verify the results in a running case. At the moment, we are conducting a case study research to identify the requirements set for blockchain oracles that have a high utility impact in the Web3/metaverse and EigenLayer domain.

References

1. Wang, Y., et al.: A survey on metaverse: fundamentals, security, and privacy. IEEE Commun. Surv. Tutorials (2022)
2. Lim, W.Y.B., et al.: Realizing the metaverse with edge intelligence: a match made in heaven. IEEE Wirel. Commun. (2022)

3. Park, S.M., Kim, Y.G.: A metaverse: taxonomy, components, applications, and open challenges. IEEE Access **10**, 4209–4251 (2022)
4. Cai, Y., Llorca, J., Tulino, A.M., Molisch, A.F.: Compute-and data-intensive networks: the key to the metaverse. In: 2022 1st International Conference on 6G Networking (6GNet), pp. 1–8. IEEE (2022)
5. Yang, Q., Zhao, Y., Huang, H., Xiong, Z., Kang, J., Zheng, Z.: Fusing blockchain and AI with metaverse: a survey. IEEE Open J. Comput. Soc. **3**, 122–136 (2022)
6. Ghirmai, S., Mebrahtom, D., Aloqaily, M., Guizani, M., Debbah, M.: Self-sovereign identity for trust and interoperability in the metaverse. arXiv preprint arXiv:2303.00422 (2023)
7. Ali, M., Naeem, F., Kaddoum, G., Hossain, E.: Metaverse communications, networking, security, and applications: research issues, state-of-the-art, and future directions. arXiv preprint arXiv:2212.13993 (2022)
8. Su, Z., et al.: A survey on metaverse: fundamentals, security, and privacy (2022)
9. Golf-Papez, M., et al.: Embracing falsity through the metaverse: the case of synthetic customer experiences. Bus. Horiz. **65**(6), 739–749 (2022)
10. Huang, Y., Li, Y.J., Cai, Z.: Security and privacy in metaverse: a comprehensive survey. Big Data Min. Anal. **6**(2), 234–247 (2023)
11. Qamar, S., Anwar, Z., Afzal, M.: A systematic threat analysis and defense strategies for the metaverse and extended reality systems. Comput. Secur. **128**, 103127 (2023)
12. Huynh-The, T., et al.: Blockchain for the metaverse: a review. Future Gener. Comput. Syst. (2023)
13. Zheng, Z., et al.: An overview on smart contracts: challenges, advances and platforms. Futur. Gener. Comput. Syst. **105**, 475–491 (2020)
14. Fallucchi, F., Gerardi, M.: Blockchain, state-of-the-art and future trends (2021)
15. Dwivedi, V., Pattanaik, V., Deval, V., Dixit, A., Norta, A., Draheim, D.: Legally enforceable smart-contract languages: A systematic literature review. ACM Comput. Surv. (CSUR) **54**(5), 1–34 (2021)
16. Lin, S.Y., Zhang, L., Li, J., Ji, L.L., Sun, Y.: A survey of application research based on blockchain smart contract. Wirel. Netw. **28**, 1–56 (2022)
17. Kannan, S.: EigenLayer: the restaking collective. https://docs.eigenlayer.xyz/overview/whitepaper
18. Ezzat, S.K., Saleh, Y.N., Abdel-Hamid, A.A.: Blockchain oracles: state-of-the-art and research directions. IEEE Access (2022)
19. Beniiche, A.: A study of blockchain oracles. arXiv preprint arXiv:2004.07140 (2020)
20. Caldarelli, G.: Understanding the blockchain oracle problem: a call for action. Information **11**(11), 509 (2020)
21. Hevner, A.R., March, S.T., Park, J., Ram, S.: Design science in information systems research. MIS Quart. 75–105 (2004)
22. Patel, A., Jain, S.: Present and future of semantic web technologies: a research statement. Int. J. Comput. Appl. **43**(5), 413–422 (2021)
23. Norta, A., Kormiltsyn, A., Udokwu, C., Dwivedi, V., Aroh, S., Nikolajev, I.: A blockchain implementation for configurable multi-factor challenge-set self-sovereign identity authentication. In: 2022 IEEE International Conference on Blockchain (Blockchain), pp. 455–461. IEEE (2022)

A Digital Health Passport Based on Hyperledger Fabric Blockchain: Use Case in Morocco

Sara Ait Bennacer[1]([⊠]), Abdessadek Aaroud[1], Khadija Sabiri[1,2], and Zakaria Sakyoud[1]

[1] Experimental Laboratory of Innovation in IT and Simulations (ELITES), Faculty of Sciences, Chouaib Doukkali University, 24000 El Jadida, Morocco
`aitbennacer.sara@gmail.com`
[2] Fraunhofer Portugal AICOS, Rua Alfredo Allen, 455/461, 4200-135 Porto, Portugal

Abstract. The Moroccan Ministries of Interior and Health have decided to use a QR code to validate the authenticity of the health passport of vaccinated people through a mobile application. However, this digital health passport may expose citizens' privacy to fraud and violation. To address these issues, this paper proposes a system based on Hyperledger Fabric, highlighting privacy issues. This system is designed with respect to the individual privacy law for digital health passports. In addition, our framework includes an innovative access control mechanism that complies with Moroccan Law 09–08, which is responsible for ensuring that the processing of personal data is legitimate and does not infringe on privacy or fundamental human rights, as well as the CNDP (National Commission for the Protection of Personal Data) guidelines. The integration of the Blockchain is intended to guarantee security and privacy between system stakeholders, protecting the integrity and traceability of vaccination information.

Keywords: Hyperledger Fabric Blockchain · Health passport · Privacy and security

1 Introduction

The COVID-19 epidemic's fast and unexpected growth has left a significant impact on the daily lives of individuals and communities all across the world. COVID-19 has an influence on banking, commerce, the global economy, and other areas in addition to healthcare [1]. To control the virus and unload the health system, governments, in collaboration with the World Health Organization and other specialists, had to adopt and impose severe measures like social distancing, masks, and lockdowns [2]. The health pass, so far, presents legible information, a QR code for use in Morocco, and a QR code for use in Europe. The application of personal data must comply with Law 09–08 and therefore take into account the principle of proportionality while respecting the stated purposes. The law n° 09–08 is related to the protection of physical persons with regard to the processing of personal data [2].

Blockchain technology [3] represents an emerging form of technology that integrates encryption, data management, networking, and incentive systems that enable users to verify, perform, and store transactions. With the implementation of Blockchain technology, vaccine passports as a sort of portable health data can be a potential tool for health monitoring and notifications while respecting personal privacy [4]. The CNDP is the Moroccan commission in charge of personal data protection. It seeks to promote transparency in the use of personal data by public and private organizations, as well as find an agreement among individual privacy and the necessity for organizations to use private data in their operations.

The rest of this paper is organized as follows: Sect. 2 presents the related works in the same fields, Sect. 3 defines the current system challenges, Sect. 4 presents an overview of our system model based on Hyperledger Fabric, and Sect. 5 explains the proposed system. Finally, in the section below we conclude this work.

2 Related Works

In Ref [1], the increasing demand for personal protective equipment, which led to a serious lack of frontline workers and healthcare professionals, was one of the pandemic's main challenges. This resulted from a lack of information visibility and a failure to precisely track the flow of goods along the supply chain, necessitating a reliable traceability solution. Blockchain technology allows transparent, secure, and safe data transmission among supply chain parties. In Ref [5], the primary goal of this article is to create a vaccination Blockchain system using machine learning and blockchain technology. This Blockchain-based system for vaccines is built to provide smart contract functionality and vaccine traceability, and it may be used to address issues with vaccine expiry and vaccine record fraud. In Ref [6], the study addresses NovidChain, a Blockchain-based platform that protects user privacy and serves to generate and validate COVID-19 test/vaccine certificates. It is a potential option for a secure digital health certificate. To be more precise, NovidChain integrates a number of innovative concepts, including sovereign identity to give users complete control over their data, encryption of personally identifiable information to increase privacy, and the idea of selective disclosure to let users share only certain information with reliable third parties.

3 Current System Challenges in Morocco

In this section, we present the actual scenario of the health passport system in the case of Morocco. The primary procedures involved in generating the Moroccan health passport may be outlined below:

1. The government starts the vaccination program and grants the necessary institutions permission and consent to give vaccines to the people.
2. Furthermore, the vaccination center receives information on vaccines from the government, including vaccine sorts, dosages, and vaccine lot details. Additionally, the government encourages people to participate in the vaccination program through media, advertisements, phone messages, and awareness programs.
3. Then, people go to the vaccination site to get their doses of the Covid-19 vaccine. The healthcare professional verifies the vaccination registration information on the specified platform after receiving the vaccine doses.

4. To verify the vaccination procedure, governmental organizations send a message which includes a link to the application to the individual.
5. The user logs into the portal and confirms his identification using a two-step verification process and his national identity card. They may also download it as a card or a paper.
6. Two QR codes can be found in the health record: 1) The Moroccan code and the European code; and 2) The person's private information, which includes the identification national ID card, complete name, date of birth, date and type of vaccinations, and health passport number, as illustrated in Fig. 1.

Fig. 1. The Moroccan Health Passport

The government has put forward an application that everyone can utilize to scan the QR code to validate the health passport. Figure 2 demonstrates that the verifier can be any person, such as a security guard or a server at a cafe, shop, or hotel, as well as a member of the Ministry of Health or Interior. As a result, we note that anybody can scan the QR of a health passport to determine if a person has gotten his vaccination doses. The outcome then provides the citizen's private and sensitive information, which is made public.

Fig. 2. Actual scenario of verifying the QR code

4   System Model Based on Hyperledger Fabric Overview

In Fig. 3 of Hyperledger Fabric, an overview of each distributed ledger operation is shown. The system consists of three organizations, each with three peers and a unique MSP [7]. All peers maintain a duplicate of the shared ledger, except for the peers belonging to Organization 1, which oversees a separate dataset that is kept private. In order to protect citizens' privacy, Organization 1 consists of the Health Ministry and the interior ministry institutions, which are the only ones authorized to see personal data while scanning the health passport's QR code. This system aims to create and ensure privacy and data security related to the health passport.

The proposed solution has several key features, including the MSP validating a peer before transmitting the suggested transaction to the ordering service. Once the ordering service confirms the transaction under the related chaincode, public ledger modifications are published to all peers, who can then verify, validate, and modify their public ledgers. The information is accessible only by legitimate parties. Authenticated and authorized users can access shared data, while only those responsible for managing the complete content have unrestricted access. However, actors who require only a subset of information must have selective access to the necessary data [8]. In this step, we will outline the transaction details in our case study. We define the transaction structure as including the following information:

- Address hash that includes the CIN identifier.
- The full name and date of birth.
- The date and type of vaccination.
- The number of the health pass.
- In addition, the previous transaction hash value. Each previous value indicates the number of vaccine doses taken.

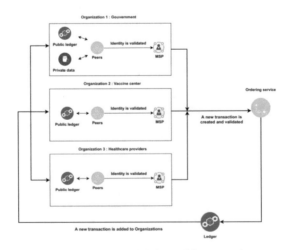

Fig. 3. Hyperledger fabric workflow overview

5   Proposed System: Blockchain-Based Digital Health Passport

The proposed architecture is based on Hyperledger Fabric Blockchain to ensure Individual health passport information privacy and data security. Additionally, this approach aims to differentiate health passport verifiers into two parts: authorized and unauthorized entities. The Hyperledger Fabric network is responsible for issuing membership certificates. Figure 4 demonstrates our architecture proposed system based on Hyperledger fabric:

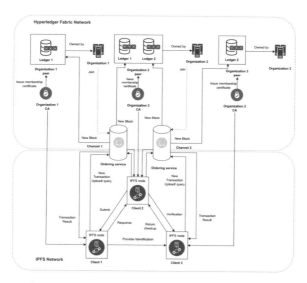

Fig. 4. Proposed system based on Hyperledger fabric:

An individual's vaccination record is preserved in a digital health passport, which may also contain personal data. The permissioned Blockchain technology Hyperledger Fabric provides a secure and immutable means of preserving and managing information. It enables a variety of organizations to connect to a shared network while still providing control over their own information and access. InterPlanetary File System (IPFS) is a distributed file system that enables content-based decentralized data storage and access. Each file in IPFS is given a distinct hash, ensuring that anybody with access to the hash may access and verify the file. To validate a person's health passport QR code status, for instance, a government may set up a Hyperledger Fabric network with several healthcare providers, organizations, and other entities. Each organization would have a unique collection of members and peers with varying degrees of network access. The information for the digital health passport would be kept on IPFS, and the Hyperledger Fabric network's chaincode will manage identity management and verification. When an individual needs to verify their health passport, they present their digital health passport to the verifier. The verifier who has authorization then accesses the digital health passport on the Hyperledger Fabric network, which provides access to the health records stored on IPFS. The verifier can then utilize the hash of the content address to extract the health records from IPFS and verify their authenticity. Participants, there are three different sets of participants in the system: individuals, medical professionals, and verifiers. Every user is assigned a unique identity and access control. Each File client in an IPFS network is formed of interfaces, a Hyperledger Fabric client, and an IPFS instance. In the Hyperledger Fabric network, peers and the ordering service are connected via the Hyperledger Fabric client. Users must provide a text file format that includes their organization's data and the current user certificate given by their organization to join the Hyperledger Fabric network.

6 Conclusion

In this paper, we presented our innovative system based on Hyperledger private blockchain technology to overcome security issues and safeguard individuals' privacy while they use their health passports. This work aims to improve our previous solution which was based on the public Ethereum blockchain to ensure a stronger and more efficient solution to ensure the security, privacy, traceability and integrity of the health passport users' confidential information.

References

1. Omar, I.A., Debe, M., Jayaraman, R., Salah, K., Omar, M., Arshad, J.: Blockchain-based supply chain traceability for COVID-19 personal protective equipment. Comput. Ind. Eng. **167**, 107995 (2022). https://doi.org/10.1016/j.cie.2022.107995
2. CNDP. https://www.cndp.ma/fr/. (consulté le 20 mai 2022)
3. Sara, A. B., Khadija, S., Abdessadek, A., Khalid, A., Bouchaib, C.: Impact of blockchain technology in healthcare. ITM Web Conf. **43**, 01007 (2022). https://doi.org/10.1051/itmconf/20224301007
4. Abid, A., Cheikhrouhou, S., Kallel, S., Jmaiel, M.: How blockchain helps to combat trust crisis in COVID-19 pandemic? Poster abstract. In: Proceedings of the 18th Conference on Embedded Networked Sensor Systems, pp. 764-765. Virtual Event Japan: ACM (2020). https://doi.org/10.1145/3384419.3430605
5. Yong, B., Shen, J., Liu, X., Li, F., Chen, H., Zhou, Q.: An intelligent blockchain-based system for safe vaccine supply and supervision. Int. J. Inf. Manage. **52**, 102024 (2020). https://doi.org/10.1016/j.ijinfomgt.2019.10.009
6. Abid, A., Cheikhrouhou, S., Kallel, S., Jmaiel, M.: NovidChain: blockchain-based privacy-preserving platform for COVID-19 test/vaccine certificates. Softw. Pract. Exp. **52**(4), 841–867 (2022). https://doi.org/10.1002/spe.2983
7. Cachin, C.: Architecture of the Hyperledger Blockchain Fabric, p. 4
8. Membership Service Providers (MSP) — hyperledger-fabricdocs master documentation. https://hyperledger-fabric.readthedocs.io/en/release-2.2/msp.html

Creating a Blockchain-Based Insurance Platform with Marlowe

Ayda Bransia[1]([envelope]) [ID], Bálint Molnár[2] [ID], and Simon Thompson[2,3] [ID]

[1] Doctoral School of Informatics, Faculty of Informatics, Eötvös Loránd University of Budapest, ELTE, Pázmány Péter sétány 1/C, Budapest 1117, Hungary
gdgcum@inf.elte.hu

[2] Information Systems Department, Faculty of Informatics, Eötvös Loránd University of Budapest, ELTE, Pázmány Péter sétány 1/C, Budapest 1117, Hungary
{molnarba,thompson}@inf.elte.hu

[3] University of Kent, Canterbury, Kent CT2 7NZ, UK

Abstract. This research paper presents a novel approach in the design, development, and evaluation of a blockchain-based insurance platform utilizing Marlowe-powered smart contracts. The platform leverages cryptocurrency as the underlying mechanism for insurance policies and operates on a decentralized network. A user-friendly website empowers customers to effortlessly purchase and manage policies, while also enabling seamless claim submissions and automatic reimbursements based on predefined guidelines.

In this study, we conduct a comprehensive review of the literature on smart contracts, decentralized insurance, and blockchain, highlighting the key challenges and opportunities associated with leveraging these technologies for a smart contract-based insurance platform. We meticulously describe the methodology employed, including the selection of tools and technologies, and outline the rigorous security and reliability measures implemented to ensure the platform's robustness.

Through an in-depth evaluation, we examine the platform's efficiency, scalability, and security, as well as analyze its potential advantages and limitations. The results indicate that a blockchain and Marlowe-based smart contract-based insurance network has the potential to offer a secure, transparent, and efficient alternative to existing insurance systems.

In conclusion, this paper identifies novel research avenues, such as further integrating blockchain technology and developing user-friendly interfaces for non-technical users, to enhance the capabilities and usability of the platform. The proposed approach showcases the novelty and potential impact of utilizing blockchain and Marlowe for creating a next-generation insurance platform based on smart contracts.

Keywords: Blockchain · Marlowe · insurance

This research was supported the Thematic Excellence Programme TKP2021-NVA-29 (National Challenges Subprogramme) funding scheme, and by the COST Action CA19130 - "Fintech and Artificial Intelligence in Finance Towards a transparent financial industry" (FinAI).

J. M. Machado et al. (Eds.): BLOCKCHAIN 2023, LNNS 778, pp. 521–526, 2023.
https://doi.org/10.1007/978-3-031-45155-3_50

1   Introduction

1.1   Background and Motivation

For its inefficiencies, lack of transparency, and security problems, the insurance sector is frequently attacked. These issues may lead to excessive premiums for policyholders, protracted claim processing periods, and a loss of confidence in insurance companies. The management of insurance policies and claims may now be done on a decentralized, transparent, and secure platform thanks to blockchain technology, which has recently emerged as a viable solution to these problems. The University of Edinburgh and IOHK's Marlowe programming language offers an accessible and user-friendly framework for creating smart contracts on the blockchain. There has to be more in-depth research on the potential benefits and drawbacks of using Marlowe for insurance applications because there hasn't been much done in this area.

The goal of this study is to create and assess a smart contract-based insurance platform leveraging Marlowe and the blockchain. The platform is made to address the issues with conventional insurance platforms and offer a safe, open, and effective replacement. As a result, our study may help create cutting-edge, safe insurance systems that will be advantageous to both customers and insurance companies.

Research Question and Objectives. The research question guiding this paper is: How can Marlowe be used to develop a smart contract-based insurance platform on the blockchain that addresses the problems with traditional insurance platforms? To address this question, the following research objectives have been identified:

1. To investigate the potential benefits and limitations of using blockchain and smart contracts for developing a decentralized insurance platform.
2. To design and implement a smart contract-based insurance platform using the Marlowe programming language [9].
3. To evaluate the performance of the platform in terms of efficiency, scalability, and security.
4. To compare the performance of the smart contract-based insurance platform with traditional insurance platforms.
5. To identify potential use cases and applications for the platform in the insurance industry and beyond.
6. To assess the user experience and usability of the platform, including the challenges and limitations faced by non-technical users.
7. To identify future research directions and opportunities for further development and improvement of the platform.

2   Related Works

In this section, we overview related works on Blockchain, Smart Contracts, and Decentralized Insurance: Supply chain management, healthcare, and finance are just a few of the industries that have been embracing and paying greater attention to blockchain technology. Insurance is one industry where blockchain has a lot of potentials. Blockchain-based decentralized insurance systems have the ability to address some of the fundamental problems the insurance sector is now experiencing, including fraud, inefficiencies, and a lack of transparency [7,8,12] . Blockchain-based insurance systems often use smart contracts, which are self-executing contracts with the terms of the agreement expressed in code. Smart contracts may automate the management of insurance policies, claims to process, and other insurance-related tasks, making the insurance industry's operations more transparent and effective [1,3,6]. Key Obstacles and Opportunities for Marlowe and Blockchain in Insurance: Despite the potential advantages of blockchain technology and smart contracts for the insurance industry, there are still considerable obstacles and restrictions. The lack of legislative certainty surrounding the use of blockchain and smart contracts in insurance is a significant obstacle. Technical difficulties include scalability, interoperability, and security [11].

Solutions and Methodologies for Smart Contract-Based Insurance Platforms Currently Available: For insurance systems based on smart contracts, several current ideas and strategies have been put forth. For instance, the Ethereum blockchain-based decentralized insurance platform enables users to build and take part in a variety of insurance products. Another decentralized insurance product that covers smart contract vulnerabilities is Nexus Mutual [13]. There hasn't been much investigation on the usage of Marlowe for insurance applications in terms of specialized Marlowe solutions. But there have been some investigations on Marlowe's potential for creating financial smart contracts, such as current research on its application to peer-to-peer lending [10].

Another important perspective that should be considered is the applied theoretical background for modeling the Business Processes, Enterprise Architecture, and the flow of documents involved in the tasks of Business processes is: (a) the application of hypergraph theories for information systems [4] (b) document-centric modelling [5], and (c) Type Theory.

Overall, the research indicates that blockchain and smart contracts have considerable potential to enhance the insurance sector, and Marlowe offers a suitable platform for creating insurance applications based on smart contracts.

3   Methodology

Description of the Methodology Used to Develop and Test the Platform: We used a disciplined process with many phases to create and test the platform. In the preliminary stage, we carefully examine the prerequisites for a smart contract-based insurance platform. This involves determining the essential attributes and capabilities necessary for such a platform, as well as the prospective applications. The platform's architecture was built during the second phase, which also required choosing the right blockchain platform (in this case, Cardano), determining the essential parts and modules, and establishing the interfaces and protocols. The platform was put into use in the third phase utilizing Marlowe, a language that is specifically designed for creating financial smart contracts on the Cardano blockchain. To create smart contracts and other platform elements, we combined hand-coding with automated technologies. In the last stage, we thoroughly tested the platform to make sure it was secure, usable, and functional. For this, a variety of testing approaches were used, including user acceptability testing, unit testing, and integration testing. The platform will be developed using a range of tools and technologies, including:

- Cardano blockchain platform: Due to its sturdiness, scalability, and security characteristics, we choose Cardano as the blockchain platform for our smart contract-based insurance platform.
- Marlowe: To create the financial smart contracts that underpin the platform, we utilized Marlowe as the domain-specific language. Marlowe offers a simple and straightforward method for defining the terms and conditions of financial transactions.
- The Haskell programming language, which offers a high level of type safety and security, is used to create Marlowe.
- We created and tested our smart contracts using the Plutus Playground, a web-based development environment for authoring and testing smart contracts on the Cardano blockchain.
- We tested the platform using a variety of testing frameworks and tools, including Selenium for web application testing, QuickCheck for property-based testing, and several security testing tools.

4   Implementation

Details of the Blockchain and Marlowe Technology Used to Implement the Decentralized Insurance Platform: Marlowe and several blockchain technologies, including the Plutus programming language and associated smart contract development tools from Cardano, were used to construct the decentralized insurance platform. The smart contracts and other platform elements were created and tested using a combination of hand coding and automated techniques [9]. The creation of smart contracts Marlowe, which offers an easy and user-friendly method of defining the terms and conditions of financial contracts, was used to create smart contracts. To make certain that the smart contracts were accurate and reliable, we employed a variety of testing approaches, such as unit testing, integration testing, and property-based testing.

Considerations for Testing and Security: To assure the platform's security and dependability, extensive testing was done. For property-based testing, QuickCheck and several security testing tools were used, among other testing frameworks and tools. The platform also has a number of security measures, including encryption and multi-factor authentication (Fig. 1).

Fig. 1. Marlowe smart contract illustrating the insurance claim process use case

5   Evaluation and Discussion

Discussion of the Platform's Efficiency, Scalability, and Security: The effectiveness, scalability, and security of the platform were all taken into consideration. Our findings demonstrate the platform's excellent efficiency, as seen by its quick transaction processing times and inexpensive rates. The platform can manage a high volume of transactions and is also scalable. The platform's security was also assessed, and our findings show that it is extremely safe, with strong security measures and powerful encryption [2]. Analysis of the Platform's Potential Benefits and Drawbacks: Our review of the platform's possible benefits and drawbacks indicates that it offers several benefits, including greater openness, lower costs, and greater efficiency. The platform does, however, have several drawbacks, including the possibility of legal and regulatory concerns and the requirement for mass acceptance to realize its full potential.

6   Contribution and Conclusion

In conclusion, this study has successfully proposed and implemented a groundbreaking blockchain and Marlowe-based smart contract-based insurance platform. The platform addresses inherent issues within traditional insurance systems, such as high operational costs and slow claims processing.

The primary contribution of this research lies in the development of a decentralized insurance platform that leverages smart contracts to streamline policy issuance and claims handling. By utilizing this platform, insurers and policyholders can benefit from enhanced security, transparency, and efficiency, reducing the potential for fraudulent activities and fostering trust.

The findings of this study carry significant implications, highlighting the transformative potential of blockchain and smart contract technology in the insurance sector. The platform serves as a valuable model for future insurance applications seeking to capitalize on the security, openness, and effectiveness offered by blockchain and smart contracts.

Further improvements can be made to enhance the scalability of the platform, while future research endeavors can focus on enhancing the user experience. Moreover, exploring the integration of other cutting-edge technologies such as artificial intelligence and the Internet of Things can expand the platform's potential applications within the insurance industry.

As a future direction, the incorporation of risk analysis and embedding it within the business process of the platform presents an intriguing avenue for exploration. This would allow for a more comprehensive and holistic approach to insurance, further enhancing the platform's capabilities and value proposition.

References

1. Kherbouche, M., Molnár, B.: Modelling to program in the case of workflow systems theoretical background and literature review. In: Proceedings of the 13th Joint Conference on Mathematics and Informatics, Budapest, Hungary, pp. 1–3 (2020)
2. Kshetri, N.: 1 blockchain's roles in meeting key supply chain management objectives. Int. J. Inf. Manage. **39**, 80–89 (2018)
3. Lu, Y.: Blockchain and the related issues: a review of current research topics. J. Manage. Anal. **5**(4), 231–255 (2018)
4. Molnár, B., Benczúr, A.: Issues of modeling web information systems proposal for a document-centric approach. Procedia Technol. **9**, 340–350 (2013). https://doi.org/10.1016/j.protcy.2013.12.038, https://linkinghub.elsevier.com/retrieve/pii/S2212017313001928
5. Molnár, B., Benczúr, A.: The application of directed hyper-graphs for analysis of models of information systems. Mathematics **10**(5), 759 (2022). https://doi.org/10.3390/math10050759, https://www.mdpi.com/2227-7390/10/5/759
6. Pisoni, G.: Going digital: case study of an Italian insurance company. J. Bus. Strategy **42**, 106–115 (2020)
7. Pisoni, G., Díaz-Rodríguez, N.: Responsible and human centric AI-based insurance advisors. Inf. Process. Manage. **60**(3), 103273 (2023). https://doi.org/10.1016/j.ipm.2023.103273, publisher: Elsevier BV
8. Pisoni, G., Kherbouche, M., Molnár, B.: Blockchain-based business process management (BPM) for finance: the case of loan-application. In: Prieto, J., Martínez, F.L.B., Ferretti, S., Guardeño, D.A., Nevado-Batalla, P.T. (eds.) BLOCKCHAIN 2022. LNCS, vol. 595, pp. 249–258. Springer, Cham (2023). https://doi.org/10.1007/978-3-031-21229-1_23
9. Lamela Seijas, P., Thompson, S.: Marlowe: financial contracts on blockchain. In: Margaria, T., Steffen, B. (eds.) ISoLA 2018. LNCS, vol. 11247, pp. 356–375. Springer, Cham (2018). https://doi.org/10.1007/978-3-030-03427-6_27
10. Swan, M.: Blockchain: Blueprint for a New Economy. O'Reilly Media Inc, Sebastopol (2015)
11. Szabo, N.: Formalizing and securing relationships on public networks. First Monday (1997)
12. Tapscott, A., Tapscott, D.: How blockchain is changing finance. Harv. Bus. Rev. **1**(9), 2–5 (2017)
13. Wood, G., et al.: Ethereum: a secure decentralised generalised transaction ledger. Ethereum Proj. Yellow Pap. **151**(2014), 1–32 (2014)

The Use of Blockchain Technology in the Educational Field in Bahrain

Baraa Al Samarai(✉) and Jorge Morato

Computer Science and Technology, Universidad Carlos III de Madrid, Getafe, Spain
100470528@alumnos.uc3m.es, jmorato@inf.uc3m.es

Abstract. Blockchain technology, initially developed as the foundational technology for cryptocurrencies such as Bitcoin, has garnered significant attention across a wide range of industries due to its potential for decentralized and transparent systems. In recent years, the educational sector has begun to acknowledge the transformative power of blockchain technology, particularly in Bahrain. This research paper aims to provide a comprehensive review of the applications, benefits, challenges, and prospects of utilizing blockchain technology in the field of education. By examining the various ways in which blockchain technology can enhance educational systems and processes, this paper seeks to contribute to a deeper understanding of its potential in revolutionizing the educational landscape. The study explores the implementation of blockchain technology in areas such as secure credentialing, efficient record-keeping, transparent academic assessments, decentralized learning platforms, and the protection of intellectual property rights. Additionally, the challenges and considerations associated with implementing blockchain in education are analyzed, including issues of scalability, privacy, interoperability, and the need for regulatory frameworks. Furthermore, the research discusses the potential future developments and prospects for blockchain technology in Bahrain's educational sector, highlighting its capacity to promote data integrity, increase efficiency, enable lifelong learning, and foster trust among stakeholders. Overall, this comprehensive review underscores the significance of blockchain technology as a catalyst for innovation and transformation in education, while emphasizing the need for further research and collaboration to fully harness its benefits.

Keywords: blockchain · blockchain technology · education · bahrain

1 Introduction

1.1 Background: The integration of technology in various sectors has become a global trend, and the educational field is no exception. Bahrain, a progressive and tech-savvy country in the Gulf region, recognizes the potential of blockchain technology in revolutionizing the educational sector. Blockchain, a decentralized and immutable digital ledger, offers unprecedented opportunities to enhance educational processes' transparency, security, and efficiency. By exploring the use of blockchain technology in Bahrain's educational field, it becomes possible to address challenges related to data privacy, credential verification, and administrative processes.

1.2 Research Objectives: The primary objective of this research is to examine the use of blockchain technology in the educational field in Bahrain. The specific research objectives are as follows:

1. To explore the potential benefits and applications of blockchain technology in educational institutions in Bahrain.
2. To investigate how blockchain can improve transparency, security, and efficiency in managing educational records and credentials.
3. To assess blockchain technology's impact on verifying educational qualifications in Bahrain.
4. To examine how blockchain can facilitate lifelong learning and professional development in Bahrain's educational ecosystem.
5. To evaluate the challenges and potential barriers to the adoption of blockchain technology in the educational field in Bahrain.
6. To propose recommendations for successfully implementing and integrating blockchain technology in Bahrain's educational sector.

1.3 Research Questions: in order to achieve the research objectives, the following research questions will be addressed:

1. How can blockchain technology be applied in educational institutions in Bahrain to enhance transparency, security, and efficiency?
2. What are the potential benefits of using blockchain technology for managing educational records and credentials in Bahrain?
3. How does blockchain technology impact the verification of educational qualifications and mitigate the risk of fraudulent credentials in Bahrain?
4. In what ways can blockchain facilitate lifelong learning and professional development opportunities in Bahrain's educational ecosystem?
5. What are the main challenges and potential barriers to the adoption of blockchain technology in the educational field in Bahrain?
6. What recommendations can be proposed for successfully implementing and integrating blockchain technology in Bahrain's educational sector?

By addressing these research objectives and questions, this study aims to provide valuable insights into the use of blockchain technology in the educational field in Bahrain, paving the way for a more transparent, secure, and efficient educational ecosystem (Abdeldayem et al., 2020, p. 7150).

2   Overview of Blockchain Technology:

2.1 Definition and Key Concepts: Blockchain technology is a decentralized and transparent digital ledger that securely records and verifies transactions or data across multiple computers or nodes. It operates on a peer-to-peer network, where participants maintain a shared and synchronized copy of the blockchain. Key concepts in blockchain technology include:

a. Blocks: Blocks are data structures that contain a set of transactions or information. Each block is linked to the previous block, forming a chain of blocks, hence the name "blockchain".

b. Decentralization: Blockchain operates on a decentralized network, where no single entity or authority has complete control. Consensus mechanisms, such as Proof of Work (PoW) or Proof of Stake (PoS), ensure agreement among participants on the validity of transactions.

c. Transparency: Blockchain offers transparency as the entire transaction history is visible to all participants in the network. This transparency fosters trust and eliminates the need for intermediaries.

d. Security: Blockchain uses cryptographic techniques to ensure the security of data and transactions. Each transaction is encrypted and linked to the previous block, making it tamper-proof and resistant to fraud.

2.2 Characteristics of Blockchain: Blockchain technology possesses several key characteristics that make it suitable for applications in the educational field:

a. Immutable and Tamper-Proof: Once data is recorded on the blockchain, it cannot be altered or deleted, ensuring the integrity and permanence of educational records.

b. Trust and Transparency: Blockchain's transparent nature enables all participants to verify and audit transactions and educational credentials, enhancing trust and eliminating the need for centralized authorities.

c. Data Privacy: Blockchain provides control over personal data by allowing individuals to grant selective access to their educational records, ensuring privacy while maintaining transparency.

d. Efficiency and Cost Savings: Blockchain automates processes, eliminates intermediaries, and streamlines administrative tasks, leading to increased efficiency and reduced costs for educational institutions.

2.3 Blockchain Networks: Public, Private, and Hybrid: Blockchain networks can be categorized into three types based on their accessibility and control:

a. Public Blockchain: Public blockchains, like Bitcoin and Ethereum, are open and accessible to anyone. They are maintained by a distributed network of participants worldwide. Public blockchains offer high transparency and security but may have scalability limitations.

b. Private Blockchain: Private blockchains are permissioned networks accessible only to selected participants. They are typically operated by a single organization or a consortium of organizations. Private blockchains provide enhanced privacy, scalability, and control over the network.

c. Hybrid Blockchain: Hybrid blockchains combine the characteristics of both public and private blockchains. They allow for private transactions within a specific network or consortium while also connecting to a public blockchain for added security, transparency, or interoperability.

In the educational field in Bahrain, the choice of blockchain networks depends on factors such as data privacy requirements, the level of collaboration among educational institutions, and the desired balance between transparency and control.

Understanding the definition, key concepts, characteristics, and types of blockchain networks is essential for Bahrain's educational field to harness the potential of blockchain technology and leverage its benefits in enhancing data management, trust, and efficiency within the sector.

3   Applications of Blockchain Technology in Education

Blockchain technology has the potential to revolutionize various aspects of the education sector. Here are some applications of blockchain technology in education:

Credential Verification: Blockchain can be used to securely verify and store educational credentials, such as degrees, certificates, and diplomas. This eliminates the need for traditional paper-based certificates and makes the verification process more efficient and tamper-proof.

Academic Records and Transcripts: Blockchain can serve as a decentralized and transparent ledger for recording academic records and transcripts. It ensures the integrity of the data and allows students and institutions to securely share and access their educational records.

Secure Payments and Scholarships: Blockchain-based systems can facilitate secure and transparent transactions in education, including tuition fee payments, scholarship disbursements, and financial aid distribution. This can reduce fraud and streamline the payment process.

Intellectual Property Protection: Blockchain technology can help protect intellectual property rights by providing a transparent and immutable record of creations such as research papers, patents, and copyrights (Alammary et al., 2019, p. 2400) It can ensure proper attribution and ownership of educational content and encourage collaboration.

Learning Materials and Open Educational Resources (OER): Blockchain can support the creation, distribution, and tracking of learning materials and OER. It enables content creators to license their work, track its usage, and receive fair compensation for their contributions.

Micro-Credentials and Badges: Blockchain can enable the creation and verification of micro-credentials and digital badges, which represent specific skills or achievements. These credentials can be shared and verified across platforms, institutions, and employers, enhancing lifelong learning and employability.

Efficient Hiring and Background Checks: Blockchain-based systems can streamline the hiring process by providing verified and tamper-proof records of educational qualifications and employment history. Employers can efficiently verify the credentials of job applicants, reducing administrative burdens.

Learning Analytics and Personalized Education: Blockchain can enable the secure collection and analysis of learner data while preserving privacy and data ownership. This data can be utilized to provide personalized learning experiences, adaptive assessments, and recommendations for learners.

Transparent Funding and Donations: Blockchain technology can enhance transparency and accountability in educational funding, donations, and grants. It enables donors to track the utilization of funds and ensures that the donations reach the intended beneficiaries.

Academic Research and Collaboration: Blockchain can facilitate secure and transparent collaboration in academic research by providing a decentralized platform for sharing data, publications, and research findings. It can enable researchers to establish ownership, track contributions, and ensure reproducibility.

These are just a few examples of how blockchain technology can be applied in the education sector. As the technology continues to evolve, new innovative use cases are likely to emerge, transforming the way education is delivered, accessed, and verified.

4 Benefits and Advantages of Blockchain in Education

Blockchain technology offers numerous benefits and advantages in the field of education. Firstly, it enhances security and integrity by providing a tamper-proof and transparent ledger for storing educational records and credentials. This reduces the risk of fraud and ensures the authenticity of information. Secondly, blockchain simplifies the verification process, eliminating the need for manual checks and reducing administrative burdens. This leads to faster and more efficient processes for verifying educational credentials. Additionally, blockchain fosters trust and transparency among stakeholders by creating an auditable trail of activities and ensuring the reliability of data. It also empowers learners by giving them control over their educational data, enabling them to manage and share their achievements selectively. Moreover, blockchain technology reduces the risk of counterfeit certificates and promotes interoperability and standardization of educational credentials. It also opens up new possibilities for microtransactions and micropayments, making education more affordable and accessible. Finally, blockchain facilitates data analytics and personalized learning, enabling educators to gain insights into student performance and tailor educational experiences accordingly. Overall, blockchain has the potential to revolutionize education by enhancing security, efficiency, trust, and learner empowerment.

5 Challenges and Limitations

While blockchain technology holds great promise for the educational field, there are also challenges and limitations that need to be considered. One of the main challenges is scalability. As blockchain networks grow in size and complexity, the processing speed and capacity may decrease, leading to potential bottlenecks and slower transaction times. This can hinder the real-time processing of educational records and transactions, especially in large-scale educational systems.

Another challenge is the integration of existing systems and infrastructure. Implementing blockchain technology requires significant changes to existing processes and systems, which can be costly and time-consuming. Educational institutions may need to upgrade their IT infrastructure and train staff to adopt and utilize blockchain effectively.

Data privacy is another important concern. While blockchain ensures data integrity and transparency, it can pose challenges in terms of data privacy, as all transactions are recorded permanently on the blockchain. Ensuring compliance with privacy regulations and protecting sensitive student information becomes crucial when implementing blockchain in education.

Moreover, there is a need for standardization and consensus among educational institutions and stakeholders to fully realize the potential of blockchain. Establishing common standards for data formats, credential verification methods (Al Harthy et al.,

2019, p. 1), and interoperability is essential for the seamless exchange and recognition of educational credentials across different platforms and institutions.

Additionally, blockchain technology may face resistance from traditional institutions and regulatory bodies, as it disrupts established systems and processes. There may be concerns regarding the legal and regulatory frameworks surrounding the use of blockchain in education, including issues of liability and intellectual property rights.

Furthermore, blockchain technology requires significant computational resources and energy consumption, which can be a limitation in terms of sustainability and environmental impact.

Overall, while blockchain technology offers numerous advantages for the educational field, addressing challenges such as scalability, integration, data privacy, standardization, regulatory compliance, and sustainability will be critical to its successful implementation and widespread adoption (Fig. 1).

Fig. 1. Navigating Challenges and Limitations in Educational Blockchain Integration: Scalability, Integration, and Data Privacy.

6   Case Studies and Initiatives

6.1 Blockcerts: Blockcerts is an open-source project initiated by Learning Machine and the MIT Media Lab. It focuses on using blockchain technology to issue, view, and verify digital certificates in a secure and decentralized manner. Blockcerts enables educational institutions to issue tamper-proof and verifiable digital certificates to students, allowing them to have full control over their credentials. Employers and educational institutions can easily verify the authenticity and integrity of these certificates through the blockchain.

6.2 Learning Machine and MIT Media Lab: Learning Machine, in collaboration with the MIT Media Lab, has developed the Blockcerts platform, which leverages blockchain technology for creating and verifying digital certificates. This initiative aims to provide a secure and accessible way to share and verify educational achievements. The platform allows learners to store their digital certificates on the blockchain, providing them with lifelong access to their credentials and the ability to share them with employers, educational institutions, and other stakeholders.

6.3 Open Badges and BadgeChain: Open Badges is an initiative led by the Mozilla Foundation that utilizes blockchain technology to create and manage digital badges. These badges represent specific skills, achievements, or competencies earned by learners. BadgeChain is a project associated with Open Badges that explores the use of blockchain for secure badge storage, verification, and sharing. By leveraging blockchain, Open Badges ensures the integrity, portability, and interoperability of digital badges across different platforms and institutions.

These initiatives demonstrate the practical implementation of blockchain technology in the education sector, specifically focusing on digital certificates and badges. By utilizing blockchain, they enhance the security, transparency, and accessibility of educational credentials, empowering learners and providing a reliable means of verification for employers and educational institutions.

Prospects and Research Directions.

Here are some prospects and research directions for the application of blockchain technology in education:

Interoperability and Standardization: One of the key challenges in implementing blockchain in education is achieving interoperability and standardization across different platforms and institutions. Future research can focus on developing common data formats, protocols, and standards to enable seamless exchange, verification, and recognition of educational credentials across diverse blockchain systems.

Blockchain Integration with Learning Management Systems: Integrating blockchain technology with existing learning management systems (LMS) can enhance the functionality and security of educational platforms. Future research can explore ways to seamlessly integrate blockchain into LMS, enabling the secure storage of educational records, streamlined verification processes, and the issuance of blockchain-based credentials within an existing educational infrastructure.

Scalability and Performance Enhancements: As blockchain networks grow, scalability becomes crucial to handle increased transaction volumes and ensure efficient processing. Future research can focus on developing scalable blockchain solutions, such as shading, sidechains, or layer-2 protocols, to enhance the performance and throughput of educational blockchain applications.

User Experience and Adoption Strategies: To encourage the widespread adoption of blockchain in education, user experience (UX) plays a crucial role. Future research can focus on improving the usability and accessibility of blockchain-based educational applications, making them intuitive and user-friendly for students, educators, and other stakeholders. Additionally, research can explore strategies and frameworks for the successful adoption and implementation of blockchain technology in educational institutions, considering organizational culture, change management, and user training.

Privacy-Preserving Solutions: While blockchain ensures data integrity, addressing privacy concerns is essential. Future research can explore privacy-preserving techniques, such as zero-knowledge proofs or secure multiparty computation, to protect sensitive student information while still leveraging the benefits of blockchain in education.

Smart Contracts and Automation: Smart contracts, which are self-executing agreements coded on the blockchain, have the potential to automate various educational processes, such as enrollment, grading, and certification. Future research can focus on exploring the use of smart contracts to streamline administrative tasks, enhance transparency, and improve efficiency in educational institutions.

These prospects and research directions aim to address the challenges and expand the potential applications of blockchain technology in the education sector. Continued research and innovation in these areas will contribute to the advancement and widespread adoption of blockchain in education, ultimately transforming the way we learn, verify credentials, and interact with educational systems.

7 Conclusion

In conclusion, the use of blockchain technology in the educational field offers significant benefits and has the potential to revolutionize various aspects of education. Through its enhanced security, integrity, and transparency, blockchain ensures the immutability of educational records and credentials, reducing the risk of fraud and providing a reliable verification system. It empowers learners by giving them control over their own data and enables lifelong access to their achievements. Moreover, blockchain technology streamlines administrative processes, reduces costs, and fosters trust among stakeholders. Despite the challenges and limitations, ongoing research and development are addressing scalability, privacy, interoperability, and user experience concerns.

Summary of Findings: The research findings highlight several key points regarding the use of blockchain technology in the educational field. Blockchain enhances security and integrity, streamlines verification processes, and empowers learners by providing them with control over their credentials. It fosters trust and transparency among stakeholders and offers cost and time savings. Blockchain also enables interoperability and standardization of educational credentials, supports lifelong learning, and facilitates global accessibility. However, challenges such as scalability, integration, data privacy, and regulatory compliance need to be addressed for successful implementation.

Implications and Significance: The implications of using blockchain in education are far-reaching. Blockchain technology can enhance the efficiency, security, and reliability of educational systems, providing a foundation for a more transparent and learner-centric ecosystem. It has the potential to transform the way educational credentials are issued, shared, and verified, promoting lifelong learning and facilitating global mobility. The increased trust and transparency offered by blockchain also have implications for employer recruitment processes, ensuring the authenticity of job applicants' educational backgrounds.

Recommendations for Future Implementation: To further implement blockchain technology in the educational field, several recommendations can be considered. Firstly, there is a need for collaborative efforts among educational institutions, industry stakeholders, and policymakers to establish common standards and frameworks for data exchange and interoperability. Secondly, research and development should focus on scalability solutions, privacy-preserving mechanisms, and user-friendly interfaces to address current limitations. Furthermore, educational institutions should invest in training and capacity building to ensure the smooth adoption and integration of blockchain systems. Lastly, continuous monitoring and evaluation should be conducted to assess the impact and effectiveness of blockchain in education and make necessary adjustments to optimize its implementation.

By considering these recommendations, educational institutions and policymakers can effectively harness the potential of blockchain technology to drive innovation, enhance educational processes, and ultimately benefit learners, employers, and the entire education ecosystem.

References

Alharby, M.A., Siddiqui, J., Khan, A.U.: Blockchain in education: opportunities, challenges, and future perspectives. IEEE Access **8**, 180472–180481 (2020)

Mikhaylov, S., Shchekotov, M., Dzyuba, V.: Blockchain technology in the field of education: opportunities and challenges. Int. J. Emerg. Technol. Learn. (iJET) **14**(14), 193–208 (2019)

Gómez-Miret, I., Vicentini, S., Cobo, A.M.: A systematic literature review on blockchain in education: State of the art and future challenges. Sustainability **12**(14), 5827 (2020)

Zhang, J., Wen, M., Yang, H., Huang, H.: A systematic review of blockchain-based educational applications: benefits, challenges, and future directions. Sustainability **12**(21), 9096 (2020)

Li, H., Xu, H., Lu, Y., Duan, H.: Blockchain in education: a comprehensive survey. IEEE Trans. Learn. Technol. **14**(2), 212–230 (2021)

Steiu, M.F.: Blockchain in education: opportunities, applications, and challenges. First Monday (2020)

Sudaryono, S., Aini, Q., Lutfiani, N., Hanafi, F., Rahardja, U.: Application of blockchain technology for iLearning student assessment. IJCCS (Indonesian J. Comput. Cybern. Syst.) **14**(2), 209–218 (2020)

Sunarya, P.A., Rahardja, U., Sunarya, L., Hardini, M.: The role of blockchain as a security support for student profiles in technology education systems. InfoTekJar: J. Nasional Inform. dan Teknologi Jaringan, **4**(2), 203–207 (2020)

Themistocleous, M., Christodoulou, K., Iosif, E., Louca, S., Tseas, D.: Blockchain in academia: where do we stand and where do we go? In: HICSS, pp. 1–10 (2020)

Turkanović, M., Hölbl, M., Košič, K., Heričko, M., Kamišalić, A.: EduCTX: a blockchain-based higher education credit platform. IEEE Access **6**, 5112–5127 (2018)

Ullah, N., Mugahed Al-Rahmi, W., Alzahrani, A.I., Alfarraj, O., Alblehai, F.M.: Blockchain technology adoption in smart learning environments. Sustainability **13**(4), 1801 (2021)

UntungRahardja, S.K., EkaPurnamaHarahap, Q.: Authenticity of a diploma using the blockchain approach. Int. J. **9**(1.2), 2 (2020)

Vidal, F., Gouveia, F., Soares, C.: Analysis of blockchain technology for higher education. In: 2019 International Conference on Cyber-Enabled Distributed Computing and Knowledge Discovery (CyberC), pp. 28–33. IEEE (2019)

Al Harthy, I., Ramadan, R., Hussain, A.: Non-destructive techniques for linking methodology of geochemical and mechanical properties of rock samples. J. Petrol. Sci. Eng. **195**, 107804, 6–8 (2019)

Business Processes and Patterns of Blockchain Technology

Yossra Zghal[1]([envelope]) [ID], Bálint Molnár[2] [ID], and Simon Thompson[2,3] [ID]

[1] Doctoral School of Informatics, Faculty of Informatics, Eötvös Loránd University of Budapest, ELTE, Pázmány Péter sétány 1/C, Budapest 1117, Hungary
`p599ba@inf.elte.hu`
[2] Information Systems Department, Faculty of Informatics, Eötvös Loránd University of Budapest, ELTE, Pázmány Péter sétány 1/C, Budapest 1117, Hungary
`{molnarba,thompson}@inf.elte.hu`
[3] University of Kent, Canterbury, Kent CT2 7NZ, UK

Abstract. In the context of digital transformation in the financial industry, blockchain technology and smart contracts can play a significant role in streamlining and enhancing business processes. The research plan outlined in this paper aims to investigate the integration of blockchain to improve the digitization of business processes in the financial sector. By leveraging blockchain's decentralized and secure nature, financial enterprises can increase the transparency and efficiency of their operations. Smart contracts can further enhance this by automating contract execution and eliminating the need for intermediaries, leading to cost savings and reduced processing time. Overall, the integration of blockchain and smart contracts can offer a powerful solution for digital transformation in the financial industry.

Keywords: Business Processes · Blockchain Technology · Smart Contract · Marlowe

1 Introduction

Due to its capacity to deliver safe and transparent transactions, the decentralized digital ledger system blockchain technology has grown in popularity in a variety of sectors in recent years. One of blockchain's advantages is that it uses a network of computers to verify and authenticate transactions, making it more secure than previous systems. Furthermore, smart contracts the self-executing digital contract that operates on a blockchain technology may greatly enhance corporate operations by removing intermediaries and increasing efficiency. Smart contracts may automate numerous operations, such as payments and contracts,

This research was supported by the Thematic Excellence Programme TKP2021-NVA-29 (National Challenges Subprogramme) funding scheme, and by the COST Action CA19130 - "Fintech and Artificial Intelligence in Finance Towards a transparent financial industry" (FinAI).

J. M. Machado et al. (Eds.): BLOCKCHAIN 2023, LNNS 778, pp. 536–541, 2023.
https://doi.org/10.1007/978-3-031-45155-3_52

reducing the need for human interaction and resulting in faster and more efficient transactions.

Blockchain technology has made a significant impact across various industries, including banking, real estate, insurances, supply chain management, and healthcare. It has the potential to enhance transaction security, reduce fraud, lower costs, and enable transparency in these sectors.

In this paper, our focus is on the insurance industry, specifically exploring how blockchain technology can improve the insurance claims process. We will examine the benefits of utilizing smart contracts powered by the Marlowe domain-specific functional programming language that is specifically designed for creating and executing smart contracts on blockchain platforms to automate claim verification, fraud detection, and payments. By leveraging blockchain technology, insurance companies can enhance efficiency, transparency, and accuracy while minimizing the risk of errors and fraud. To lay a strong theoretical foundation, the next section will delve into the theoretical background of business processes, blockchain technology, and smart contracts.

In order to provide a solid theoretical foundation for our subsequent analysis, we will delve in the next sections into the theoretical background of business processes, blockchain technology, and smart contracts. We will delve into the practical implementation of blockchain technology in Insurance.

2   Theoretical Background

This section provides a foundational understanding of business processes, blockchain technology, and smart contracts. By examining their theoretical implications, potential applications, current research, and future directions, we can better analyze their practical implementations.

2.1   Business Process

A business process is a collection of interconnected actions or activities carried out by individuals, teams, or departments within an organization to achieve a specific business objective. These tasks or activities are typically designed to create value for customers, suppliers, and other stakeholders, and may involve the use of technology, information, and other resources.

2.2   Blockchain Technology and Smart Contract

Blockchain Technology. Blockchain technology, as described by Nakamoto in the seminal Bitcoin whitepaper [2], refers to a specific type of digital ledger system characterized by its distribution and decentralization.

Its main purpose is to enable the recording of data in a way that is secure, transparent, and unchangeable. It achieves this through the use of cryptographic algorithms and consensus mechanisms, which work together to create a database that is tamper-proof and accessible to everyone in the network without the need for intermediaries or third parties. The chain of blocks in the system is linked by a unique identifier called a hash, allowing for easy tracing of the data back to its original source. This makes blockchain technology an excellent tool for a variety of applications, including cryptocurrency, supply chain management, and digital identity verification.

Smart Contract. A smart contract, as defined by Szabo [3], is a digital contract that is self-executing and facilitates secure transaction execution between parties without the need for intermediaries. This computer program operates on a blockchain platform and carries out the conditions of the agreement automatically when certain predefined conditions are fulfilled. Typically coded in software, smart contracts can perform a diverse range of functions, including validating product authenticity, managing asset transfers, and enforcing agreement terms.

3   Objectives

The main objective of this research is to enhance the Insurance claim process by implementing a blockchain smart contract with the Marlowe functional programming language [4]. To achieve this goal, we utilized a model-to-text process that focused on improving the UML activity diagram's XML meta-model. This involved using the Papyrus tool to create an activity diagram and then applying Acceleo to extract and enhance the non-optimized XML. The improved XML was then used to create an enhanced version of the activity diagram. Finally, we integrated the blockchain smart contract to enhance the activity diagram's efficiency, transparency, and security.

3.1   The Use of Model to Program M2P in Business Process Improvement and Optimization

Before delving into the implementation of smart contracts, it is important to acknowledge the existing research on the use of business process modeling for the improvement and optimization of business processes [1].

Figure 1 represent the summary of the insurance claim improvement process in which we apply the improvement algorithm in order to improve the business process within the insurance process.

To begin with, we utilized Papyrus, a graphical editing tool for UML diagrams, to model the insurance claim process activity diagram, which includes editors for all UML diagrams and complete support for SysML to facilitate a model-based system. Subsequently, we extracted the corresponding XML and developed an algorithm that uses Acceleo, a template-based technology that features authoring tools for customized code generators, to convert the non-enhanced XML file into an improved version. Acceleo the template based technology can generate any type of source code from any available data source in EMF format. The improvement algorithm extracts the keyword from the XML format, and we analyze the non-enhanced diagram and apply the transformation

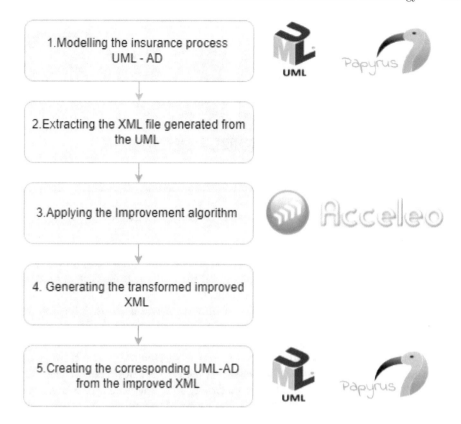

Fig. 1. Summary of the insurance claim improvement process

based on the keyword using Acceleo and the Java programming language. The outcome is an improved UML activity diagram.

3.2 The Implementation of Blockchain Smart Contract After applying the Improvement Algorithm

After using M2T in business process improvement and optimization, the next step is to implement blockchain smart contract technology for insurance to ensure better security, transparency, and efficiency. Additionally, to spot problem areas in Natural Language Processing (NLP), pattern matching can be utilized to discover irregular patterns. NLP can also be used to identify suspicious denotation and constructions, which can improve the business process automatically. This can be especially useful in the insurance industry where accuracy and security are paramount.

The implementation of smart contracts in this work involves designing and developing code that reflects the specific insurance processes and requirements. This includes identifying the conditions, triggers, and actions that need to be

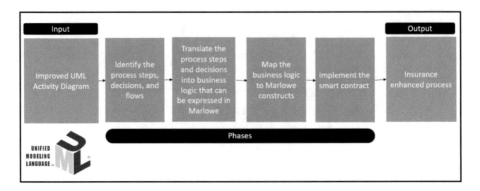

Fig. 2. Steps of the implementation of blockchain

programmed into the smart contract. The Marlowe functional programming language can be utilized to create the smart contract, allowing for automated execution based on predefined conditions (Fig. 2).

To implement this new approach, we undergo several steps. The first step is to identify the insurance process that requires improvement. We choose the claims process that is deemed to be inefficient or insecure. In the second step, we translate the process steps and decisions into business logic that can be expressed in Marlowe. After that, we map the business logic to Marlowe constructs. Then we develop and implement the smart contract. This involves creating a secure and automated system that can execute the terms of the agreement when certain conditions are met. Finally, we integrate the smart contract with the insurance process. The figure below demonstrates the step-by-step process for implementing this new approach. In the upcoming section, we will present and analyze the results obtained from our research, shedding light on the findings and insights gained through our investigation. The implementation of blockchain smart contracts in the insurance sector becomes a viable solution. This approach enables automated and secure execution of insurance processes, ensuring transparency, efficiency, and reduced risks of fraud or errors.

3.3   Results

The implementation of blockchain and smart contract technology within the insurance claim process is a significant step toward achieving better security, transparency, and efficiency in the finance sector.

In this work, we have demonstrated how the use of modeling can facilitate the improvement and optimization of business processes within an insurance company. By identifying the areas of the insurance process that require improvement and translating the process steps into business logic that can be expressed in Marlowe.

The implementation of blockchain and smart contract technology within the insurance claim process, combined with the use of modeling and NLP techniques, can help to streamline the insurance process and improve accuracy and security.

4   Conclusion and Future Direction of the Research

In this work, we have demonstrated the effectiveness of using a combination of modeling, blockchain, and smart contract technology to improve the insurance claim process. By identifying the areas that require improvement and translating the process steps into business logic. This research also highlights the importance of utilizing NLP techniques, such as pattern matching, to spot problem areas and improve the business process automatically.

There are several avenues for future work in this area. One possible direction is to investigate the potential use of artificial intelligence (AI) and machine learning (ML) techniques to further optimize the insurance claim process. For example, ML algorithms could be used to predict claim approval or rejection, reducing the workload for insurance assessors and improving the efficiency of the process.

Another area for future work is to explore the potential benefits of integrating the blockchain-based insurance claim process with other areas of the insurance industry, such as underwriting and policy management. This could create a more streamlined and efficient insurance ecosystem, improving the customer experience and reducing costs for insurers.

This research establishes a robust foundation for future studies and advancements in utilizing blockchain, smart contract technology, and NLP techniques to optimize the insurance process. By fostering innovation and exploring cutting-edge technologies, we can contribute to the creation of an insurance industry that is characterized by enhanced security, transparency, and efficiency. Ultimately, this will result in widespread benefits for all stakeholders involved.

References

1. Kherbouche, M., Zghal, Y., Molnár, B., Benczúr, A.: The use of M2P in business process improvement and optimization. In: Chiusano, S., et al. (eds.) ADBIS 2022. CCIS, vol. 1652, pp. 109–118. Springer, Cham (2022). https://doi.org/10.1007/978-3-031-15743-1_11
2. Nakamoto, S.: Bitcoin: a peer-to-peer electronic cash system. Decent. Bus. Rev. 21260 (2008)
3. Szabo, N.: Smart contracts: building blocks for digital markets. EXTROPY: J. Transhumanist Thought (16) **18**(2), 28 (1996)
4. Taherdoost, H.: Smart contracts in blockchain technology: a critical review. Information **14**(2), 117 (2023)

A Modelling Approach for a High Utility Decentralized Autonomous Organization Development

Sowelu Avanzo[1(✉)], Alex Norta[2], and Claudio Schifanella[1]

[1] Computer Science Department, University of Torino, via Pessinetto, 12, 10149 Turin, Italy
{soweluelios.avanzo,claudio.schifanella}@unito.it
[2] BFM, Tallinn University, Narva tee 27, 10120 Tallinn, Estonia
alex.norta.phd@ieee.org

Abstract. Decentralized Autonomous Organizations (DAO) are decentralized applications (Dapps) enabling governance processes. These face limitations caused by the underlying blockchain complexity, usability and scalability issues. In addition, the design and validation of the governance mechanisms lack appropriate methods and tools, resulting in a misalignment of DAO policies to the stakeholders' needs. This work proposes a software engineering methodology for developing DAOs with a high degree of utility. In order to do so, we will extend the Trusted-Dapp Modeling concepts to facilitate specification of the incentive mechanisms involved in the token economy. These are an essential component of DAO-based governance, ultimately contributing to the project's success or failure. As a proof-of-concept prototype, a support tool will be developed and evaluated in vivo field studies in the context of community currency-system governance.

Keywords: Decentralized Autonomous Organization · DAO · Agent-Oriented Modeling · Trusted Dapp Modelling · T-DM · Blockchain

1 Problem Statement

Decentralized Autonomous Organizations (DAOs) are systems enabling decentralised governance enacted by a set of self-executing rules deployed on a public blockchain [1]. Despite the increasing scientific interest in the topic, DAOs face obstacles limiting their utility in several domains [2]. DAOs are affected by the main limitations of decentralized applications (Dapps), including a lack of scalability and usability, and complexity [3]. Furthermore, the presence of governance protocols such as Holographic Consensus, or Conviction Voting [4], relying on token-based incentive mechanisms, adds a further layer of complexity in developing such a DAO system. The current lack of support to token-economy design

A. Norta and C. Schifanella—Contributing authors.

and validation causes misalignment of the system behavior to the stakeholders' goals. Some consequences of this are the high centralization of governance tokens in several DAOs [5], which contradicts the decentralization goal, and lack of economic security in DAO projects [2]. Furthermore, in [4], insufficient testing of DAOs is reported. Therefore, a holistic software engineering approach that addresses the above issues assuring a high utility in the developed DAOs is needed. In this work we aim to address this gap by extending Trusted Dapp Modeling (T-DM) [3], a state-of-the-art model-driven approach for Dapp development to meet the needs of DAOs.

2   Related Works

Even though no modeling approach is specifically focused on DAOs, diverse proposals facilitating Dapp development exist. A process reduction method was proposed in [6], which enables generation of smart contracts with significantly optimized gas costs. Still, this doesn't solve the usability concerns, as mentioned in [3]. The software engineering approach proposed in [7] involves design and development support and facilitates requirement modeling and stakeholder communications. This approach, however, fails to deal with the transaction scalability problem. T-DM extends Agent-Oriented Modeling concepts and UML [8]. Profile diagrams integrate traditional UML with *ad hoc* elements allowing to distinguish between actions performed *on-chain* and *off-chain*. This addresses the issue of transaction cost scalability. Moreover, citation [9] provides a design support tool, based on T-DM, reducing complexity and usability issues. None of the methodologies above, however, support token economy design and validation needed to implement DAOs with high utility.

Among solutions for DAO development, frameworks, such as Aragon[1], Colony[2] or DAOstack[3] enable DAO deployment as a service based on a no-code approach [4]. These platforms also enable customization of voting mechanisms and other functionalities of DAOs. Despite the advantages provided by such tools in implementing DAOs, a high degree of complexity remains, as users might not be able to effectively choose the features and platforms that best meet their needs, as discussed in [10]. The solution proposed by Baninemeh et al. consists in a multi-criteria decision model facilitating the selection of platform and features [10] based on the requirements elicited. Yet, features and platforms might be updated or change over time, without the stakeholders fully realizing the impact of each feature on the resulting DAO project. These aspects limit the utility of the decision model. Other approaches address the design and validation of specific functionalities of DAOs. Some examples are discussed in [11], where mechanism design optimization is applied to Conviction Voting, and citation [12], in which an agent-based simulation supports the validation of the Furtarchy voting protocol. These approaches take into account and recognize the importance of token

---

[1] https://aragon.org/.
[2] https://colony.io/.
[3] https://daostack.io/.

economy-related concepts and assumptions involved in the governance systems considered. Still, a holistic approach for modeling and validating DAO policies is missing.

3   Hypothesis and Proposal

To confirm the hypothesis below, we will respond to the main research question: *How to develop a software engineering methodology for a high utility DAOs design?* To increase DAO utility, it is necessary to overcome the structural limitations of DAOs related to complexity, usability and scalability. Since all other approaches for Dapp development fail to address these three issues [3], we will focus on extending the T-DM approach. This still does not take into account the complexity of DAO token economies design. The hypothesis of the research is as follows: *Extending the T-DM approach by including elements that facilitate the specification of incentive mechanisms involved in token economies of DAOs increases the utility of the corresponding systems developed.* To achieve a separation of concerns, we deduce a set of sub-research questions. Hence, the first sub-research question is: *How to extend the T-DM approach to model DAOs based on sound token economies?* (RQ1). The reference requirements set for DAOs will be elicited based on the data of the running case (described below). Based on this input, a set of heuristics will be developed enabling the specification of DAO models with sound token economies in the Extended T-DM. Subsequently, an ontology will be provided for defining the modeling language elements and their relations. These aspects aim to enable the specification of incentive mechanisms involved in governance processes. The next sub-research question addresses the need to implement adequate design and validation tools for the approach developed: *How to develop a support tool for designing DAOs based on Extended T-DM?* (RQ2). We will, therefore, develop tool support to design DAOs meeting the requirements of the running case, based on the ontology previously modeled. Its initial design will be based on the T-DM tool support described in [9]. Finally, we will enable the validation of DAO design models. This entails responding to the sub-research question *How to support Extended T-DM DAO model validation via simulations?* (RQ3). To do so, we will propose a heuristics mapping from T-DM model elements to DAO simulation elements. Hence, we will extend the support tool to facilitate generation of simulations. The Action Design Science (ADS) methodology based on three IT-dominant building, intervention and evaluation cycles [13] will be adopted. The first cycles will lead to the implementation and evaluation of the ontology and design support tool proof-of-concept (RQ1,2). The subsequent, to the heuristic-mapping enabling the generation of DAO simulations (RQ3). As a running case, the Extended T-DM will be adopted in developing and implementing a smart contract framework enabling the decentralized governance of the Circles[4] community currency system (CCS). A CCS is a blockchain-based system supporting circulation of tokens used as local means of exchange complementary to national currencies. These are accepted by voluntary agreement and

---

[4] https://github.com/CirclesUBI/whitepaper.

managed by social economy organizations aiming to bootstrap local trade and provide humanitarian aid. In this context, DAOs shall manage minting and token-distribution policies based on collective governance. Their development, however, is particularly challenging due to the complexity of the token economy involved. The main aspects of the research execution are outlined below. This will follow the timeline reported in Table 1, and adhere to the principles of ADS (in italics) [13]. The PhD candidate will collaborate with the cooperative managing the Circles CCS to develop high utility DAOs for the governance of the CCS. The related challenges provide opportunities for knowledge creation concerning approaches to DAO development (*Practice-Inspired Research*). The T-DM framework will provide the preliminary design of the artifact (*Theory-Ingrained Artifact*). The T-DM extension will be shaped to solve the concrete problems faced by developers in implementing the smart contract framework (*Reciprocal Shaping*). The PhD candidate will be employed at the Cooperative during the research. He will contribute knowledge of DAOs and approaches to Dapp development, while the developers will provide in-depth knowledge of CCSs (*Mutually Influential Roles*). The approach will be evaluated progressively and concurrently with the building and intervention(*Authentic and Concurrent Evaluation*). Through building, intervention and evaluation a novel version of the T-DM will be proposed, ensuring high utility of DAOs developed (*Guided Emergence*). As a result, the modeling approach will improve the utility of DAOs (*Generalized Outcomes*).

4   Evaluation Plan

The evaluation of the utility of the ontology and support tool for the Extended T-DM will be performed concurrently with building and intervention via participant observation and in-depth interviews with the developer team and other stakeholders. Additionally, the XML-based model-checking method proposed in [14] will be adopted to evaluate the syntactic correctness and consistency of the modeling language. The user interactions with the CCS will be monitored based on the network analysis methods proposed in [15]. This will support the evaluation of the utility of the DAO deployment framework to CCS users.

Table 1. Time line for PhD research.

Activity	2022 Jan-Dec	2023 Jan-Dec	2024 Jan-Dec
Problem Formulation: systematic literature review on DAO design and applications and survey of existing development approaches	✓	✓	
Building, Intervention and Evaluation: development of the extended T-DM approach and support tool. Adoption and evaluation by implementing DAO deployment framework for the governance of Circles UBI		✓	
Reflection and Learning: the results of the project will be compared to the goals of facilitating DAO development and validation of DAOs			✓
Fromalization of learning: thesis and scientific articles will be written to formalize the contributions made to the state-of-the-art.			✓

5   Conclusive Remarks

DAOs inherit the complexity, scalability and usability constraints affecting all Dapps. An additional limitation of DAOs is related to the complexity in incentive mechanism design and validation, aspect particularly relevant in governance protocols, on which little research has been done so far. We propose to extend T-DM [3] in order to provide a holistic approach to DAO development that targets all issues mentioned. The approach shall enable the modeling of token economies with the aim to build DAOs with high utility. This approach will be evaluated through in vivo field studies concerning its adoption to develop DAOs for the governance of CCSs. In this domain, DAOs could achieve positive socio-economic impact, if adequate development support is provided. Further research is needed to develop the approach and the relative support tool.

References

1. Hassan, S., De Filippi, P.: Decentralized autonomous organization **10**(2). https://doi.org/10.14763/2021.2.1556. Accessed 13 Dec 2022
2. Liu, L., Zhou, S., Huang, H., Zheng, Z.: From technology to society: an overview of blockchain-based DAO. **2**, 204–215 (2021). https://doi.org/10.1109/OJCS.2021.3072661. Accessed 23 Dec 2022
3. Udokwu, C., Anyanka, H., Norta, A.: Evaluation of approaches for designing and developing decentralized applications on blockchain. In: Proceedings of the 2020 4th International Conference on Algorithms, Computing and Systems, pp. 55–62. ACM (2020). https://doi.org/10.1145/3423390.3426724, https://dl.acm.org/doi/10.1145/3423390.3426724. Accessed 13 Dec 2022
4. Faqir-Rhazoui, Y., Arroyo, J., Hassan, S.: A comparative analysis of the platforms for decentralized autonomous organizations in the ethereum blockchain. **12**(1), 9 (2021). https://doi.org/10.1186/s13174-021-00139-6. Accessed 13 Dec 2022
5. Fritsch, R., Müller, M., Wattenhofer, R.: Analyzing voting power in decentralized governance: Who controls DAOs? (2022). https://doi.org/10.48550/ARXIV.2204.01176. Accessed 14 Mar 2023
6. García-Bañuelos L., Ponomarev, A., Dumas, M., Weber, I.: Optimized execution of business processes on blockchain. https://doi.org/10.48550/ARXIV.1612.03152. Accessed 09 Mar 2023
7. Lallai, G., Pinna, A., Marchesi, M., Tonelli, R.: Software engineering for DAPP smart contracts managing workers contracts. In: DLT@ITASEC (2020)
8. Sterling, L.S., Taveter, K.: The Art of Agent-Oriented Modeling. The MIT Press (2009). OCLC: 1178943942
9. Udokwu, C., Brandtner, P., Norta, A., Kormiltsyn, A., Matulevičius, R.: Correction to: implementation and evaluation of the DAOM framework and support tool for designing blockchain decentralized applications (2021). https://doi.org/10.1007/s41870-022-01026-4. Accessed 09 Mar 2023
10. Baninemeh, E., Farshidi, S., Jansen, S.: A decision model for decentralized autonomous organization platform selection: three industry case studies (2023). https://doi.org/10.48550/ARXIV.2107.14093. Accessed 23 Dec 2022
11. Emmet, J.: Mathematically formalizing the conviction voting algorithm. https://github.com/1Hive/conviction-voting-cadcad/blob/master/algorithmoverview.ipynb. Accessed 14 Jan 2023

12. Ding, W.W., et al.: Parallel governance for decentralized autonomous organizations enabled by blockchain and smart contracts. In: 2021 IEEE 1st International Conference on Digital Twins and Parallel Intelligence (DTPI), pp. 1–4. IEEE (2021). https://doi.org/10.1109/DTPI52967.2021.9540069, https://ieeexplore.ieee.org/document/9540069/. Accessed 13 Dec 2022

13. Sein, M.K., Henfridsson, O., Purao, S., Rossi, M., Lindgren, R.: Action design research. **35**(1), 37 (2011). https://doi.org/10.2307/23043488. Accessed 13 Dec 2022

14. Udokwu, C., Norta, A.: Deriving and formalizing requirements of decentralized applications for inter-organizational collaborations on blockchain. **46**(9), 8397–8414 (2021). https://doi.org/10.1007/s13369-020-05245-4. Accessed 10 Apr 2023

15. Criscione, T., Guterman, E., Avanzo, S., Linares, J.: Community currency systems: basic income, credit clearing, and reserve-backed. Models and Design Principles (2022). http://www.econstor.eu/dspace/Nutzungsbedingungen. Accessed 13 Dec 2022

Towards an Enterprise Blockchain Interoperability Framework: A Case of the Banking Sector

Senate Sylvia Mafike$^{(\boxtimes)}$ (ID) and Tendani Mawela (ID)

University of Pretoria, Hatfield, Pretoria 0002, South Africa
senatemafike@gmail.com, tendani.mawela@up.ac.za

Abstract. Blockchain technology offers attractive benefits to many industries, however, realizing these benefits is hindered by the lack of interoperability between different blockchain systems and other non-blockchain systems. The lack of standards, models and frameworks that guide the development of blockchain systems has led to fragmented systems that are unable to share information seamlessly without employing a central party. This PhD study proposes a generic blockchain interoperability framework to assist organizations in making the appropriate design choices for interoperable blockchain systems. The study tackles blockchain interoperability from a General Systems Theory perspective and focuses on the four levels of interoperability (legal, organizational, semantic and technical) and focuses on the banking sector as a case study. This paper presents the preliminary findings of the study, and the proposed frameworks and highlights the next steps.

Keywords: Blockchain · Blockchain Interoperability · Framework · Banking sector

1 Introduction

Blockchain is a technology developed to enable collaborative validation and verification of financial transactions. Its initial intent was to facilitate decentralized governance of transactions as an alternative to the traditional centralized banking system. However, recently, the scope of application of the technology has extended within the financial industry and in other industries such as supply chain, health, and real estate. In particular, the banking sector globally, is exploring blockchain in payment systems to reduce costs and enhance the security and efficiency of their operations [1].

The current industrial and global central bank efforts in developing blockchain applications are highly fragmented and has led to the emergence of various blockchain platforms, that differ in consensus mechanisms, protocols and data models [2]. This is because, different industries have different requirements, therefore, the choice of blockchain varies across industries. Consequent to these variations, the blockchain-based systems emerging from the industries are also different and incompatible. Existing blockchain platforms are inherently designed to operate only within their networks and lack mechanisms to directly interoperate with other platforms. Incompatibilities in the consensus mechanisms, data models and protocols complicate data and asset sharing between these blockchain platforms [2]. The inability of different blockchain platforms to share information among themselves and with other non-blockchain systems inhibits collaboration between enterprises and is an obstacle to the mainstream adoption of the technology [3].

For organizations to fully reap the benefits of blockchain, new blockchain solutions must communicate with each other and with existing technologies within incumbent organizations. This is particularly critical for collaborative business models and industries such as the financial sector, which consists of a large network of participants. The global banking sector is exploring the application of blockchain to improve payment systems. Payment systems by nature involve multiple participants to enable parties to transfer funds. These participants may use different technologies, legal frameworks, and may be located in different geographical areas as is the case of cross-border payments.

However, due to the lack of appropriate standards and models for developing blockchain systems [4], these blockchain-based payment systems are created in silos, thus leading to new payments systems that are not interoperable with each other and with existing legacy payment systems. It is therefore pertinent to develop models and guidelines for organizations to develop blockchain systems that are interoperable across and within organizations.

1.1 Problem Statement

The problem is that blockchain interoperability is a complex process [5]. Achieving interoperability between different blockchain systems does not only require data to be shared but requires the validation of activities that occurred in a foreign blockchain. Validating events that occurred on a separate blockchain is a challenging undertaking. This challenge is amplified when communication is required between centralized and decentralized networks which may have different security, privacy, efficiency and performance requirements. Traditional interoperability protocols, models, and frameworks were designed for centralised systems, and are therefore incompatible with blockchain, which by nature is decentralised [6]. Furthermore, emerging solutions addressing the blockchain interoperability challenge are still immature and mostly rely on centralised third-party-based approaches and are not interoperable [7]. These challenges, coupled with the lack of appropriate standards and frameworks [4] hinder the ability of organizations to create standardized and interoperable blockchain systems. To our knowledge, the literature lacks a clear and comprehensive framework that identifies key legal, organizational, semantic and technical design considerations, features/elements and requirements for enabling cross-blockchain interoperability and blockchain to non-blockchain interoperability.

1.2 Research Objectives

The main purpose of this PhD study is to address the blockchain interoperability challenge by developing and evaluating an enterprise technology-agnostic interoperability framework, that includes technical, semantic, organizational and legal design choices, considerations and characteristics for enabling interoperability across blockchain systems. The study addresses the following research question:

How can a blockchain interoperability framework be developed to contribute to the effective communication between heterogeneous blockchain systems and existing legacy systems?

2 Related Works

As part of this research study, we have analyzed the state of the art of blockchain interoperability. The analysis indicates that there have been efforts relating to the interoperability of blockchain systems from different industrial domains. The existing literature gives particular focus to the design and comparison of cross-chain technologies such as sidechains, Hash-time-locks, notaries, and industry solutions. For instance, [8, 9] have provided comparisons of these technologies. Few studies [10–14] have proposed frameworks relating to blockchain interoperability. Belcior et al. [10], proposed an assessment framework for selecting interoperability solutions. Their work provides an overview of legal, organizational, semantic and technical considerations for selecting an interoperability solution. Pillai et al. [12], proposed a cross-blockchain integration design decision framework. Their framework highlights some considerations for enabling interoperability between different blockchains systems. Their work also identifies security assumptions relating to different integration modes. Another study [14], presented an enterprise blockchain design framework including architectural elements for designing enterprise blockchain systems. The framework includes an interoperation layer to support connectivity between the blockchain and external systems.

While the above-stated papers provide some invaluable insights regarding the different considerations and design choices for enabling interoperability between blockchains and other systems, they are not comprehensive. Most of the studies, focus on semantic and technical interoperability solutions and overlook key considerations for enabling other forms of interoperability such as legal and organizational interoperability. In addition, pertinent requirements that interoperability solutions should fulfil have been ignored in the studies (Table 1).

Table 1. Existing literature on blockchain interoperability frameworks

Interoperability concern	Blockchain to Blockchian	Blockchian to Legacy
Technical & semantic	[10, 12–14]	[14]
Legal	[10]	
Organizational	[10, 11]	[14]

3   Methodology

The study follows the Design Science Research Methodology (DSRM) proposed by Peffers [15] to address the research question. The DSRM is a problem-solving approach that generates new knowledge about a particular problem through the construction of an Information Systems (IS) artefact [15]. IS artefacts include; software, frameworks, methods, design principles and requirements. This study proposes a framework as the artefact to address blockchain interoperability challenges and to guide organizations on the design of interoperable blockchain systems. The core purpose of DSRM is to develop knowledge that can be used by professionals in a practical context to provide solutions to contextual problems [16]. From this perspective, DSRM is a good fit for this study. A preliminary overview of the proposed framework is provided in the proceeding section. This framework will be evaluated and refined through expert questionnaires, expert interviews, and a banking case study.

4   Preliminary Results

The study is currently in the first iteration of the design and development phase of the DSRM process described above. The problem identification was carried out as the first step. A systematic literature review [17] was performed to identify open research gaps relating to the topic of blockchain adoption. The purpose of the SLR was to identify challenges and techniques concerning blockchain implementation in the banking sector. The results indicate several challenges including scalability, lack of relevant regulation and interoperability as critical obstacles to the implementation of the technology in banking. The interoperability challenge was selected as the topic for this study as it was observed from the literature that addressing interoperability also helps to address scalability issues. In the second step, expert interviews and a second SLR were conducted to identify limitations and requirements that blockchain interoperability solutions should fulfil. The current and third step involves designing and developing the proposed framework. A preliminary framework was proposed as shown in Fig. 1. The design of the framework was informed by the results of the second SLR and will be refined to include the results of the expert interviews. The General System theory is applied as the theoretical basis for the research study. Systems theory is suitable for the study of blockchain interoperability because it helps to understand the wholeness of a system by evaluating interactions and relationships between its parts and has the advantage of providing a general perspective for understanding systems, their environment and interactions [18] and therefore can be applied to provide generic and domain-independent frameworks.

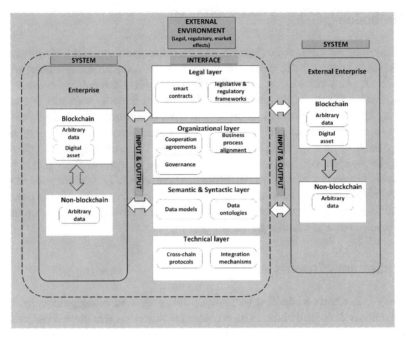

Fig. 1. Preliminary enterprise blockchain interoperability framework (Source: Author's own)

5 Conclusion and Way Forward

This PhD study aims to address the lack of interoperability between blockchain systems by developing and evaluating a generic blockchain interoperability framework. The first phase of the study was to identify the problem and propose a potential solution and its objectives. A systematic literature review [17] was published as part of this phase. A second systematic literature review focusing on identifying requirements for interoperable blockchain systems has also been submitted for publication. Currently, the study is in the framework development phase, which involves designing and developing a preliminary framework. The next phase will focus on refining the proposed framework. The framework will be refined and evaluated through an iterative process involving expert interviews and a case study. Details behind the framework and its elements will also be provided in the proceeding phases of the study.

References

1. Sakho, S., et al.: Improving banking transactions using blockchain technology. In: 2019 IEEE 5th International Conference on Computer and Communications (ICCC). IEEE (2019). https://doi.org/10.1109/ICCC47050.2019.9064344
2. Ghaemi, S., et al.: A pub-sub architecture to promote blockchain interoperability. arXiv preprint arXiv:2101.12331 (2021)
3. Liu, Z., et al.: HyperService: interoperability and programmability across heterogeneous blockchains. In: Proceedings of the 2019 ACM SIGSAC Conference on Computer and

Communications Security, pp. 549–566. Association for Computing Machinery, London (2019). https://doi.org/10.1145/3319535.3355503

4. Belchior, R., et al.: A survey on blockchain interoperability: past, present, and future trends. ACM Comput. Surv. (CSUR) **54**(8), 1–41 (2021). https://doi.org/10.1145/3471140

5. Hardjono, T., Lipton, A., Pentland, A.: Toward an interoperability architecture for blockchain autonomous systems. IEEE Trans. Eng. Manage. **67**(4), 1298–1309 (2019). https://doi.org/10.1109/TEM.2019.2920154

6. Abebe, E., et al.: Enabling enterprise blockchain interoperability with trusted data transfer (industry track). In: Proceedings of the 20th International Middleware Conference Industrial Track, pp. 29–35. Association for Computing Machinery, Davis (2019). https://doi.org/10.1145/3366626.3368129

7. Pillai, B., Biswas, K., Muthukkumarasamy, V.: Cross-chain interoperability among blockchain-based systems using transactions. Knowl. Eng. Rev. (2020). https://doi.org/10.1017/S0269888920000314

8. Bhatia, R.: Interoperability solutions for blockchain. In: 2020 International Conference on Smart Technologies in Computing, Electrical and Electronics (ICSTCEE). IEEE (2020). https://doi.org/10.1109/ICSTCEE49637.2020.9277054

9. Qasse, I.A., Talib, M.A., Nasir, Q.: Inter blockchain communication: a survey. In: ACM International Conference Proceeding Series (2019). https://doi.org/10.1145/3333165.3333167

10. Belchior, R., et al.: Do you need a distributed ledger technology interoperability solution? Distrib. Ledger Technol. (2022). https://doi.org/10.1145/3564532

11. Llambias, G., González, L., Ruggia, R.: Blockchain interoperability: a feature-based classification framework and challenges ahead. CLEI Electron. J. **25**(3), 4:1–4:29 (2022). https://doi.org/10.19153/cleiej.25.3.4

12. Pillai, B., et al.: Cross-blockchain technology: integration framework and security assumptions. IEEE Access **10**, 41239–41259 (2022). https://doi.org/10.1109/ACCESS.2022.3167172

13. Pillai, B., et al.: Blockchain interoperability: performance and security trade-offs. In: Proceedings of the 20th ACM Conference on Embedded Networked Sensor Systems, pp. 1196–1201. Association for Computing Machinery, Boston (2023). https://doi.org/10.1145/3560905.3568176

14. Nodehi, T., et al.: EBDF: The enterprise blockchain design framework and its application to an e-Procurement ecosystem. Comput. Ind. Eng. **171**, 108360 (2022). https://doi.org/10.1016/j.cie.2022.108360

15. Peffers, K., et al.: A design science research methodology for information systems research. J. Manag. Inf. Syst. **24**(3), 45–77 (2007). https://doi.org/10.2753/MIS0742-12222240302

16. Van Aken, J.E.: Management research as a design science: articulating the research products of mode 2 knowledge production in management. Br. J. Manag. **16**(1), 19–36 (2005). https://doi.org/10.1111/j.1467-8551.2005.00437.x

17. Mafike, S.S., Mawela, T.: Blockchain design and implementation techniques, considerations and challenges in the banking sector: a systematic literature review. Acta Inform. Pragensia **2022**(3), 396–422 (2022). https://doi.org/10.18267/j.aip.200

18. Naudet, Y., et al.: Towards a systemic formalisation of interoperability. Comput. Ind. **61**(2), 176–185 (2010). https://doi.org/10.1016/j.compind.2009.10.014

Author Index

© The Editor(s) (if applicable) and The Author(s), under exclusive license
to Springer Nature Switzerland AG 2023
J. M. Machado et al. (Eds.): BLOCKCHAIN 2023, LNNS 778, pp. 555–557, 2023.
https://doi.org/10.1007/978-3-031-45155-3

Printed in the United States
by Baker & Taylor Publisher Services